Non-wettable Surfaces

Theory, Preparation and Applications

RSC Soft Matter Series

Titles in this Series:

1: Functional Molecular Gels
2: Hydrogels in Cell-based Therapies
3: Particle-stabilized Emulsions and Colloids: Formation and Applications
4: Fluid–Structure Interactions in Low-Reynolds-Number Flows
5: Non-wettable Surfaces: Theory, Preparation and Applications

How to obtain future titles on publication:
A standing order plan is available for this series. A standing order will bring delivery of each new volume immediately on publication.

For further information please contact:
Book Sales Department, Royal Society of Chemistry, Thomas Graham House, Science Park, Milton Road, Cambridge, CB4 0WF, UK
Telephone: +44 (0)1223 420066, Fax: +44 (0)1223 420247
Email: booksales@rsc.org
Visit our website at www.rsc.org/books

Non-wettable Surfaces

Theory, Preparation, and Applications

Edited by

Robin H. A. Ras
Aalto University, Espoo, Finland
Email: robin.ras@aalto.fi

and

Abraham Marmur
Technion – Israel Institute of Technology, Haifa, Israel
Email: marmur@technion.ac.il

ROYAL SOCIETY OF CHEMISTRY

RSC Soft Matter No. 5

Print ISBN: 978-1-78262-154-6
PDF eISBN: 978-1-78262-395-3
EPUB eISBN: 978-1-78262-968-9
ISSN: 2048-7681

A catalogue record for this book is available from the British Library

Published by The Royal Society of Chemistry,
Thomas Graham House, Science Park, Milton Road,
Cambridge CB4 0WF, UK

Registered Charity Number 207890

For further information see our web site at www.rsc.org

Printed in the United Kingdom by CPI Group (UK) Ltd, Croydon, CR0 4YY, UK

Preface

This book is about a topic that has been known for many decades. However, it has become extremely popular only during the last two decades. We do not know what the reason is—maybe the very successful association with the purity and cleanliness of the lotus. However, we are happy it happened, since it is a challenging as well as rewarding topic, both theoretically and practically.

This book attempts to cover the whole spectrum, from the theoretical fundamentals to the practical applications of non-wettable surfaces. Although thousands of papers have been published, mainly on various production methods, many pieces of the puzzle are still missing. The most obvious missing part is the problem of long-term durability, which may be the main reason why superhydrophobic consumer products are not yet common. There are also some differences of opinion with regard to theoretical aspects, and even terminology. We very much hope that this book will be not only a source of knowledge, but also a catalyst for future development.

Robin Ras
Abraham Marmur

RSC Soft Matter No. 5
Non-wettable Surfaces: Theory, Preparation, and Applications
Edited by Robin H. A. Ras and Abraham Marmur

Published by the Royal Society of Chemistry, www.rsc.org

Contents

RSC Soft Matter No. 5
Non-wettable Surfaces: Theory, Preparation, and Applications
Edited by Robin H. A. Ras and Abraham Marmur

Published by the Royal Society of Chemistry, www.rsc.org

CHAPTER 1

Non-Wetting Fundamentals

ABRAHAM MARMUR[a]

[a]Chemical Engineering Department, Technion – Israel Institute of Technology, Haifa 3200003, Israel
*E-mail: marmur@technion.ac.il

1.1 Introduction

Wetting is a ubiquitous process that occurs in a huge variety of everyday biological and industrial systems. It is a macroscopic process that is very sensitive to surface properties on the nano or molecular scale. In most wetting situations the solid surface is wet only to some extent, depending on its chemical and physical nature. As is well known, the common quantitative measure of wettability is the contact angle (CA), which in most cases is greater than 0° and much less than 180°. However, the extreme cases of either complete wetting (CA = 0°) or non-wetting (very high CA and additional possible criteria to be discussed below) offer interesting scientific challenges as well as practical applications. Actually, nature has been using non-wetting to solve a variety of important needs, and the main scientific principle has been known for about half a century.[1] However, it is only about two decades ago that it started to become a very popular topic in science and engineering.[2–46] The paper by Neinhuis and Barthlott[3] served as an important trigger to the vast interest in non-wetting. It introduced the term "lotus effect" that refers to the self-cleaning of the lotus leaf (and many others), achieved by water drops easily rolling off the surface of the leaf, carrying with them dust and dirt particles.

RSC Soft Matter No. 5
Non-wettable Surfaces: Theory, Preparation, and Applications
Edited by Robin H. A. Ras and Abraham Marmur

Published by the Royal Society of Chemistry, www.rsc.org

However, non-wettability is relevant not only for self-cleaning of leaves and not only for drops. For example, some aquatic animals breathe air from an air film on their body even when they are under water. This air film is re-created each time the animal goes back into the water.*e.g.*[24,36] In addition, while natural systems are predominantly aqueous, the non-wettability of solid surfaces by oils, or organic liquids in general, is also of great practical importance in daily life and in industry.[25,26]

At this point it is important to discuss terminology,[25,41] since there is no standard one and the variety of terms may lead to confusion. A surface that is not wetted by water drops in air, or may sustain an air film under water, is in many cases classified as "water repellent". This usage is unfortunate, because there is nothing active in this process that repels water. The adjective "non-wettable" (or the noun "non-wetting"), on the other hand, appears to be more true to the facts. Moreover, the so-called "water repellent" surfaces are usually classified as "superhydrophobic". However, when a surface is not wetted even by liquids of lower surface tension than of water, this term cannot be used, since "hydro" specifically means water. For this purpose, other terms are used, seemingly at random. One term is "superoleophobic". This is a problematic term, since a surface that is "superoleophobic" is usually also superhydrophobic, so "oleophobic" refers only to a part of the picture. On the other hand, a term such as "omniphobic", which means "fearing everything", is far too wide, since, after all, the discussion is about liquids, not about everything. Some time ago I suggested[25,42] using the term "superhygrophobic" to imply non-wetting, because "hygro" in Greek means "liquid". Thus, the terms "hygrophobic" and "superhygrophobic" exactly express various degrees of non-wetting by liquids in general. In summary, "non-wetting" is a generic term that may be specifically complemented by "superhydrophobic" or "superhygrophobic" when it is important to know what the specific case is.

In order to develop useful non-wettable surfaces, it is important to understand the fundamental theory and apply it in choosing the chemical and physical properties of the surfaces. The objective of this chapter is to present the thermodynamic fundamentals of non-wetting, as they are derived from the general theory of wetting equilibrium. An important aspect that has not been sufficiently noticed and is emphasized here is that of thermodynamic stability. In general, qualitative aspects are stressed in this chapter, with only a few necessary equations, in order to give the general picture rather than the mathematical details.

1.2 Wetting Equilibrium

As is well known, minimizing the energy of a system (internal, Gibbs, or Helmholtz energy, depending on the conditions at the system boundary) leads to a few indicators of equilibrium. First, for all systems, irrespective of the existence of interfaces, the temperature as well as the generalized chemical potential of each species must be uniform throughout the whole system. Then, there are two equations that govern the equilibrium state of an interface: the Young equation and the Young–Laplace equation. The former determines the boundary condition for the shape of the liquid–gas interface, in terms of the local CA that must equal the Young CA, θ_Y. For solid–liquid–gas systems it is given by

$$\cos\theta_Y = (\sigma^s - \sigma^{sl})/\sigma \qquad (1.1)$$

here, σ and σ^s are the surface tension of the liquid and of the solid, respectively, and σ^{sl} is the solid–liquid interfacial tension. This equation is correct for radii of curvature much above the nano scale, for which line tension is negligible *e.g.* ref. 47.

The Young–Laplace equation determines the shape of the interface, in terms of the local curvature that is determined by the local pressure difference across the interface:

$$P^d - P^c = \sigma(1/R_1 + 1/R_2) \qquad (1.2)$$

in this equation, P^d and P^c are the local pressure in the drop and in the continuous phase, respectively, and R_1 and R_2 are the local radii of curvature. In the absence of gravity (or other external fields), the pressure difference is constant across the interface. This implies that the average curvature is also constant across the interface. This well-known fact is important for understanding the behaviour of liquids inside roughness grooves, as will be discussed later.

Eqn (1.1) and (1.2) completely determine the equilibrium behaviour of an interface. When the solid surface is ideal (*i.e.* rigid, smooth, chemically uniform, non-reactive, and insoluble) there is only one solution to these equations, which requires the apparent, namely macroscopically measured CA, to equal the Young CA. However, when the surface is rough or chemically non-uniform, there are many possible solutions. Each solution is characterized by its own apparent CA. Naturally, it is important and interesting to find out (a) which of these solutions has the lowest energy, namely which is the thermodynamically most stable CA, and (b) what are the lowest (receding) and highest (advancing) apparent CAs. The difference between the advancing and receding CAs is called the CA hysteresis range.

When a mathematical function has multiple minima, the only way to identify these minima is to search for them one by one. To find out the global minimum, it is necessary to compare all of them and identify the lowest. There is no general mechanism for this. Luckily, for wetting on rough or chemically heterogeneous surfaces, we have approximate equations for the most stable CA.[1,47] The accuracy of these equations improves as the ratio of the radius of curvature to the heterogeneity scale increases.[49] For rough but chemically uniform surfaces we have the Wenzel equation,[48] which assumes the liquid to penetrate completely into the roughness grooves. This state will be referred to as the W state. The apparent CA associated with this global minimum, θ_W, is given by

$$\cos\theta_W = r\cos\theta_Y \qquad (1.3)$$

in this equation, r is the roughness ratio, defined as the ratio between the true area of the solid surface and its projection on a horizontal surface. The above discussion of eqn (1.3) also holds for chemically heterogeneous surfaces. The most stable minimum in energy occurs at the angle that is given by[1]

$$\cos\theta_c = x_1\cos\theta_{Y1} + x_2\cos\theta_{Y2} \tag{1.4}$$

here, x_1 and x_2 are the ratios of contact area of the solid with each chemistry to the projection of total area of the solid, and θ_{Y1} and θ_{Y2} are the Young CAs corresponding to the two chemistries. If the heterogeneous solid surface is flat, then $x_1 + x_2 = 1$; however it is >1 if the heterogeneous solid surface is also rough. We can easily generalize this equation to a higher number of chemistries, using the principle of linear averaging.

When the surface is rough, there may also be equilibrium positions associated with partial penetration of the liquid into the roughness grooves. This case was first studied by Cassie and Baxter,[1] therefore it is referred to as the CB state. The equation for the apparent CA in this case can be derived from eqn (1.3) and (1.4), assuming the solid surface to be represented by θ_{Y1}, and air (or an inert gas in general) to be represented by θ_{Y2}. Because of the perfect hydrophobicity of air, θ_{Y2} is taken to be 180°. The solid–liquid area per unit projection area is $r_f f$, where f is the area fraction of the projection of the wetted part of the solid surface, and r_f is the roughness ratio of the wetted solid. The liquid–gas interface within the roughness is assumed to be flat, therefore its true area fraction is well approximated by its projected area fraction, $(1 - f)$. The apparent flatness of the liquid–gas interface stems from the fact that the pressure inside the liquid is very nearly uniform (if the effect of gravity is small), therefore the radius of curvature around the liquid body must be uniform too. Since this radius of curvature is usually very large compared with the distance between the protrusions of the roughness, the liquid–gas interface inside the grooves appears to be almost flat. This theoretical conclusion[12,25,29] has recently been demonstrated experimentally.[50] Substituting the above information into eqn (1.4), the CB equation reads

$$\cos\theta_{CB} = r_f f\cos\theta_{Y1} + (1 - f)(-1) = -1 + f(1 + r_f\cos\theta_{Y1}) \tag{1.5}$$

A common problem in publications is the omission of r_f. Assuming $r_f = 1$ is correct only if the roughness protrusions have flat tops that are parallel to the surface.

1.3 Mechanism and Definition of Non-Wettability

The essential characteristic of a non-wettable surface is the ease of removal of a drop from the surface by applying a small force, such as a small fraction of the drop weight. This is usually tested by tilting the surface, similarly to the natural slight tilting of leaves, and measuring the angle at which the drop rolls off. The currently existing quantitative definition, which requires

CA > ~150° and roll-off angle < ~5°, has only an empirical justification. For fundamental understanding and ability to design successful new non-wettable surfaces, it is essential to study this point in more detail. Because of the prevalence of drop-related non-wetting applications, it makes sense to first reach a full understanding of these cases. However, the definition must be made more general. Easy removal of a drop from a solid surface appears to depend on two main factors: (a) the ability of a weak external force to get the drop out of equilibrium, and (b) high rate of removal from the surface. The factor that may keep a drop in equilibrium under the effect of an external force (say, gravity) is contact angle hysteresis, namely the existence of a range of metastable CAs. This allows the drop to assume a non-axisymmetric equilibrium shape as required by the external force. In contrast, on ideal surfaces the drop must be axisymmetric by definition, so it cannot stay in equilibrium even under the influence of a very small force.

Regarding the rate of removal, it is intuitively appealing to assume that the lower is the solid–liquid contact area, the higher is the rate of removal of the liquid from the solid surface.[12,25,29] If this is true, then the crux of the matter is to find a way to reduce the wetted area as much as possible. The first idea that comes to mind is making the CA as high as possible. However, a simple geometrical calculation indicates that by increasing the Young CA from 90° (considered usually as the lower limit of hydrophobicity) to 120° (the highest available Young CA in practice), the reduction in the area wetted by a drop is only by a factor of about 2. Thus, a different mechanism, capable of much bigger increase in the CA, is required.

Actually, the above two factors that characterize non-wettability can be translated into the following two objectives: (a) achieving a very small hysteresis range (by making the surface as uniform as possible); and (b) making the CA as high as possible. In principle, both objectives can be attained if the surface that is in contact with the liquid consists mostly of a gas, *e.g.* air trapped in roughness grooves. A gas is the most hydrophobic "surface" we can have, and is also the most uniform. Therefore, a CB state, where a liquid is supported by relatively few solid peaks, certainly answers the need. This statement leads to a possible unified definition of all types of non-wettable surfaces. Qualitatively, this definition may simply state that the wetted area has to be sufficiently small. Some initial calculations[14] showed that the wetted area in the CB state may be orders of magnitude lower than that in the W state, even for the same CA. Further quantitative work is required, but it is clear that non-wettability has to be associated with the CB state, as was qualitatively concluded above and also by Quéré.[8]

Whatever the exact definition, from a practical point of view it is clear that in order to be non-wettable the solid surface must be either rough or porous. The grooves of a rough surface are interconnected and open to the atmosphere. In a porous surface, the pores may be either interconnected or isolated. In the latter case it may be much easier to keep the air in the pores in a stable state, but structural constraints may limit the reduction of the wetted area. Therefore, the following discussion is limited only to structures with interconnected grooves or pores.

1.4 Stability Considerations

As previously discussed, roughness of the solid surface is a necessary condition for non-wettability; however, it is not a sufficient condition. As shown below, the geometric characteristics of the roughness may have a major influence. In general, there may be more than one equilibrium position for the liquid–air interface (*i.e.* minima in the Gibbs energy) within the roughness grooves. The most stable is, of course, the one that has the lowest Gibbs energy. Identifying equilibrium positions is easy: the two equilibrium indicators, namely the Young and the Young–Laplace equations, have to be fulfilled. The latter is fulfilled by the curvature of the liquid–gas interface inside the roughness grooves being the same as that of the outer liquid–air interface, as explained above. This is achieved by assuming that this interface is practically flat. Thus, the only question that needs to be considered is whether the local CA can equal the Young CA at the position that is tested. Then, the identified CB states as well as the W state (that is always a potential equilibrium position) have to be compared to find out the most stable state. In the following we discuss first the case of a drop on a solid, non-wettable surface and then that of a non-wettable surface beneath a liquid.

1.4.1 A Drop on a Non-Wettable Surface

To make the above analysis clearer it is best to study some examples. For a drop, it is technically easy to compare energies, since the energy varies monotonically with the apparent CA that the drop makes with the solid surface.[12,25] Thus, all that is needed in order to decide which state is more stable is to find out which is associated with a lower apparent CA. One of the simplest forms of roughness is that of straight pillars with a square cross-section. Let us assume that the height of the pillars is h, and that they have flat, horizontal tops of width $\sqrt{f}$ that cover an area fraction of f (see Figure 1.1(a)). In this case, there are only two possible equilibrium positions. One is the W state, and the other is the CB state with the liquid–gas interface attached to the top of the pillars. This is so, because it is only at the upper corner of the pillar that the liquid–gas interface can locally attain the Young CA when it is >90° (see Figure 1.1(b)). The roughness ratio is given by

$$r = 1 + 4h\sqrt{f} \tag{1.6}$$

Therefore,

$$\cos\theta_W = r\cos\theta_Y = (1 + 4h\sqrt{f})\cos\theta_Y \tag{1.7}$$

The local roughness ratio of the top of the pillar equals 1, therefore

$$\cos\theta_{CB} = -1 + f(1 + \cos\theta_Y) \tag{1.8}$$

The CB state is more stable if $\theta_{CB} < \theta_W$, namely if $\cos\theta_{CB} > \cos\theta_W$. When $\cos\theta_Y < 0$, this leads to

$$r > 1 + (f - 1)(1 + 1/\cos\theta_Y) \quad (1.9)$$

Thus, for this simple type of roughness, for a given chemistry ($\cos\theta_Y$) and surface density of protrusions (f), the only parameter that determines the stability of the non-wetting state is the roughness ratio that depends on the protrusion height, h. The wetting state turns from W to CB when the roughness ratio, namely height of protrusion, is sufficiently high.

For roughness features that are not flat at the top, the situation is more complex and interesting.[12,25] A simple example of two-dimensional roughness with a circular cross-section clearly demonstrates the phenomena that may be observed. For convex roughness features (see Figure 1.2(a)) it is

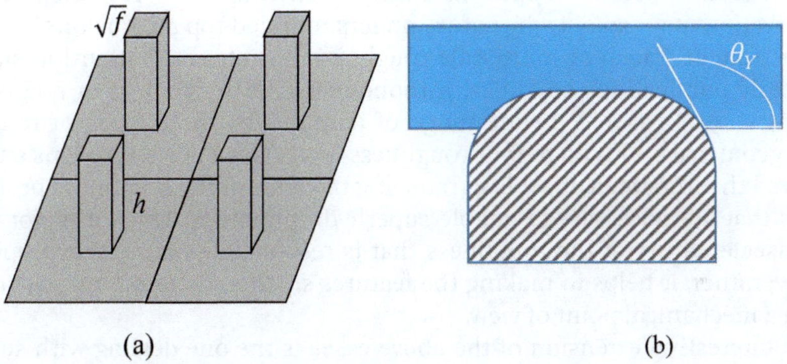

Figure 1.1 (a) A simple form of roughness, for which the transition from the Wenzel regime to the Cassie–Baxter regime depends only on the height of the protrusions, for a given chemistry ($\cos\theta_Y$) and surface density of protrusions (f). (b) The liquid–air interface may find a position that enables the local contact angle (CA) to equal the Young CA at the upper corner of the protrusion.

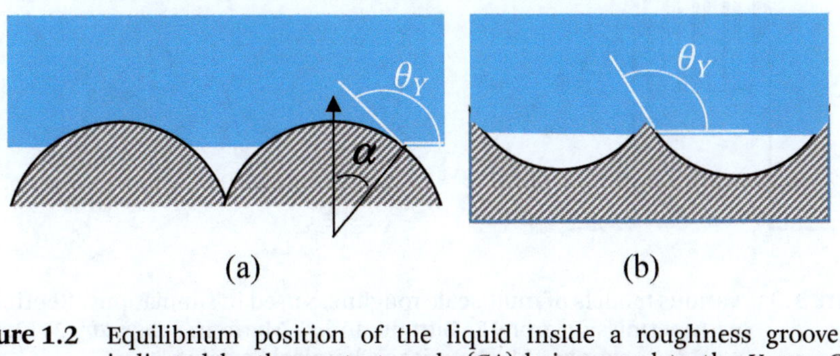

Figure 1.2 Equilibrium position of the liquid inside a roughness groove, as indicated by the contact angle (CA) being equal to the Young CA:[12] (a) convex roughness features; (b) concave roughness features.

possible to get a stable CB state above a certain roughness ratio, as explained in the following. First, a position of the liquid–air interface for which the CA equals the Young CA has to be identified. This is feasible only if the maximum position angle, α, (see Figure 1.2(a)) is bigger than $(180° - \theta_Y)$. Once this condition is fulfilled, we need to check if this position is a minimum in the Gibbs energy. It turns out[12] that indeed it is a minimum, and that above a certain roughness ratio (determined by the maximum value of α) the CB state is more stable than the W state.

The picture is reversed when the roughness features are concave (Figure 1.2(b)). In this case, the Gibbs energy keeps going down as the liquid penetration into the grooves advances until the W state is reached. Thus, although there exists a position where the CA equals the Young CA (Figure 1.2(b)), the system is unstable and must get to the W state. As concluded from additional studies,[29,43] it turns out that the specific protrusion shape within the group of convex shapes exerts a major effect. Rounded-top protrusions seem to be more effective than flat-topped ones with a sharp edges.[29,43] This theoretical observation may explain why nature prefers rounded-top protrusions.

The role of fractal or multiscale roughness has attracted attention since the early publications on superhydrophobicity.[2,15,21,23,28,30,33,34,39] A relatively recent study[43] covered a wide range of parameters: three types of roughness geometries with up to four roughness levels (see Figure 1.3). This study showed that the main effect is in reducing the sizes of the roughness protrusions that are necessary for stable superhydrophobicity. Thus, it is not the multiscale nature of the roughness that is responsible for superhydrophobicity; rather, it helps in making the features smaller, therefore more stable from a mechanical point of view.

An interesting extension of the above cases is the one dealing with superhygrophobic surfaces, namely non-wettable surfaces, for which the CAs of the wetting liquid is less than 90°. This case appears at first sight to contradict the common requirement of hydrophobicity for non-wettable surfaces. However, if we look at the CB eqn (1.5), there is no *a priori* reason that prevents $\cos\theta_{CB}$ from being negative, even if $\theta_Y < 90°$. For example, the Young CA may be acute at the equilibrium positions shown in Figure 1.4. However,

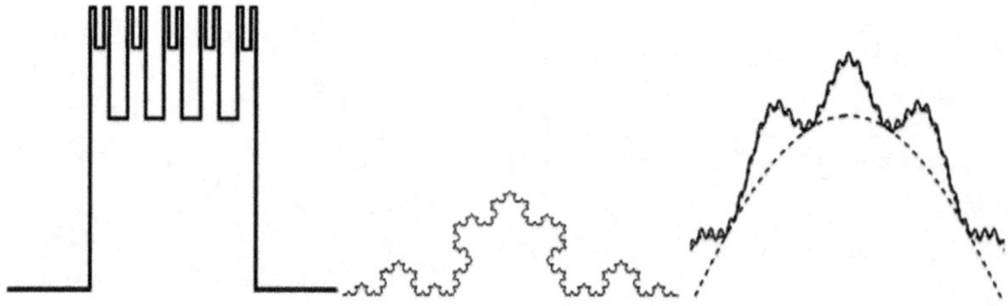

Figure 1.3 Various models of multiscale roughness used in simulations. Reprinted with permission from E. Bittoun and A. Marmur, *Langmuir*, 2012, **28**, 13933. Copyright 2012 American Chemical Society.[43]

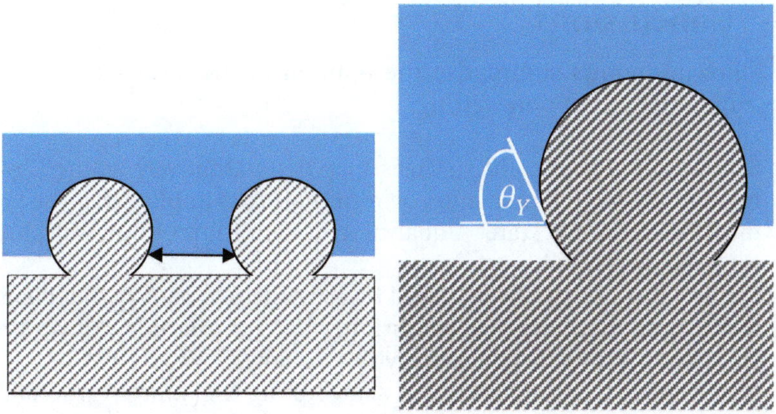

Figure 1.4 Superhygrophobic surface: the liquid–air interface is at equilibrium with a solid rough surface, the Young CA of which is <90°.[25]

stability also needs to be checked for this case. The results of this test lead to conclusions that are similar to those of a drop on a hydrophobic rough surface: it may be stable for convex protrusions, and unstable otherwise.[25] Nevertheless, it is important to realize that a superhygrophobic state is necessarily metastable, for the following reason. For a hygrophilic surface, the W state is characterized by a CA that is lower than the Young CA. On the other hand, the whole point in making a super-hygrophobic surface is to increase the CA beyond the Young CA. Thus, the W state, by definition, has a lower CA than the CB state, *i.e.* it is more stable. The special type of roughness that enables superhygrophobicity has been called by several names, such as "multivalued topography" or "re-entrant".

1.4.2 Underwater Superhydrophobicity

As mentioned in the introduction, there are important reasons for keeping a stable air film on a solid surface under water. This situation is not explicitly defined by apparent CAs related to the W or CB state. However, the concepts of the W and CB states remain valid in terms of the contact between the liquid and the solid.

The equilibrium criteria for the CB state turn out to be the same as for a drop.[18] The local CA between the liquid and the roughness protrusion must be the Young CA, and the curvature of the liquid–air interfaces must appear to be approximately zero, since it equals the curvature of the outside surface of the liquid. The condition that differentiates unstable equilibrium from metastable or stable ones, in terms of the roughness geometry, turns out to be the same as for a drop.[18] The stable CB state in this system is determined by eqn (1.9), which gives a minimum roughness ratio above which the CB state is stable.[18]

1.5 Conclusions

The following points summarize the opinion of the author regarding the main fundamentals of non-wetting:

- Non-wettability of solid surfaces may be qualitatively defined by stating that the wetted area must be minimal. This implies that the system must be in the CB state. A quantitative definition of non-wettability is yet to be developed.
- Stable CB states can be achieved by roughness geometry that conforms to a certain mathematical condition.[12,18,25] For example, convex protrusions enable it while concave cavities do not.
- Non-wetting in systems with an acute Young CA (superhygrophobicity) is feasible, but it is always metastable.
- Multiscale roughness is not essential for non-wettability; however, it improves the mechanical stability of the surface by lowering the required protrusion size.
- The detailed optimal topography of non-wettable surfaces has yet to be elucidated. Moreover, it is likely that there is more than one solution to the problem, depending on specific constraints.

References

1. A. B. D. Cassie and S. Baxter, *Trans. Faraday Soc.*, 1944, **40**, 546.
2. T. Onda, S. Shibuichi, N. Satoh and K. Tsujii, *Langmuir*, 1996, **12**, 2125.
3. C. Neinhuis and W. Barthlott, *Ann. Bot.*, 1997, **79**, 667.
4. S. Herminghaus, *Europhys. Lett.*, 2000, **52**, 165.
5. D. Oner and T. J. McCarthy, *Langmuir*, 2000, **16**, 7777.
6. M. Thieme, R. Frenzel, S. Schmidt, F. Simon, A. Hennig, H. Worch, K. Lunkwitz and D. Scharnweber, *Adv. Eng. Mater.*, 2001, **3**, 691.
7. Z. Yoshimitsu, A. Nakajima, T. Watanabe and K. Hashimoto, *Langmuir*, 2002, **18**, 5818.
8. D. Quéré, *Nat. Mater.*, 2002, **1**, 14.
9. S. Li, H. Li, X. Wang, Y. Song, Y. Liu, L. Jiang and D. Zhu, *J. Phys. Chem. B*, 2002, **106**, 9274.
10. J. Kijlstra, K. Reihs and A. Klamt, *Colloids Surf., A*, 2002, **206**, 521.
11. A. Duparre, M. Flemming, J. Steinert and K. Reihs, *Appl. Opt.*, 2002, **41**, 3294.
12. A. Marmur, *Langmuir*, 2003, **19**, 8343.
13. N. A. Patankar, *Langmuir*, 2003, **19**, 1249.
14. A. Marmur, *Langmuir*, 2004, **20**, 3517.
15. N. A. Patankar, *Langmuir*, 2004, **20**, 8209.
16. A. Otten and S. Herminghaus, *Langmuir*, 2004, **20**, 2405.
17. Y. Cong, G.-h. Chen, Y. Fang and L.-q. Ren, *J. Bionic. Eng.*, 2004, **1**, 249.
18. A. Marmur, *Langmuir*, 2006, **22**, 1400.

19. E. Bormashenko, Y. Bormashenko, T. Stein, G. Whyman and E. Bormashenko, *J. Colloid Interface Sci.*, 2007, **311**, 212.
20. A. Solga, Z. Cerman, B. F. Striffler, M. Spaeth and W. Barthlott, *Bioinspiration Biomimetics*, 2007, **2**, 126.
21. M. Nosonovsky and B. Bhushan, *Ultramicroscopy*, 2007, **107**, 969.
22. F. Peter, *Intrface*, 2007, **4**, 637.
23. Y. Yu, Z.-H. Zhao and Q.-S. Zhengu, *Langmuir*, 2007, **23**, 8212.
24. J. Genzer and A. Marmur, *MRS Bull.*, 2008, **33**, 742.
25. A. Marmur, *Langmuir*, 2008, **24**, 7573.
26. A. Tuteja, W. Choi and J. M. Mabry, *et al.*, *Proc. Natl. Acad. Sci. U. S. A.*, 2008, **105**, 18200.
27. D. Quéré, *Annu. Rev. Mater. Res.*, 2008, **38**, 71.
28. W. Li and A. Amirfazli, *Soft Matter*, 2008, **4**, 462.
29. E. Bittoun and A. Marmur, *J. Adhes. Sci. Technol.*, 2009, **23**, 401.
30. K. Koch, H. F. Bohn and W. Barthlott, *Langmuir*, 2009, **25**, 14116.
31. N. J. Shirtcliffe, G. McHale and M. I. Newton, *Langmuir*, 2009, **25**, 14121.
32. Y. Su, B. Ji, K. Zhang, H. Gao, Y. Huang and K. Hwang, *Langmuir*, 2010, **26**, 4984.
33. S. H. Sajadinia and F. Sharif, *J. Colloid Interface Sci.*, 2010, **344**, 575.
34. T. Liu, W. Sun, X. Sun and H. Ai, *Langmuir*, 2010, **26**, 14835.
35. W. Barthlott, T. Schimmel and S. Wiersch, *et al.*, *Adv. Mater.*, 2010, **22**, 2325.
36. G. McHale, M. I. Newton and N. J. Shirtcliffe, *Soft Matter*, 2010, **6**, 714.
37. Y. Su, B. Ji, K. Zhang, H. Gao, Y. Huang and K. Hwang, *Langmuir*, 2010, **26**, 4984.
38. S. H. Sajadinia and F. Sharif, *J. Colloid Interface Sci.*, 2010, **344**, 575.
39. H. H. Liu, H. Y. Zhang and W. Li, *Langmuir*, 2011, **27**, 6250.
40. C. W. Extrand, *Langmuir*, 2011, **27**, 6920.
41. S.-H. Hsu, K. Woan and W. Sigmund, *Mater. Sci. Eng., R*, 2011, **72**, 189.
42. A. Marmur, *Soft Matter*, 2012, **8**, 6867.
43. E. Bittoun and A. Marmur, *Langmuir*, 2012, **28**, 13933.
44. N. R. Geraldi, F. F. Ouali and R. H. Morris, *et al.*, *Appl. Phys. Lett.*, 2013, **102**, 214104.
45. P. R. Jones, X. Hao and E. R. Cruz-Chu, *et al.*, *Sci. Rep.*, 2015, **5**, 12311.
46. R. Ramachandran and M. Nosonovsky, *Phys. Chem. Chem. Phys.*, 2015, **17**, 24988.
47. G. Wolansky and A. Marmur, *Langmuir*, 1998, **14**, 5292.
48. R. N. Wenzel, *J. Ind. Eng. Chem.*, 1936, **28**, 988.
49. A. Marmur and E. Bittoun, *Langmuir*, 2009, **25**, 1277.
50. B. Haimov, S. Pechook, O. Ternyak and B. Pokroy, *J. Phys. Chem. C*, 2013, **117**, 6658.

CHAPTER 2

Non-Wetting, Stabilization, and Phase Transitions Induced by Vibrations and Spatial Patterns

RAHUL RAMACHANDRAN[a] AND MICHAEL NOSONOVSKY*[a]

[a]University of Wisconsin-Milwaukee, 3200 N Cramer St, Milwaukee, WI 53211, USA
*E-mail: nosonovs@uwm.edu

2.1 Introduction

Small fast vibrations constitute a temporal periodic pattern, while surface microtopography often introduces spatial patterns. Despite the fact that both types of patterns are small, they can significantly affect and alter the bulk properties of materials. We show in this chapter how small fast vibrations can be substituted by an effective force, which stabilizes an inverted pendulum or bouncing droplets. We call this effective force a "levitation" force, given that it provides support to suspended objects, such as liquid droplets, due to the effect of the vibro-levitation.

Levitation is the process by which an object is suspended by a physical force against gravity. Historically, levitation was claimed by many ancient spiritual or occult teachings, but the possibility of levitation as a physical phenomenon was also studied by many scholars including Isaac Newton, who secretly

RSC Soft Matter No. 5
Non-wettable Surfaces: Theory, Preparation, and Applications
Edited by Robin H. A. Ras and Abraham Marmur

Published by the Royal Society of Chemistry, www.rsc.org

investigated the possibility of levitation as an opposite force to gravitation.[1,2] Today physicists investigate several means of levitation including magnetic, electrostatic, acoustic, and aerodynamic forces.[3] Acoustic levitation is one interesting possibility. The phenomenon is based on the non-linear nature of intense sound waves resulting in the acoustic radiation pressure creating an average positive force on a suspended object which resists the weight of the object.

Of particular interest is the acoustic levitation of a small droplet. Droplets, despite their apparent simplicity, are quite complex objects involving such effects as surface tension, Laplace pressure, capillary waves, and non-linear viscosity.[4–11] Droplet transport, coalescence, and bouncing off solid and liquid surfaces is still not completely understood, since these processes involve complex interactions and lead to complicated scenarios of droplet evolution. This complexity of droplet behaviour makes droplets suitable for various applications. In the past, it has been suggested that the droplets could be used for microfluidic applications; for instance, they could serve as microreactors for various chemical compounds carried by coalescent water droplets. For example, it has been shown that droplet coalescence can realize Boolean logic and thus a "droplet computer" can in principle be created.[12]

Recent studies have shown experimentally that incoming droplets can bounce off a vibrating liquid surface, thus leading to the "walking droplets", which, in a sense, combine the properties of waves and particles and serve as an illustration of the particle–wave duality.[13,14] The effect of bouncing droplets is thought to be similar to the acoustic levitation due to non-linear viscosity in a thin film. However, a detailed model of such an effect remains quite complex, and several ideas have been suggested in the literature.

It has been suggested[15] that the classical stability problem of an inverted pendulum on a vibrating foundation has relevance to a diverse class of non-linear effects involving dynamic stabilization of statically unstable systems ranging from the vibrational stabilization of beams to novel "dynamic materials," the transport and separation of granular material, soft matter, bubbles and droplets, to synchronization of rotating machinery. In these problems, the small fast vibrational motion can be excluded from consideration and substituted by effective slow forces acting on the system causing the stabilizing effect.[16]

In this chapter we suggest a simple analogy between levitating droplets over a vibrating liquid surface and a well-known mechanical system consisting of an inverted pendulum on a vibrating foundation. This analogy sheds light on the necessary conditions for droplet levitation. We further discuss the relation of the phenomenon to other non-linear vibration-caused effects, such as the vibro-levitation of a flexible stiff rope ("Indian rope trick"), the shear-thickening of non-Newtonian fluids ("cornstarch monsters"), and vibration-induced phase transitions, as well as possible applications for "smart" dynamic nanocomposite materials.[17]

2.2 Effective Force Corresponding to Small Fast Vibrations

In this section, we study pure mechanical systems undergoing fast vibrations in a time-independent potential field. Two different approaches have been developed to separate the fast small vibrations from the overall motion of the system; the Mathieu equation approach and Kapitza's method of the separation of motions. The latter method was further developed by Blekhman who suggested an interesting interpretation with two observers. One observer can see the small vibrations while the other one, who does not see the vibrations (*e.g.* due to a specially designed stroboscopic light), nonetheless observes their effect as a fictitious force, similar to the force of inertia. We will discuss various examples and derive a mathematical expression for an effective stabilizing "vibro-levitation" force. First we discuss a general case of Kapitza's separation of fast and slow motions. Kapitza thought of such systems being in a state of slow oscillation with a fast vibration superimposed upon it. The effect of fast vibrations can be isolated as a change in the effective potential energy of the system. Following Blekhman, an observer in a vibrating frame of reference will perceive the stability as a result of an additional fictitious force, which we refer to as the vibro-levitation force. We apply this method to the classic example of an inverted pendulum, and calculate the vibro-levitation force. This is followed by a review of the Mathieu equation approach to studying the stability of an inverted pendulum. Then, we look at the stabilization of multiple pendulums, and a continuous system involving a rope. Replacing fast vibrations with an effective force can not only be applied to systems described above, but also to non-coalescing droplets on a vibrating bath and other liquid systems which are discussed in later sections.

2.2.1 Motion Subjected to a Rapidly Oscillating Force

The method of separation of motions was first suggested by Kapitza[16] to study the stability of a pendulum on a vibrating foundation and then generalized for the case of an arbitrary motion in a rapidly oscillating field by Landau and Lifshitz.[18]

Consider a material point with mass m in a potential field $\Pi(x)$, where x is a spatial coordinate, with the minimum corresponding to the stable equilibrium. One can think about a mechanical spring-mass system as shown in Figure 2.1a. The restoring "spring force" acting on the mass is given by $-\mathrm{d}\Pi/\mathrm{d}x$, therefore, the equation of motion of the system is $m\ddot{x} = -\mathrm{d}\Pi/\mathrm{d}x$. In addition to the time-independent potential field $\Pi(x)$, a "fast" external periodic force $f\cos\Omega t$ acts upon the mass with a small-amplitude f and high frequency $\Omega \gg \sqrt{\left(\mathrm{d}^2\Pi/\mathrm{d}x^2\right)/m}$, which is much higher than the natural frequency (Figure 2.1b). The equation of motion then becomes

$$m\ddot{x} = -\left(\mathrm{d}\Pi/\mathrm{d}x\right) + f\cos\Omega t \tag{2.1}$$

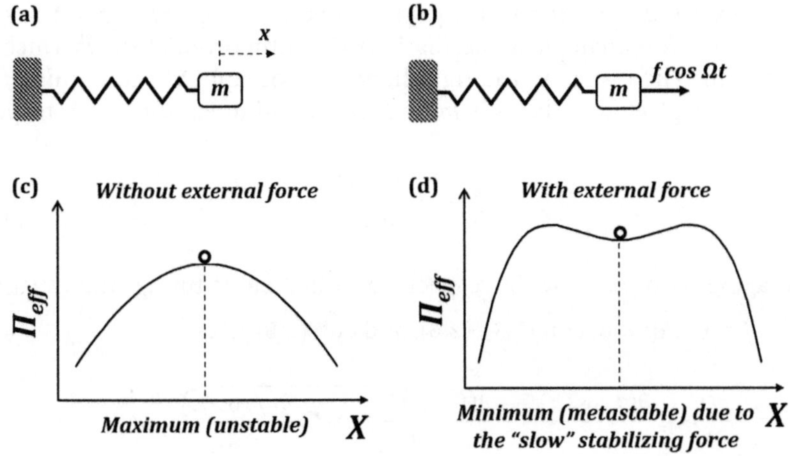

Figure 2.1 (a) Motion in one dimension of a mass m connected to a spring. (b) Application of an external force $f\cos\Omega t$ to the mass m. (c) Unstable equilibrium corresponding to the maximum potential energy. (d) A metastable equilibrium due to the stabilizing effect of the external force.

The location of the mass can be written as a sum of the "slow" oscillations $X(t)$ due to the "slow" force and small "fast" oscillations $\xi(t)$ due to the "fast" force:

$$x(t) = X(t) + \xi(t) \tag{2.2}$$

The mean value $\bar{\xi}(t)$ of this fast oscillation over its period $2\pi/\Omega$ is zero, whereas $X(t)$ changes slightly during the same period:

$$\bar{\xi}(t) = \frac{\Omega}{2\pi}\int_0^{2\pi/\Omega} \xi(t)\mathrm{d}t = 0 \tag{2.3}$$

$$\bar{X}(t) \approx X(t) \tag{2.4}$$

Therefore the mean location of the mass can be written as

$$\bar{x}(t) = \bar{X}(t) + \bar{\xi}(t) \approx X(t) \tag{2.5}$$

and the second derivative as

$$\ddot{\bar{x}}(t) \approx \ddot{X}(t) \tag{2.6}$$

Substituting eqn (2.2) in eqn (2.1) and using the Taylor series first-order terms in powers of ξ,

$$m\ddot{X} + m\ddot{\xi} = -\frac{\mathrm{d}\Pi}{\mathrm{d}x} - \xi\frac{\mathrm{d}^2\Pi}{\mathrm{d}x^2} + f\cos\Omega t + \xi\frac{\partial(f\cos\Omega t)}{\partial X} \tag{2.7}$$

The "slow" and "fast" terms in eqn (2.7) must separately be equal. The second derivative of small "fast" oscillations $\ddot{\xi}$ is proportional to Ω^2 which is a large term. The terms on the right-hand side of eqn (2.7) containing the small ξ are neglected. This gives $m\ddot{\xi} = f\cos\Omega t$, and integrating with respect to time t,

$$\xi = -\frac{f\cos\Omega t}{m\Omega^2} \tag{2.8}$$

Averaging eqn (2.7) with respect to time, substituting the relation $\frac{\Omega}{2\pi}\int_0^{2\pi/\Omega} f\cos\Omega t \mathrm{d}t = 0$, eqn (2.3)–(2.6), and eqn (2.8) gives

$$m\ddot{X} = -\frac{\mathrm{d}\Pi}{\mathrm{d}X} + \overline{\xi\frac{\partial(f\cos\Omega t)}{\partial X}} = -\frac{\mathrm{d}\Pi}{\mathrm{d}X} - \frac{1}{m\Omega^2}\overline{f\cos\Omega t\frac{\partial(f\cos\Omega t)}{\partial X}}$$

$$m\ddot{X} = -\frac{\mathrm{d}\Pi}{\mathrm{d}X} - \frac{1}{2m\Omega^2}\frac{\partial\overline{(f\cos\Omega t)^2}}{\partial X} \tag{2.9}$$

This can be written as $m\ddot{X} = -\frac{\mathrm{d}\Pi_{\mathrm{eff}}}{\mathrm{d}X}$ where Π_{eff} is an effective potential energy given by

$$\Pi_{\mathrm{eff}} = \Pi + \frac{\overline{(f\cos\Omega t)^2}}{2m\Omega^2} = \Pi + \frac{f^2}{4m\Omega^2} = \Pi + \frac{m}{2}\overline{\dot{\xi}^2} \tag{2.10}$$

Thus the effect of "fast" vibrations ξ when averaged over the time period $2\pi/\Omega$ is equivalent to the additional term $m\overline{\dot{\xi}^2}/2$ on the right-hand side in eqn (2.10). This term is the mean kinetic energy of the system under "fast" oscillations. Thus small "fast" vibrations can be substituted by an additional term in the potential energy resulting in the same effect that oscillations have on the system. The most interesting case is when this term affects the state of the equilibrium of a system. Let us say that in the absence of vibrations a system has an effective potential energy $\Pi_{\mathrm{eff}} = \Pi$ (a local maximum of the potential energy, Figure 2.1c). Vibrations can bring this system to a stable equilibrium due to the additional term discussed before (a local minimum of the potential energy, Figure 2.1d). In such cases the small "fast" vibrations have a stabilizing effect on the state of equilibrium.

Blekhman[15] has applied the method of separation of motions to many mechanical systems and suggested what he called "vibrational mechanics" as a tool to describe a diverse range of effects in the mechanics of solid and liquid media, from effective liquefying of granular media, which can flow through a hole like a liquid when on a vibrating foundation, to the opposite effect of solidifying liquid by jamming a hole in a vessel on a vibrating foundation, to vibro-synchronization of the phase of two rotating shafts on a vibrating foundation.

Blekhaman[15] has also suggested an elegant interpretation of the separation of motions. According to his interpretation, there are two different observers who can look at the vibrating system. One is an ordinary observer in an inertial frame of reference who can see both small, ζ, and large, X, oscillations. The other one is a "special" observer in a vibrating frame of reference, who does not see the small-scale motion, ζ, possibly due to a stroboscopic effect or just because his vision is not sensitive enough to see the small-scale motion. As a result, what the ordinary observer sees as an effect of the fast small vibrations is perceived by the special observer as an effect of some new effective force. This fictitious force is similar to the inertia force which is observed by observers in a non-inertial frame of reference. Furthermore, when the stabilizing effect occurs, the special observer attributes the change in effective potential energy to fictitious "slow" stabilizing forces or moments. The additional "slow" stabilizing force V for the system can be written as

$$V = -\frac{\partial}{\partial X}\left(\frac{f^2}{4m\Omega^2}\right) \tag{2.11}$$

The most common example of the stabilizing effect of the small vibrations is the inverted pendulum, which is studied in the following section.

2.2.2 Inverted Pendulum

We now consider the classic problem of stability of an inverted pendulum to apply the method of the separation of motions and determine a stabilizing vibro-levitation force. A simple pendulum is a common example used in mechanics to introduce the fundamentals of simple harmonic motion. Consider a pendulum with a point mass m connected to the end of a pivoted link of length L. The angular position of the pendulum about its pivot is described by the angle ψ. It has its stable equilibrium at its vertical lower position, $\psi = 0^\circ$ where the potential energy is minimum as shown in Figure 2.2. Any small perturbations from this position results in oscillations about the equilibrium with natural frequency $\omega = \sqrt{g/L}$ where g is the acceleration due to gravity. Eventually the pendulum returns to its equilibrium due to the restoring force $-\frac{\mathrm{d}(mgL\cos\psi)}{\mathrm{d}\psi}$.

A pendulum also has an unstable equilibrium that corresponds to the point of inflection at $\psi = 180^\circ$ (Figure 2.2). When the foundation of the pendulum is subjected to vertical harmonic oscillations $A\cos\Omega t$, where A is the amplitude and $\Omega \gg \omega$ is the frequency, the equilibrium at $\psi = 180^\circ$ can, under certain conditions, become stable. A pendulum on a vibrating foundation is called "Kapitza's pendulum" after Peter Kapitza. The equation of motion can be written as

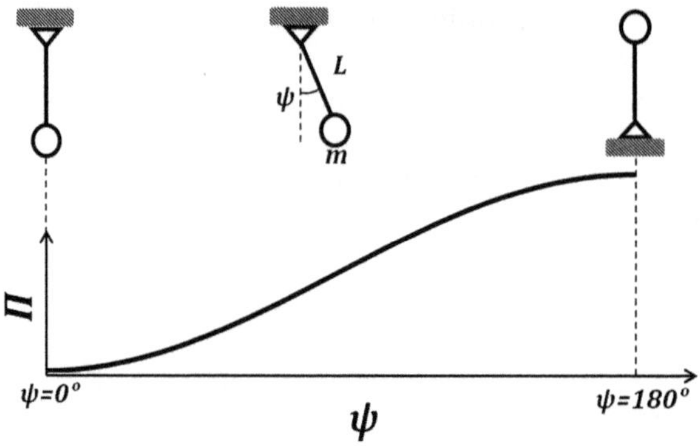

Figure 2.2 The potential energy Π of a pendulum as a function of its angular displacement ψ.

$$L\ddot{\psi} = g\sin\psi - A\Omega^2 \sin\psi\cos\Omega t \tag{2.12}$$

The form of eqn (2.12) is similar to that of eqn (2.1) with $f = -mA\Omega^2 \sin\psi$. Substituting f into eqn (2.10), the effective potential energy can be obtained as

$$\Pi_{\text{eff}} = mgL\left(-\cos\psi + \frac{A^2\Omega^2}{4gL}\sin^2\psi\right) \tag{2.13}$$

Now if we look at the stabilized inverted pendulum, it appears upright and stationary. By differentiating the effective potential energy in eqn (2.13) we obtain the generalized force (with the dimension of torque) acting upon the pendulum. In addition to the term involving $\sin\psi$, this generalized force now involves the term given by eqn (2.11):

$$V = \frac{\partial}{\partial\psi}\left(-\frac{mA^2\Omega^2}{4}\sin^2\psi\right) = -\frac{mA^2\Omega^2}{4}\sin 2\psi \tag{2.14}$$

Note that V is dimensionally a torque, because the spatial coordinate ψ is angular displacement. This additional effective force can have a stabilizing effect on the unstable equilibrium. The effect of this force is equivalent to that of a spring with spring constant $k = \dfrac{mA^2\Omega^2}{2}$ when the angle ψ is close to 180°. This is equivalent to the upright pendulum supported by a spring (Figure 2.3).

The equilibrium is stable when the effective potential energy in eqn (2.13) is a positive-definite function near the state of equilibrium, which yields the stability criterion

$$A^2\Omega^2 > 2gL \tag{2.15}$$

Thus, when the amplitude and frequency of the small fast vibrations of the foundation satisfy eqn (2.15), the otherwise unstable equilibrium at

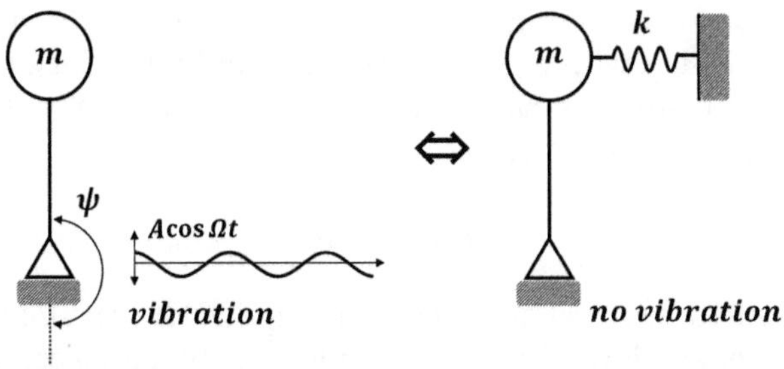

Figure 2.3 The figure on the left shows an inverted pendulum stabilized by a foundation vibrating with a periodic displacement $A\cos\Omega t$. The same system can be represented as shown in the figure on the right with the pendulum being stabilized by a spring of effective spring constant k. Reproduced from ref. 11 with permission from the Royal Society of Chemistry.

$\psi = 180°$ can correspond to a local minimum for the effective potential energy, *i.e.* it can become a stable equilibrium. Thus, we have derived the expression for the stabilizing force (eqn (2.14)) and a stability criterion for the inverted pendulum (eqn (2.15)) using the separation of motion method. We will apply this method to study non-coalescing liquid droplets later in this chapter.

2.2.3 Mathieu Equation Method

The result in eqn (2.15) has been historically obtained using a different method, namely, the parametric resonance Mathieu equation analysis suggested by A. Stephenson in 1908.[19,20] The motion of a pendulum on a vibrating foundation is an example of parametric oscillation. The differential equation of motion of such a pendulum contains time-varying coefficients and is called the Mathieu equation. Stephenson found that when the pivot of a pendulum is subjected to a vertical periodic motion at a frequency $2\omega/n$ where n is any integer, then the oscillations of the pendulum are gradually amplified. The pendulum eventually becomes highly unstable. Stephenson used the Mathieu equation approach to study the conditions for stability and instability of the pendulum. In this section we briefly describe the Mathieu equation approach to determining the stability criteria of an inverted pendulum.

The equation of motion of a pendulum on a vibrating foundation (eqn (2.12)) can be rewritten as

$$\ddot{\psi} - \left(\frac{g}{L} - \frac{A\Omega^2}{L}\cos\Omega t\right)\sin\psi = 0 \tag{2.16}$$

which has the form of the Mathieu equation. To study the stability of a solution of eqn (2.16) using the perturbation technique, the variables $z = \psi$, $\delta = 4g/L\Omega^2$, $\varepsilon = 2A/L$, where $\varepsilon \ll 1$ and $\tau = \Omega t$, are introduced. For small values of z, $\sin z \approx z$ and the equation of motion for a pendulum reduces to

$$\ddot{z} - \frac{\Omega^2}{4}(\delta - 2\varepsilon\cos\tau)z = 0 \tag{2.17}$$

The stability of a pendulum with vibrating foundation is studied in the parameter plane (δ,ε), with regions of stability and instability, the graphical representation of which is called the Ince–Strutt diagram. For an inverted pendulum the stability criterion is

$$-\frac{1}{2}\varepsilon^2+\frac{7}{8}\varepsilon^4\ldots<\delta<1-\varepsilon-\frac{1}{8}\varepsilon^2\ldots \tag{2.18}$$

and is represented by the shaded region in Figure 2.4. For stability at any δ, there is an upper and lower bound for ε. It follows that for a certain length of the inverted pendulum there exists a stability range of frequencies $\Omega_1 < \Omega < \Omega_2$. From eqn (2.18), the stability criterion can be obtained as follows. Since we are concerned with an inverted pendulum, we restrict ourselves to the set of negative values of δ in the vicinity of zero and we can write

$$-\frac{1}{2}\varepsilon^2<-\delta \tag{2.19}$$

Substituting $\delta = 4g/L\Omega^2$ and $\varepsilon = 2A/L$ into eqn (2.19) we obtain the same stability criteria as in eqn (2.15). The Mathieu equation approach is another way of analysing the vibro-levitation of an inverted pendulum.

We see that the Mathieu equation approach provides the same stability criterion as the method of separation of motion. However, the latter has a more general application and is not limited to the parametric excitation of a pendulum. We can therefore apply the method of separation of motion to more complex problems of the multiple pendulum, the continuous (flexible stiff beam) pendulum, and liquid systems like non-coalescing droplets. We also draw an analogy between mechanical systems undergoing vibration and non-linear behaviour in vibrating fluids that leads to non-wetting and phase transition.

2.2.4 Multiple Pendulums and the Indian Rope Trick

We have discussed the stabilization of a single inverted pendulum by small-amplitude fast vibration of the pendulum's foundation. Inverted multiple pendulums consisting of a number of freely jointed links can also be stabilized by applying a harmonic oscillation at the foundation as long as the frequency of the oscillation is sufficiently large. The theoretical proof was put

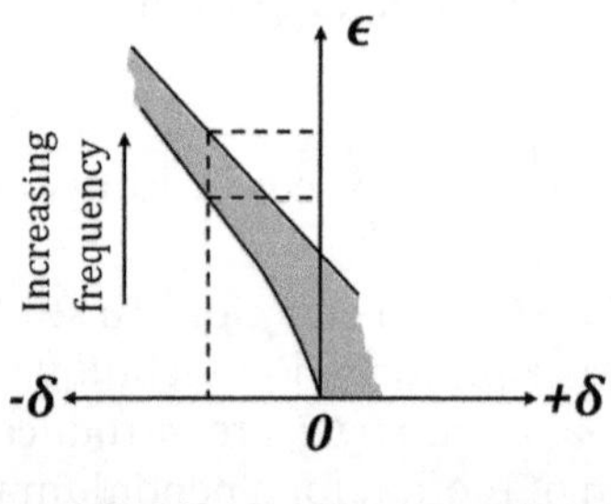

Figure 2.4 The region of stability for an inverted pendulum as seen in the Ince–Strutt diagram. Adapted from ref. 11 with permission from the Royal Society of Chemistry.

forward by Stephenson[21] who derived the stability criteria. Acheson derived the stability criterion for a multiple pendulum using the Mathieu equation approach. He showed that the region of stability in the Ince–Strutt diagram diminishes as the number of links in the pendulum increases. As the number of links approaches infinity, as in the case of a perfectly flexible string, the region of stability vanishes.[22] Acheson and Mullin later experimentally demonstrated the stability of double and triple inverted pendulums.[23]

An even more complex, albeit related, case is a continuous system consisting of a flexible beam. Since it has been shown that the limiting case of multiple pendulums, *i.e.* a string, cannot be stabilized in the upside-down position, flexural stiffness must be introduced.

Interestingly, some researchers have suggested that stabilization by a vibrating foundation can explain the so-called Indian rope trick. This trick involves a magician (traditionally an Indian fakir) throwing one end of a flexible rope vertically upwards, which under certain conditions levitates like a vertical rod. In certain versions of the trick a small animal (an ape) could even climb the rope, leaving the audience in awe. This defies the empirical observation that an upright column exceeding a critical length will buckle under its own weight. Although accounts of the trick remain controversial, it has been shown that a rope with bending stiffness can be stabilized at sufficiently high frequencies. A piece of steel curtain wire longer than its critical buckling length was able to stay upright when its pivot was vibrated within a certain frequency range $\Omega_1 < \Omega < \Omega_2$. When the frequencies were reduced below Ω_1 the wire fell over, while increasing the frequencies above Ω_2 resulted in instabilities in the wire.[24] Ramachandran and Nosonovsky[11] demonstrated instabilities in a plastic rope when its pivot was oscillated at a certain range of frequencies. The rope, which was initially in a buckled state, became unstable at a certain frequency (Figure 2.5). The instabilities grew with increase in frequency till an upper limit

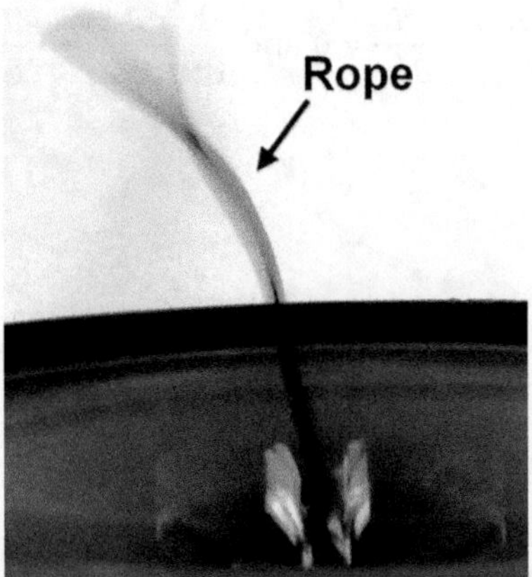

Figure 2.5 Instabilities in a plastic rope on a foundation vibrating at 130 Hz. Reproduced from ref. 11 with permission from the Royal Society of Chemistry.

was reached, beyond which the instabilities gradually decreased and the rope returned to its buckled state.

Now we derive the expression for the stabilizing force for multiple pendulums and a flexible stiff rope. First let us consider a double pendulum as shown in Figure 2.6 with point masses m_1 and m_2 attached to links of lengths L_1 and L_2 respectively. The foundation of the pendulum is subjected to a harmonic oscillation $A\cos\Omega t$. Let the angular displacements of masses m_1 and m_2 be ψ_1 and ψ_2 respectively.

For m_1 we can write the horizontal and vertical displacements as $x_1 = L_1 \sin\psi_1$ and
$y_1 = L_1\cos\psi_1 + A\cos\Omega t$. Similarly for m_2, $x_2 = L_1\sin\psi_1 + L_2\sin\psi_2$ and
$y_2 = L_1\cos\psi_1 + L_2\cos\psi_2 + A\cos\Omega t$. The x and y components of velocities are
$\dot{x}_1 = L_1\dot{\psi}_1\cos\psi_1$,
$\dot{y}_1 = -L_1\dot{\psi}_1\sin\psi_1 - A\Omega\sin\Omega t$, $\dot{x}_2 = L_1\dot{\psi}_1\cos\psi_1 + L_2\dot{\psi}_2\cos\psi_2$ and
$\dot{y}_2 = -L_1\dot{\psi}_1\sin\psi_1 - L_2\dot{\psi}_2\sin\psi_2 - A\Omega\sin\Omega t$.

The kinetic energy of the system is given by $K = \frac{1}{2}m_1\left(\dot{x}_1^2 + \dot{y}_1^2\right) + \frac{1}{2}m_2\left(\dot{x}_2^2 + \dot{y}_2^2\right)$.
The potential energy of the system is given by $\Pi = m_1 g y_1 + m_2 g y_2$. The Lagrangian of the system can be written in terms of the angular displacements and their derivatives as $L = K - \Pi$:

$$
\begin{aligned}
L &= \frac{1}{2}m_1\left(L_1^2\dot{\psi}_1^2 + A^2\Omega^2\sin^2\Omega t + 2L_1\dot{\psi}_1 A\Omega\sin\psi_1\sin\Omega t\right) \\
&+ \frac{1}{2}m_2\begin{pmatrix} L_1^2\dot{\psi}_1^2 + L_2^2\dot{\psi}_2^2 + 2L_1L_2\dot{\psi}_1\dot{\psi}_2\cos(\psi_1 - \psi_2) \\ + 2A\Omega\sin\Omega t\left(L_1\dot{\psi}_1\sin\psi_1 + L_2\dot{\psi}_2\sin\psi_2\right) + A^2\Omega^2\sin^2\Omega t \end{pmatrix} \\
&- m_1 g\left(L_1\cos\psi_1 + A\cos\Omega t\right) - m_2 g\left(L_1\cos\psi_1 + L_2\cos\psi_2 + A\cos\Omega t\right)
\end{aligned}
\tag{2.20}
$$

The equations of motion are then given by the Lagrange equations

$$
\frac{\mathrm{d}}{\mathrm{d}t}\left(\frac{\partial L}{\partial\dot{\psi}_1}\right) - \frac{\partial L}{\partial\psi_1} = 0 \text{ and } \frac{\mathrm{d}}{\mathrm{d}t}\left(\frac{\partial L}{\partial\dot{\psi}_2}\right) - \frac{\partial L}{\partial\psi_2} = 0 \tag{2.21}
$$

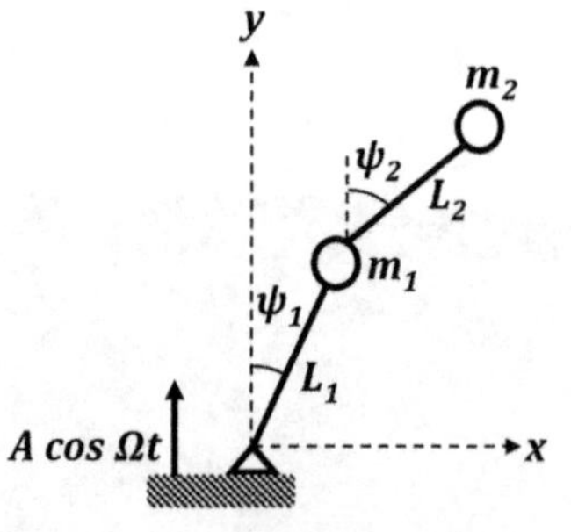

Figure 2.6 An inverted double pendulum whose foundation is subjected to a sinusoidal vibration $A\cos\Omega t$.

Substituting for L and simplifying we obtain

$$(m_1+m_2)L_1\ddot{\psi}_1 + m_2L_2\ddot{\psi}_2\cos(\psi_1-\psi_2) + m_2L_2\dot{\psi}_2^2\sin(\psi_1-\psi_2) - m_1gL_1\sin\psi_1 \\ - m_2gL_1\sin\psi_1 + A\Omega^2\cos\Omega t(m_1\sin\psi_1 + m_2\sin\psi_1) = 0$$

and

$$m_2L_2\ddot{\psi}_2 + m_2L_1\ddot{\psi}_1\cos(\psi_1-\psi_2) - m_2L_1\dot{\psi}_1^2\sin(\psi_1-\psi_2) - m_2gL_2\sin\psi_2 \\ + m_2A\Omega^2\sin\psi_2\cos\Omega t = 0$$

Rewriting the equations of motion in the form of eqn (2.1), we have

$$(m_1+m_2)L_1\ddot{\psi}_1 + m_2L_2\ddot{\psi}_2\cos(\psi_1-\psi_2) + m_2L_2\dot{\psi}_2^2\sin(\psi_1-\psi_2) \\ = m_1gL_1\sin\psi_1 + m_2gL_1\sin\psi_1 - (m_1+m_2)A\Omega^2\sin\psi_1\cos\Omega t$$

and

$$m_2L_2\ddot{\psi}_2 + m_2L_1\ddot{\psi}_1\cos(\psi_1-\psi_2) - m_2L_1\dot{\psi}_1^2\sin(\psi_1-\psi_2) \\ = m_2gL_2\sin\psi_2 - m_2A\Omega^2\sin\psi_2\cos\Omega t$$

Comparing these with eqn (2.1), we see

$$f_1 = -(m_1+m_2)A\Omega^2\sin\psi_1 \text{ and } f_1 = -m_2A\Omega^2\sin\psi_2 \tag{2.22}$$

Using eqn (2.11), the effective generalized forces on m_1 and m_2 can be written as

$$\begin{aligned} V_1 &= \frac{\partial}{\partial\psi_1}\left(\frac{f_1^2}{4m_1\Omega^2}\right) = -\frac{(m_1+m_2)^2}{4m_1}A^2\Omega^2\sin 2\psi_1 \\ V_2 &= \frac{\partial}{\partial\psi_2}\left(\frac{f_2^2}{4m_2\Omega^2}\right) = -\frac{m_2}{4}A^2\Omega^2\sin 2\psi_2 \end{aligned} \tag{2.23}$$

For any mass m_i in a system of n connected pendulums as shown in Figure 2.7,

$$f_i = -A\Omega^2\left(\sum_{j=i}^{n} m_j\right)\sin\psi_i \tag{2.24}$$

and the stabilizing effective generalized force is

$$V_i = \frac{\partial}{\partial\psi_i}\left(\frac{f_i^2}{4m_i\Omega^2}\right) = -\frac{A^2\Omega^2}{4m_i}\left(\sum_{j-i}^{n} m_j\right)^2\sin 2\psi_i \tag{2.25}$$

The multiple pendulums are stabilized due to the system of effective generalized forces$\{V_1,V_2,...V_n\}$ as shown in Figure 2.7. For small angular

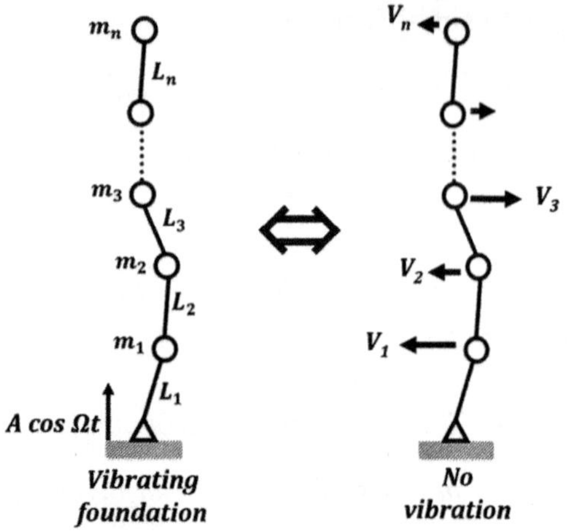

Figure 2.7 A multiple pendulum which is being stabilized by vibrating its foundation is equivalent to a multiple pendulum which is stabilized by a system of generalized vibro-levitation forces $V_1, V_2, ..., V_n$.

displacements of the system of n connected pendulums, the equivalent spring constant at the first link is

$$k_1 = \frac{A^2\Omega^2}{2m_1}\left(\sum_{i=1}^{n} m_i\right)^2 \tag{2.26}$$

Studies of the Indian rope trick usually approximate the rope or wire to continuum objects such as a rod or column with appreciable stiffness. For example, Champneys and Fraser[25] studied the Indian rope trick for a linearly elastic rod. The equation of motion in terms of the lateral displacement u at arc length s is

$$\eta\frac{\partial^2 u}{\partial t^2} + (1 - \eta\varepsilon\cos t)\frac{\partial}{\partial s}\left((1 - s)\frac{\partial u}{\partial s}\right) + b\frac{\partial^4 u}{\partial s^4} = 0 \tag{2.27}$$

where η, ε, and b are the dimensionless acceleration, amplitude, and stiffness respectively. Comparing this with eqn (2.1) we can write

$$f = \eta\varepsilon\frac{\partial}{\partial s}\left((1 - s)\frac{\partial u}{\partial s}\right)$$

and formulate the effective vibro-levitation force using eqn (2.11).

Shishkina *et al.* investigated a rope treated as a flexible Euler beam with the stiffness k subjected to the gravity and an axial load oscillating near the constant value of c^2 with amplitude εa^2 and frequency Ω. The transversal deflection of the beam $u(x,t)$ is governed by

$$\frac{\partial^2 u}{\partial t^2} + k\frac{\partial^4 u}{\partial x^4} + (c^2 + \varepsilon\Omega^2\sin\Omega t)\frac{\partial u}{\partial x} + (c^2 + \varepsilon\Omega^2\sin\Omega t)x\frac{\partial^2 u}{\partial x^2} = 0 \tag{2.28}$$

They showed that effect of the oscillating load is equivalent to the increase of the effective flexural stiffness of the rope k, which becomes equal to

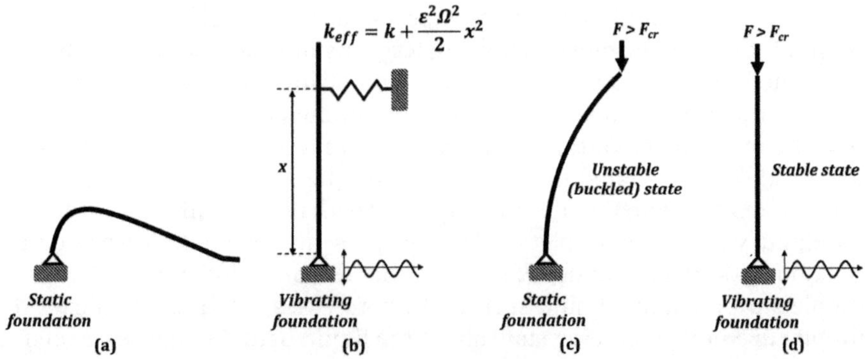

Figure 2.8 (a) A rope which is subject to no vibration buckles under its own weight. (b) Vertical vibrations results in an increased effective stiffness which prevents buckling. (c) For any beam there is a critical force (F_{cr}) that depends on the beam material and geometry. Any load (F) greater than this will cause the beam to buckle. (d) Vibrating the foundation leads to an increase in the effective stiffness of the beam, and the beam is able to resist buckling.

$k_{eff} = k + \frac{\varepsilon^2 \Omega^2}{2} x^2$, where x is the distance along the rope (Figure 2.8a and b). This increase can be sufficient to exceed the critical value of the stiffness and prevent buckling of the beam (Figure 2.8c and d).[26]

For a multiple pendulum of n connected links, as $n \to \infty$ the system becomes more flexible and its stiffness decreases. Now the system is similar to a limp rope. From eqn (2.25), the vibro-levitation force is proportional to the mass. Therefore as $n \to \infty$, the vibro-levitation force becomes infinite. It follows that the Indian rope trick cannot be performed if the rope does not have sufficient inherent stiffness.

In the previous sections we introduced the method of separation of motions, applied it to various mechanical systems undergoing vibration, and derived an effective stabilizing force for each case. In the next section, we study non-coalescing droplets stabilized by vibrations. We also apply the method of separation of motions to formulate an expression for the effective force that causes their non-coalescing, non-wetting behaviour. We also draw parallels with the vibration-induced stability of an inverted pendulum.

2.3 Vibro-Levitation of Droplets

Water droplets are seen to float momentarily on the surface of water and then coalesce into the bulk fluid. Sometimes they emit a smaller droplet as a result of coalescence, which then undergoes the same fate as the parent droplet.[27] This phenomenon is called coalescence cascade. Such non-coalescing droplets were noticed as early as 1881 when Reynolds studied the influence of surface impurities on this peculiar behaviour of droplets. He concluded that a pure liquid surface is required for droplets to float over it.[28]

Walker demonstrated by a simple experiment that droplets of an aqueous soap solution can levitate in a non-coalescent state above a vibrating bath of the same bulk solution. The droplets could levitate indefinitely if standing waves (Faraday instabilities) were set up on the bulk liquid surface.[29] Recently this phenomenon has attracted the attention of researchers once again. Couder *et al.*[30] demonstrated that silicone oil droplets could be levitated indefinitely over a sinusoidally vibrating ($A \cos \Omega t$) bath of oil. While Walker noticed indefinitely levitating droplets only in the presence of standing waves on the bulk liquid surface, Couder *et al.* were able to obtain indefinitely levitating droplets over a stable liquid surface. In both cases, vibration stabilizes the droplet in a non-coalescing state above the liquid bath. Therefore we refer to such a droplet as a vibro-levitating droplet.

A vibro-levitating droplet is in a repetitive cycle of impact and bounce-off at the liquid surface. If its radius is larger than the capillary length ($\sqrt{\gamma / \rho g}$, where γ and ρ are the liquid surface tension and density respectively) the droplet undergoes continuous deformation from spherical to oblate and prolate shapes, which may setup oscillations along the droplet surface.[31] When the droplet impacts the liquid surface, the kinetic energy of the droplet is dissipated into surface energy by flattening of the droplet, oscillations of the droplet, and viscous damping in the air film between the droplet and the liquid surface.[32] The droplet does not coalesce with the bulk liquid surface so long as the thin air film is replenished and stabilized due to the applied vibrations.

The vibro-levitating droplets produced weak surface waves every time they bounced off the liquid surface. These surface waves grew larger in amplitude when the amplitude A of the applied vibration was increased. At a critical value of A near the onset of Faraday instabilities, the levitating droplets started to move in seemingly random horizontal trajectories over the vibrating liquid surface. This motion is due to the interaction between the surface wave and the levitating droplet on each impact. Couder *et al.* called the system of the droplet and its associated wave a "walker".[13] These walkers can interact and orbit with each other, and can also form self-assembled ordered patterns.[14,33,34] Within a certain range of frequencies, the vibro-levitating droplets can roll over the liquid bath due to internal rotation.[35]

Vibro-levitating droplets have some parallels with the wave–particle duality from quantum mechanics.[36] The droplets illustrate several quantum mechanical phenomena such as single-particle diffraction, quantized orbits, and tunnelling.[37–39] But this comes with a caveat that there is a great difference between the physics in the macro and subatomic domains. Discussion of these topics is beyond the scope of this chapter.

There are models which describe the levitation and horizontal motion of these non-coalescent droplets.[30,40–42] The effect of bouncing droplets is thought to be similar to the acoustic levitation due to non-linear viscosity in a thin film which leads to hysteresis. However, a detailed model of such effects remains quite complex. In the following section, we suggest a simple analogy between the vibro-levitating droplets and the inverted pendulum.

2.3.1 Vibro-Levitating Droplets and Inverted Pendulum

The potential energy function of a droplet as it moves from a non-coalescing state to a completely coalesced state is similar to that of a pendulum as it swings from an inverted state to stable state. Consider a liquid droplet above a bath of the same liquid. The droplet radius (R_0) is small (compared to the capillary length) so that gravitational effects can be neglected. Assume that the droplet takes the shape of a truncated sphere (Figure 2.9a) as it coalesces into the bath. The droplet can be characterized by the radius R, height h, and the radius of the foundation x (Figure 2.9a and b). The volume, surface area, and the position of the centre of mass above the foundation of the truncated sphere are given by

$$V = \frac{1}{3}\pi h^2 (3R - h) = \frac{1}{3}\pi R^3 (2 - 3\cos\theta + \cos^3\theta)$$

and

$$A_S = 2\pi Rh = \pi(x^2 + h^2),\ z = \frac{3(2R - h)^2}{4(3R - h)}$$

respectively, where θ is the contact angle of the droplet, $\sin\theta = x/R$, and $x^2 = 2Rh - h^2$. As the droplet spreads from the initial spherical shape along the flat surface, the total volume of the droplet remains constant. Therefore $R = R_0 \sqrt[3]{4/(2 - 3\cos\theta + \cos^3\theta)}$. The change in the net surface free energy

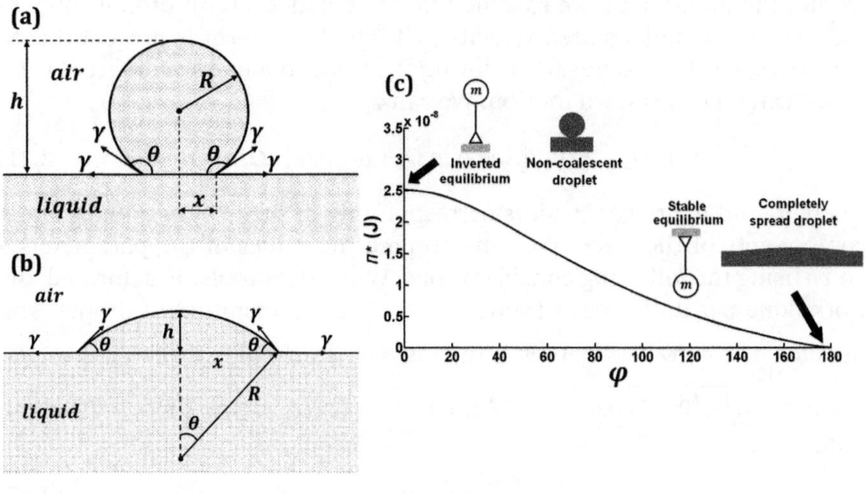

Figure 2.9 (a) and (b) The droplet as it spreads from a full sphere to a spherical cap of radius R. (c) Energy of a droplet (corn oil, $R_0 = 0.25$ mm, $\gamma = 0.032$ N m^{-1}) as it coalesces with the bulk liquid, and the similarity of this energy function to that of an inverted pendulum. Reproduced from ref. 11 with permission from the Royal Society of Chemistry.

during spreading is given by the free surface energy (γ) times the area of the droplet minus the foundation area:

$$\Pi = \gamma\pi R^2\left(1 - \cos\theta\right)^2 = \gamma\pi R_0^2\left(\frac{4}{2 - 3\cos\theta + \cos^3\theta}\right)^{\frac{2}{3}}\left(1 - \cos\theta\right)^2 \qquad (2.29)$$

The plot of energy as a function of θ for a corn oil droplet of R_0 = 0.25 mm and γ = 0.032 N m^{-1} is shown in Figure 2.9c, and it can be seen that θ = 180° corresponds to the unstable equilibrium, similar to an inverted pendulum. Therefore, it is convenient to introduce the variable $\varphi = 180^\circ - \theta$ to characterize the shape of the droplet so that φ is equal to zero at the unstable equilibrium.[11]

Now consider a vibro-levitating droplet over the sinusoidally vibrating liquid bath, whose vertical displacement is $u = A\sin\Omega t$. The dynamic equation of motion of the droplet in the vicinity of the unstable equilibrium can be written as

$$\chi\ddot{\varphi} + \beta\dot{\varphi} + \frac{\partial\Pi}{\partial\varphi} = Q_\varphi \qquad (2.30)$$

where χ is the inertial coefficient associated with droplet's shape change, β is the viscous coefficient, and Q_φ is the periodic force from the substrate affecting the droplet shape change. The force Q_φ includes a term proportional to the area of contact, and a term proportional to the length of the contact line ($2\pi R_0\varphi$); however, for small φ, the second term prevails. Furthermore, assuming that a non-linear viscous force acts in the thin air film between the droplet and the liquid bath, we assume that Q_φ includes a term proportional to the velocity ($\dot{u}$) and squared velocity ($\dot{u}^2$). The latter term is present due to hysteresis, *i.e.*, the viscous force during the forward motion is different from that during the backward motion. We can write

$$Q_\varphi = 2\pi R_0\varphi[\alpha_1 A\Omega\sin\Omega t + \alpha_2(A\Omega\sin\Omega t)^2] \qquad (2.31)$$

where α_1 and α_2 are coefficients corresponding to the linear and non-linear components of the force. We can estimate the values of the parameters χ and β using the following considerations. When the droplet is deformed, the work done per unit time is proportional to the momentum of droplet and thus $m\frac{\mathrm{d}x}{\mathrm{d}t}\mathrm{d}\dot{x} = \chi\dot{\varphi}\mathrm{d}\dot{\varphi}$, where m is the mass of the droplet. From Figure 2.9a, $x = R_0\sin\theta\sqrt[3]{4/\left(2 - 3\cos\theta + \cos^3\theta\right)}$. For $\theta \approx 180^\circ$(or $\varphi \approx 0^\circ$), $\mathrm{d}x/\mathrm{d}\varphi = R_0$ which yields

$$\chi = mR_0^2 \qquad (2.32)$$

Similarly, one can argue that the viscosity of the liquid, μ, is related to β as

$$B = \mu R_0 \qquad (2.33)$$

Using $(\sin \Omega t)^2 = (1 - \cos 2\Omega t)/2$ in eqn (2.31) and substituting the amplitudes $f_1 = 2\pi R_0 \varphi \alpha_1 A\Omega$ and $f_1 = 2\pi R_0 \varphi \alpha_2 A^2 \Omega^2$ into eqn (2.11) yields the following expression for the effective vibro-levitation force:[11]

$$V = \frac{4\pi^2 A^2 R_0^2}{\chi}\left(4\alpha_1^2 + \alpha_2^2 A^2 \Omega^2\right)\varphi \tag{2.34}$$

In the following section a simple experiment to study the vibro-levitation of droplets is described.

2.3.2 Experimental Study

The levitation of oil droplets over a vibrating oil bath was investigated experimentally using a setup similar to the one used by Walker.[29] In this study a 6.5 inch (16.5 cm) speaker cone (Pyle Company) formed the vibrating foundation. Sinusoidal waves at a desired frequency (10 Hz < Ω < 1000 Hz) were generated using a Matlab code, and were then amplified using a 20 W amplifier (Lepai) and fed to the speaker (Figure 2.10a). The vibration of the speaker cone was of the form $A \sin \Omega t$. Since the amplitude of the sound wave was not a controlled parameter, the loudness setting was kept constant during the experiment.

The liquids studied were water, corn oil, SAE 30 engine oil, and 10W40 engine oil. The working liquid was placed at the centre of the speaker cone to form a bath. Once the speaker was excited by the sound wave, a small drop of the same liquid was dropped on to the surface of the liquid bath using a syringe. This produced satellite droplets which levitated at certain frequencies of vibration of the speaker cone. Levitating droplets could also be produced by pinching and lifting off the liquid surface using a pipette tip/needle.[11] The results from this experimental study are discussed in the following section.

2.3.3 Results

Pure water did not produce levitating droplets in the tested frequency range. However, the higher-viscosity liquids, corn oil (Figure 2.10b), SAE 30, and 10W40, all produced levitating drops in the frequency ranges listed in

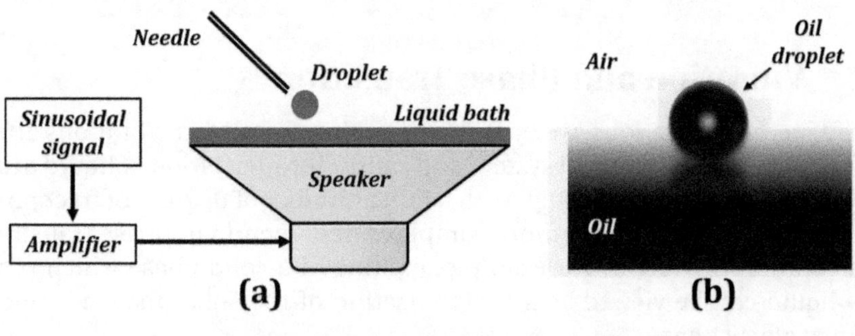

Figure 2.10 (a) Experimental setup. (b) A droplet of corn oil levitating on the surface of corn oil vibrating at 150 Hz. Reproduced from ref. 11 with permission from the Royal Society of Chemistry.

Table 2.1 Frequency range where stable levitating droplets were observed.[a]

Liquid	Viscosity (Pa s)	Surface tension (N m^{-1})	Frequency range (Hz)
Water	0.001	0.072	NA
Corn oil	0.052	0.032	35–350
10W40	0.160	0.031	30–400
SAE 30	0.400	0.031	30–400

[a]Reproduced from ref. 11 with permission from the Royal Society of Chemistry.

Table 2.1. Note, however, that in each case a certain frequency range was observed. In other words, besides a lower frequency limit corresponding to the stability onset there was an upper frequency limit above which the droplet was not stable. Using the analogy with the inverted pendulum, this may be due to the fact that for high frequency the assumptions of small vibration may not be valid.

The droplets were seen to levitate for several minutes. However, they coalesced with the bath as soon as the sound generation stopped. Outside the specified frequency range the levitating droplets were highly unstable, coalescing with the bath after a short while. At low frequencies it was clearly visible that the interaction between the levitating droplet and the bulk liquid surface created a surface wave. It was also possible to have multiple droplets levitating at the same time. The dependence of the stability of multiple levitating droplets on the frequency was not conclusive from the experiments conducted. Increasing the amplitude of vibration by increasing the loudness resulted in the levitating droplet "walking" on the surface of the liquid bath. Again, the dependence of horizontal motion of droplets on the amplitude could not be conclusively studied since the loudness could not be precisely regulated.[11]

In this section we saw a form of vibro-levitation in which liquid droplets can remain in a non-coalescent state above a vibrating liquid bath under certain conditions. In the following section we discuss vibration-induced phase transitions in continuum systems. We suggest an analogy between shear-thickening of fluids and a mechanical system like the inverted pendulum. We also discuss liquid-like properties manifested in vibrating granular matter, and solid-like properties manifested in vibrating fluids.

2.4 Vibration and Phase Transition

In the preceding sections, we have discussed how small fast vibrations affect the stability of a mechanical system and cause vibro-levitation of liquid droplets. Destabilization of a system with a finite number of degrees of freedom is closely related to the much more complex phenomenon of phase transition in a continuum system. For example, melting of a solid phase which turns into liquid can be viewed as a destabilization of the solid phase *via* nucleation of a new phase.

Phase transitions may be of the first kind (with energy released or consumed during the phase transition), such as melting and boiling, or of the second kind (when no additional energy is released or consumed), such as the transition between elastic deformation and plastic flow. The shifting of the stability region in a beam can be viewed as a transition from a soft to a hard "quasi-phase" analogous to the elastic–plastic transition in continuous medium. Thus, it has been suggested by K. Lurie[17] that a dynamic composite material with fibre reinforcement can be created using tuneable dielectrics, optical pumping with high-energy pulse compression, or electromagnetic stealth technology in such a way that stiffness of the reinforcing fibres can be controlled by an external fast oscillating electric field, thus controlling the phase transition in the composite (the parametric stiffness modulation).

In this section we discuss the effective freezing of bouncing droplets and other vibration-induced phenomena which can be interpreted as effective or apparent phase transitions.

2.4.1 Effective Freezing

We saw in previous sections how liquid droplets are effectively confined to a spherical shape by vibrations under certain conditions. The vibro-levitating droplets are effectively "frozen" in the spherical shape due to the vibro-levitation force. As soon as the exciting vibration is turned off, the droplet "melts" and coalesces into the bulk liquid. We also saw how vibration results in the increased stiffness of a rope. A soft rope effectively becomes stiff due to the exciting vibrations, making the Indian rope trick possible. Turning off the vibrations once again results in the rope going limp. These vibration-induced stabilizations can be viewed as analogous to the latent heat induced solid–liquid phase transition.

2.4.2 Cornstarch Monsters

A colloidal suspension of cornstarch in water is a common example of dilatant or shear-thickening fluid. If the cornstarch suspension is taken in the hand and squeezed, it can be observed that the suspension turns solid and its surface feels powdery. As soon as the pressure is released, it returns to its initial flowing state.

The Péclet number (Pe), which is the ratio of hydrodynamic to diffusion transport rates, governs the behaviour of colloids:

$$\mathrm{Pe} = r^2 \frac{\partial u}{\partial y} \Big/ D \tag{2.35}$$

where r is the particle radius, D is the diffusion coefficient, and $\partial u/\partial y$ is the shear rate. At high Pe (high shear rates) the hydrodynamic forces are too strong for the diffusion transport to restore the equilibrium of colloidal particles in the suspension. This non-equilibrium state consists of particles clustering together, called hydroclusters.[43]

The hydroclusters are an unstable state, returning to the equilibrium state of randomness and fluidity once the shear stress is removed. For the system of cornstarch colloidal suspension in water, it is observed that the "cornstarch monsters" levitate on a vibrating surface like an inverted pendulum or the rope in the Indian rope trick within a certain range of frequencies. The harmonic vibration of foundation is again seen to stabilize this system in its otherwise unstable equilibrium state.

The hydroclusters formed in cornstarch on application of stress can be simplified into a system of multiple pendulums as shown in Figure 2.7. High strain rates due to the harmonically vibrating foundation causes formation of hydroclusters of cornstarch particles in water. The hydroclusters of masses m_1, m_2,..., m_n (separated by distances l_1, l_2,..., l_n) are assumed to be held together by the viscous forces in the surrounding medium. This reduces the phenomenon of "cornstarch monsters" into a problem of stability of a chain of inverted pendulums.[11]

2.4.3 Effective Liquid Properties and Surface Tension of Granular Materials

Small-amplitude fast vibrations have an important effect on the properties of granular materials. For instance, vibrations can overcome jamming of the granular material due to friction. This is because vibrational acceleration creates an inertia force which can overcome dry Coulomb friction between the grains of the granular medium. As a result, a granular medium can flow into a narrow pipe, demonstrating an effective liquid-like behaviour, which is used in certain industrial applications.[15]

Note that from the viewpoint of rheological models, dry friction represents the key mechanisms of the plasticity. Therefore, the vibration-induced effective "melting" of the granular flow can be interpreted as an elastic–plastic transition rather than a true melting (which is a phase transition of the first kind).

An opposite effect of the "vibrational injection" of gas into liquid and effective locking or jamming of a valve in a vibrating vessel containing a liquid (thus preventing leaking of the liquid through the valve) has also been reported in the literature and studied both theoretically and experimentally by Blekhman.[44] This "vibro-jet effect", when applied at the microscale, can have broad consequences for such phenomena as multiphase flow separation and the control of liquid penetration through a semipermeable membrane (osmosis).

Another effect of vibration on granular media is the emergence of apparent surface tension. Clewett *et al.*[45] studied the vertical vibration of a layer of bronze spheres with diameter 150–180 μm placed between flat glass substrates. The vibrated particles formed 2D clusters demonstrating behaviour similar to 3D liquid droplets, thus suggesting the presence of an effective surface tension consistent with Laplace's equation, demonstrating the existence of an actual surface tension. The surface tension results predominantly from an anisotropy in the kinetic energy part of the pressure tensor, in contrast to thermodynamic systems where it arises from either the attractive interaction between particles or entropic considerations. The spheres inside the cluster had on average more collisions with neighbouring spheres than those at the border of the clusters. Since the collisions are not pure elastic and some energy is dissipated during the collisions, the average energy at the border of the clusters is larger than that inside the clusters, and the trend to minimize energy results in the clusters attaining a circular shape.

2.4.4 Locomotion in a Viscous Liquid

Another vibrational effect which is worth mentioning is the propulsion in a viscous medium due to small-amplitude fast vibrations, which is believed to be a principle of aquatic locomotion of many aquatic microorganisms.[46] Due to the small size of these microorganisms, viscosity prevails over inertia, so the way of swimming normally practised by larger organisms would result in just a back-and-forth motion rather than successful locomotion.

The so-called "scallop theorem" states that to achieve propulsion at low Reynolds number in Newtonian fluids a swimmer must deform in a way that is not invariant under time-reversal. Similarly to what we have observed in the preceding sections, such a motion (fast vibration) results in the effective propulsive force which drags the microorganism forward, facilitating aquatic locomotion.

The effect of vibrational locomotion is not limited to microorganisms and is widespread among aquatic animals, including whales.[44] In general, a system should involve an asymmetry to realize this effect. According to Blekhman's classification,[44] there are six main types of such asymmetry caused by force, kinematic, structural, gradient, wave, and initial conditions asymmetry which can lead to an effective propulsive force. In general, the effects of small fast vibrations and patterns are summarized in Table 2.2.

2.5 Surface Texture-Induced Phase Transitions

Vibrations are temporal periodic structures, whereas surface microstructure provides spatial patterns. It is remarkable that, similarly to small fast vibrations, surface micropatterns can affect the bulk properties of a liquid phase; in particular, they result in effective phase transitions of the material. In this section, we start by considering of the so-called Kirchhoff analogy between the dynamics of motion of a rigid body and bending of a beam, which establishes parallelism between time and spatial coordinates. After that we review recent findings in the area of surface texture-induced phase behaviour.

Table 2.2 Summary of the effects of small-amplitude fast vibrations and patterns on the mechanical and phase state equilibrium.

Effect	Large-scale system	Small-scale effect	Effective manifestation
Inverted pendulum stabilization, single or multiple	Mathematical pendulum	Fast vibrating foundation	Effective stabilizing "levitation" force
"Indian rope trick"	Flexible beam	Fast vibrating foundation	Effective stabilizing "levitation" force, strengthening of the soft rope
Stabilization of a column for buckling	Beam subjected to compressing load	Fast vibrating foundation	Increased critical load for buckling (soft becomes hard)
Dynamic materials	Composite material reinforced by fibres	Induced fast vibrating electric field	Soft fibres become hard (or plastic becomes elastic)
Vibro-levitation of droplet	Oil droplet over oil bath	Fast vibrations of the oil bath	Vibro-levitation, effective "freezing"
Melting of the granular flow	Granular material	Vibration	Liquid-like flow
"Cornstarch monsters"	Non-Newtonian liquid	Vibration	Rising of the figurines
Effective surface tension in granular flow	Granular material	Vibration	Formation of clusters minimizing the surface area
Vibrational injection of gas into liquid	Liquid in a vessel with a valve or a semipermeable membrane	Vibration	Effective closing of a valve; gas is injected into the liquid. Permeability properties of a membrane are changed
Locomotion in viscous liquid and the scallop theorem	Viscous liquid	Vibration (asymmetric)	Effective propulsion force
Kirchhoff analogy between dynamics of a pendulum/gyroscope and bending of a rod	Elastic rod or beam	Pattern in elastic properties	Effective shift of the critical load for buckling destabilization
Surface texture phase transition	Liquid	Microstructured profile	Suppression of the boiling point
Surface texture-induced propulsion	Liquid droplet	Asymmetric profile	Effective propulsion force

2.5.1 Kirchhoff's Analogy

Gustav Kirchhoff (1824–1887) was a German physicist who made a significant contribution to mechanics by developing a theory of bending of deformable elastic rods and beams. One particular result of Kirchhoff's theory was to establish an isomorphism between the bending shape of a beam and the dynamics of motion of a rigid body, such as a pendulum or a hydroscope, in 3D space.[47,48] This isomorphism, referred to as the Kirchhoff analogy, is due to the fact that the differential equations describing the bending of an elastic rod are the same as the differential equations describing the dynamics of the rigid body with the local orientation of the rod corresponding to the position of the pendulum and the length of the rod corresponding to the time variable.

Consider a slender beam of area moment of inertia I, and modulus of elasticity E, whose end points are loaded by an axial force F as shown in Figure 2.11a. At any point (x,y) the deflection in the beam is denoted by the angle ψ. From geometry of any small element ds on the beam, where s is the length along the beam, we can write $\mathrm{d}y/\mathrm{d}s = \sin\psi$. The bending moment at (x,y) is given by $EI\ \mathrm{d}\psi/ds = -Fy$. Combining these two relations,

$$\frac{\mathrm{d}^2\psi}{\mathrm{d}s^2} + \frac{F}{EI}\sin\psi = 0 \tag{2.36}$$

which describes the spatial pattern formed in the beam. Note its similarity to the equation of motion of a simple pendulum (of length L and angular displacement ψ)

$$\frac{\mathrm{d}^2\psi}{\mathrm{d}t^2} + \frac{g}{L}\sin\psi = 0 \tag{2.37}$$

which in turn describes the temporal patterns of the pendulum.

The time variable, t, in eqn (2.37) corresponds to the spatial length variable, s, in eqn (2.36). We should also note the difference that the dynamic problem of motion of a pendulum in the time domain constitutes an initial value problem, whereas bending of the beam in space constitutes a boundary value problem. However, despite this difference, an analogy exists between the motion of a pendulum and the shape of a bended elastic rod.

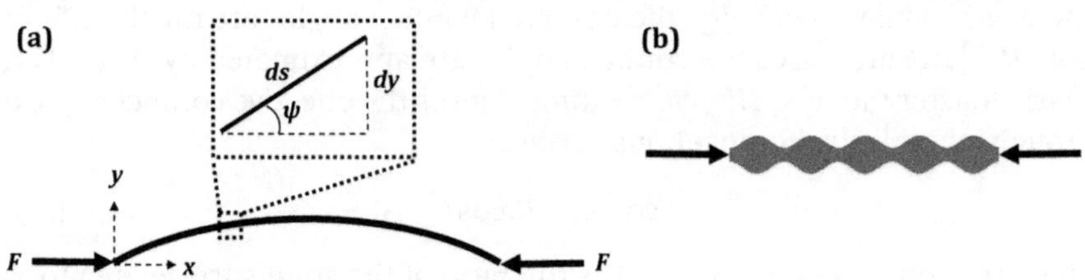

Figure 2.11 (a) Spatial patterns in a beam due to an axial force F. (b) The critical buckling load can be suppressed by introducing patterns in the surface profile of the beam.

Furthermore, if the properties of the rod are changed in a periodic manner with small amplitude $h \ll 1$ and frequency Ω about the stationary value EI_0 such that

$$EI = EI_0\left(1 + h\cos\Omega s\right) \approx \frac{EI_0}{\left(1 - h\cos\Omega s\right)} \tag{2.38}$$

we find that the equation for bending of the beam attains the form of eqn (2.39), which is similar to eqn (2.12):

$$\frac{d^2\psi}{ds^2} + \frac{F}{EI_0}\sin\psi = \frac{Fh}{EI_0}\sin\psi\cos\Omega s \tag{2.39}$$

We conclude that a pattern on the surface profile of a rod affects the critical load of buckling destabilization of the rod (Figure 2.11b).

While the effect of the vibration or surface patterns is to shift the stability region, it can also be viewed as affecting the transition between different regimes (hard *vs.* soft). For continuous systems this transition corresponds to the phase transitions of the second kind (plastic *vs.* elastic) or even of the first kind (liquid *vs.* vapour). As we have already seen, the small-amplitude fast vibrations can lead to the effective "freezing" of liquid droplets or "melting" of granular material. In the following sections we discuss surface texture-induced phase transitions.

2.5.2 Surface Texture-Induced Superhydrophobicity

Wettability of a surface is usually characterized by the contact angle (θ) a droplet of water makes with the surface.[49] The surface is hydrophobic if $\theta > 90^\circ$, and hydrophilic if $\theta < 90^\circ$. For an ideally smooth homogenous surface, the equilibrium contact angle (θ_0) of a liquid droplet (say, of water) is given by the Young equation

$$\cos\theta_0 = \frac{\gamma_{SA} - \gamma_{SW}}{\gamma_{WA}} \tag{2.40}$$

where γ_{SA}, γ_{SW}, and γ_{WA} are the surface free energies of the solid–air, solid–water, and water–air interfaces. However, on real surfaces with roughness and chemical heterogeneity, the observed contact angles are much different from θ_0.[50] In such cases the contact angles are approximated by Wenzel and Cassie–Baxter models. The Wenzel model gives the effective contact angle on a rough, chemically homogenous surface:

$$\cos\theta_W = R_f\cos\theta_0 \tag{2.41}$$

where the roughness factor $R_f \geq 1$ is the ratio of the solid surface area to the projected area. We can see from eqn (2.41) that roughening a hydrophobic surface makes it more hydrophobic, while roughening a hydrophilic surface makes it more hydrophilic. On a superhydrophilic surface, the droplet spreads out into a thin film. If a rough surface harbours pockets of air, thus creating chemical heterogeneities, then the contact angle is given by the Cassie–Baxter model:

$$\cos\theta_{CB} = r_f f_{SW}\cos\theta_0 - 1 + f_{SW} \tag{2.42}$$

where r_f is the roughness factor of the wet area, and $0 \leq f_{SW} \leq 1$ is the fractional solid–liquid interfacial area. The air pockets can lead to the surface being superhydrophobic. On superhydrophobic surfaces water beads up into a near-spherical shape.[51]

In both the above cases we see that surface texture (roughness) is an essential parameter in determining the wettability (or non-wettability) of a surface. On a superhydrophobic surface, a water droplet effectively "freezes" into a spherical shape. This observation is analogous to the non-coalescing state in a vibro-levitating droplet. The roughness features on the superhydrophobic surface also harbour and stabilize pockets of air. On a superhydrophilic surface, a water droplet effectively "melts" into a thin film, just like the coalescence of a droplet into a liquid bath.

From eqn (2.40) and (2.41), the increase in apparent contact angle is the result of the modified surface free energy term $R_f(\gamma_{SA} - \gamma_{SW})$. When surface texturing results in an increase in the apparent contact angle (Figure 2.12), the centre of gravity of a droplet placed on the surface is also displaced vertically. This is analogous to the effect of an effective stabilizing force.

2.5.3 Surface Texture-Induced Phase Transitions

Marmur[52] suggested that a textured surface with sufficiently high roughness value can harbour thermodynamically stable air pockets under water resulting in underwater superhydrophobicity. Later on, surface texture induced phase transitions were studied by Patankar and coworkers.[53–55] They studied how surface texture affects the Leidenfrost effect[56] manifested by water droplets levitating over a sufficiently hot skillet due to the presence of an evaporating vapour film (Figure 2.13a). Such a film is formed only when the hot surface is above a critical temperature: at lower temperatures the vapour film

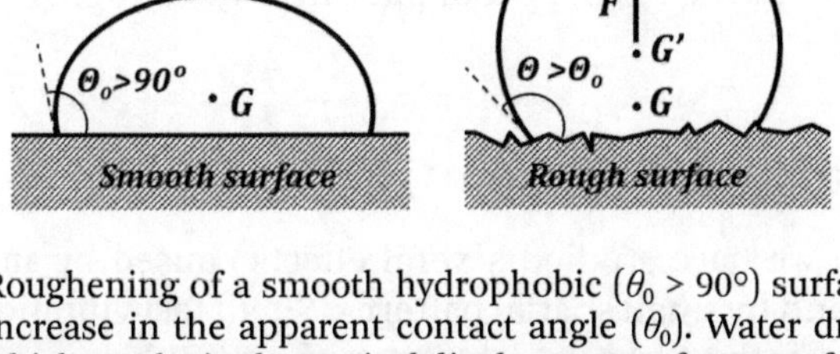

Figure 2.12 Roughening of a smooth hydrophobic ($\theta_0 > 90°$) surface results in an increase in the apparent contact angle (θ_0). Water droplets bead up, which results in the vertical displacement of centre of gravity (G) of the droplets. This can be visualized as the effect of a force F.

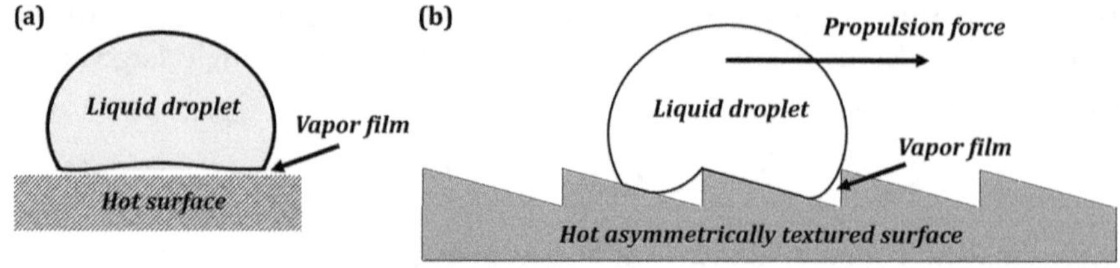

Figure 2.13 (a) A levitating liquid droplet over a sufficiently hot surface due to the Leidenfrost effect. The liquid that comes into contact with the hot surface vaporizes. This thin vapour film slows down further boiling and evaporation of the liquid droplet. (b) Self-propelled Leidenfrost droplets on a sawtooth surface.

collapses. However, the critical temperature can be reduced, and the vapour film collapse can even be completely suppressed when microtextured superhydrophobic surfaces are used.[53] Their result demonstrated that the surface texturing can potentially be applied to control other phase transitions, such as ice or frost formation and to the design of low-drag surfaces at which the vapour phase is stabilized in the grooves of textures without heating.

Jones *et al.*[54] later demonstrated that rough-textured surfaces may be used to manipulate the phase of water since the nanoscale roughness pattern stabilizes the vapour phase of water, even when liquid is the thermodynamically favourable phase. Furthermore, the reverse phenomenon exists when patterned hydrophilic surfaces keep a liquid-phase layer of water under conditions for boiling. They used molecular dynamics simulations to demonstrate the stability of the vapour and liquid phases of water adjacent to textured surfaces. Patankar[55] has also identified the critical roughness scale below which it is possible to sustain the vapour phase of water and/or trapped gases in roughness valleys, thus keeping the immersed surface dry.

Linke *et al.*[57] demonstrated that hot surfaces with slight asymmetric texture (sawtooth profile) can induce self-propulsion in Leidenfrost droplets, which in the process climb over the steep sides of the surface texture.[58] The effect of the asymmetric surface texture is to cause a propulsive force as shown in Figure 2.13b.

In general, the phenomenon of surface texture-based phase transition can be described as suppressing the boiling point and thus it is similar to superheating or subcooling of water. Similarly to the vibration-induced phase transitions, the effect of the small spatial pattern is in changing the phase state of the material.

2.6 Conclusions

In this chapter, we have studied several effects caused by small-amplitude fast vibrations and by small spatial patterns. Small fast vibrations can be substituted by an effective force, which affects the equilibrium of a mechanical system. This is a stabilizing vibro-levitation force in the case of an inverted single or multiple pendulum on a vibrating foundation or in the case of a flexible elastic beam or rope (the Indian rope trick). Vibrations affect the critical load corresponding to the buckling loss of stability of an axially loaded elastic beam, thus leading to the "soft–hard" regime transition.

A similar effect of vibration is found in liquids, including bouncing droplets of a liquid upon the bath of the same liquid, and in granular material which effectively "melts" leading to liquid-like behaviour. For a continuous system in space, the "soft–hard" regime transition would correspond to the elastic–plastic phase transition leading to the plastic flow which is observed as effective melting or freezing of the liquid. Thus, small vibrations do not just affect the stability of a mechanical equilibrium, but can also cause effective phase transitions. The same effect of effective melting or freezing is also observed in granular media and in non-Newtonian liquids, leading to the rise of quasi-solid rising figurines. Vibrations also result in effective surface tension of granular media, which is a liquid-like characteristic.

A mathematical technique to study the effect of small-amplitude fast vibrations is the method of separation of the fast and slow motions, as discussed in this chapter. In addition, the effect is related to the parametric vibrations. Although the mathematical implications have not been discussed here in detail, it is noted that the method of separating the fast and slow motion is related to the novel mathematical field of inertial manifolds as well as to the well-established technique of the renormalization group method in theoretical physics.[59,60]

Since there is an analogy between vibrations in time and patterned surfaces in space, surface patterns were expected to affect the phase stability. This was indeed found and has been reported in the literature in the case of micropatterned superhydrophobic surfaces, where the surface micropattern preserves a vapour phase or delays boiling. To summarize, the mathematical techniques for studying small vibrations developed in non-linear mechanics provide an important tool to investigate various effects related to phase transition and phase manipulation in liquids and droplets.

References

1. F. E. Manuel, *The Religion of Isaac Newton*, Clarendon Press, Oxford, 1974.
2. J. D. Black, *Comp. Lit.*, 1980, **32**, 42–62.
3. D. L. Price, *High-Temperature Levitated Materials*, Cambridge University Press, New York, 2010.
4. P. de Gennes, F. Brochard-Wyart and D. Que're', *Capillarity and Wetting Phenomena: Drops, Bubbles, Pearls, Waves*, Springer, New York, 2004.
5. E. Bormashenko, R. Pogreb, T. Stein, G. Whyman, M. Erlich, A. Musin, V. Machavariani and D. Aurbach, *Phys. Chem. Chem. Phys.*, 2008, **10**, 4056–4061.
6. M. Nosonovsky and B. Bhushan, *J. Phys.: Condens. Matter*, 2008, **20**, 395005.
7. R. Tadmor, P. Bahadur, A. Leh, H. E. N'guessan, R. Jaini and L. Dang, *Phys. Rev. Lett.*, 2009, **103**, 266101.

8. R. Tadmor, *Soft Matter*, 2011, 7, 1577–1580.
9. M. Nosonovsky, *Nature*, 2011, **477**, 412–413.
10. E. Y. Bormashenko, *Wetting of Real Surfaces*, De Gruyter, Berlin, Boston, 2013.
11. R. Ramachandran and M. Nosonovsky, *Soft Matter*, 2014, **10**, 4633–4639.
12. I. R. Epstein, *Science*, 2007, **315**, 775–776.
13. Y. Couder, S. Protiere, E. Fort and A. Boudaoud, *Nature*, 2005, **437**, 208.
14. Y. Couder, A. Boudaoud, S. ProtiÃ¨re and E. Fort, *Europhys. News*, 2010, **41**, 14–18.
15. I. I. Blekhman, *Vibrational Mechanics*, World Scientific, Singapore, 2000.
16. P. L. Kapitza, *Usp. Fiz. Nauk*, 1951, **44**, 7–15.
17. K. A. Lurie, *An Introduction to the Mathematical Theory of Dynamic Materials*, Springer, Berlin, 2007.
18. L. D. Landau and E. M. Lifshitz, *Mechanics Volume 1 of Course of Theoretical Physics*, Pergamon Press, 1969.
19. A. Stephenson, *Mem. Proc.–Manchester Lit. Philos. Soc.*, 1908, **52**, 1–10.
20. A. Stephenson, *Philos. Mag.*, 1908, **15**, 233–236.
21. A. Stephenson, *London, Edinburgh Dublin Philos. Mag. J. Sci.*, 1909, **17**, 765–766.
22. D. J. Acheson, *Proc. R. Soc. London, Ser. A*, 1993, **443**, 239–245.
23. D. J. Acheson and T. Mullin, *Nature*, 1993, **366**, 215–216.
24. T. Mullin, A. Champneys, W. B. Fraser, J. Galan and D. Acheson, *Proc. R. Soc. London, Ser. A*, 2003, **459**, 539–546.
25. A. R. Champneys and W. B. Fraser, *Proc. R. Soc. London, Ser. A*, 2000, **456**, 553–570.
26. E. V. Shishkina, I. I. Blekhman, M. P. Cartmell and S. N. Gavrilov, *Nonlinear Dyn.*, 2008, **54**, 313–331.
27. F. Blanchette and T. P. Bigioni, *Nat. Phys.*, 2006, **2**, 254–257.
28. O. Reynolds, *Proc.–Manchester Lit. Philos. Soc.*, 1881, **21**, 413–414.
29. J. Walker, *Sci. Am.*, 1978, **238**, 151–158.
30. Y. Couder, E. Fort, C. H. Gautier and A. Boudaoud, *Phys. Rev. Lett.*, 2005, **94**, 177801.
31. A. L. Biance, C. Clanet and D. Quere, *Phys. Fluids*, 2003, **15**, 1632–1637.
32. J. Qian and C. K. Law, *J. Fluid Mech.*, 1997, **331**, 59–80.
33. N. Vandewalle, D. Terwagne, K. Mulleners, T. Gilet and S. Dorbolo, *Phys. Fluids*, 2006, **18**, 091106.
34. S. I. Lieber, M. C. Hendershott, A. Pattanaporkratana and J. E. Maclennan, *Phys. Rev. E*, 2007, **75**, 056308.
35. S. Dorbolo, D. Terwagne, N. Vandewalle and T. Gilet, *New J. Phys*, 2008, **10**, 113021.
36. J. W. M. Bush, *Proc. Natl. Acad. Sci. U. S. A.*, 2010, **107**, 17455–17456.
37. Y. Couder and E. Fort, *Phys. Rev. Lett.*, 2006, **97**, 154101.
38. A. Eddi, E. Fort, F. Moisy and Y. Couder, *Phys. Rev. Lett.*, 2009, **102**, 240401.
39. E. Fort, A. Eddi, A. Boudaoud, J. Moukhtar and Y. Couder, *Proc. Natl. Acad. Sci. U. S. A.*, 2010, **107**, 17515–17520.
40. S. Protiere, A. Boudaoud and Y. Couder, *J. Fluid Mech.*, 2006, **554**, 85–108.

41. J. Molacek and J. W. M. Bush, *J. Fluid Mech.*, 2013, **727**, 612–647.
42. A. U. Oza, R. R. Rosales and J. W. M. Bush, *J. Fluid Mech.*, 2013, **737**, 552–570.
43. N. J. Wagner and J. F. Brady, *Phys. Today*, 2009, **62**, 27–32.
44. I. I. Blekhman, *Selected Topics in Vibrational Mechanics*, World Scientific, New Jersey, 2004.
45. J. P. D. Clewett, K. Roeller, R. M. Bowley, S. Herminghaus and M. R. Swift, *Phys. Rev. Lett.*, 2012, **109**, 228002.
46. E. M. Purcell, *Am. J. Phys.*, 1977, **45**, 3–11.
47. Y. M. Shi and J. E. Hearst, *J. Chem. Phys.*, 1994, **101**, 5186–5200.
48. E. L. Starostin, *Philos. Trans. R. Soc., A*, 2004, **362**, 1317–1334.
49. A. Marmur, *Soft Matter*, 2006, **2**, 12–17.
50. A. Marmur, *Langmuir*, 2003, **19**, 8343–8348.
51. A. Marmur, *Langmuir*, 2004, **20**, 3517–3519.
52. A. Marmur, *Langmuir*, 2006, **22**, 1400–1402.
53. I. U. Vakarelski, N. A. Patankar, J. O. Marston, D. Y. C. Chan and S. T. Thoroddsen, *Nature*, 2012, **489**, 274–277.
54. P. Jones, A. Kirn, D. Rich, A. Elliot and N. A. Patankar, *Bull. Am. Phys. Soc.*, 2014, **59**, BAPS.2014.DFD.R35.11.
55. N. A. Patankar, , DOI: arXiv:1505.06233 [cond-mat.soft].
56. J. G. Leidenfrost, *Int. J. Heat Mass Transfer*, 1966, **9**, 1153–1166.
57. H. Linke, B. J. Aleman, L. D. Melling, M. J. Taormina, M. J. Francis, C. C. Dow-Hygelund, V. Narayanan, R. P. Taylor and A. Stout, *Phys. Rev. Lett.*, 2006, **96**, 154502.
58. D. Quere, *Annu. Rev. Fluid Mech.*, 2013, **45**, 197–215.
59. C. Foias, G. R. Sell and R. Temam, *J. Differ. Equations*, 1988, **73**, 309–353.
60. E. Kirkinis, *SIAM Rev.*, 2012, **54**, 374–388.

CHAPTER 3

Superoleophobic Materials

THIERRY DARMANIN[a] AND FRÉDÉRIC GUITTARD*[a]

[a]Univ. Nice Sophia Antipolis, CNRS, LPMC, UMR 7336, 06100 Nice, France
*E-mail: Frederic.guittard@unice.fr

3.1 Introduction

The study of superoleophobic materials, which are not wetted by low surface tension liquids such as oils, is very much in demand not only for understanding the underlying wetting theories but also for a large range of potential applications such as in anti-stain textiles,[1] anti-fingerprint screens,[2] liquid displacement in microfluidic devices,[3] enhancement of printing technologies,[4] or oil/water separation membranes.[5] In this chapter, we report on superoleophobic materials when the media is air; underwater superoleophobic materials[6,7] are not reported here. Due to their extremely low surface tension (typically $\gamma_{LV} < 35$ mN m^{-1}), in comparison to water ($\gamma_{LV} = 72.8$ mN m^{-1}), it is extremely difficult to prevent wetting by oils.

In nature, superhydrophobic properties are present in many structures of plant and animal origin, such as leaves, feet, or wings.[8] For example, the famous lotus leaves possess superhydrophobic properties and are able to resist water wetting during rainfall. Recently it has been shown that some insects also possess superoleophobic properties.[9–11] Werner *et al.* demonstrated that many species of *Collembola* (springtails) have superoleophobic cuticles, allowing these arthropods to live in soils (Figure 3.1). To resist wetting by oils, these insects have developed highly ordered hexagonal or rhombic comb-like patterned structures on their cuticles. Here, the negative overhang in the profile

RSC Soft Matter No. 5
Non-wettable Surfaces: Theory, Preparation, and Applications
Edited by Robin H. A. Ras and Abraham Marmur

Published by the Royal Society of Chemistry, www.rsc.org

Figure 3.1 Cuticle patterns of different life forms and different orders of superoleophobic *Collembola* (A epedaphic Entomobryomorpha, B hemiedaphic Isotomidae, C euedaphic Poduromorpha and D hemiedaphic Symphypleona). Reprinted from ref. 9, copyright (2013) with kind permission from Springer Science and Business Media.

of the ridges and granules induces a strong pinning of the three-phase contact line of an oil droplet put on the surface. Moreover, springtails can also resist immersion in oil by forming a very stable plastron, *i.e.* a trapped air layer, even at elevated pressures, and can also resist bacterial adhesion.

Hence, superoleophobic properties can be produced by mimicking phenomena observed in nature. Different processes have been developed in order to obtain superoleophobic properties. In this chapter, we present a relatively comprehensive list of strategies used by researchers.

3.2 Superoleophobicity Theories

The apparent contact angle (θ^{Y}) of a liquid deposited on a "smooth" substrate is given by the Young equation: $\cos\theta^{Y} = (\gamma_{SV} - \gamma_{SL})/\gamma_{LV}$. θ^{Y} is dependent on the liquid surface tension (γ_{LV}) and the surface free energy (γ_{SV}).[12] More precisely, the capacity of a liquid to wet a substrate increases as its γ_{LV} decreases or the γ_{SV} increases. To characterize the adhesion of a liquid to a substrate, it is also necessary to determine the advancing (θ_{adv}) and receding (θ_{rec}) contact angles and as a consequence the contact angle hysteresis ($H = \theta_{adv} - \theta_{rec}$).[13] The maximum surface inclination before a droplet put on the substrate moves, called the sliding angle or tilt angle (α), also provides information on the surface adhesiveness.

For rough surfaces, using water as the probe liquid, many properties can be obtained as described by Marmur,[14] but usually a substrate is said to be superhydrophobic if the apparent contact angle (θ_{water}) > 150° and the value of H is low.[8] But what is the limit for the superoleophobicity? Indeed, if the superhydrophobic properties are determined using water, different oils of various surface tensions can be used to describe the superoleophobic properties. For example, using a microbalance to measure the adhesive forces, Kock-Yee Law showed that the adhesion of a hexadecane droplet (γ_{LV} = 27.4 mN m^{-1}), used as oil, is extremely low and as a consequence the substrate can be considered as truly superoleophobic if $\theta_{adv,hexadecane}$ > 165°.[15]

Two equations are often used to predict the wettability of rough substrates: the Wenzel and the Cassie–Baxter equations.[16,17] In the Wenzel equation, a liquid droplet placed on a rough substrate enters into all the surface roughness, leading to only a solid–liquid interface.[16] The Wenzel equation is $\cos\theta = r\cos\theta^Y$ where r is a roughness parameter. It is possible to have $\theta > \theta^Y$ but the only condition is $\theta^Y > 90°$. However, although many materials are intrinsically hydrophobic ($\theta^Y_{water} > 90°$) such as polytetrafluoroethylene (PTFE) or polydimethylsiloxane (PDMS), to our knowledge all materials are intrinsically oleophilic ($\theta^Y_{oils} < 90°$), especially to low surface tension oils such as hexadecane. Hence, the use of the Wenzel equation on rough substrates leads to highly oleophilic or superoleophilic properties. Only the Cassie–Baxter equation can predict superoleophobic properties.[17] In the Cassie–Baxter equation, a liquid droplet placed on a substrate stays on top of the surface roughness and on air trapped between the droplet and the substrate. The Cassie–Baxter equation is $\cos\theta = r_f f\cos\theta^Y + f - 1$. In this equation, r_f corresponds to the roughness ratio of the substrate wetted by the liquid, f to the solid fraction or $(1 - f)$ to the air fraction.[18] Hence, to obtain superoleophobic properties it is necessary to have surface structures that can trap a large amount of air. Moreover, the stability of the oil droplet increases if the energy difference between the Cassie–Baxter and the Wenzel state increases (Figure 3.2). For example, it was shown that the presence of surface structures with re-entrant curvature such as convex microstructures, pillars with overhangs, or mushroom-like structures allows greatly increased surface oleophobicity.[19–23] Indeed, an oil droplet placed on substrates with re-entrant structures is strongly pinned to these structures. Moreover, it is possible to calculate the energies corresponding to the Wenzel and Cassie–Baxter models as a function of geometrical parameters of surface structures, such as the

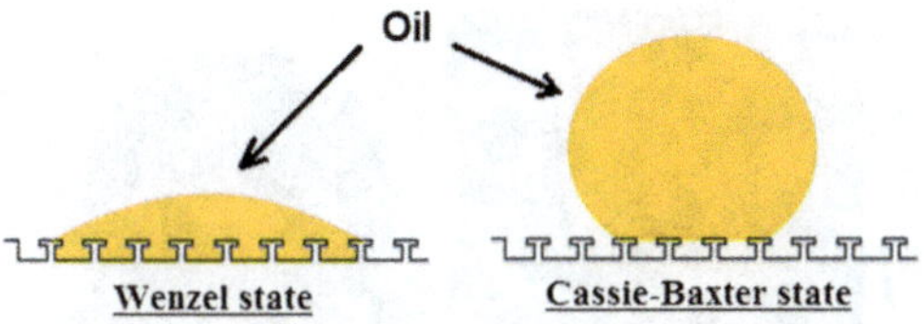

Figure 3.2 Schematic representation of an oil droplet on a substrate with re-entrant curvatures in the Wenzel state and the Cassie–Baxter state.

size of the structures, the size of the overhangs, and the distance between the structures. For example, McKinley's group implemented design parameters and robustness parameters in order to predict superoleophobic properties and their robustness as a function of the geometrical parameters and the liquid surface tension.[23]

After this brief introduction, we now summarize most of the published examples from the literature describing the fabrication of superoleophobic materials.

3.3 Fabrication of Superoleophobic Materials

3.3.1 Plasma Etching/Reactive Ion Etching

During plasma treatment, a substrate is put in contact with ionized species produced by creating an electric field between two electrodes. The plasma treatment can have different effects such as surface cleaning, the formation of chemical groups, or the formation of surface structures.[24–29] These effects can be controlled by many parameters such as the gas used, the pressure, or the power. For example, the plasma treatment of polyvinylidene fluoride (PVDF) using argon at atmospheric pressure made it possible to create microporosities inducing an increase of $\theta_{diiodomethane}$ from 51.7° to 115.2°.[24] Several authors have reported the use of low-pressure plasma treatments in order to attain superoleophobic properties. For example, Gogolides's group showed that an oxygen plasma treatment of various polymer substrates such as polymethylmethacrylate (PMMA), polyether ether ketone (PEEK), or polydimethylsiloxane (PDMS) can lead to the formation of surface nanofibrous structures.[25–27] After a post-treatment with perfluorinated compounds, the characteristics of the fibres (diameter, length, distance between fibres, *etc.*) and as a consequence the superoleophobic properties are highly dependent on plasma parameters such as the treatment time. For example, after 20 min O_2 plasma treatment, the authors reported $\theta_{hexadecane}$ = 142° and $H_{hexadecane}$ = 10° for PMMA and $\theta_{hexadecane}$ = 138° and $H_{hexadecane}$ = 11° for PEEK. Polybenzoxazine materials with papillae-like nanostructures were also reported after Ar plasma exposure in a reactive ion etching (RIE) system.[28] The authors reported $H_{diiodomethane}$ = 1° after a subsequent CF_4 plasma treatment. Silicon

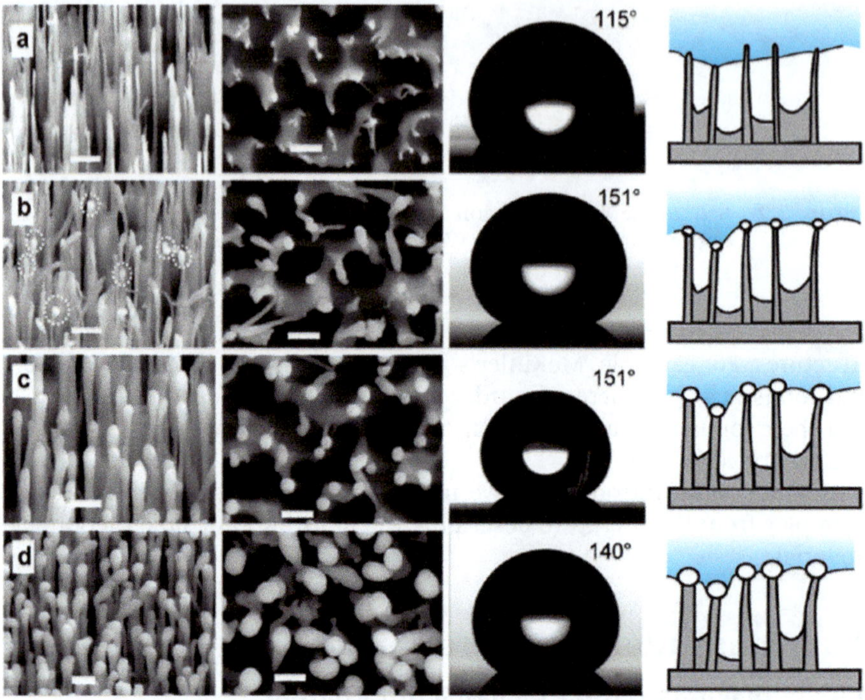

Figure 3.3 Silicon nanograss obtained by anisotropic RIE using SF_6 and O_2 as gas (a–d represent the different results obtained with different silicon oxide thicknesses). Reprinted with permission from R. T. R. Kumar, K. B. Mogensen and P. Bøggild, *J. Phys. Chem. C*, 2010, **114**, 2936[29], Copyright (2010) American Chemical Society.

nanograss was also reported by anisotropic RIE using SF_6 and O_2 (Figure 3.3). The authors reported the possibility of obtaining overhanging nanostructures by increasing the O_2 flow with enhanced oleophobic properties.[29]

3.3.2 Chemical Etching

Depending on the oxidation potential of the M^{n+}/M couple, a metal substrate can be oxidized when it is in contact with O_2. In aqueous solution, a metal substrate can react differently depending on the pH.

- In acidic medium:

$$4n\,M + nO_2 + 4nH^+ \rightarrow 4M^{n+} + 2nH_2O$$

The substrate is etched.

- In neutral or alkaline medium:

$$4n\,M + nO_2 + 2nH_2O \rightarrow 4M^{n+} + 4nOH^-$$

The M^{n+} ions produced during the reaction can react with OH^- to form metal oxides or alkoxides.

3.3.2.1 Etching in Acidic Media

The etching of aluminium substrates in aqueous HCl solutions led to micro/nanoscale cubic structures with a high porosity content close to "building block" architectures.[30,31] After surface modification with a fluorinated silane, it is possible to attain superoleophobic properties. In order to increase the apparent oil contact angle and decrease the oil sliding angle, the aluminium substrate can be pretreated by grinding[32] or post-treated by deposition of nanoparticles.[33]

When an aluminium substrate is immersed in boiling water, petal-like crystalline nanostructures may be formed, depending on the immersion time (Figure 3.4).[34,35] The crystalline phase was found to be boehmite ($\alpha.Al_2O_3.H_2O$). After modification with a fluorinated silane, superhydrophobic substrates (θ_{water} = 163°) were reported.[35] In order to obtain superoleophobic properties, one of the strategies is to induce microstructures, for example by sanding or etching.[36–42] It is also possible to induce the formation of γ-alumina (γ-Al_2O_3) after a heat treatment at 400 °C. Zhou's group studied the superoleophobic properties of aluminium substrates after etching in HCl and immersion in boiling water.[39] They observed that the re-entrant structures obtained after etching in HCl were necessary to obtain superoleophobic properties while the formation of petal-like nanostructures after immersion in boiling made it possible to decrease the contact angle hysteresis. They obtained $H_{hexadecane}$ = 8.0° and $\alpha_{hexadecane}$ = 7.2° or H_{decane} = 45.1° and α_{decane} = 40.1°, for example. Deng *et al.* also showed that these substrates are chemically stable toward corrosive liquids, solvents and also mechanically

Figure 3.4 Aluminium substrate obtained after HCl etching (a–c) following by immersion in boiling water (d–f). Reprinted with permission from S. Peng, X. Yang, D. Tian and W. Deng, *ACS Appl. Mater. Interfaces*, 2014, **6**, 15188[36], Copyright (2014) American Chemical Society.

stable toward abrasion, scratching for example.[36] In order to control the adhesion forces against different oils, it is possible to used mixed carboxylic acids ($C_7F_{15}COOH$ and $C_7H_{15}COOH$).[37] Using a high-sensitivity microelectromechanical balance, the authors observed that the substrate is not superoleophobic if the $C_7F_{15}COOH$ fraction (x_F) is below 0.8. As x_F increased from 0.8 to 1.0, the adhesion force of a hexadecane droplet changed greatly, from 67 μN (sticky: $\alpha_{hexadecane} > 90°$) to 10 μN ($\alpha_{hexadecane} = 4°$). The formation of petal-like crystalline nanostructures on stainless steel substrates was also possible if a shot-blasting process was used first to roughen the substrates before etching in HCl.[43] The maximum apparent contact angle was found to be 140° with peanut oil.

In microelectronics and microfabrication, there are numerous approaches to the etching of silicon wafers.[44–49] Etching involves the oxidation of silicon by oxidizing agents such as HNO_3 or H_2O_2 and the formation of water-soluble silicon complexes (*e.g.* SiF_6^{2-}, $HSiF_6^-$, SiF_4, $HSiF_3$) by F^-.[44,45] For example, the reaction in the presence of HF and H_2O_2 is:

$$Si + 2H_2O_2 + 6HF \rightarrow H_2SiF_6 + 4H_2O$$

Lee's group used this process to produce roughness and nanoholes on silicon wafers with superoleophobic properties.[45] In order to produce nanostructures on a silicon wafer, a metal-assisted etching process is recommended. A noble metal is usually used to induce anisotropic etching.[46] Indeed, the silicon beneath the noble metal is etched much faster than the areas without noble metal, generating pores in the silicon substrate. Coffinier *et al.* reported the fabrication of silicon etching with different nanostructures by using $AgNO_3$ and $NaBF_4$.[47,48] Values of $\theta_{hexadecane}$ up to 125° were measured on these substrates, depending on the $AgNO_3$ concentration.

Other groups also replaced the use of oxidizing agents by an electrochemical system to etch a substrate. The process is called electrochemical etching. Xu *et al.* reported the use of this process to produce rough titanium substrates. Superoleophobic substrates could be obtained by controlling electrochemical parameters.[50]

By replacing the inorganic acids used for etching substrates by organic acids such as carboxylic acids, *e.g.* $C_9F_{19}COOH$, the released M^{n+} can react to form nanoclusters by complexation, following the reaction:

$$M^{n+} + nR\text{–}COOH \rightarrow M[R\text{–}COO]_n + nH^+$$

Various nanostructures such as petal-like structures or nanosheets could be deposited by modifying the substrate (copper, zinc, aluminium, iron, nickel, and alloys) or the carboxylic acid.[51,52] Here, superoleophobic properties were directly obtained using perfluorinated carboxylic acids.

3.3.2.2 *Etching in Basic Media*

When a metal substrate is immersed in basic solutions, it can react to form metal oxides or alkoxides following the reaction:

$$4n\,M + nO_2 + 2nH_2O \rightarrow 4M^{n+} + 4nOH^-$$

For example, the immersion of copper substrates in ammonia solution induces the formation of $Cu(OH)_2$ ribbed nanoneedle arrays. The density of nanoneedles increases with immersion time. Superoleophobic properties were achieved after surface modification with a perfluorinated thiol. The authors also reported that the increase in nanoneedle density reduced the adhesive force of an oil droplet. These surfaces could be used for oil transportation in microreactors.[53] Other works also reported that the superoleophobic properties can be switched by plasma treatments.[54]

The oxygen necessary for the surface oxidation can be furnished using powerful oxidizing agents.[55–60] For example, the immersion of copper substrates in solutions of NaOH and $(NH_4)_2S_2O_8$ made it possible to change the surface morphology from nanoneedles or nanorods to microflowers composed of nanosheets as the immersion time increased from 3 min to 50 min.[56] After fluorination, the substrates with the highest superoleophobic properties ($\theta_{\text{hexadecane}}$ = 153°, $H_{\text{hexadecane}}$ = 29°, $\alpha_{\text{hexadecane}}$ = 23°) were observed with a mixture of both nanoneedles and microflowers. Relatively similar results were also reported replacing $(NH_4)_2S_2O_8$ by antiformin solutions of various concentrations.[55] In order to develop multilevel hierarchical nanostructures with a high degree of re-entrant curvatures from CuO nanowires, the deposition of hydrocarbon or fluorocarbon wax crystals by thermal evaporation (Figure 3.5) has been reported.[58,59] The surface

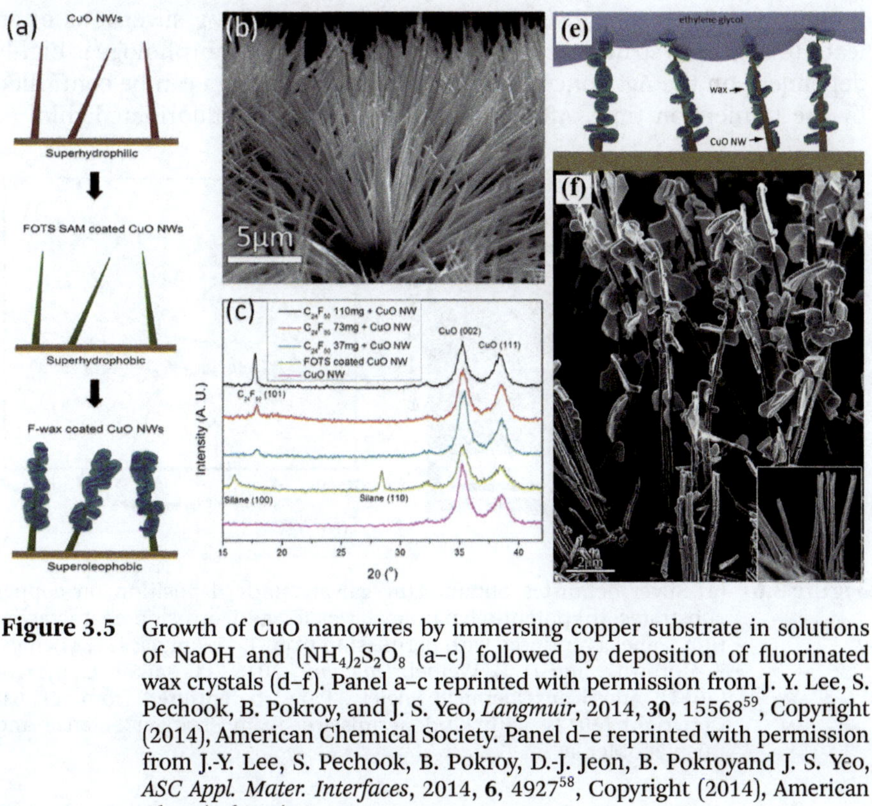

Figure 3.5 Growth of CuO nanowires by immersing copper substrate in solutions of NaOH and $(NH_4)_2S_2O_8$ (a–c) followed by deposition of fluorinated wax crystals (d–f). Panel a–c reprinted with permission from J. Y. Lee, S. Pechook, B. Pokroy, and J. S. Yeo, *Langmuir*, 2014, **30**, 15568[59], Copyright (2014), American Chemical Society. Panel d–e reprinted with permission from J.-Y. Lee, S. Pechook, B. Pokroy, D.-J. Jeon, B. Pokroyand J. S. Yeo, *ASC Appl. Mater. Interfaces*, 2014, **6**, 4927[58], Copyright (2014), American Chemical Society.

properties could be controlled by tuning the CuO nanowires, changing the type or amount of wax used. The best results were obtained with these surfaces with $\theta_{hexadecane} \approx 150°$ and $H_{hexadecane} < 10°$. In order to obtain both micro- and nanostructures, micropits on copper substrates were first formed by etching in HNO_3 and using cetyl trimethylammonium bromide (CTAB) micelles as a soft template.[57] The authors observed that superoleophobic surfaces using hexadecane can be obtained when densely distributed micropits are combined to copper oxide nanorods. Superoleophobic properties ($\theta_{rapeseed\ oil} = 151°$, $\alpha_{rapeseed\ oil} = 16°$) with flower-like microstructures were also reported on iron substrates after etching in acetic acid and immersion in H_2O_2.

3.3.3 Galvanostatic Deposition

A metal substrate immersed in a solution containing metal ions of a different metal can react following the reaction

$$m{M_1}^{n+} + nM_2 \rightarrow mM_1 + n{M_{2m}}^{+}$$

The reaction is spontaneous if oxidation potential of the metal substrate (here, M_2) is lower than that of the metal ions (here, ${M_1}^{n+}$). This reaction is called galvanostatic deposition or electroless deposition. Frequently this reaction induces the deposition of crystalline structures. For example, the immersion of copper substrates in a solution containing silver ions often leads to dendritic structures (Figure 3.6).[61] The surface morphology is highly dependent on the Ag^+ concentration while the roughness can be controlled by the immersion time. After modification with a perfluorinated thiol or

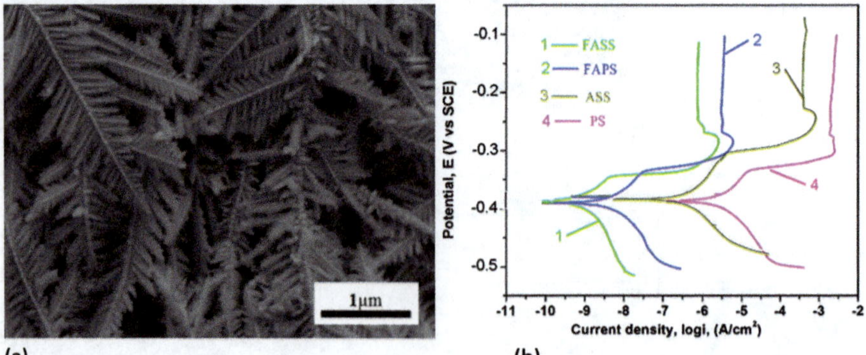

Figure 3.6 (a) Silver dendrites obtained by galvanostatic deposition on copper substrates; (b) potentiodynamic curves for anticorrosion characterization. Panel a reprinted with permission from T. C. Rangel, A. F. Michels, F. Horowitz, and D. E. Weibel, *Langmuir*, 2015, **31**, 3465[62], Copyright (2015), American Chemical Society. Panel b reprinted from ref. 61, Copyright (2012), with kind permission from Springer Science and Business Media.

silane the surface became superhydrophobic, but to reach superoleophobicity two strategies have been reported in the literature: the addition of a pretreatment by sandblasting to form surface microstructures or post-treatment with a PVDF coating.[62] Moreover, the substrates were corrosion resistant and could easily be repaired after damage. Similar results were also reported by immersion of zinc substrates which could induce the deposition of horn-like silver rods with superoleophobic properties ($\theta_{hexadecane} = 138°$) by adjusting the Ag^+ concentration and immersion time.[63]

3.3.4 Anodization

Tsujii's group was the first to report the use of anodization of aluminium substrates to produce superoleophobic properties.[64,65] The substrates were obtained after 3 h in 0.5 M H_2SO_4 and using a current density of 10 A cm^{-2}, leading to rough surfaces with fractal structures. After treating with a perfluorinated phosphate, the substrates displayed superoleophobic properties with $\theta_{hexadecane} = 135.5°$.

Now, the conventional anodization (in sulfuric acid) of aluminium substrate is known to produce hexagonally packed nanopore arrays on the surface. In this reaction, a metal substrate is oxidized in acidic media. The resulting metal ions react with water to form metal oxides but, unlike natural oxide, highly structured (often on the nanoscale) and/or adherent oxide layers can be obtained. In the case of aluminium, the reaction is:

$$Al \rightarrow Al^{3+} + 3e^-$$

$$H_2O \rightarrow O_2 + 2e^- + 2H^+$$

$$2Al + 3H_2O \rightarrow Al_2O_3 + 6H^+ + 6e^-$$

The oxidation of Al to Al_2O_3 is in competition with the dissolution of Al_2O_3 by H^+ and the electrical field. The pore size, the interpore distance, or the growth rate of an oxide layer can be tuned by modulating the anodizing conditions such as voltage, time, or temperature. However, such structures were found to be not sufficient to reach superoleophobic properties. Hopefully, it is possible to both widen the nanopores and to induce the growth of nanopillars from the nanopores using mild (*e.g.* in phosphoric acid) or hard (*e.g.* in oxalic acid) anodization processes (Figure 3.7) in order to enhance the oleophobic properties.[66,67]

The presence of densely packed nanopillars made it possible to produce superoleophobic surfaces with $\theta_{hexadecane} = 153.2°$ and $\alpha_{hexadecane} = 3°$ after post-treatment with a perfluorinated phosphate. In order to enhance the superoleophobic properties, aluminium substrates can be first treated to form microstructures before anodizing processes.[68–74] This could be done by acid etching to form "building block" microstructures[71] or by DC magnetron sputtering[72,73] to form submicrometre pillar structures. After anodizing and pore widening, it is possible to reduce the H_{octane} up to 4° by acid etching and H_{octane} up to 6° after DC magnetron sputtering.

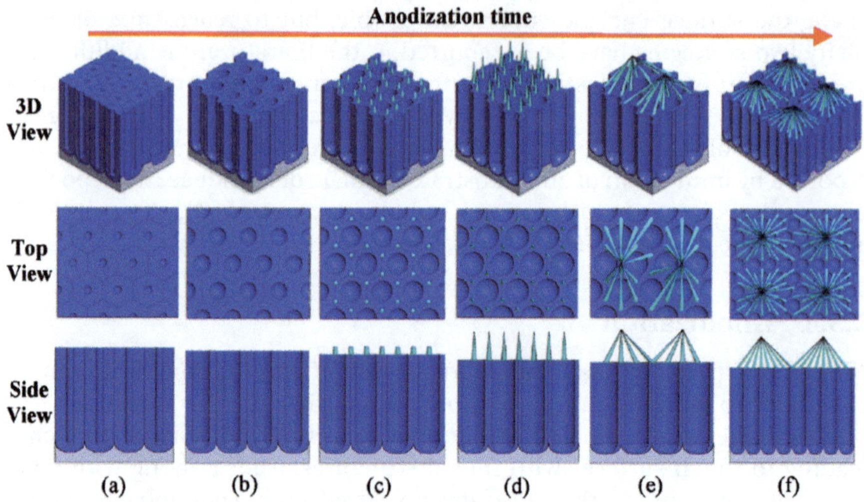

Figure 3.7 Different morphology obtained by aluminum anodizing as the anodization time increases (a–f). Reprinted with permission from C. Jeong and C.-H. Choi, *ACS Appl. Mater. Interfaces*, 2012, **4**, 842[66], Copyright (2012) American Chemical Society.

The anodization process is also well known to induce the formation of nanotubes on titanium substrates.[75–81] The main difference is the necessity to use F^- ions to induce the process. The characteristics of the nanotubes are dependent on several parameters such as voltage, pH, water content, or anodization time. The reaction for titanium oxidation is

$$Ti + 2H_2O \rightarrow TiO_2 + 4H^+ + 4e^-$$

but in the presence of F^- ions in the electrolyte, the formation of nanopits and the resulting nanopores is induced following the reaction

$$TiO_2 + 6F^- + 4H^+ \rightarrow TiF_6^{2-} + 2H_2O$$

Lim's group studied the superoleophobic properties of titanium substrates after electrochemical etching in NaCl to form microstructure and anodization (Figure 3.8).[76] The best results were obtained in ethylene glycol containing 0.25 wt% NH_4F and 2 wt% water after 2 h at 50 V. At this voltage, the nanotube wall thickness was less than 10 nm and the pore diameter 95–105 nm, leading to superoleophobic properties with $\theta_{\text{olive oil}} = 151°$. In order to obtain anisotropic superoleophobic properties, Zhou's group also reported the use of pre-patterning (micropatterned lines) by laser micromachining.[78] They observed that in order to obtain superoleophobic properties the distance between the patterned lines should not be too wide or too narrow, while $\theta_{\text{hexadecane}}$ also increased from 137° to 157° as the numbers of nanotubes increased from 195 to 492 per mm^2. For the best surfaces, $\alpha_{\text{hexadecane}}$ was 3.5° along the lines and 7.7° vertical to the line direction.

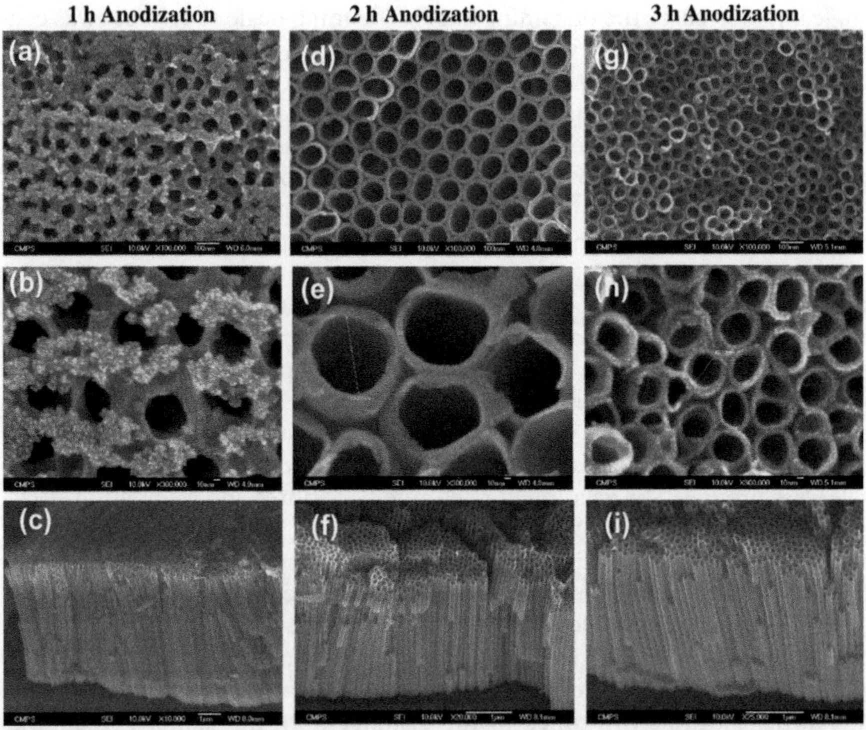

Figure 3.8 Different morphology obtained by titanium anodizing as the anodization time increases ((a–c) 1 h, (d–f) 2 h and (g–i) 3 h). Reprinted from *Journal of Colloid and Interfaces Science*, 400, S. Barthwal *et al.*, Fabircation of amphiphobic surface by using titanium anodization for large-area three-dimensional substrates, 123–129, Copyright (2013) with permission from Elsevier.[76]

3.3.5 Use of Nanoparticles

One of the easiest ways to produce structured substrates is to use nanoparticles. Nanoparticles can be produced in solution and the deposition can be performed using various methods such as dip-coating, spin-coating, and spray-coating. In the literature, silica (SiO_2) nanoparticles are often used to obtain superoleophobic properties. Indeed, spherical silica nanoparticles of different sizes can be easily obtained by the Stöber method using a silica precursor and different compounds for hydrolysis, crosslinking, and particle stabilization. Here, the superoleophobic properties are obtained either by introducing fluorinated compounds as a post-treatment or during the Stöber method.

The use of fluorinated silica nanoparticles has been reported by several research groups.[82–91] Motlagh *et al.* showed that to obtain superoleophobic properties using spherical nanoparticles it is preferable to mix nanoparticles of different sizes.[85,86] They showed that a multi-scale roughness induced by mixing nanoparticles of different sizes greatly increases the oil apparent contact angle and decreases the oil sliding angle. The multi-scale roughness was confirmed by He's group,[87,88] who mixed different kind of silica particles including 20 nm silica nanoparticles, 60 nm hollow silica nanoparticles, and silica nanosheets to obtain superoleophobic properties ($\theta_{hexadecane}$ = 132.4°) but with high adhesion with hexadecane. In order to enhance the coating robustness, a silica layer was applied by a chemical vapour deposition (CVD) post-treatment.[89] An O_2 plasma treatment just before the CVD treatment was also reported to enhance the wear resistance. In order to obtain superoleophobic and translucent properties ($\theta_{hexadecane}$ = 151°, $H_{hexadecane}$ = 10°, $\alpha_{hexadecane}$ = 9°) on various substrates, spray-coating of fluorinated silica nanoparticles has also been reported. Mixtures of silica nanoparticles and fluorinated silica nanoparticles were reported to enhance the superoleophobic properties.[90,91] Similar results were obtained using titania (TiO_2) nanoparticles.[92,93] One of the advantages of titania nanoparticles is their photocatalytic activity, which is of interest for photolithography. Ultralow-density boehmite (γ- and α-Al_2O_3) nanofibre aerogels were produced by gelation of boehmite nanofibres and calcination.[94] The resulting materials were superoleophobic with $\theta_{hexadecane}$ = 155°, $H_{hexadecane}$ = 10° and $\alpha_{hexadecane}$ = 9°. To obtain superoleophobic properties with carbon nanoparticles, a substrate can be held in the flame of a candle to obtain a soot layer.[95–99] Here, the superoleophobic properties ($\theta_{hexadecane}$ = 156°, $\alpha_{hexadecane}$ = 5°) could be enhanced by CVD of silica nanoparticles or by annealing above 1100 °C, which resulted in the formation of more rod-like nanostructures. The size of the particles can be controlled by the distance from the flame. The optimal contact angles were obtained after sooting in the middle of the flame.[96] Due to the possibility of moving droplets easily on these substrates, they were investigated in digital microfluidics.[3] The use of mixtures of carbon nanotubes and silica or titania nanoparticles has also been described in the literature.[100–102]

Superoleophobic surfaces were also reported using metal or polymer nanoparticles.[103–112] Robust superoleophobic surfaces ($\theta_{hexadecane}$ = 154.7°, $H_{hexadecane}$ = 21.8°, $\alpha_{hexadecane}$ = 18.9°) with multi-scale roughness were obtained by spraying copper perfluorooctanoate suspension.[103,104] This method is very interesting because many metals and carboxylates can be used. Silver nanowires were also produced using a two-step procedure.[105] After preparation of silver seeds, the silver nanowires were obtained using the polyol process by reduction of silver ions by ethylene glycol at high temperature and by microwave. The height of the nanowires could be controlled using different treatment times. The longest silver nanowires gave rise to the best oleophobic properties ($\theta_{ethylene\ glycol}$ = 146.2°, $H_{ethylene\ glycol}$ = 4.3°, $\alpha_{ethylene\ glycol} \approx$ 18°). Moreover, these silver materials can be used as an antibacterial coating and for surface-enhanced Raman scattering (SERS).[106]

Fluorinated polymer micro- and nanospheres were produced by conventional dispersion polymerization using different perfluoroalkyl methacrylates.[107] The spheres were deposited on glass slides using double-sided carbon adhesive tape. The coatings were superoleophobic with $\theta_{diiodomethane}$ = 159–160°. The formation of silicone nanofilaments directly on glass substrates by condensation of trichloromethylsilane in toluene and in the presence of water has also been reported.[108–112] The nanofilaments were 50–90 nm in diameter and several micrometres in length. To obtain superoleophobic properties ($\theta_{hexadecane}$ = 174.4°, $\alpha_{hexadecane}$ = 2°), an O_2 plasma treatment was performed to convert the methyl groups into hydroxyl groups before introducing a perfluorinated silane. The substrates were also stable against ozone, strong UV light, heat treatments, or immersion in basic solution. It was also possible to obtain superoleophobic nanofilaments in one step using a perfluorinated trichlorosilane, but the use of tetraethylorthosilicate (TEOS) was necessary to induce the formation of nanofilaments with this silane derivative.[109] To increase the sliding speed of liquid drops, a perfluoropolyether was spread on the perfluorinated silicone nanofilaments.[110] Moreover, it was possible to form micropatterns composed of parallel lines with contrasting wettability by exposing the substrates with silicone nanofilaments to a near-UV Nd/YAG laser.[112]

In order to improve the mechanical properties of the coating obtained by deposition of nanoparticles, one of the most widely used methods is to form a nanocomposite.[113–119] In the case of oxide nanoparticles such as silica, a monomer can be introduced during the formation of the nanoparticles in order to form a polymer/silica nanocomposite. A perfluoroether containing two terminal carboxylic acid groups was used in the Stöber reaction in the presence of TEOS.[113] Surprisingly, the resulting materials displayed both superoleophobic ($\theta_{dodecane}$ = 129°) and superhydrophilic properties, which is extremely rare in the literature. These materials could be used in oil/water separation membranes. Cho's group first elaborated silica nanoparticles with methacrylate groups on their surface, following by polymerization in the presence of fluorinated methacrylate to obtain superoleophobic surfaces.[114] In order to produce a polyhedral oligomeric silsesquioxane/poly(methyl methacrylate) (POSS/PMMA) nanocomposite, the surface of a fluorinated POSS was also modified by methacrylate groups and the composite was obtained by reversible addition–fragmentation chain transfer (RAFT) polymerization in the presence of methyl methacrylate monomers and a chain transfer agent.[115] The best properties were obtained for 25 wt% fluorinated POSS. Bromide functions were also introduced at the surface of multiwalled carbon nanotubes (MWCNTs).[116] Then, a polymer containing ammonium groups was grown at the surface of the MWCNTs by surface-initiated atom transfer radical polymerization (ATRP). The resulting materials displayed reversible wettability from superoleophobicity and superoleophilicity by successive anion exchanges using thiocyanate and perfluorooctanoate anions. The ATRP was also used to induce the growth of polymer corona at the surface of functionalized polymer nanoparticles.[117] Patton *et al.* introduced silica nanoparticles with trimethylsilyl groups, a formulation containing a photoinitiator, a trifunctional alkene, and fluorinated thiols (Figure 3.9).[118] After spray-coating in the presence of a UV lamp, superoleophobic thiol-ene resins ($\theta_{hexadecane}$ = 155.3°, $H_{hexadecane}$ = 9.5°, $\alpha_{hexadecane}$ = 4° using 30 wt% of silica) were obtained. Following this strategy, a polyacrylate resin obtained by photopolymerization was also used.[119]

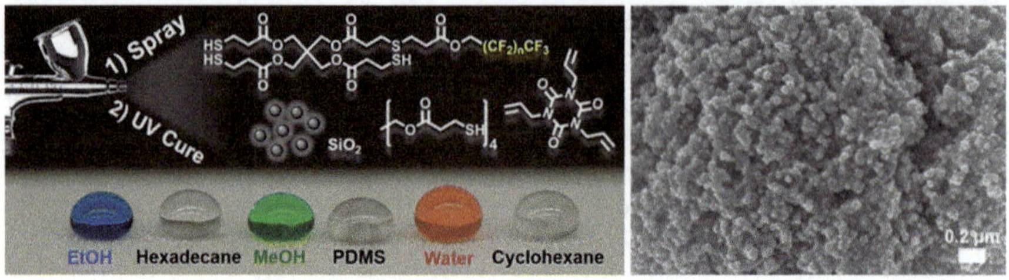

Figure 3.9 Spray-deposition and photopolymerization of organic–inorganic thiol-ene resins. Reprinted with permission from L. Xiong, *et al.*, *ACS Appl. Mater. Interfaces*, 2014, **6**, 10763.[118] Copyright (2014) American Chemical Society.

Otherwise, one of the easiest ways to form a nanocomposite is by introduction of nanoparticles into a polymer matrix.[120–137] Highly stable and transparent coatings were obtained using silica nanoparticles and polydimethylsiloxane (PDMS).[120–122] To obtain superoleophobic properties ($\theta_{\text{diiodomethane}}$ = 140.7°), the coatings were immersed in piranha solution in order to create hydroxyl functions in the PDMS backbone before functionalization by a perfluorinated silane.[120] Similar results were also obtained using fluorinated graphene oxide.[123] Polyurethane has also been used as a polymer matrix.[124] Spraying of polyurethane/molybdenum disulfide (MoS_2) was used to obtain coatings with wear resistance. After modification with a fluorinated silane, the coatings showed superoleophobic properties ($\theta_{\text{hexadecane}}$ = 151°, $H_{\text{hexadecane}}$ = 30°). In order to produce fibre structures by spray-coating, a "solution blow spinning" technique was also developed.[125] This process consists of the introduction of a polymer solution through a nozzle using a syringe pump and a high-velocity gas flow. Superoleophobic PMMA/fluorinated POSS nanofibres were obtained with this strategy.

In order to obtain superoleophobic properties in one step, fluoropolymers can also be used.[126–137] Using this process, superoleophobic properties ($\theta_{\text{hexadecane}}$ = 152°, $\alpha_{\text{hexadecane}}$ = 40°) were obtained using fluorinated MWCNTs and fluorinated polyurethane.[126] By mixing fluorinated silica and with poly(vinylidene fluoride-hexafluoropropylene) as fluoropolymer, superoleophobic properties with low adhesion were obtained ($\theta_{\text{hexadecane}}$ = 158°, $\alpha_{\text{hexadecane}}$ = 5.1°).[127] Moreover, these coatings showed enhanced corrosion resistance. Mabry's group used fluorinated silica and the fluoropolymer Viton ETP-600S (DuPont) to obtain superoleophobic properties.[128] However, these authors observed that rather than being present in the interstices between the nanoparticles, the binder is widely distributed across the surface roughness. Fluorinated polyacrylates were also widely used as binder for nanoparticles because of their higher substrate adhesion in comparison to PTFE. Using 20 nm silica nanoparticles, Hsieh *et al.* showed that the highest superoleophobic properties ($\theta_{\text{ethylene glycol}}$ = 165.2°, $H_{\text{ethylene glycol}}$ = 2.5°) are obtained for a F/Si ratio = 2.13.[129,130] Other authors also showed the importance of having an optimal particle/binder ratio to remain in the Cassie–Baxter state.[133] Fluorinated polyacrylates were used with 50 nm ZnO nanoparticles.[134] Rough substrates were obtained by spraying using acetone as cosolvent to obtain a "dry" coating. The best properties ($\theta_{hexadecane}$ = 154°, $H_{hexadecane}$ = 6°) were obtained for a ZnO:polyacrylate mass fraction of 3.3. In order to obtain conductive coatings, mixtures of fluorinated polyacrylates and carbon nanofibres (diameter ~100 nm, length ~130 μm) were employed.[135–137] Superoleophobic properties and electrical conductivity could both be controlled by the carbon nanofibre content. Superoleophobic properties with low adhesion were obtained from a carbon nanofibre content of 60%.

In order to greatly enhance the stability of the properties, an excellent strategy is to graft a polymer on the nanoparticle surfaces.[138–142] For this purpose, Liu *et al.* prepared a diblock copolymer containing both fluorinated chains and triisopropyloxysilyl group and used it to coat the surface of silica nanoparticles uniformly.[138–141] Values of $\theta_{hexadecane}$ = 149°, $H_{hexadecane}$ = 13° were measured using these materials, as well as a high resistance in basic solution. Raspberry-like polymer particles were also produced by grafting.[142] Here, small polymer particles with glycidyl groups were mixed to larger polymer particles containing amine groups.

Another strategy used in the literature was to coat nanoparticles with polymer using a layer-by-layer approach. For example, titania nanoparticles were coated with successive layers of poly(acrylic acid) and a perfluorooalkyl methacrylic copolymer to obtain superoleophobic properties.[143]

3.3.6 Hydrothermal and Solvothermal Processes

The hydrothermal process consists of the growth of crystals (especially oxides and hydroxides), using an autoclave at high temperature and high vapour pressure, from substances that are insoluble at atmospheric pressure and low temperature.[144] The crystalline structures often being fractal, they can lead to superoleophobic properties. Modification using a perfluorinated agent is necessary to obtain these properties. Using this process, titania nanotubes were obtained by hydrothermal process using titanium substrates directly immersed in aqueous NaOH solution and heated in an autoclave at 120 °C. To obtain superoleophobic properties ($\theta_{glycerol}$ = 165°, $H_{glycerol}$ = 5°, $\alpha_{glycerol}$ = 7°), a pretreatment by electrochemical etching and a post-treatment by annealing at 500 °C to obtain the anatase phase were required. Yu's group reported the formation of ZnO nanorods using a hydrothermal process (in aqueous solution containing ammonia and ethanol) on zinc plates previously etched in HCl to increase the surface roughness.[145] The surface morphology was highly dependent on the etching time and also on the hydrothermal temperature. The best properties ($\theta_{peanut\ oil}$ = 138°) were obtained at 95 °C. On these surfaces, the ZnO nanorods were about 4 μm in length, 0–1 μm in gap distance, and 0.1–0.4 μm in diameter. In order to induce the formation of ZnO nanostructures on any substrates, Sun'group first coated a substrate with ZnO and then obtained ZnO nanoflowers by hydrothermal process in aqueous solution containing zinc nitrate ($Zn(NO_3)_2$), hexamethylenetetramine, and sodium dodecyl sulfonate.[146]

The surfaces displayed superoleophobic properties with $\theta_{hexadecane} = 150°$ and $\alpha_{hexadecane} = 18°$. Various ZnO nanostructures were also reported using $ZnSO_4$, monoethanolamine, and ammonia. By changing the deposition time or the amine used, different structures such as nanorods, flower-like structures or spherical particles were obtained with different superoleophobic properties.[147–149]

Non-aqueous solvents are often necessary to solubilize non-oxide materials such as nitrides or chalcogenides, in a so-called solvothermal process.[150–152] $FeSe_2$ particles of various crystalline structures were obtained by mixing an iron source ($FeCl_2$, $FeSO_4$, or Fe_2O_3 powder), selenium, and oleylamine at 200–220 °C. The different structures are obtained by formation of $FeSe_2$ seeds and either growth of these seed crystals or self-assembly among different seed crystals. Using this process pompon-like, chip-like, or flower-like structures were produced.[150,151] The addition of HF or CTAB also induced a change in the crystalline structures (Figure 3.10).[152] Superoleophobic properties ($\theta_{diiodomethane} > 150°$, $\alpha_{diiodomethane} < 8°$) with ultralow reflectance in the range 300–1800 nm were obtained with these particles.

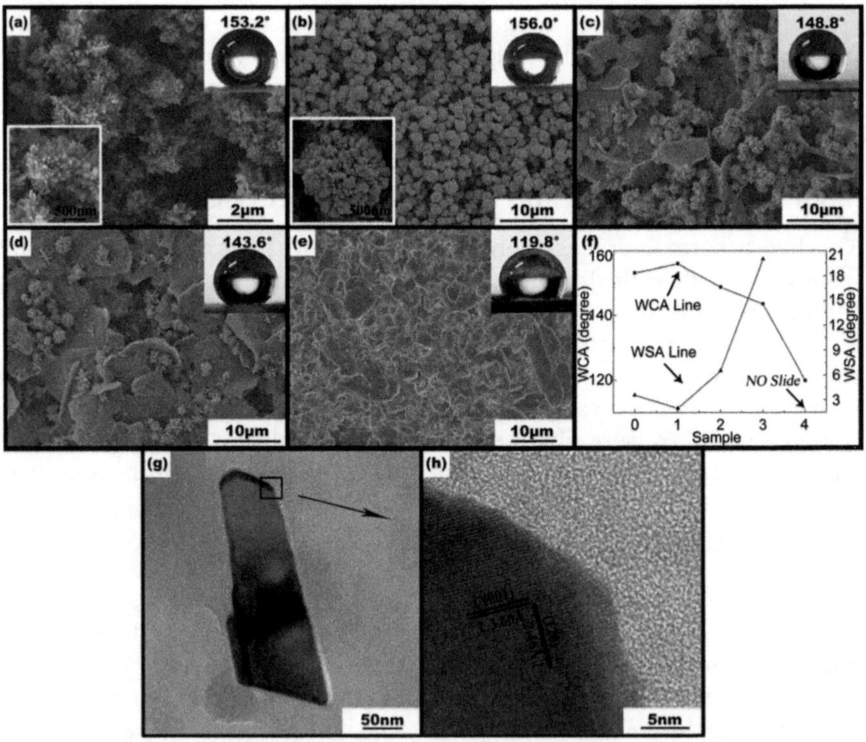

Figure 3.10 Pompon-like and chip-like $FeSe_2$ particles using HF-assisted solvothermal process.[152] (a–e) are SEM images using different amounts of HF. The insets represent water droplets. (f) represents the apparent contact angles and sliding angles of the different samples. (g and h) are TEM and HRTEM images.

3.3.7 Chemical Vapour Deposition

The CVD process is widely used in the fabrication of semiconductors. It allows the growth of thin films on a substrate from volatile precursors and using a gas flow. The crystallinity of the materials is highly dependent on the precursors used, or the CVD method and parameters. For example, Brunet *et al.* reported the growth silicon nanowires from SiH_4 (Figure 3.11).[153] Silicon nanowires of high aspect ratios (from 30 to 100) and 50–100 nm in diameter were obtained depending on the pressure and the treatment time. The substrates were superoleophobic for $\gamma_{LV} > 26$ mN m^{-1}. Using a relatively similar process, aligned carbon nanotubes were produced on quartz glass substrates by metal phthalocyanines FePc/YPc.[154] Their length was about 3 µm and their external diameter 60 nm. The substrates were superoleophobic with $\theta_{\text{rapeseed oil}} = 161°$.

3.3.8 Electrodeposition

The electrodeposition of conducting polymers makes it possible to obtain structured conducting polymer films by monomer oxidation in an electrochemical cell.[155] In this process, polymerization and polymer deposition are obtained in one step. This process is extremely interesting because the layer thickness can be easily controlled while the surface morphology is highly governed by electrochemical parameters, and especially by the chemical structure of the monomer. The polymerizable core of the monomer is fundamental because both the optoelectronic properties and the surface morphology are dependent on it. To date, the highest superoleophobic properties ($\theta_{\text{hexadecane}} = 148.0°$, $H_{\text{hexadecane}} = 15°$, $\alpha_{\text{hexadecane}} = 6°$) have been obtained using 3,4-ethylenedioxypyrrole (EDOP) as the polymerizable core.[156–165] Indeed, although the synthesis of these monomers is extremely long and complex, fluorinated PEDOP can have extremely high intrinsic oleophobicity (for

Figure 3.11 (a) Drop of a dried polymer on a carpet of Si nanowires; (b) pressure threshold for liquid impalement. Reprinted with permission from T. P. N. Nguyen, P. Brunet, Y. Coffinier and R. Boukherroub, *Langmuir*, 2010, **26**, 18369.[153] Copyright (2010) American Chemical Society.

the best PEDOP polymers, $\theta^{Y}_{hexadecane}$ = 78°, $\theta^{Y}_{sunflower\ oil}$ = 90.8°, $\theta^{Y}_{diiodomethane}$ = 99.5°),[156] comparable to fluorinated POSS.[166] Moreover, these monomers often lead to micro/nanostructured surfaces with nanoporosity (Figure 3.12), which highly enhances the superoleophobic properties. However, it was found that the presence of nanoporosity is due to a complex doping process and that the polymer has to be sufficiently rigid to induce the formation of nanopores.[157] For example, increased alkylenedioxy bridge length,[157] replacement of the two oxygen atoms of EDOP by sulfur,[161] or the presence of bulky or mobile substituents such as short fluorinated chains have a negative effect on the presence of surface nanoporosity.[162,163] However, it was possible to obtain superoleophobic nanoporous surfaces ($\theta_{hexadecane}$ = 138.0° but sticky) with short perfluorobutyl (C_4F_9) chains using a long alkyl spacer to reduce their mobility by van der Waals interactions.[164]

However, it is also extremely important to find a way to obtain superoleophobic properties using fluorinated 3,4-ethylenethiophene (EDOT) derivatives. Indeed, EDOT derivatives have faster polymerization capacity and can lead to various surface morphologies including nanofibres, nanosheets, or nanoflowers.[167–169] Moreover, the EDOT monomers are much easier to synthesize. However, whatever the fluorinated chain length, all the first attempts to obtain superoleophobic properties from EDOT derivatives failed because PEDOT polymers have much lower θ^{Y}_{oils} than PEDOP ones. In order to greatly increase θ^{Y}_{oils} and obtain superoleophobicity, the best strategy was found to be the use of extremely polar linkers such as amide, urea, carbamate, or thiocarbamate (Figure 3.13).[170,171] Here, for the best PEDOT polymers using carbamate linkers, $\theta^{Y}_{hexadecane}$ = 71.2°, $\theta^{Y}_{sunflower\ oil}$ = 80.3°, and $\theta^{Y}_{diiodomethane}$ = 92.9°.

Figure 3.12 Surface morphologies obtained by electrodeposition of a PEDOP conducting polymer containing a short fluorinated chains and as a function of the electrolyte used. Reprinted from ref. 165 with permission from John Wiley and Sons. Copyright © 2013 Wiley-VCH Verlag GmbH & Co. KGaA, Weinheim.

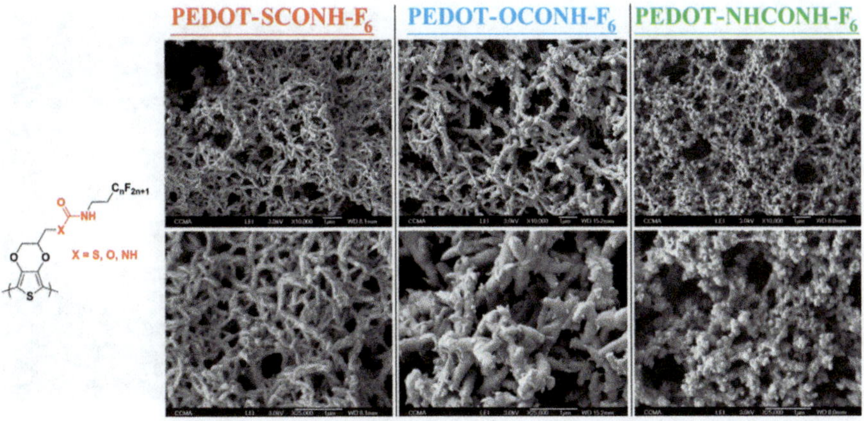

Figure 3.13 Surface morphologies obtained by electrodeposition of PEDOT conducting polymers as a function of the fluorinated chain length and the linker. Reprinted from ref. 170 with permission from John Wiley and Sons. Copyright © 2015 Wiley-VCH Verlag GmbH & Co. KGaA, Weinheim.

By electrodeposition of conducting polymers, it was also possible to obtain both oleophobic and hydrophilic properties by introduction of metal ions in the polymer structures or by copolymerization with a polymer with extremely polar pyridinium groups.[172,173]

Otherwise, Tsujii *et al.* also reported the obtaining of highly oleophobic films by electrodeposition of fluorinated pyrrole or the electrodeposition of hydrocarbon pyrrole followed by a post-treatment with a fluorinated silane.[174,175]

Electrodeposited conducting polymers could also be used as a sacrificial template. Using porous PEDOT substrates, transparent and highly stable superoleophobic films ($\theta_{hexadecane} = 128.6°$ but sticky) were obtained by depositing of a silica layer and removing the template by annealing at 500 °C.[176]

Electrodeposition can also be used to deposit metal films by reducing a metal salt. The surface morphology is also highly dependent on electrochemical parameters. For example, gold pyramidal nanostructures were obtained by reducing $HAuCl_4$ in $HClO_4$ aqueous solutions. The surfaces displayed superoleophobic properties with $\theta_{rapeseed\ oil} = 146°$.[177]

3.3.9 Electrospinning

The electrospinning technique is based on the deposition of horizontally aligned nanofibres (nanofibre mats), especially polymers although other materials can also be used.[178] By applying a high voltage, electrically charged jets of material are expelled through a needle and a collected on a substrate (Figure 3.14a). To obtain superoleophobic properties in one step, Lim's group electrospun the fluorinated polymer poly(2,2,2-trifluoroethyl

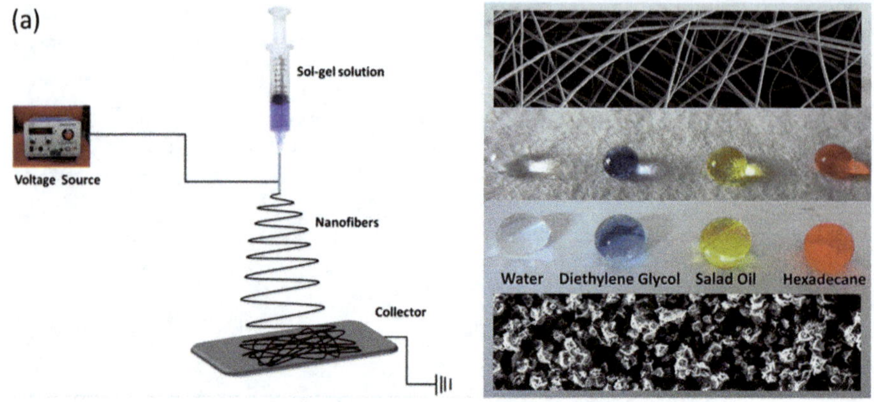

Figure 3.14 Electrospinning process (a) with microphase separation using block copolymers with 3,3,3-trifluoropropyl substituted siloxane segments (b).[180] Panel a reprinted with permission from V. A. Ganesh, S. S. Dinachali, A. S. Nair and S. Ramakrishna, *ACS Appl. Mater. Interfaces*, 2013, **5**, 1527.[181] Copyright (2013) American Chemical Society. Panel b reprinted with permission from L. Yi, X. Meng, X. Tian, W. Zhou and R. Chen, *J. Phys. Chem.* C, 2014, **118**, 26671.[180] Copyright (2014) American Chemical Society.

methacrylate).[179] Different fibre diameters were obtained by varying the concentration of the polymer solution. The highest values, $\theta_{hexadecane} > 150°$, were reached with the smallest gap distance between the fibres, which also corresponds to the highest robustness parameter.

In order to obtain multi-scale surface roughness, Yi *et al.* electrospun block copolymers with 3,3,3-trifluoropropyl-substituted siloxane segments to induce microphase separation during the electrospinning process (Figure 3.14b).[180] The surface properties could be adjusted by the copolymer used and its concentration. The surface with the highest properties exhibited $\theta_{hexadecane} = 135.2°$. To obtain robust superoleophobic properties, rice-shaped titania nanostructures were created by electrospinning of a solution of polyvinyl acetate and titania sol.[181] After heating at 500 °C to remove the polyvinyl acetate, the nanostructures were highly porous and rough. The resulting films were mechanically and thermally stable, highly adhesive, and superoleophobic with $\theta_{hexadecane} = 138.5°$, $H_{hexadecane} = 12°$, and $\alpha_{hexadecane} = 15°$. Silica nanofibres were also reported by electrospinning of silica sol and vapour deposition of silica coating to obtain robust superoleophobic properties ($\theta_{hexadecane} = 146.5°$, $H_{hexadecane} = 6°$, $\alpha_{hexadecane} = 9°$).[182] Superoleophobic substrates were also obtained by electrospinning of fluorinated POSS and PMMA to form a highly porous morphology with re-entrant structures.[183] Ultralow hysteresis ($H_{heptane} = 10°$) was reported on these substrates, and by exposing them to O_2 plasma treatment their surface wettability changed from superoleophobic to superoleophilic. Moreover, using a mask, after O_2 plasma treatment it was possible to obtain patterned substrates with both superoleophobic and superoleophilic zones. Such substrates are excellent candidates for water harvesting applications.

3.3.10 Layer-by-Layer Deposition

Layer-by-layer assembly involves the deposition of alternating layers by molecular interactions. The molecular interactions can be electrostatic, hydrophobic, covalent, or coordination reactions, for example.[184,185] After multiple depositions, the surface may be sufficiently structured to yield to superoleophobic properties. Yang's group used layer-by-layer assembly on aluminium substrates through electrostatic interactions by successive immersion in poly(diallyldimethylammonium chloride) (PDDA) and poly-(sodium 4-styrene-sulfonate) (PSS) (Figure 3.15).[186,187] To obtain superoleophobic properties, a pretreatment by etching in HCl solution was used to form microstructured substrates as well as a post-treatment to exchange the chloride ions by perfluorinated ions. By adjusting the number of bilayers, superoleophobic properties from sticky to non-sticky could be obtained.

3.3.11 Lithography

3.3.11.1 Photolithography

Photolithography is the most widely used lithographic technique and allows the design of extremely well-defined patterns on a substrate.[188] Because of the homogeneity of the patterns produced, this technique is often used to determine the influence of geometric parameters of the pattern on the surface hydrophobicity or oleophobicity. In this process, a light-sensitive polymer and a photomask are used to form the desired patterns after UV illumination. The minimum feature size is approximately 2–3 μm. However, the pattern obtained using "classic" photolithography is not sufficient to obtain superoleophobicity. These microstructured surfaces can be combined with another technique to obtain micro/nanostructured surfaces with superoleophobic properties. For example, nanoporous fluorinated PEDOP polymers were electrodeposited on micropatterned substrates made of cylindrical arrays (diameter 13 μm, height 25 μm, distance between cylinders or pitch

Figure 3.15 Aluminium substrates after layer-by-layer assembly of PDDA and PSS at different magnifications (a–c).[186]

40 μm).[189] Here it was shown that, with the patterning, only a thin layer of fluorinated PEDOP is necessary to reach superoleophobic properties ($\theta_{hexadecane}$ = 145°, $H_{hexadecane}$ = 38°, $\alpha_{hexadecane}$ = 25°). Similar results were obtained after spraying of fluorinated POSS,[190,191] which has among the highest oleophobic properties ($\theta^{Y}_{hexadecane}$ = 80° for fluorodecyl POSS, for example),[166] on micropatterned substrates made of cylindrical arrays. The authors reported that the highest oleophobic values were obtained from a spray time of 120 s. Bhushan's group also used micropatterned substrates made of cylindrical arrays (diameter 14 μm, height 30 μm) with different pitch values (21–210 μm), but they deposited n-perfluoroeicosane nanosheets by CVD.[192,193] The nanosheets were 50–100 nm thick and 500–1000 nm long. This group found that the highest oleophobic properties were obtained for a pitch value of 23 μm ($\theta_{hexadecane}$ = 133°). Hence, using cylindrical arrays it is necessary to deposit structured materials in order to obtain superoleophobic properties. Otherwise, it is possible to reach superoleophobicity by creating microstructures with re-entrant curvatures, often using Bosch deep reactive ion etching (DRIE).[194–220] In this process, the parameters of the re-entrant structures are highly dependent on plasma process parameters such the gas flow and the etching time of the gas used. For example, cylindrical arrays (diameter 3 μm, height 7 μm, pitch 6 μm) were fabricated but with 500 nm re-entrant structures on the straight sidewall pillars. Here, the structures were created by DRIE using a different etching process in order to remove the passivation layer at the bottom of the trench or to etch the silicon isotropically. The height of the pillar can be controlled by the number of etching cycles. Superoleophobic properties ($\theta_{hexadecane}$ = 158°, $H_{hexadecane}$ = 40°, $\alpha_{hexadecane}$ = 10°) were obtained and compared to cylindrical arrays of pillars with an overhang re-entrant structure ($\theta_{hexadecane}$ = 145°, $H_{hexadecane}$ = 47°, $\alpha_{hexadecane}$ = 17°) as well as to smooth, straight sidewall pillars ($\theta_{hexadecane}$ = 120°, sticky) (Figure 3.16).[4,199,200] The authors also evaluated the influence of the pillar diameter, height, and spacing. They observed

Figure 3.16 Cylindrical arrays (a) with 500 nm re-entrant structures on the straight sidewall pillars, (b) with an overhang re-entrant structure, and (c) without re-entrant structures. Reprinted with permission from H. Zhao, K.-Y. Law and V. Sambhy, *Langmuir*, 2011, **27**, 5927.[199] Copyright (2011) American Chemical Society.

that the receding contact angle, the hysteresis, and the sliding angle increase as the solid area fraction increases, which is attributable to an increase in pinning. This observation is in agreement with another publication.[201] They also investigated the presence of nanostructures on top of the pillars and concluded that it is preferable to have flat-topped pillars in order to increase the stability of the Cassie–Baxter state.[200] This is in contradiction to the work of Coffinier *et al.*[203] Indeed, they have also developed cylindrical arrays with re-entrant structures on the straight sidewall pillars (diameter 10 μm, height 10 μm, pitch 5 μm) and measured superoleophobic properties ($\theta_{hexadecane} \approx 140°$) with liquids of $\gamma_{LV} > 27.4$ mN m^{-1} (hexadecane); after nanostructuring using silicon nanowires, it was possible to have superoleophobic properties ($\theta_{hexadecane} \approx 135°$) with liquids of $\gamma_{LV} > 21.6$ mN m^{-1} (octane). Zhu *et al.* also investigated the influence of the shape of various re-entrant structures as well as the presence of nanostructures and obtained the best results with nanostructures.[204]

The superoleophobic properties of cylindrical arrays of pillars with an overhang re-entrant structure have been extensively studied in the literature by other research groups. Tuteja's group developed such substrates with superoleophobic properties for all liquids $\gamma_{LV} > 15.1$ mN m^{-1} (pentane).[23,206] They have shown that following the geometrical parameter of the structure, it is possible to calculate design parameters and as a consequence to determine the robustness of the Cassie–Baxter state as a function of γ_{LV}. In order to impede the wetting of liquids of $\gamma_{LV} > 10$ mN m^{-1} (perfluorohexane), Kim *et al.* also developed cylindrical arrays of pillars with an overhang but with doubly re-entrant structures (Figure 3.17).[207] This is the first artificial superoleophobic substrate without any hydrophobic materials, confirming the

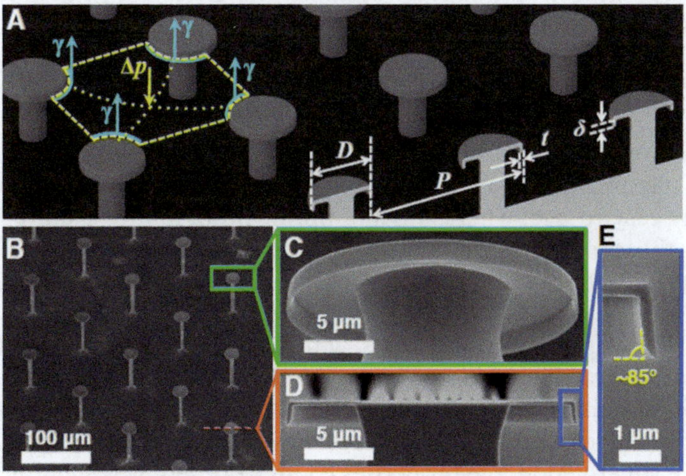

Figure 3.17 Cylindrical arrays with doubly re-entrant structures (A–E represent SEM images at different positions and inclinations). From T. Liu, C.-J. Kim, *Science*, 2014, **346**, 1096.[207] Reprinted with permission from AAAS.

superoleophobic properties observed in nature, and in particular on springtails (*Collembola*).[9–11]

Cylindrical arrays of pillars with an overhang but in a polymeric material were also reported by using the substrate after DRIE as a mould to introduce a liquid polymer before replication.[208–215] This process was also used to create oblique walls (trapezoidal microstructures) with superoleophobic properties ($\theta_{methanol}$ = 135°).[212] Lee's group also fabricated trapezoidal microstructures but with nanopatterns (cone-, pillar-, hole-, and line-shaped) on top of each microstructure in order to create new overhang angles.[214]

Honeycomb patterns composed of nanonails with re-entrant structures were also fabricated.[216–218] The nail head diameter was 405 nm, the nail head thickness 125 nm, the stem diameter 280 nm, and the pitch 900 or 2000 nm, and 400 nm re-entrant structures were obtained on the straight sidewall pillars. Superoleophobic properties for liquids of $\gamma_{LV} > 21.8$ mN m^{-1} (ethanol) were obtained whatever the pitch. Here, the stem was made of conductive silicon and the head of dielectric silicon oxide. By applying a high voltage, it was possible to change the surface properties from non-wetting to wetting. Indeed, the electric energy supplied was sufficient to exceed the energy barrier of the Cassie–Baxter state.

In order to have directional (anisotropic) superoleophobic properties, microgrooves with re-entrant structures were fabricated.[219,220] Law *et al.* obtained microgrooves (width 3 μm, height 4 μm, pitch 3 μm) with various re-entrant structures on the sidewalls.[219] They observed that the surfaces were superoleophobic in the orthogonal direction but with high adhesion ($\theta_{hexadecane}$ = 162°, $H_{hexadecane}$ = 66°, $\alpha_{hexadecane}$ = 34°). Conversely, the apparent contact angles were much lower in the parallel direction but droplet mobility was much easier ($\theta_{hexadecane}$ = 113°, $H_{hexadecane}$ = 19°, $\alpha_{hexadecane}$ = 4°). Microgrooves with anisotropic superoleophobic properties but with oblique walls have also been reported in the literature.[220]

3.3.11.2 *Soft Lithography and Nanoimprint Lithography*

In soft lithography, a pattern is transferred on to a substrate by imprinting using a relatively soft polymer stamp.[188] This technique is inexpensive and relatively easy to use, whereas nanoimprint lithography or hot embossing requires the use of a hard mould. In this technique, a thermoplastic is heated above its glass transition temperature in order to fill the structures of the mould.

Anodized aluminium substrates with a "building block" architecture were used to transfer patterns to polyurethane (PU) and PDMS substrates and enhance their surface oleophobicity.[221] Carbon nanotubes were also imprinted on stretchable PDMS substrates by soft lithography.[222] Interestingly, the superoleophobic properties could be tuned by stretching or bending the substrate. Superoleophobic properties were also obtained by soft lithography on flexible and transparent substrates.[223,224] In order to make the re-entrant structures and obtain pillars with overhangs, isotropic wet etching was performed by simple immersion in tetramethylammonium hydroxide solution. The $\theta_{toluene}$ measured on these substrates was 141°.

Otherwise, Lee's group fabricated pyramidal structures by hot embossing.[225,226] Then, ZnO nanorods of various lengths (250–1000 nm) were induced on these substrates using a hydrothermal process (Figure 3.18). The best oleophobic properties ($\theta_{hexadecane}$ = 131.4°) were measured with 500 nm ZnO nanorods. The hot embossing technique was also used with microscale steel gauzes to form hairy microstructures.[227] To induce nanoroughness of these substrates, a cold radiofrequency CF_4 plasma treatment was used. The authors measured $\theta_{diiodomethane}$ = 143°, $\alpha_{diiodomethane}$ = 50° on these substrates.

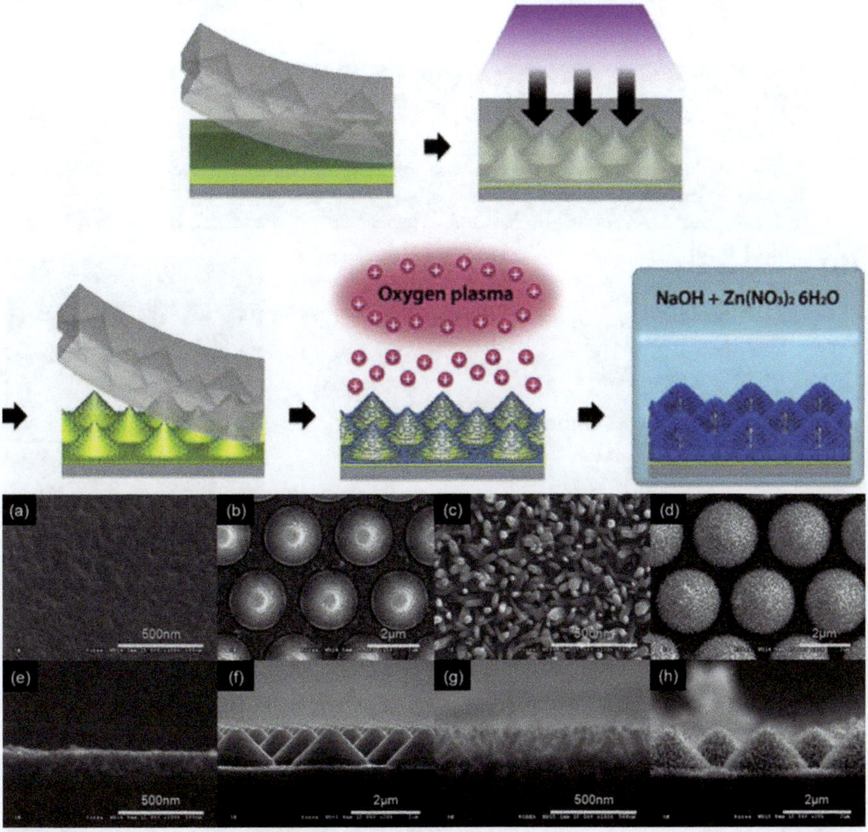

Figure 3.18 Pyramidal structures fabricated by hot embossing and growth of ZnO nanorods using a hydrothermal process. SEM images of (a,e) ZnO nanoparticle resin, (b,f) micropattern, (c,g) nanorods and (d,h) hierarchical structure. Reprinted from *Microelectronic Engineering*, **116**, H-B Jo *et al.*, Superhydrophobic and superleophobic surfaces using ZnO nano-in-micro hierarchical structures, Copyright 2014 with permission from Elsevier.[225]

3.3.11.3 Colloidal Lithography

Colloidal lithography uses highly packed nanoparticles as template, and various post-treatments are possible (Figure 3.19). Hsieh *et al.* used 300 nm silica nanoparticles to form a highly packed layer of a substrate on which small nanoparticles were deposited.[228] The substrates displayed superoleophobic properties for liquids of γ_{LV} > 23.4 mN m^{-1} (isopropanol). Otherwise, highly packed silica nanoparticles could be used a template to embed a polymer in the spaces between the particles.[229] Afterwards, RIE could be performed to form re-entrant structures and the nanoparticles could be removed in basic solution to form an inverse opal with photonic applications.

Polymer spherical nanoparticles can also be used.[230–235] Using these nanoparticles, it is possible to change the morphology of the pattern by

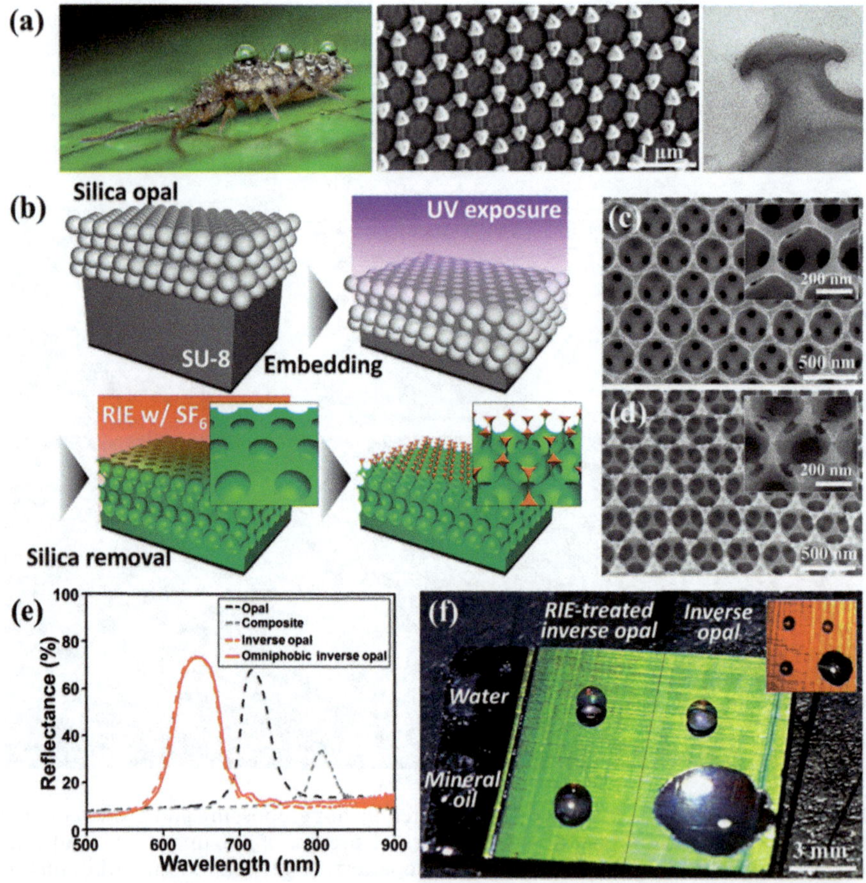

Figure 3.19 Formation of inverse opal structures using silica nanospheres and RIE etching. Reprinted from ref. 230 with permission from John Wiley and Sons. Copyright © 2015 Wiley-VCH Verlag GmbH & Co. KGaA, Weinheim.

applying a plasma etching post-treatment. For example, Gogolides's group demonstrated the possibility of changing the surface morphology from nanospheres to vertically aligned nanocones as the etching time increased, with the additional possibility of obtaining pillars with overhangs.[25,232,233] The patterning can be controlled by the size of the polymer nanoparticles and the plasma parameters. With optimized parameters $\theta_{\text{soya oil}} = 134°$, $H_{\text{soya oil}} = 15°$ were obtained.

A multi-step strategy was also used to form micronails on a substrate.[235] This strategy includes the use of an anodized aluminium oxide membrane as sacrificial templates to form nickel nanowires. Here, the nail heads were obtained by colloidal lithography on the anodized aluminium oxide membrane.

3.3.12 Use of Textured Substrates

In order to easily obtain superoleophobic properties, patterned substrates such as membranes and textiles can be used.

3.3.12.1 Membranes

Superoleophobic paper membranes, consisting of cellulose-based materials with microfibrous morphology, have been reported in the literature.[236–246] This fibrous morphology allows enhanced oleophobic properties. Indeed, Whitesides *et al.* just treated paper membranes by vapour-phase silanization of fluoroalkyltrichlorosilanes and observed that the membranes could support wetting by liquids having $\gamma_{LV} > 27.4$ mN m^{-1} (hexadecane).[236–238] Moreover, this treatment makes it possible to keep the gas permeability and mechanical properties of the paper membranes. In order to graft perfluorinated silanes on to cellulose materials, Silvestre's group used (3-isocyanatopropyol)triethoxysilane to do the bonding.[239] To enhance the superoleophobic properties of microfibrous paper membranes, one strategy is to form nanostructures on the substrates.[240–246] With this aim, Hess *et al.* etched paper membranes by plasma following by a pentafluoroethane plasma to obtain membranes with $\theta_{\text{hexadecane}} = 154°$.[240] Paper membranes were also etched in alkaline solution and the surface roughness increased by titania nanoparticles to obtain $\theta_{\text{hexadecane}} = 146.5°$.[241]

In order to fabricate polymer membranes with microfibrous morphology, the electrospinning technique can be used. Fluorinated polyurethane membranes with superoleophobic properties ($\theta_{\text{octane}} = 136°$) were fabricated by incorporation of silica nanoparticles in order to induce nanoroughness.[247,248] These membranes exhibited high water resistance, good air permeability, and water vapour transmittance. Fluorinated polyurethane and polyacrylonitrile composites as well as poly(*m*-phenylene isopthalimide) membranes were also reported.[249,250] Using a similar approach, glass fibre membranes were also modified with silica nanoparticles to enhance the oleophobic properties.[251,252] Smart polymer membranes by counterion exchange were also reported by electrospinning of a polymer containing ammonium groups.[253] Here, the oleophobic properties of the membranes could be reversibly modified by exchanging thiocyanate ions with bis(trifluoromethylsulfonyl)imide ions, for example. Nanocellulose aerogel membranes with a fibrillar network were fabricated by drying native nanocellulose hydrogels and fluorination by CVC.[254] The $\theta_{\text{mineral oil}}$ was 158° but oil droplets remained stuck to these membranes, indicating high oil adhesion. The membranes also showed buoyant properties on water surface and oils, gas permeability, and drag reduction. These authors also studied silica aerogels.[255] They showed that the applying of a surfactant can also highly enhance the mechanical properties, including abrasion resistance, due to self-replenishing.

Stainless steel meshes are excellent substrates for controlling the geometrical parameters of the superoleophobic properties of membranes.[256–265] These properties can be easily controlled by adjusting the mesh opening and the wire diameter as well as by the formation of surface structures around the mesh wires. Differently structured conducting polymers were electrodeposited on stainless steel meshes (Figure 3.20a).[256,257] The best properties ($\theta_{\text{hexadecane}} = 155°$, $H_{\text{hexadecane}} = 29°$, $\alpha_{\text{hexadecane}} = 30°$) were obtained using 100 µm mesh opening and using fluorinated PEDOP, which allowed the formation of nanoporous structures. It is also noteworthy that a lower deposition charge ($Q_s = 25$ mC cm^{-2}) was necessary to reach the best results on the nanoporous surfaces than on smooth substrates ($Q_s = 225$ mC cm^{-2}). Moreover, it was also observed that it is very important to keep the mesh openings free after deposition. As a consequence, polymers with 2D rather than 3D growth are preferable for this application.[256] Tuteja's group also reported the formation of superoleophobic meshes by deposition of fluorinated POSS (Figure 3.20b).[258–261] The oil hysteresis and sliding angles were extremely low on these substrates even for $\gamma_{LV} < 25$ mN m^{-1}. The coated substrates also resisted acid and basic solutions. These authors also reported

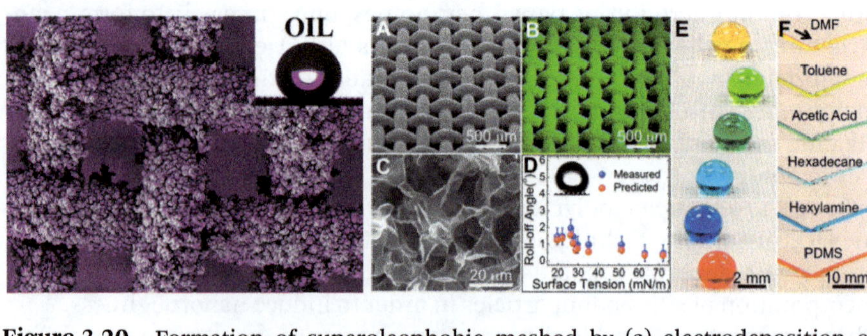

Figure 3.20 Formation of superoleophobic meshed by (a) electrodeposition of fluorinated PEDOP and (b) deposition of fluorinated POSS. Panel a reprinted from ref. 256 with permission from John Wiley and Sons. Copyright © 2014 Wiley-VCH Verlag GmbH & Co. KGaA, Weinheim. Panel b reprinted with permission from S. Pan, A. K. Kota, J. M. Mabry and A. Tuteja, *J. Am. Chem. Soc.*, 2013, **135**, 578.[258] Copyright (2013) American Chemical Society.

that the robustness of the superoleophobic properties is higher for an opening mesh of 100 μm than for an opening mesh of 235 μm, in agreement with other work.[255,256] These results were also supported by hysteresis calculations in the Cassie–Baxter regime and Wenzel regime as a function of the mesh opening and wire diameter.[260] Membranes with both superoleophobicity and superhydrophilicity were also reported in the literature, which is extremely rare because $\gamma_{LV\ oil} < \gamma_{LV\ water}$.[262] This condition is possible if there are specific interactions between water and the substrate in order to lower the solid–liquid interface. In all the examples reported in the literature, there is a combination of oleophobic materials (fluorinated compounds) with highly hydrophilic ones (materials with highly polar or charged species). Tuteja *et al.* reported superoleophobic and superhydrophilic membranes by coating with a blend of fluorinated POSS and cross-linked poly(ethylene glycol) diacrylate.[263] When a water droplet is deposited on the surface, the poly(ethylene glycol) chains are reconfigured, increasing their interfacial area with water. Superoleophobic and superhydrophilic membranes were also obtained by deposition of poly(diallyldimethylammonium perfluorooctanoate).[5]

3.3.12.2 Textiles

Finally, textile materials such as fabrics can be used as textured substrates for enhancing superoleophobic properties. The principle of a see-through fabric is the same than that of a membrane.[266] The superoleophobic properties can be controlled by the dimensions of the fibres, the presence of surface structures on the fibres, or the spacing between the fibres. For example, superoleophobic ($\theta_{hexadecane} > 140°$) see-through fabrics were reported by spraying of silica nanoparticles.[267]

Otherwise, the principle of a "classical woven" fabric is the same than that of a see-though fabric but without spacing between the fibres. Superoleophobic fabric can be obtained by a simple fluorination process.[268–271] However, in order to obtain both durable and robust superoleophobic fabrics, it is preferable to modify the roughness of the fibres. A triple-length-scale surface roughness was achieved by grafting both silica micro- and nanoparticles on cotton fabrics, using the Stöber reaction.[1] Superoleophobic properties ($\theta_{hexadecane} = 152°$) with low $\alpha_{hexadecane}$ were reported with these fabrics. Mesoporous silica nanoparticles were also reported to obtain superoleophobicity.[272] Other inorganic nanoparticles such as ZnO[273] and CuO[274] as well as polymer nanoparticles[275,276] such as silicone or polyaniline and composites[277] also showed remarkable superoleophobic properties. In particular, the use of fluorinated POSS allowed not only superoleophobic properties ($\theta_{hexadecane} = 157°$, $\alpha_{hexadecane} \approx 18°$) but also self-healing properties (Figure 3.21). Moreover, the fabrics were able to withstand acid, UV light, machine wash (>200 laundry cycles), and abrasion (>5000 Martindale abrasion cycles).[278–281] It was also possible to form electrically conductive fabrics by incorporating conducting polymers.[282,283] Finally, smart fabrics with reversible oleophobicity could be

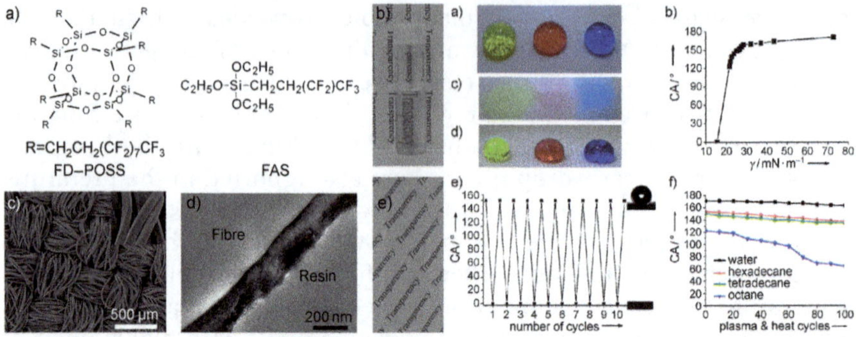

Figure 3.21 Formation of inverse opal structures using silica nanospheres and RIE etching. Reprinted from ref. 280 with permission from John Wiley and Sons. Copyright © 2011 Wiley-VCH Verlag GmbH & Co. KGaA, Weinheim.

achieved by using polymers with ammonium groups.[284] On these fabrics, the reversibility was obtained by exchanging chloride counterions by perfluorooctanoate ones.

3.4 Conclusion

In this review, we have summarized most of the methods reported in the literature to reach superoleophobic properties. Such materials can be obtained by plasma etching/RIE, etching in acidic and basic media, galvanostatic deposition, anodization, the use of nanoparticles, hydrothermal and solvothermal processes, chemical vapour deposition, electrodeposition, electrospinning, layer-by-layer deposition, lithography (photolithography, soft lithography, nanoimprint lithography, colloidal lithography), or the use of textured substrates (membranes, textiles). In all the processes reported in the literature, fluorinated compounds are used in order to enhance the intrinsic oleophobic properties. However, in nature species such as *Collembola* demonstrate superoleophobic properties without using fluorinated compounds.[9–11] Hence, the next step will be to synthesize non-fluorinated superoleophobic materials. Indeed, the first example of such a material has just been reported using cylindrical arrays with doubly re-entrant structures.[207]

References

1. B. Leng, Z. Shao, G. de With and W. Ming, *Langmuir*, 2009, **25**, 2456.
2. S. Jin and C. Choi, *PCT Int. Appl.*, WO 2012088209 A2 20120628, 2012.
3. S. L. S. Freire and B. Tanner, *Langmuir*, 2013, **29**, 9024.
4. H. Zhao and K.-Y. Law, *ACS Appl. Mater. Interfaces*, 2012, **4**, 4288.
5. J. Yang, Z. Zhang, X. Xu, X. Zhu, X. Men and X. Zhou, *J. Mater. Chem.*, 2012, **22**, 2834.
6. K. Liu, Y. Tian and L. Jiang, *Prog. Mater. Sci.*, 2013, **58**, 503.

7. L.-P. Xu, J. Zhao, B. Su, X. Liu, J. Peng, Y. Liu, H. Liu, G. Yang, L. Jiang, Y. Wen, X. Zhang and S. Wang, *Adv. Mater.*, 2013, **25**, 606.
8. T. Darmanin and F. Guittard, *Mater. Today*, 2015, **18**, 273.
9. J. Nickerl, R. Helbig, H.-J. Schulz, C. Werner and C. Neinhuis, *Zoomorphology*, 2013, **132**, 183.
10. R. Helbig, J. Nickerl, C. Neinhuis and C. Werner, *PLoS One*, 2011, **6**, e25105.
11. R. Hensel, R. Helbig, S. Aland, H.-G. Braun, A. Voigt, C. Neinhuis and C. Werner, *Langmuir*, 2011, **6**, e25105.
12. T. Young, *Philos. Trans. R. Soc. London*, 1805, **95**, 65.
13. H. Zhu, Z. Guo and W. Liu, *Chem. Commun.*, 2014, **50**, 3900.
14. A. Marmur, *Soft Matter*, 2012, **8**, 6867.
15. K.-Y. Law, *J. Phys. Chem. Lett.*, 2014, **5**, 686.
16. R. N. Wenzel, *Ind. Eng. Chem.*, 1936, **28**, 988.
17. A. B. D. Cassie and S. Baxter, *Trans. Faraday Soc.*, 1944, **40**, 546.
18. A. Marmur, *Langmuir*, 2003, **19**, 8343.
19. A. Marmur, *Langmuir*, 2008, **24**, 7573.
20. G. Whyman and E. Bormashenko, *Langmuir*, 2011, **27**, 8171.
21. L. Cao, H.-H. Hu and D. Gao, *Langmuir*, 2007, **23**, 4310.
22. M. Nosonovsky, *Langmuir*, 2007, **23**, 3157.
23. A. Tuteja, W. Choi, M. Ma, J. M. Mabry, S. A. Mazzella, G. C. Rutledge, G. H. McKinley and R. E. Cohen, *Science*, 2007, **318**, 1618.
24. X. Liu, H.-S. Choi, B.-R. Park and H.-K. Lee, *Appl. Surf. Sci.*, 2011, **257**, 8828.
25. K. Ellinas, S. P. Pujari, D. A. Dragatogiannis, C. A. Charitidis, A. Tserepi, H. Zuilhof and E. Gogolides, *ACS Appl. Mater. Interfaces*, 2014, **6**, 6510.
26. A. K. Gnanappa, D. P. Papageorgiou, E. Gogolides, A. Tserepi, A. G. Papathanasiou and A. G. Boudouvis, *Plasma Processes Polym.*, 2012, **9**, 304.
27. D. Kontziampasis, G. Boulousis, A. Smyrnakis, K. Ellinas, A. Tserepi and E. Gogolides, *Microelectron. Eng.*, 2014, **121**, 33.
28. C.-F. Wang, S.-F. Chiou, F.-H. Ko, C.-T. Chou, H.-C. Lin, C.-F. Huang and F.-C. Chang, *Macromol. Rapid Commun.*, 2006, **27**, 333.
29. R. T. R. Kumar, K. B. Mogensen and P. Bøggild, *J. Phys. Chem. C*, 2010, **114**, 2936.
30. S. Li, M. Jin, C. Yu and M. Liao, *Colloids Surf., A*, 2013, **430**, 46.
31. A. Starostin, V. Valtsifer, V. Strelnikov, E. Bormashenko, R. Grynyov, Y. Bormashenko and A. Gladkikh, *Adv. Eng. Mater.*, 2014, **16**, 1127.
32. S. Ji, P. A. Ramadhianti, T.-B. Nguyen, W.-d. Kim and H. Lim, *Microelectron. Eng.*, 2013, **111**, 404.
33. J. Song, S. Huang, K. Hu, Y. Lu, X. Liu and W. Xu, *J. Mater. Chem. A*, 2013, **1**, 14783.
34. K. Tadanaga, N. Katata and T. Minami, *J. Am. Ceram. Soc.*, 1997, **80**, 3213.
35. G. Zhang, X. Zhang, M. Li and Z. Su, *ACS Appl. Mater. Interfaces*, 2014, **6**, 1729.

36. S. Peng, X. Yang, D. Tian and W. Deng, *ACS Appl. Mater. Interfaces*, 2014, **6**, 15188.
37. Z. Cheng, H. Lai, Y. Du, R. Hou, C. Li, N. Zhang and K. Sun, *ChemPlusChem*, 2015, **80**, 151.
38. J. Yang, Z. Zhang, X. Xu, X. Men, X. Zhu and X. Zhou, *New J. Chem.*, 2011, **35**, 2422.
39. G. Ren, Z. Zhang, X. Zhu, B. Ge, K. Wang, X. Xu, X. Men and X. Zhou, *Appl. Phys. A*, 2014, **114**, 1129.
40. J. Yang, H. Song, H. Tang, H. Ji and C. Li, *Appl. Surf. Sci.*, 2013, **280**, 940.
41. Z. S. Saifaldeen, K. R. Khedir, M. F. Cansizoglu, T. Demirkan and T. Karabacak, *J. Mater. Sci.*, 2014, **49**, 1839.
42. M. Raimondo, M. Blosi, A. Caldarelli, G. Guarini and F. Veronesi, *Chem. Eng. J.*, 2014, **258**, 101.
43. H. Li, S. Yu, E. Liu and Y. Zhao, *Microsyst. Technol.*, 2015, **21**, 2003.
44. C. Gondek, M. Lippold, I. Roever, K. Bohmhammel and E. Kroke, *J. Phys. Chem. C*, 2014, **118**, 2044.
45. T.-i. Kim, D. Tahk and H. H. Lee, *Langmuir*, 2009, **25**, 6576.
46. Z. Huang, N. Geyer, P. Werner, J. de Boor and U. Goesele, *Adv. Mater.*, 2011, **23**, 285.
47. T. P. N. Nguyen, R. Dufour, V. Thomy, V. Senez, R. Boukherroub and Y. Coffinier, *Appl. Surf. Sci.*, 2014, **295**, 38.
48. N. T. P. Nguyen, Y. Coffinier, V. Thomy and R. Boukherroub, *Phys. Status Solidi A*, 2013, **210**, 2178.
49. Y. Liu, Y. Xiu, D. W. Hess and C. P. Wong, *Langmuir*, 2010, **26**, 8908.
50. Y. Lu, J. Song, X. Liu, W. Xu, Y. Xing and Z. Wei, *ACS Sustainable Chem. Eng.*, 2013, **1**, 102.
51. J. Xi, L. Feng and L. Jiang, *Appl. Phys. Lett.*, 2008, **92**, 053102/1.
52. H. Meng, S. Wang, J. Xi, Z. Tang and L. Jiang, *J. Phys. Chem. C*, 2008, **112**, 11454.
53. X. Yao, J. Gao, Y. Song and L. Jiang, *Adv. Funct. Mater.*, 2011, **21**, 4270.
54. J. Yang, X. Wu, Q. Zhang and H. Song, *J. Adhes. Sci. Technol.*, 2014, **28**, 1687.
55. H. Wang, J. Yu, Y. Wu, W. Shao and X. Xu, *J. Mater. Chem. A*, 2014, **2**, 5010.
56. X. Zhu, Z. Zhang, X. Xu, X. Men, J. Yang, X. Zhou and Q. Xue, *J. Colloid Interface Sci.*, 2015, **367**, 443.
57. J. Ou, W. Hu, S. Liu, M. Xue, F. Wang and W. Li, *ACS Appl. Mater. Interfaces*, 2013, **5**, 10035.
58. J.-Y. Lee, S. Pechook, B. Pokroy, D.-J. Jeon, B. Pokroy and J. S. Yeo, *ASC Appl. Mater. Interfaces*, 2014, **6**, 4927.
59. J. Y. Lee, S. Pechook, B. Pokroy and J. S. Yeo, *Langmuir*, 2014, **30**, 15568.
60. S. Barthwal and S.-H. Lim, *Appl. Surf. Sci.*, 2015, **328**, 296.
61. Z. Zhang, X. Zhu, J. Yang, X. Xu, X. Men and X. Zhou, *Appl. Phys. A*, 2012, **108**, 601.
62. T. C. Rangel, A. F. Michels, F. Horowitz and D. E. Weibel, *Langmuir*, 2015, **31**, 3465.

63. X. Xu, Z. Z. Zhang, F. Guo, J. Yang, X. T. Zhu, X. Y. Zhou and Q. J. Xue, *Colloids Surf., A*, 2012, **396**, 90.
64. K. Tsujii, T. Yamamoto, T. Onda and S. Shibuichi, *Angew. Chem., Int. Ed.*, 1997, **36**, 1011.
65. S. Shibuichi, T. Yamamoto, T. Onda and K. Tsujii, *J. Colloid Interface Sci.*, 1998, **208**, 287.
66. C. Jeong and C.-H. Choi, *ACS Appl. Mater. Interfaces*, 2012, **4**, 842.
67. T. Fujii, Y. Aoki and H. Habazaki, *Langmuir*, 2011, **27**, 11752.
68. W. Wu, X. Wang, D. Wang, M. Chen, F. Zhou, W. Liu and Q. Xue, *Chem. Commun.*, 2009, 1043.
69. Y. Wu, M. Cai, Z. Li, X. Song, H. Wang, X. Pei and F. Zhou, *J. Colloid Interface Sci.*, 2014, **414**, 9.
70. R. Weng, H. Zhang and X. Liu, *Microelectron. Eng.*, 2013, **119**, 183.
71. S. Barthwal, Y. S. Kim and S.-H. Lim, *Langmuir*, 2013, **29**, 11966.
72. K. Nakayama, E. Tsuji, Y. Aoki and H. Habazaki, *RSC Adv.*, 2014, **4**, 30927.
73. T. Fujii, H. Sato, E. Tsuji, Y. Aoki and H. Habazaki, *J. Phys. Chem. C*, 2012, **116**, 23308.
74. T. T. Y. Tan, M. R. Reithofer, E. Y. Chen, A. G. Menon, T. S. A. Hor, J. Xu and J. M. Chin, *J. Am. Chem. Soc.*, 2013, **135**, 16272.
75. B. Li, X. Gao, H.-C. Zhang and C. Yuan, *ACS Sustainable Chem. Eng.*, 2014, **2**, 404.
76. S. Barthwal, Y. S. Kim and S.-H. Lim, *J. Colloid Interface Sci.*, 2013, **400**, 123.
77. S. Farsinezhad, P. R. Waghmare, B. D. Wiltshire, H. Sharma, S. Amiri, S. K. Mitra and K. Shankar, *RSC Adv.*, 2014, **4**, 33587.
78. D. Wang, X. Wang, X. Liu and F. Zhou, *J. Phys. Chem. C*, 2010, **114**, 9938.
79. Y. Sun, L. Wang, Y. Gao and D. Guo, *Appl. Surf. Sci.*, 2015, **324**, 825.
80. Q. Ye, X. Wang, S. Li and F. Zhou, *Macromolecules*, 2010, **43**, 5554.
81. J.-Y. Huang, Y.-K. Lai, F. Pan, L. Yang, H. Wang, K.-Q. Zhang, H. Fuchs and L.-F. Chi, *Small*, 2014, **10**, 4865.
82. Y.-C. Sheen, Y.-C. Huang, C.-S. Liao, H.-Y. Chou and F.-C. Chang, *J. Polym. Sci., Part B: Polym. Phys.*, 2008, **46**, 1984.
83. Y.-C. Sheen, W.-H. Chang, W.-C. Chen, Y.-H. Chang, Y.-C. Huang and F.-C. Chang, *Mater. Chem. Phys.*, 2009, **114**, 63.
84. O.-U. Nimittrakoolchai and S. Supothina, *J. Nanosci. Nanotechnol.*, 2012, **12**, 4962.
85. N. V. Motlagh, F. C. Birjandi, J. Sargolzaei and N. Shahtahmassebi, *Appl. Surf. Sci.*, 2013, **283**, 636.
86. N. V. Motlagh, J. Sargolzaei and N. Shahtahmassebi, *Surf. Coat. Technol.*, 2013, **235**, 241.
87. Z. Geng, J. He, L. Xu and L. Yao, *J. Mater. Chem. A*, 2013, **1**, 8721.
88. Z. Geng and J. He, *J. Mater. Chem. A*, 2014, **2**, 16601.
89. Y. Wang and B. Bhushan, *ACS Appl. Mater. Interfaces*, 2015, **7**, 743.
90. J. Li, L. Yan, Q. Ouyang, F. Zha, Z. Jing, X. Li and Z. Lei, *Chem. Eng. J.*, 2014, **246**, 238.

91. D. Ge, L. Yang, Y. Zhang, Y. Rahmawan and S. Yang, *Part. Part. Syst. Charact.*, 2014, **31**, 763.
92. J. Yang, Z.-Z. Zhang, X.-H. Men and X.-H. Xu, *J. Dispersion Sci. Technol.*, 2011, **32**, 485.
93. G. Soliveri, R. Annunziata, S. Ardizzone, G. Cappelletti and D. Meroni, *J. Phys. Chem. C*, 2012, **116**, 26405.
94. G. Hayase, K. Nonomura, G. Hasegawa, K. Kanamori and K. Nakanishi, *Chem. Mater.*, 2015, **27**, 3.
95. X. Deng, L. Mammen, H.-J. Butt and D. Vollmer, *Science*, 2012, **335**, 67.
96. M. Paven, P. Papadopoulos, L. Mammen, X. Deng, H. Sachdev, D. Vollmer and H.-J. Butt, *Pure Appl. Chem.*, 2014, **86**, 87.
97. C.-J. Liang, J.-D. Liao, A.-J. Li, C. Chen, H.-Y. Lin, X.-J. Wang and Y.-H. Xu, *Fuel*, 2014, **128**, 422.
98. J. He, H. Li, X. Liu and M. Qu, *J. Nanosci. Nanotechnol.*, 2013, **13**, 1974.
99. H. Yu, X. Tian, H. Luo and X. Ma, *Mater. Lett.*, 2015, **138**, 184.
100. Y. Li, X. Zhu, X. Zhou, B. Ge, S. Chen and W. Wu, *Appl. Phys. A*, 2014, **115**, 765.
101. X. Zhu, Z. Zhang, G. Ren, X. Men, B. Ge and X. Zhou, *J. Colloid Interface Sci.*, 2014, **421**, 141.
102. M. Zhang, T. Zhang and T. Cui, *Langmuir*, 2011, **27**, 9295.
103. J. Yang, Z. Zhang, X. Men, X. Xu and X. Zhu, *New J. Chem.*, 2011, **35**, 207.
104. J. Yang, Z. Zhang, X. Men, X. Xu and X. Zhu, *Colloids Surf., A*, 2010, **367**, 60.
105. C.-T. Hsieh, D.-Y. Tzou, C. Pan and W.-Y. Chen, *Surf. Coat. Technol.*, 2012, **207**, 11.
106. X. Li, H. K. Lee, I. Y. Phang, C. K. Lee and X. Y. Ling, *Anal. Chem.*, 2014, **86**, 10437.
107. E. Yoshida, *Colloid Polym. Sci.*, 2012, **290**, 80.
108. J. Zhang and S. Seeger, *Angew. Chem., Int. Ed.*, 2011, **50**, 6652.
109. J. Zhang, A. Wang and S. Seeger, *Polym. Chem.*, 2014, **5**, 1132.
110. J. Zhang, A. Wang and S. Seeger, *Adv. Funct. Mater.*, 2014, **24**, 1074.
111. G. R. J. Artus and S. Seeger, *Adv. Colloid Interface Sci.*, 2014, **209**, 144.
112. A. Stojanovic, G. R. J. Artus and S. Seeger, *Nano Res.*, 2010, **3**, 889.
113. E. Sumino, T. Saito, T. Noguchi and H. Sewada, *Polym. Adv. Technol.*, 2015, **26**, 345.
114. S. G. Lee, D. S. Ham, D. Y. Lee, H. Bong and K. Cho, *Langmuir*, 2013, **29**, 15051.
115. S. M. Ramirez, Y. J. Diaz, C. M. Sahagun, M. W. Duff, O. B. Lawal, S. T. Iacono and J. M. Mabry, *Polym. Chem.*, 2013, **4**, 2230.
116. C. Jiang, Q. Wang and T. Wang, *New J. Chem.*, 2013, **37**, 810.
117. D. Xiong, G. Liu, J. Zhang and S. Duncan, *Chem. Mater.*, 2011, **23**, 2810.
118. L. Xiong, L. L. Kendrick, H. Heusser, J. C. Webb, B. J. Sparks, J. T. Goetz, W. Guo, C. M. Stafford, M. D. Blanton, S. Nazarenko and D. L. Patton, *ACS Appl. Mater. Interfaces*, 2014, **6**, 10763.
119. S.-H. Hsu, Y.-L. Chang, Y.-C. Tu, C.-M. Tsai and W.-F. Su, *ACS Appl. Mater. Interfaces*, 2013, **5**, 2991.

120. Z. He, M. Ma, X. Lan, F. Chen, K. Wang, H. Deng, Q. Zhang and Q. Fu, *Soft Matter*, 2011, **7**, 6435.
121. B. J. Basu, V. D. Kumar and C. Anandan, *Appl. Surf. Sci.*, 2012, **261**, 807.
122. B. J. Basu, T. Bharathidasan and C. Anandan, *Surf. Innovations*, 2013, **1**, 40.
123. T. Bharathidasan, T. N. Narayanan, S. Sathyanaryanan and S. S. Sreejakumari, *Carbon*, 2015, **84**, 207.
124. Y. Tang, J. Yang, L. Yin, B. Chen, H. Tang, C. Liu and C. Li, *Colloids Surf., A*, 2014, **459**, 261.
125. S. Srinivasan, S. S. Chhatre, J. M. Mabry, R. E. Cohen and G. H. McKinley, *Polymer*, 2011, **52**, 3209.
126. X. Wang, H. Hu, Q. Ye, T. Gao, F. Zhou and Q. Xue, *J. Mater. Chem.*, 2012, **22**, 9624.
127. B. Ge, Z. Zhang, X. Men, X. Zhu and X. Zhou, *Appl. Surf. Sci.*, 2014, **293**, 271.
128. R. Campos, A. J. Guenthner, A. J. Meuler, A. Tuteja, R. E. Cohen, G. H. McKinley, T. S. Haddad and J. M. Mabry, *Langmuir*, 2012, **28**, 9834.
129. C.-T. Hsieh, F.-L. Wu and W.-Y. Chen, *Surf. Coat. Technol.*, 2009, **203**, 3377.
130. C.-T. Hsieh, B.-S. Chang and J.-Y. Lin, *Appl. Surf. Sci.*, 2011, **257**, 7997.
131. R. V. Lakshmi, T. Bharathidasan, P. Bera and B. J. Basu, *Surf. Coat. Technol.*, 2012, **206**, 3888.
132. P. Muthiah, B. Bhushan, K. Yun and H. Kondo, *J. Colloid Interface Sci.*, 2013, **409**, 227.
133. U. Cengiz and C. E. Cansoy, *Appl. Surf. Sci.*, 2015, **335**, 99.
134. A. Steele, I. Bayer and E. Loth, *Nano Lett.*, 2009, **9**, 501.
135. A. Das, T. M. Schutzius, I. S. Bayer and C. M. Megaridis, *Carbon*, 2012, **50**, 1346.
136. T. M. Schutzius, M. Elsharkawy, M. K. Tiwari and C. M. Megaridis, *Lab Chip*, 2012, **12**, 5237.
137. C.-F. Wang, S.-W. Hung, S.-W. Kuo and C.-J. Chang, *Appl. Surf. Sci.*, 2014, **320**, 658.
138. D. Xiong, G. Liu, L. Hong and E. J. S. Duncan, *Chem. Mater.*, 2011, **23**, 4357.
139. G. Zhang, S. Lin, I. Wyman, H. Zou, J. Hu, G. Liu, J. Wang, F. Li, F. Liu and M. Hu, *ACS Appl. Mater. Interfaces*, 2013, **5**, 2991.
140. G. Zhang, J. Hu, G. Liu, H. Zou, Y. Tu, F. Li, S. Hu and H. Luo, *J. Mater. Chem. A*, 2013, **1**, 6226.
141. D. Xiong, G. Liu and E. J. S. Duncan, *Polymer*, 2013, **54**, 3008.
142. W. Jiang, C. M. Grozea, Z. Shi and G. Liu, *ACS Appl. Mater. Interfaces*, 2014, **6**, 2629.
143. S. Nishizawa and S. Shiratori, *Appl. Surf. Sci.*, 2012, **263**, 8.
144. H. Hayashi and Y. Hakuta, *Materials*, 2010, **3**, 3794.
145. H. Li, S. Yu, X. Han, E. Liu and Y. Zhao, *Colloids Surf., A*, 2015, **469**, 271.
146. C. Jin, J. Li, S. Han, J. Wang and Q. Sun, *Appl. Surf. Sci.*, 2014, **320**, 322.

147. H. Kim, K. Noh, C. Choi, J. Khamwannah, D. Villwock and S. Jin, *Langmuir*, 2011, **27**, 10191.
148. W. Pan and J. Gong, *Key Eng. Mater.*, 2010, **434–435**, 538.
149. G. Perry, Y. Coffinier, V. Thomy and R. Boukherroub, *Langmuir*, 2012, **28**, 389.
150. G. Demazeau, *J. Mater. Sci.*, 2008, **43**, 2104.
151. J. Yu, H.-J. Wang, W.-J. Shao and X.-L. Xu, *Chin. Phys. B*, 2014, **23**, 016803/1.
152. J. Yu, H. Wang, N. Yin and X. Xu, *RSC Adv.*, 2014, **4**, 24163.
153. T. P. N. Nguyen, P. Brunet, Y. Coffinier and R. Boukherroub, *Langmuir*, 2010, **26**, 18369.
154. H. Li, X. Wang, Y. Song, Y. Liu, Q. Li, L. Jiang and D. Zhu, *Angew. Chem., Int. Ed.*, 2001, **4**, 1743.
155. T. Darmanin and F. Guittard, *Prog. Polym. Sci.*, 2014, **39**, 656.
156. T. Darmanin and F. Guittard, *J. Phys. Chem. C*, 2014, **118**, 26912.
157. T. Darmanin and F. Guittard, *J. Am. Chem. Soc.*, 2009, **131**, 7928.
158. T. Darmanin and F. Guittard, *J. Mater. Chem.*, 2009, **19**, 7130.
159. H. Bellanger, T. Darmanin, E. Taffin de Givenchy and F. Guittard, *Colloids Surf., A*, 2013, **433**, 47.
160. H. Bellanger, T. Darmanin and F. Guittard, *Langmuir*, 2012, **28**, 186.
161. H. Bellanger, T. Darmanin, E. Taffin de Givenchy and F. Guittard, *RSC Adv.*, 2012, **2**, 10899.
162. H. Bellanger, T. Darmanin, E. Taffin de Givenchy and F. Guittard, *J. Mater. Chem. A*, 2013, **1**, 2896.
163. A. Zenerino, T. Darmanin and F. Guittard, *Langmuir*, 2010, **26**, 13545.
164. H. Bellanger, T. Darmanin, E. Taffin de Givenchy and F. Guittard, *RSC Adv.*, 2013, **3**, 555.
165. T. Darmanin and F. Guittard, *Macromol. Chem. Phys.*, 2013, **214**, 2036.
166. S. S. Chhatre, J. O. Guardado, B. M. Moore, T. S. Haddad, J. M. Mabry, G. H. McKinley and R. E. Cohen, *ACS Appl. Mater. Interfaces*, 2010, **2**, 3544.
167. T. Darmanin and F. Guittard, *J. Am. Chem. Soc.*, 2011, **133**, 15627.
168. J. El-Maiss, T. Darmanin and F. Guittard, *J. Colloid Interface Sci.*, 2015, **447**, 167.
169. M. Wolfs, T. Darmanin and F. Guittard, *Soft Matter*, 2012, **8**, 9110.
170. T. Darmanin and F. Guittard, *Adv. Mater. Interfaces*, 2015, **2**, 1500081.
171. T. Darmanin and F. Guittard, *Soft Matter*, 2013, **9**, 5982.
172. T. Darmanin and F. Guittard, *J. Colloid Interface Sci.*, 2013, **408**, 101.
173. S. Taleb, T. Darmanin and F. Guittard, *ACS Appl. Mater. Interfaces*, 2014, **6**, 7953.
174. K. Chiba, K. Kurogi, K. Monde, M. Hashimoto, M. Yoshida, H. Mayama and K. Tsujii, *Colloids Surf., A*, 2010, **354**, 234.
175. H. Yan, K. Kurogi and K. Tsujii, *Colloids Surf., A*, 2007, **292**, 27.
176. L. Xu, D. Zhu, X. Lu and Q. Lu, *J. Mater. Chem. A*, 2014, **6**, 7953.
177. Y. Tina, H. Liu and Z. Deng, A review on polymer nanofibers by electrospinning and their applications in nanocomposites, *Chem. Mater.*, 2006, **18**, 5820.

178. Z.-M. Huang, Y.-Z. Zhang, M. Kotaki and S. Ramakrishna, *Compos. Sci. Technol.*, 2003, **63**, 2223.
179. G.-R. Choi, J. Park, J.-W. Ha, W.-D. Kim and H. Lim, *Macromol. Mater. Eng.*, 2010, **295**, 995.
180. L. Yi, X. Meng, X. Tian, W. Zhou and R. Chen, *J. Phys. Chem. C*, 2014, **118**, 26671.
181. V. A. Ganesh, S. S. Dinachali, A. S. Nair and S. Ramakrishna, *ACS Appl. Mater. Interfaces*, 2013, **5**, 1527.
182. V. A. Ganesh, S. S. Dinachali, H. K. Raut, T. M. Walsh, A. S. Nair and S. Ramakrishna, *RSC Adv.*, 2013, **3**, 3819.
183. S. P. R. Kobaku, A. K. Kota, D. H. Lee, J. M. Mabry and A. Tuteja, *Angew. Chem., Int. Ed.*, 2012, **51**, 10109.
184. J. Borges and J. F. Mano, *Chem. Rev.*, 2014, **114**, 8883.
185. Y. Li, X. Wang and J. Sun, *Chem. Soc. Rev.*, 2012, **41**, 5998.
186. J. Yang, H. Song, B. Chen, H. Tang and C. Li, *RSC Adv.*, 2014, **4**, 14227.
187. J. Yang, Z. Zhang, X. Men, X. Xu, X. Zhu, X. Zhou and Q. Xue, *J. Colloid Interface Sci.*, 2012, **366**, 191.
188. A. Pimpin and W. Srituravanich, *Eng. J.*, 2012, **16**, 37.
189. T. Darmanin, F. Guittard, S. Amigoni, E. Tafin de Givenchy, X. Noblin, R. Kofman and F. Celestini, *Soft Matter*, 2011, **7**, 1053.
190. K. Golovin, D. H. Lee, J. M. Mabry and A. Tuteja, *Angew. Chem., Int. Ed.*, 2013, **52**, 13007.
191. S. M. Ramirez, Y. J. Diaz, R. Campos, R. L. Stone, T. S. Haddad and J. M. Mabry, *J. Am. Chem. Soc.*, 2011, **133**, 20084.
192. Y. C. Jung and B. Bhushan, *Langmuir*, 2009, **25**, 14165.
193. B. Bhushan, *Beilstein J. Nanotechnol.*, 2011, **2**, 66.
194. Z. Pan, H. Shahsavan, W. Zhang, F. K. Yang and B. Zhao, *Appl. Surf. Sci.*, 2015, **324**, 612.
195. D. Infante, K. W. Koch, P. Mazumder, L. Tian, A. Carrilero, D. Tulli, D. Baker and V. Pruneri, *Nano Res.*, 2013, **6**, 429.
196. P. Mazumder, Y. Jiang, D. Baker, A. Carrilero, D. Tulli, D. Infante, A. T. Hunt and V. Pruneri, *Nano Lett.*, 2014, **14**, 4677.
197. F. Joki-Korpela, J. Karvinen, B. Päivänranta, A. Partanen, M. Suvanto, M. Kuittinen and T. T. Pakkanen, *Microelectron. Eng.*, 2014, **114**, 38.
198. A. Susarrey-Arce, A. G. Marín, S. Schlautmann, L. Lefferts, J. G. E. Gardeniers and A. van Houselt, *J. Micromech. Microeng.*, 2013, **23**, 025004/1.
199. H. Zhao, K.-Y. Law and V. Sambhy, *Langmuir*, 2011, **27**, 5927.
200. H. Zhao, K.-C. Park and K.-Y. Law, *Langmuir*, 2012, **28**, 14925.
201. B. M. L. Koch, J. A. W. Elliott and A. Amirfazli, *Langmuir*, 2014, **30**, 14039.
202. A. Zeniou, K. Ellinas, A. Olziersky and E. Gogolides, *Nanotechnology*, 2014, **25**, 035302/1.
203. T. P. N. Nguyen, R. Boukherroub, V. Thomy and Y. Coffinier, *J. Colloid Interface Sci.*, 2014, **416**, 280.
204. H. Hu, V. V. Swaminathan, M. R. Z. Farahani, G. Mensing, J. Yeom, M. A. Shannon and L. Zhu, *J. Micromech. Microeng.*, 2011, **24**, 095023/1.

205. R. Almeida and J. W. Kwon, *Langmuir*, 2013, **29**, 994.
206. A. Tuteja, W. Choi, J. M. Mabry, G. H. McKinley and R. E. Cohen, *Proc. Natl. Acad. Sci. U. S. A.*, 2008, **105**, 18200.
207. T. Liu and C.-J. Kim, *Science*, 2014, **346**, 1096.
208. R. Dufour, M. Harnois, Y. Coffinier, V. Thomy, R. Boukherroub and V. Senez, *Langmuir*, 2010, **26**, 17242.
209. R. Dufour, G. Perry, M. Harnois, Y. Coffinier, V. Thomy, V. Senez and R. Boukherroub, *Colloid Polym. Sci.*, 2013, **291**, 409.
210. S. M. Kang, S. M. Kim, H. N. Kim, M. K. Kwak, D. H. Tahk and K. Y. Suh, *Soft Matter*, 2012, **8**, 8563.
211. L. Yuan, T. Wu, W. Zhang, S. Ling, R. Xiang, X. Gui, Y. Zhu and Z. Tang, *J. Mater. Chem. A*, 2014, **2**, 6952.
212. M. Im, H. Im, J.-H. Lee, J.-B. Yoon and Y.-K. Choi, *Soft Matter*, 2010, **6**, 1401.
213. A. Cavalli, P. Bøggild and F. Okkels, *Langmuir*, 2012, **28**, 17545.
214. H.-J. Choi, S. Choo, J.-H. Shin, K.-I. Kim and H. Lee, *J. Phys. Chem. C*, 2013, **117**, 24354.
215. R. Hensel, A. Finn, R. Helbig, H.-G. Braun, C. Neinhuis, W.-J. Fischer and C. Werner, *Adv. Mater.*, 2014, **26**, 2029.
216. A. Ahuja, J. A. Taylor, V. Lifton, A. A. Sidorenko, T. R. Salamon, E. J. Lobaton, P. Kolodner and T. N. Krupenkin, *Langmuir*, 2008, **24**, 9.
217. C. Aulin, S. H. Yun, L. Wagberg and T. Lindstroem, *ACS Appl. Mater. Interfaces*, 2009, **1**, 2443.
218. T. Wu and Y. Suzuki, *Sens. Actuators, B*, 2011, **156**, 401.
219. H. Zhao and K.-Y. Law, *Langmuir*, 2012, **28**, 11812.
220. S. M. Kang, C. Lee, H. N. Kim, B. J. Lee, J. E. Lee, M. K. Kwak and K.-Y. Suh, *Adv. Mater.*, 2013, **25**, 5756.
221. X. Liu, W. Wu, X. Wang, Z. Luo, Y. Liang and F. Zhou, *Soft Matter*, 2009, **5**, 3097.
222. D. Kim, D. Jung, J. H. Yoo, Y. Lee, W. Choi, G. S. Lee, K. Yoo and J.-B. Lee, *J. Micromech. Microeng.*, 2014, **24**, 055018/1.
223. S. E. Lee, H.-J. Kim, S.-H. Lee and D.-G. Choi, *Langmuir*, 2013, **29**, 8070.
224. N. Ghosh, A. V. Singh and A. A. Vaidya, *ACS Appl. Mater. Interfaces*, 2013, **5**, 8869.
225. H.-B. Jo, J. Choi, K.-J. Byeon, H.-J. Choi and H. Lee, *Microelectron. Eng.*, 2014, **116**, 51.
226. Y. Luo, L. Wang, Y. Ding, H. Wei, X. Hao, D. Wang, Y. Dai and J. Shi, *Appl. Surf. Sci.*, 2013, **279**, 36.
227. E. Bormashenko, R. Grynyov, G. Chaniel, H. Taitelbaum and Y. Bormashenko, *Appl. Surf. Sci.*, 2013, **270**, 98.
228. C.-T. Hsieh, F.-L. Wu and W.-Y. Chen, *Mater. Chem. Phys.*, 2010, **121**, 14.
229. J. Zhao, B. Leng, Z. Shao, G. de With and W. Ming, *RSC Adv.*, 2013, **3**, 22332.
230. H. Kang, J.-S. Lee, W. S. Chang and S.-H. Kim, *Adv. Mater.*, 2015, **27**, 1282.
231. N. Vogel, R. A. Belisle, B. Hatton, T.-S. Wong and J. Aizenberg, *Nat. Commun.*, 2013, **4**, 2167.

232. K. Ellinas, A. Tserepi and E. Gogolides, *Langmuir*, 2011, **27**, 3960.
233. K. Ellinas, M. Chatzipetrou, I. Zergioti, A. Tserepi and E. Gogolides, *Adv. Mater.*, 2015, **27**, 2231.
234. W. Liu, X. Liu, J. Fangteng, S. Wang, L. Fang, H. Shen, S. Xiang, H. Sun and B. Yang, *Nanoscale*, 2014, **6**, 13845.
235. A. Grigoryev, Y. Roiter, I. Tokarev, I. Luzinov and S. Minko, *Adv. Funct. Mater.*, 2013, **23**, 870.
236. A. C. Glavan, R. V. Martinez, A. B. Subramaniam, H. J. Yoon, R. M. D. Nunes, H. Lange, M. M. Thuo and G. M. Whitesides, *Adv. Funct. Mater.*, 2014, **24**, 60.
237. A. C. Glavan, R. V. Martinez, E. J. Maxwell, A. B. Subramaniam, R. M. D. Nunes, S. Soh and G. M. Whitesides, *Lab Chip*, 2013, **13**, 2922.
238. J. Lessing, A. C. Glavan, S. B. Walker, C. Keplinger, J. A. Lewis and G. M. Whitesides, *Adv. Mater.*, 2014, **26**, 4677.
239. A. G. Cunha, C. S. R. Freire, A. J. D. Silvestre, C. P. Neto and A. Gandini, *Carbohydr. Polym.*, 2010, **80**, 1048.
240. L. Li, V. Breedveld and D. W. Hess, *ACS Appl. Mater. Interfaces*, 2013, **5**, 5381.
241. D. Ge, L. Yang, C. Wang, E. Lee, Y. Zhang and S. Yang, *Chem. Commun.*, 2015, **51**, 6149.
242. C. Jin, Y. Jiang, T. Niu and J. Huang, *J. Mater. Chem.*, 2012, **22**, 12562.
243. L. Tang, J. Han, Z. Jiang, S. Chen and H. Wan, *Carbohydr. Polym.*, 2015, **117**, 230.
244. W. Zhang, S. Chen, W. Hu, B. Zhou, Z. Yang, N. Yin and H. Wang, *Carbohydr. Polym.*, 2011, **86**, 1760.
245. J. Vasiljević, B. Tomšič, I. Jerman, B. Orel, G. Jakša, J. Kovač and B. Simončič, *J. Sol-Gel Sci. Technol.*, 2014, **70**, 385.
246. S. G. Im, D. Kusters, W. Choi, S. H. Baxamusa, M. C. M. van de Sanden and K. K. Gleason, *ACS Nano*, 2008, **2**, 1959.
247. J. Wang, A. Raza, Y. Si, L. Cui, J. Ge, B. Ding and J. Yu, *Nanoscale*, 2012, **4**, 7549.
248. J. Ge, Y. Si, F. Fu, J. Wang, J. Yang, L. Cui, B. Ding, J. Yu and G. Sun, *RSC Adv.*, 2013, **3**, 2248.
249. N. Wang, Z. Zhu, J. Sheng, S. S. Al-Deyab, J. Yu and B. Ding, *J. Colloid Interface Sci.*, 2014, **428**, 41.
250. J. Lin, F. Tian, B. Ding, J. Yu, J. Wang and A. Raza, *Appl. Surf. Sci.*, 2013, **276**, 750.
251. S. Lin, S. Nejati, C. Boo, Y. Hu, C. O. Osuji and M. Elimelech, *Environ. Sci. Technol. Lett.*, 2014, **1**, 443.
252. M. Guo, B. Ding, X. Li, X. Wang, J. Yu and M. Wang, *J. Phys. Chem. C*, 2010, **114**, 916.
253. C. H. Lee, S. K. Kang, J. A. Lim, Y.-J. Kwark, H. S. Lim, J. Kim and J. H. Cho, *J. Mater. Chem.*, 2012, **22**, 14656.
254. H. Jin, M. Kettunen, A. Laiho, H. Pynnoenen, J. Paltakari, A. Marmur, O. Ikkala and R. H. A. Ras, *Langmuir*, 2011, **27**, 1930.
255. H. Jin, X. Tian, O. Ikkala and R. H. A. Ras, *ACS Appl. Mater. Interfaces*, 2013, **5**, 485.

256. T. Darmanin, J. Tarrade, E. Celia, H. Bellanger and F. Guittard, *ChemPlusChem*, 2014, **79**, 382.
257. T. Darmanin, J. Tarrade, E. Celia and F. Guittard, *J. Phys. Chem. C*, 2014, **118**, 2052.
258. S. Pan, A. K. Kota, J. M. Mabry and A. Tuteja, *J. Am. Chem. Soc.*, 2013, **135**, 578.
259. A. K. Kota, Y. Li, J. M. Mabry and A. Tuteja, *Adv. Mater.*, 2012, **24**, 5838.
260. S. S. Chhatre, W. Choi, A. Tuteja, K.-C. Park, J. M. Mabry, G. H. McKinley and R. E. Cohen, *Langmuir*, 2010, **26**, 4027.
261. J. A. Kleingartner, S. Srinivasan, J. M. Mabry, R. E. Cohen and G. H. McKinley, *Langmuir*, 2013, **29**, 13396.
262. M. Paven, P. Papadopoulos, S. Schoettler, X. Deng, V. Mailänder, D. Vollmer and H.-J. Butt, *Nat. Commun.*, 2013, **4**, 2512.
263. A. K. Kota, G. Kwon, W. Choi, J. M. Mabry and A. Tuteja, *Nat. Commun.*, 2012, **3**, 1025.
264. H. Yang, X. Hu, R. Chen, S. Liu, P. Pi and Z.-r. Yang, *Appl. Surf. Sci.*, 2013, **280**, 113.
265. H. Yang, X. Zhang, Z.-Q. Cai, P. Pi, D. Zheng, X. Wen, J. Cheng and Z.-r. Yang, *Surf. Coat. Technol.*, 2011, **205**, 5387.
266. R. Saraf, H. J. Lee, S. Michielsen, J. Owens, C. Willis, C. Stone and E. Wilusz, *J. Mater. Sci.*, 2011, **46**, 5751.
267. S. Nishizawa and S. Shiratori, *J. Mater. Sci.*, 2013, **48**, 6613.
268. H. J. Lee and J. R. Owens, *J. Mater. Sci.*, 2010, **45**, 3247.
269. E. Jeong, B. H. Lee, S. J. Doh, I. J. Park and Y.-S. Lee, *J. Fluorine Chem.*, 2012, **141**, 69.
270. S. Pan, R. Guo and W. Xu, *AIChE J.*, 2014, **60**, 2752.
271. D. Satam, H. J. Lee and E. Wilusz, *AATCC Rev.*, 2010, **10**, 59.
272. C. Pereira, C. Alves, A. Monteiro, C. Magén, A. M. Pereira, A. Ibarra, M. R. Ibarra, P. B. Tavares, J. P. Araújo, G. Blanco, J. M. Pintado, A. P. Carvalho, J. Pires, M. F. R. Pereira and C. Freire, *ACS Appl. Mater. Interfaces*, 2011, **3**, 2289.
273. C.-T. Hsieh, D.-Y. Tzou, C.-C. Chen, C.-Y. Lai and Y.-Y. Liu, *Surf. Coat. Technol.*, 2012, **206**, 4639.
274. J. Zeng, B. Wang, Y. Zhang, H. Zhu and Z. Guo, *Chem. Lett.*, 2014, **43**, 1566.
275. G. R. J. Artus, J. Zimmermann, F. A. Reifler, S. A. Brewer and S. Seeger, *Appl. Surf. Sci.*, 2012, **258**, 3835.
276. X. Zhou, Z. Zhang, X. Xu, X. Men and X. Zhu, *Appl. Surf. Sci.*, 2013, **276**, 571.
277. J. Lin, C. Zheng, W. J. Ye, H. Q. Wang, D. Y. Feng, Q. Y. Li and B. W. Huan, *J. Appl. Polym. Sci.*, 2015, **132**, 41458.
278. H. Zhou, H. Wang, H. Niu, J. Fang, Y. Zhao and T. Lin, *Adv. Mater. Interfaces*, 2015, **2**, 1400559.
279. H. Zhou, H. Wang, H. Niu, A. Gestos and T. Lin, *Adv. Funct. Mater.*, 2013, **23**, 1664.

280. H. Wang, Y. Xue, J. Ding, L. Feng, X. Wang and T. Lin, *Angew. Chem., Int. Ed.*, 2011, **50**, 11143.
281. H. Wang, H. Zhou, A. Gestos, J. Fang and T. Lin, *ACS Appl. Mater. Interfaces*, 2013, **5**, 10221.
282. H. Wang, H. Zhou, A. Gestos, J. Fang, H. Niu, J. Ding and T. Lin, *Soft Matter*, 2013, **9**, 277.
283. H. Wang, Y. Xue and T. Lin, *Soft Matter*, 2011, **7**, 8158.
284. J. Yang, Z. Zhang, X. Men, X. Xu, X. Zhu and X. Zhou, *Langmuir*, 2011, **27**, 7357.

CHAPTER 4

Liquid-Repellent Nanostructured Polymer Composites

ILKER S. BAYER[a]

[a]Smart Materials, Istituto Italiano di Tecnologia, Genoa, Italy
*E-mail: ilker.bayer@iit.it

4.1 Introduction

When botanists started to investigate the chemistry and structure of certain plant surfaces such as the lotus leaf, one fact became obvious: these plants have evolved in such a way that their leaves can remain dry and dirt-free in natural habitats like murky ponds in wet climates. This was achieved by natural engineering of surface texturing and surface chemistry simultaneously. The lotus leaf, for instance, has two levels of surface structure in the form of micrometre-scale bumps decorated by nanometre-scale hair-like structures with a wax–cellulose composite surface chemistry. The surface structure allows air to be trapped under the water droplets that fall on the leaf. This composite surface is responsible for the high water contact angles and droplets cleaning their paths as they roll off the leaves. These observations sparked tremendous interest from chemists and materials and surface scientists who attempted to replicate such naturally engineered surfaces,[1] and along the way the term "biomimetic surfaces" became very popular.[2] For the rest of this chapter, it

RSC Soft Matter No. 5
Non-wettable Surfaces: Theory, Preparation, and Applications
Edited by Robin H. A. Ras and Abraham Marmur

Published by the Royal Society of Chemistry, www.rsc.org

is assumed that readers have some familiarity with theories of wetting and the concept of liquid droplet contact angle, as these subjects are beyond the scope of this chapter. Readers are referred to a number of outstanding reviews on these subjects.[3–5] Polymer nanocomposites (or polymer matrix based nanocomposites) are polymer composites in which nanoparticles or nanofillers are dispersed in the polymer or copolymer matrix. Nanofillers can be of different shapes (*e.g.* platelets, fibres, spheroids), but at least one dimension must be in the range of 1–50 nm. Polymer nanocomposites are multiphase systems, the fabrication of which requires controlled mixing/compounding, dispersion stabilization, and orientation of the dispersed phase to achieve the desired functionality. Polymer nanocomposites have become a prominent area of current research and development. At first, exfoliated clay-based polymer nanocomposites dominated the polymer literature but now there are many other significant areas of current and emerging interest. The "nanoeffect" of nanoparticle or fibre inclusion relative to their larger-scale counterparts is also very important for polymer surfaces and interfaces, eventually reflecting in wetting. The incorporation of nanoparticles instead of microparticles in the polymer matrix leads to changes in physical as well as chemical properties. A major change is the increase in the ratio of the surface area-to-volume ratio. This increase leads to an increasing dominance of the behaviour of atoms on the surface of particles, rather than in the bulk. This affects the properties of the particles when they are reacting with other particles. Hence, the interaction with the other particles within the polymer nanocomposite is greater and this enhances properties such as strength, heat resistance, electrical conductivity, and resistance to gas and liquid permeability.

4.2 Polymer Coatings

Perhaps one of the most important applications of polymer composites is the field of coatings including industrial paints and adhesives. Coatings such as paints and lacquers mostly have dual uses, protecting the substrate and being decorative. However, artists' paints are intended only for artistic and decorative purposes, and some special paints on large industrial pipe surfaces are generally applied to prevent corrosion. Functional coatings may be applied to change the surface properties of the substrate, such as adhesion, wettability, corrosion resistance, or wear resistance. In other cases, *e.g.* semiconductor device fabrication (where the substrate is a wafer), the coating adds a completely new property such as a magnetic response or electrical conductivity and forms an essential part of the finished product. A major consideration for most coating processes is that the coating must be applied at a controlled thickness. A number of different processes can be used to apply polymer coatings, ranging from simple brushes or sprays for painting large areas to specialized and expensive machinery, particularly for printing and the electronics industry. In the last few years, a number of hydrophobic and superhydrophobic coatings and paints have been commercialized. These are generally formulated by dispersing polymer binders along with micro- and nanoparticles in appropriate solvents in order to form microporous liquid-repellent top coats.

Various polymeric binders can be used for polymer composite liquid-repellent technologies such as polyurethanes, fluorinated acrylics, silicone polymers and copolymers, styrene copolymers and poly(tetrafluoroethylene) (PTFE)-based polymers, to name a few.[6–8] The fillers are generally nanostructured materials such as organo-clays, silica nanoparticles, metal oxides, carbon nanotubes or nanofibres as well as graphene and metals such as lubricated silver flakes. Figure 4.1 shows, for instance, a superhydrophobic polyurethane/MoS_2 nanocomposite coating which also displays a good degree of wear resistance.[9]

Spraying is considered as one of the easiest and most efficient ways of applying liquid-repellent polymer composites.[10,11] Even for thermosetting polymers such as polydimethylsiloxane (PDMS), spray application has been shown to be effective. For instance, abrasion resistant and self-healing superhydrophobic polymer composite coatings were fabricated by spraying a liquid solution comprising polystyrene/SiO_2 core/shell nanoparticles and PDMS in a suitable solvent (see Figure 4.2).[12] Such coatings were also shown to be repairable by various techniques such as heat curing and solvent swelling

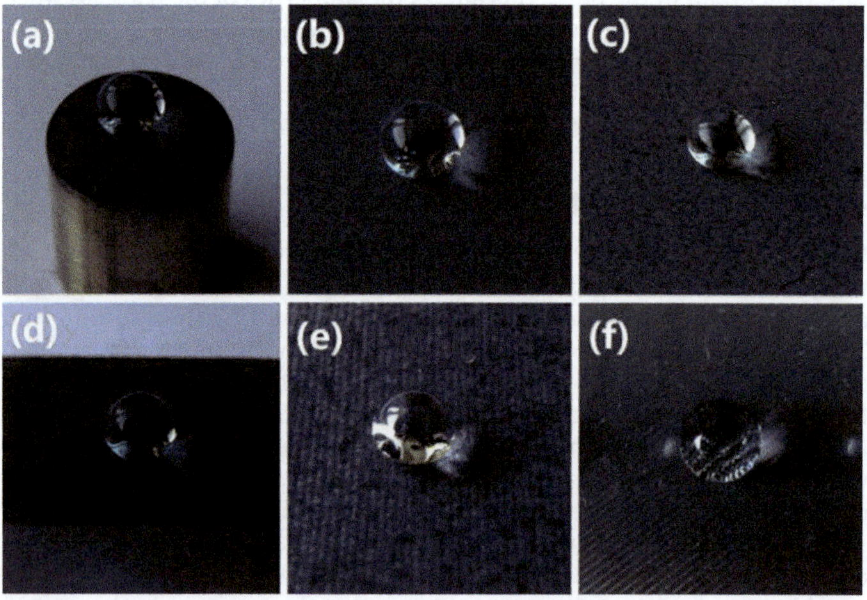

Figure 4.1 Images of a water droplet on superhydrophobic polyurethane/MoS2 nanocomposite coatings applied to (a) copper pillar, (b) glass plate, (c) paper, (d) stainless steel, (e) fabric, and (f) copper mesh. Reprinted from *Colloids and Surfaces A: Physicochemical and Engineering Aspects*, **459**, Y. Tang *et al.*, Fabrication of Super hydrophobic polyurethane/MoS_2 nonacomposite coatings with wear-resistance, 261–266, Copyright (2014) with permission from Elsevier.[9]

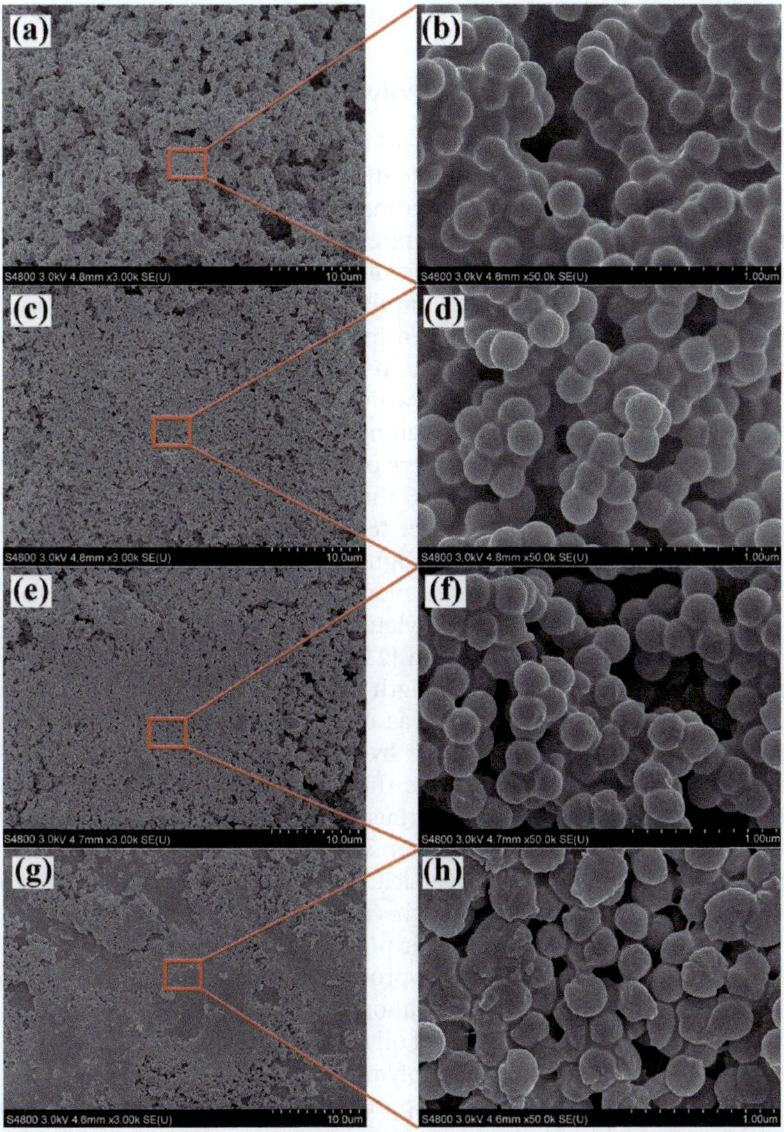

Figure 4.2 SEM images of spray-coated superhydrophobic polymer nanocomposites comprising polystyrene/SiO_2 core/shell nanoparticles dispersed in PDMS: (a) original; (b) higher magnification of (a); (c) after sand blasting one time; (d) higher magnification of (c); (e) after sand blasting 10 times; (f) higher magnification of (e); (g) after sand blasting 20 times; (h) higher magnification of (g). Reproduced from ref. 12 with permission from the Royal Society of Chemistry.

in tetrahydrofuran (THF). Such easy-to-apply strategies may find practical applications in all kinds of substrates because spray coating is a simple process, and the coatings can be formulated for lasting superhydrophobicity.

4.2.1 Fluoropolymer Matrix Polymer Composites

Fluoropolymers such as PTFE, PTFE-AF copolymer, polyvinylidene fluoride, and fluorinated acrylics have frequently been utilized as a hydrophobic matrix for liquid-repellent polymer composite formulations. Perfluorinated polymers, in general, display excellent chemical resistance, thermal stability, and low dielectric constant. Most of these polymers pose certain environmental concerns such as bioaccumulation, due to long molecules or side chains having a carbon backbone entirely surrounded by fluorine. Chain lengths of C8 or more (*i.e.* 8 carbon atoms surrounded with fluorine) degrade into perfluorooctanoic acid (PFOA), which persists indefinitely in the environment and is a toxicant and carcinogen in animals.[13] For this reason, most commercial producers and users of perfluorinated polymers now use molecules or polymers containing C6 chemistry instead of C8. This is a particular drawback for oil or low surface tension liquid repellency.[14,15] Cansoy and Cengiz[16] studied the effect of weight percent of perfluoroalkyl content and hydrocarbon chain length on the oleophobic properties of perfluoroethyl alkyl methacrylate-methyl methacrylate (Zonyl-TM-MMA) copolymers by using oils of varying surface tension (21.6–27.5 mN m^{-1}). They found that increasing the hydrocarbon chain length caused an increase in contact angle values of flat copolymer films. It was also discovered that contact angle hysteresis increased with increasing hydrocarbon chain length when low surface tension oils were used while there was no significant variation in the hysteresis values when higher surface tension oils were used. In order to demonstrate how effective the C8 chemistry is against oils, a polymer nanocomposite spray-cast from nanoparticle–polymer suspensions is shown in Figure 4.3.[17] The method involves the use of ZnO nanoparticles blended with a waterborne C8 perfluoroacrylic polymer emulsion using acetone as a cosolvent. Spray coating was used to produce self-assembling nanocomposite slurries that form hierarchical nanotextured morphology upon curing. Figure 4.3 shows the static water and oil contact angles as a function of ZnO nanoparticle concentration in this polymer nanocomposite.

In a recent publication[18] a simple approach was demonstrated to prepare a transparent superhydrophobic coating and a translucent superamphiphobic coating by spraying silica–fluoropolymer hybrid nanoparticles (SFNs) without any pre- or post-treatment of substrates; the nanoparticles create micro/nanoscale roughness, and the fluoropolymer acts as a low surface energy binder. An increase in the concentration of the nanoparticles facilitates the transition between the superhydrophobic/transparent and superamphiphobic/translucent states (see Figure 4.4). This transition results from an increase in the discontinuities in the three-phase (solid–liquid–gas) contact line and in the light-scattering properties due to micropapillae tuned by varying the concentration of the nanoparticles.

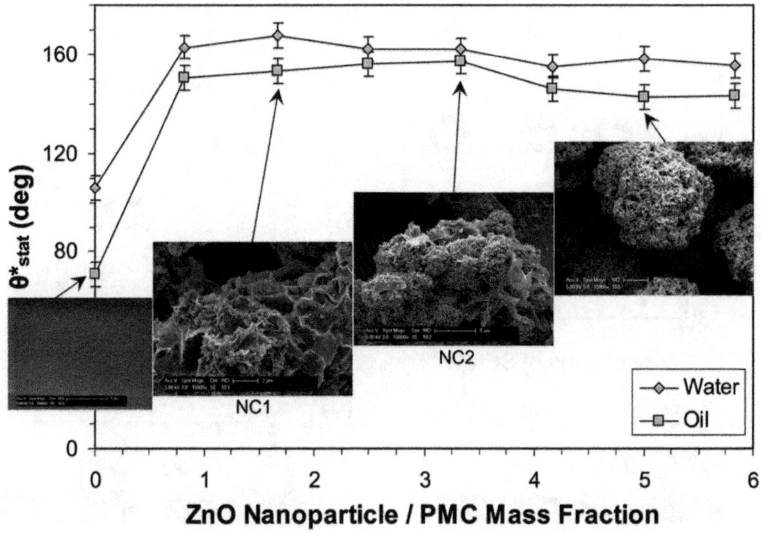

Figure 4.3 Apparent static contact angle of 10 μL droplets as a function of nanoparticle/polymer mass fraction for performance measurement using ideal acetone cosolvent concentration. NC1 (most superhydrophobic nanocomposite): 4% wt ZnO, 88% wt acetone, 2.4% wt perfluoroalkyl methacrylic copolymer, 5.6% wt distilled water. NC2 (most superoleophobic nanocomposite): 8% wt ZnO, 84% wt acetone, 2.4% wt perfluoroalkyl methacrylic copolymer, 5.6% wt distilled water. Reprinted with permission from A. Steele, I. Bayer and E. Loth, *Nano Lett.*, 2008, **9**, 501. Copyright 2009 American Chemical Society.[17]

The micro/nanoscale surface roughness of the spray-coated samples is enhanced by increasing the concentration of the SFNs in solution. When the concentration of the SFNs in solution is low, 0.05 wt%, the substrate is not fully covered with SFNs (see arrows in Figure 4.5(a)). When the concentration is increased to 0.1 wt%, the SFNs fully cover the substrate, and the coated surface contains nanostructures but no microstructures (Figure 4.5(b)). This relatively smooth surface is attributed to the remaining solvent of the sprayed liquid droplet. When the sprayed liquid droplets impact substrate, the SFNs within the liquid droplets tend to stack flatways. As the concentration is increased beyond 0.1 wt%, both the nanostructures and microstructures are roughened by numerous irregular bumps ranging in size from several to tens of micrometres, and there are a large number of nanopores on the surface (Figure 4.5(c and d)). The formation of the hierarchical structure can be explained in terms of fast solvent evaporation.

The drawbacks associated with C6 chemistry can be overcome by using polymer blends as matrices.[19] For instance, a highly efficient technique was recently presented to form novel fluoropolymer blend dispersions containing poly(vinylidene fluoride) (PVDF) and a C6 fluorinated acrylic copolymer using a cosolvent system comprising *N*-methyl-2-pyrrolidone (NMP), acetone, and water under pH control. In this process certain surface-functionalized,

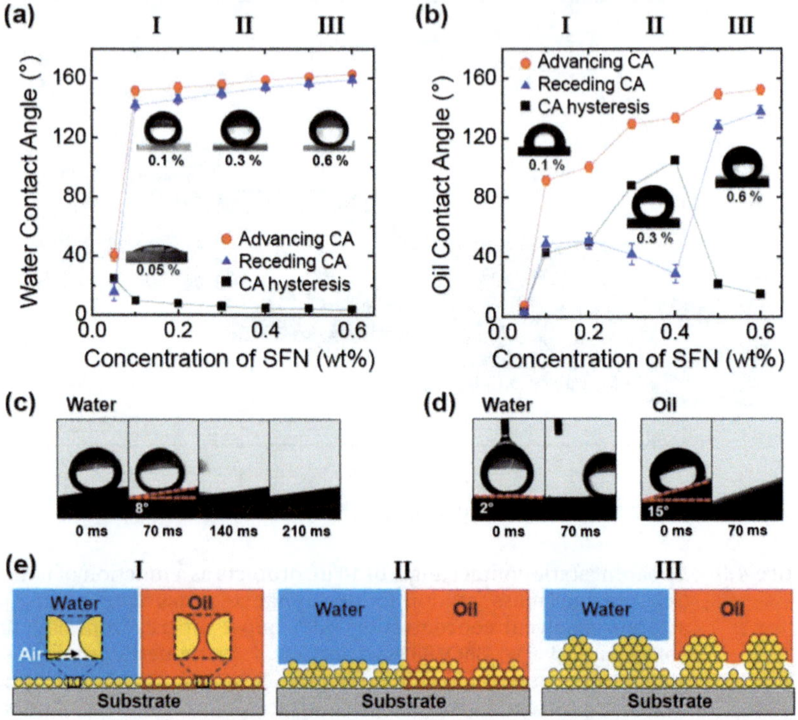

Figure 4.4 (a,b) Contact angles of water (a) and hexadecane (b) on surfaces prepared from solutions with various concentrations of silica–fluoropolymer hybrid nanoparticles (SFN). Reprinted with permission from S. G. Lee, D. S. Ham, D. Y. Lee, H. Bong, K. Cho, *Langmuir* 2013, **29**, 15 051. Copyright (2013) American Chemical Society.[18] (c,d) Series of photographs showing the rolling behaviour of a liquid droplet (5 μL) on spray-deposited substrates with 0.1 wt% solution of SFNs (c) and 0.6 wt% solution of SFNs (d). (e) Schematic illustrations of the possible solid–liquid contact modes in regimes I–III in (a) and (b).

high-aspect-ratio nanostructured materials, such as organoclay and carbon nanowhiskers (CNWs), were easily dispersed in these fluoropolymer blends to fabricate durable and functional superhydrophobic composite coatings upon spray casting. Both clay and CNW superhydrophobic coatings were also reported to repel lower surface tension liquids, such as water–alcohol mixtures (~40 mN m^{-1}). Moreover, electrical conductivity measurement of CNW composite coatings demonstrates the ability to fabricate multifunctional superhydrophobic composites using these fluoropolymer dispersions. Figure 4.6 shows phase diagrams in which miscibility (with no phase separation) of solvent-based PVDF with pure water or with the water-based fluoroacrylate polymer can be seen as the white window at the bottom.

Adding nanoclay, which has both micro- and nanoscale features, to these dispersions is intended to affect the surface texture of the resulting spray-cast composite coating. These clay platelets, after being introduced into the

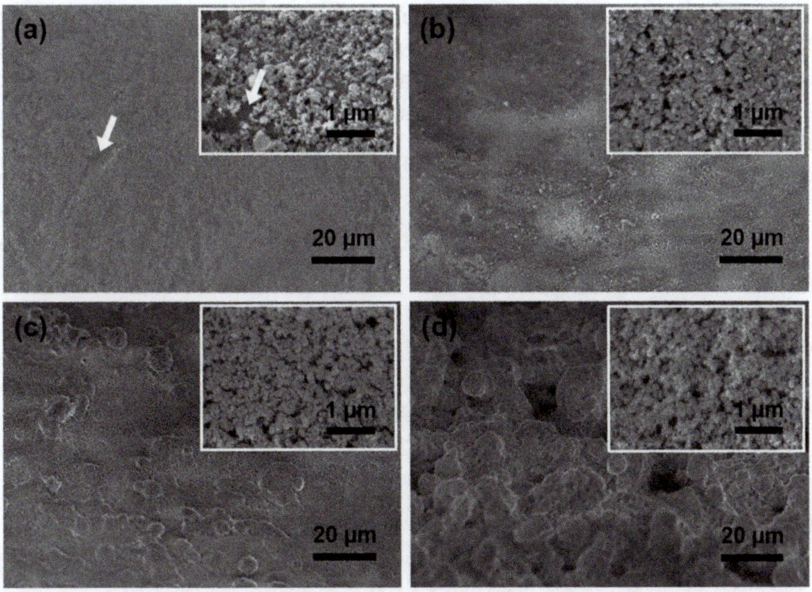

Figure 4.5 Scanning electron microscopy images of spray-deposited SFNs on Si wafers with various sol concentrations: (a) 0.05, (b) 0.1, (c) 0.3, (d) 0.6 wt%. Reprinted with permission from S. G. Lee, D. S. Ham, D. Y. Lee, H. Bong, K. Cho, *Langmuir* 2013, **29**, 15051. Copyright (2013) American Chemical Society.[18]

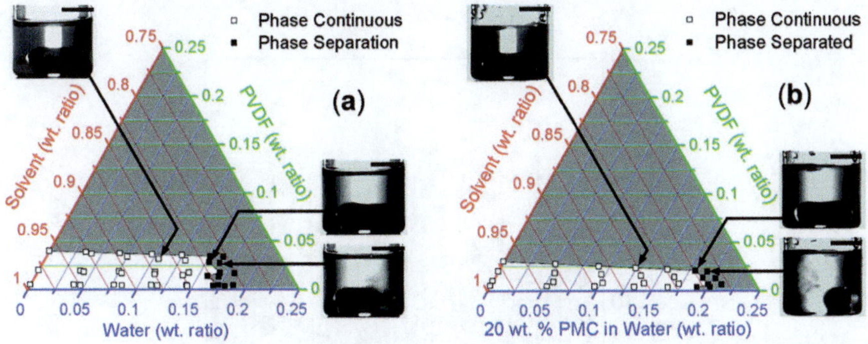

Figure 4.6 Quaternary and sexternary phase diagrams of the solutions utilized for spray without any clay fillers. Filled symbols indicate PVDF phase separation from a liquid to a solid state. The insets display photographs of vials containing solutions with the specified compositions. The scale bar is 10 mm. The magnetic stirrer bar can be seen at the bottom of each vial: (a) PVDF, solvent (acetone and *n*-methyl-2-pyrrolidone, NMP), water phase diagram; (b) PVDF, solvent (acetone, NMP, and trifluoroacetic acid, TFA.), 20 wt% PMC in water. For both parts a and b, NMP was kept at a constant 9 : 1 weight ratio with respect to PVDF. The concentration of TFA in the total solution was <0.2 wt% TFA. Grey areas indicate unstable or phase-inverted regimes. Reprinted with permission from T. M. Schutzius, I. S. Bayer, M. K. Tiwari, C. M. Megaridis. *Ind. & Eng. Chem. Res.* 2011, **50**, 11117. Copyright (2011) American Chemical Society.[19]

composite coatings, can self-assemble into hierarchical micro/nanoscale structures. Such structures are known to facilitate a high degree of water repellency (lotus effect). Figure 4.7 shows electron micrographs of two spray-cast PVDF–perfluoroalkyl methacrylate copolymer (PMC)–clay composite coatings. Repeating micro/nanoscale features (due to clay) are apparent in these composite coatings; this affects the surface wetting of the composites. Figure 4.8 shows that such polymer–clay composite coatings also resist low surface tension aqueous solutions, such as a 9 : 1 (by weight) water–isopropyl

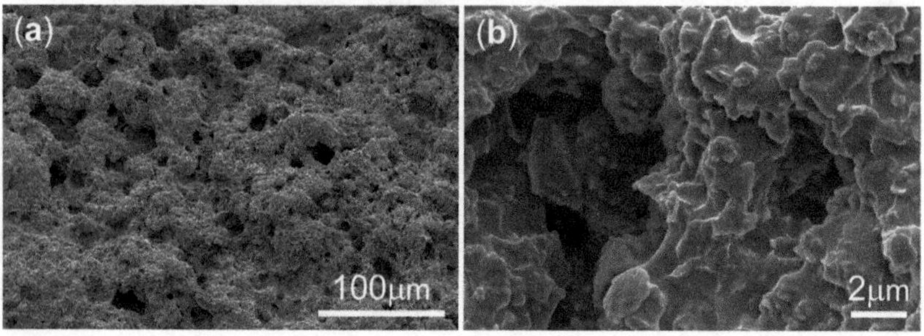

Figure 4.7 Electron microscopy images of PVDF–PMC–clay composite coatings: (a) 2 : 3 PVDF–PMC weight ratio and 0.4 clay–polymer mass ratio; (b) 1 : 1 PVDF–PMC weight ratio and 0.4 clay–polymer mass ratio. Reprinted with permission from T. M. Schutzius, I. S. Bayer, M. K. Tiwari, C. M. Megaridis. *Ind. & Eng. Chem. Res.* 2011, 50, 11117. Copyright (2011) American Chemical Society.[19]

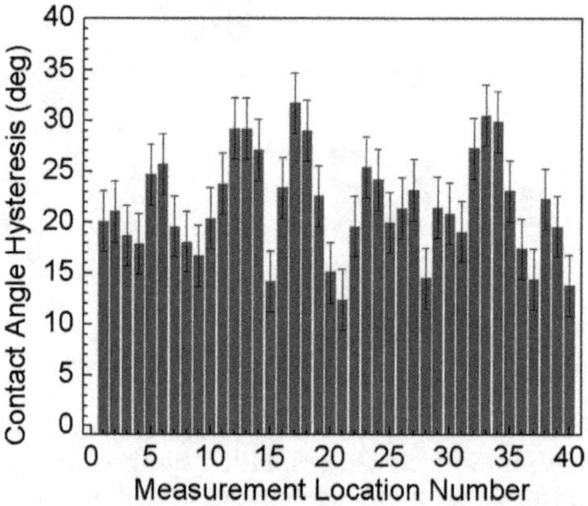

Figure 4.8 Water–isopropyl alcohol, IPA (9 : 1 weight ratio) solution droplet hysteresis measurements for a PVDF–PMC–clay composite coating spray-cast on aluminium foil. For this particular coating, the polymer is a mixture of a 1 : 1 PVDF–PMC weight ratio and the clay–polymer mass ratio is 0.6. The water–IPA sessile CA for this coating is 154 ± 4°. Reprinted with permission from T. M. Schutzius, I. S. Bayer, M. K. Tiwari, C. M. Megaridis. *Ind. & Eng. Chem. Res.* 2011, **50**, 11117. Copyright (2011) American Chemical Society.[19]

alcohol (IPA) mixture, with static contact angles exceeding 150° with reasonably low contact angle hysteresis.

Polyhedral oligomeric silsesquioxanes (POSS) are thermally robust cages consisting of a silicon–oxygen core framework possessing alkyl functionality on the periphery (Figure 4.9).[20] POSS molecules can be functionally tuned, are easily synthesized with inherent functionality, are nanomaterials, and are often commercially available. Furthermore, POSS compounds may possess a high degree of compatibility in blended polymers and can easily be covalently linked into a polymer backbone. The incorporation of POSS into polymers produces nanocomposites with improved properties, including glass transition temperature, mechanical strength, thermal and chemical resistance, and ease of processing. For instance, fluorinated polyhedral oligomeric silsesquioxanes–poly(vinylidene fluoride-*co*-hexafluoro propylene) (fluoroPOSS–PVDF-HFP) nanocomposites were recently reported by individually mixing two different fluorinated POSS materials with PVDF-HFP solution and transparent superhydrophobic coatings on a glass substrate were made by electrospinning (Figure 4.10).[21] The nanocomposites exhibited continuous, uniform and non-beaded nanofibres with a high water contact angle (157.3°) and a low sliding angle (SA < 5°). As the concentration of fluoro-POSS in the PVDF solution was increased, the amount of fluorine content increased thereby the surface energy of the coatings decreased (~10 mN m^{-1}) leading to a superhydrophobic surface with low contact angle hysteresis (<5°). Some of the nanocomposite coatings were also relatively transparent.

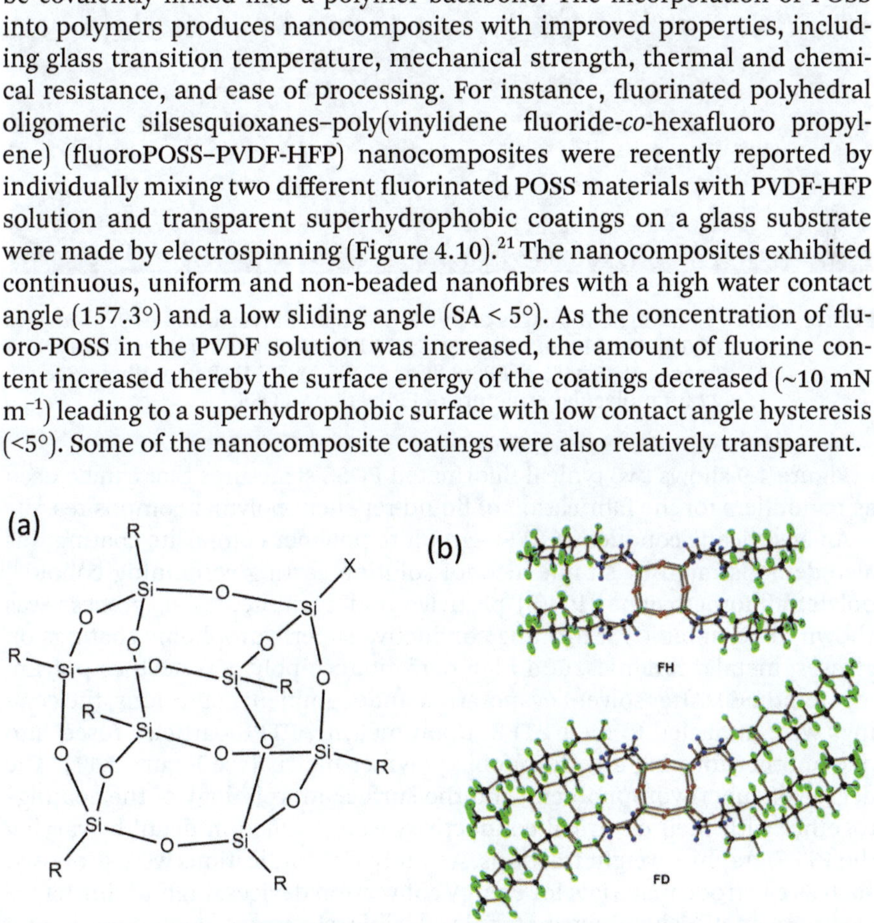

Figure 4.9 (a) General structure of POSS, hydrogen silsesquioxane, nanocage structure. (b) 3D Representation of FH and FD POSS molecules at 103 K, with thermal ellipsoids set at 50% probability. Green, F; black, C; dark blue, H; red, O; light blue, Si. FH stands for (1*H*,1*H*,2*H*,2*H*-nonafluorohexyl)$_8$Si$_8$O$_{12}$ and FD stand for (1*H*,1*H*,2*H*,2*H*-heptadecafluorodecyl)-$_8$Si$_8$O$_{12}$ nanoparticles. *Functional Molecular Silicon Compounds I: Regular Oxidation State, Silsesquioxanes*, **155**, 2013, 1–28, Guido Kickelbick, © Springer International Publishing Switzerland 2013, with permission from Springer.[22]

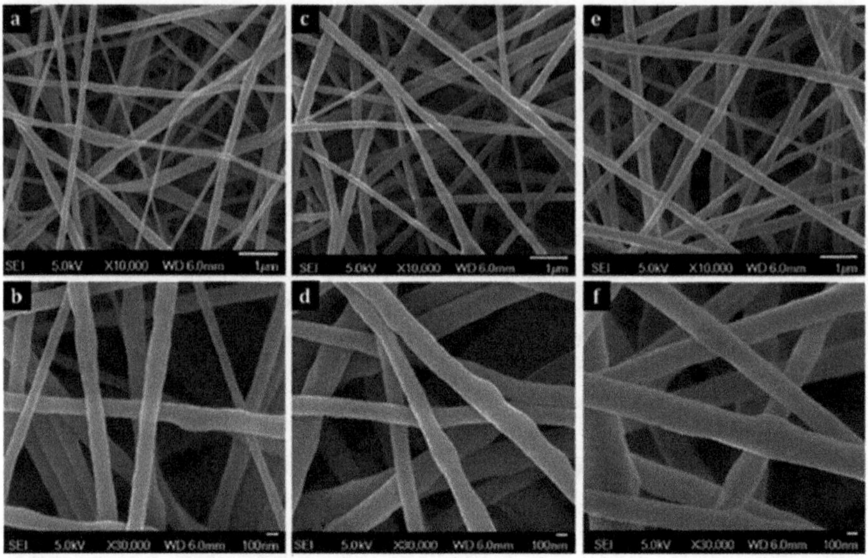

Figure 4.10 (a and b) SEM images of PVDF-HFP nanofibres; (c and d) SEM images of 5 wt% of FPSi8 fluoroPOSS–PVDF-HFP nanofibres; (e and f) SEM images of 15 wt% of FPSi8 fluoroPOSS–PVDF-HFP nanofibres. See ref. 22 for molecular structure of FPSi8 fluoroPOSS.

Figure 4.9 shows two typical fluorinated POSS structures that can be used as nanofillers for the fabrication of liquid-repellent polymer composites.[22]

An electrically conductive PTFE–graphite polymer composite coating was also demonstrated by simple alcohol solution casting containing colloidal polytetrafluoroethylene (PTFE) particles and graphite.[23] The process was shown to be suitable for forming conductive superhydrophobic coatings on glasses, metals, ceramics, and high-performance polymers such as polyimide (Kapton®). After solvent evaporation under ambient conditions, the coatings were annealed to melt PTFE. Upon melting, PTFE particles fused into one another forming a hydrophobic polymer matrix (see Figure 4.11). The degree of superhydrophobicity and the surface morphology of the coatings together with their electrical conductivity were studied in detail by varying the PTFE/graphite weight fractions. A number of applications were proposed such as electrode materials for energy conversion devices, high-performance electromagnetic shielding materials, flexible electronic components, and heat exchanger surfaces.

Superhydrophobic films could be formed only when certain PTFE/graphite weight fractions were maintained in the composite films. The most promising composites displaying self-cleaning superhydrophobicity were observed in dry films having 1 : 2 to 2 : 2 PTFE/graphite weight fractions as a result of thermal annealing. After solvent evaporation, dry films, before annealing at 350 °C to melt the PTFE, showed a mixed degree of hydrophobicity with high contact angle hysteresis depending on the PTFE/graphite weight fraction.

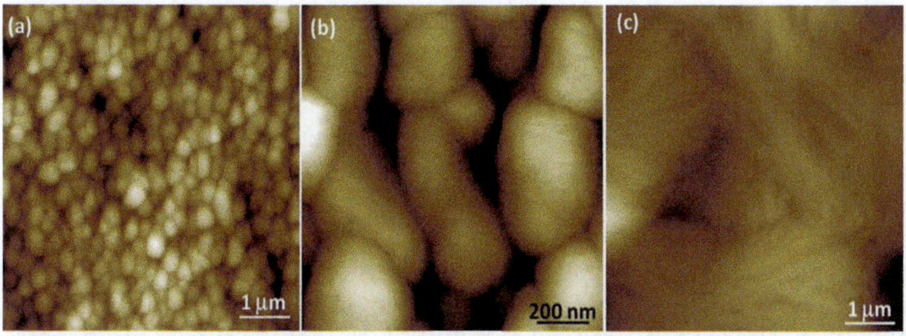

Figure 4.11 (a) AFM topography acquired in non-contact mode of a polytetrafluoroethylene (PTFE) film obtained by drop-casting 3 wt% PTFE suspensions in isopropanol. (b) A smaller area scan of the film shown in (a), showing the geometrical shape of the submicron PTFE particles. (c) AFM topography of the film in (a), after thermal annealing at 350 °C for 30 s. Annealing-induced morphological change into a fibrillar network is clearly seen. Reproduced from ref. 23 with permission from the Royal Society of Chemistry.

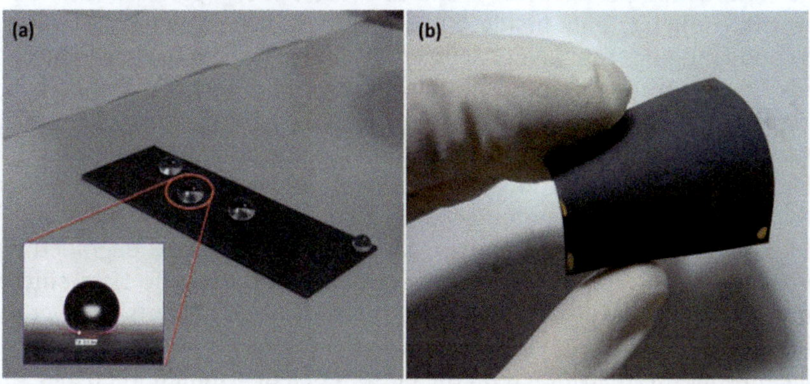

Figure 4.12 (a) Microscope glass slide coated with a conducting superhydrophobic film with a PTFE/graphite weight fraction of 2.2. Sessile water droplet contact angle on this film is 167°. (b) Kapton® film coated with the same conductive superhydrophobic composite. Both films were annealed at 350 °C for 0.5 min. Reproduced from ref. 23 with permission from the Royal Society of Chemistry.

In addition, before annealing, the films had very poor substrate adhesion for both glass and Kapton and could be easily removed from the substrates by gentle rubbing. Therefore, annealing to melt PTFE served two purposes: (a) melting induced generation of self-cleaning superhydrophobicity and (b) formation of a good degree of substrate adhesion. Figure 4.11(a) shows a photograph of a microscope glass slide coated with a superhydrophobic film having a PTFE/graphite mass fraction of 2 : 2 after annealing for 0.5 min. The sessile water droplet contact angle on this surface was 167°. Figure 4.12(b) shows the same film deposited on a piece of Kapton.

4.2.2 Silicone Matrix Polymer Composites

Most of the liquid-repellent materials fabricated by utilizing silicone polymers such as PDMS generally take advantage of moulding or replicating textures into PDMS. Thermosetting PDMS polymer composites in the form of coatings or freestanding films have so far rarely been reported. One very interesting study involving superhydrophobic silicone polymer composites, however, is based on discarded silicone products. This recyclable scrap can be used to prepare superhydrophobic powder by simply burning and compressing. The powder can be used to fabricate a superhydrophobic polypropylene (PP) composite with mechanical durability such that the superhydrophobicity was kept after 50 abrasion cycles. A robust electroconductive superhydrophobic surface can also be obtained by this simple method.[24]

The process can be briefly described as the combustion of cross-linked silicone elastomers. Commercially available Sylard-184A and Sylard-184B (RTV-2) were mixed in a mass ratio of 10 : 1 and cured at ambient temperature. Cured silicone was cut into small pieces and held in the flame of an alcohol lamp in air until the silicone pieces were completely burned. The silicone combustion product was found to consist of two parts: a white outer layer with caulk-like bulk inside. FT-IR spectra indicated a series of absorption bands ranging from 800 to 3600 cm^{-1}, implying the presence of Si–O–Si bonds, $Si–CH_3$ groups, and CH_3 groups. Compared to unburned silicone, the absorption intensities of $Si–CH_3$ groups and CH_3 groups decrease remarkably in the inside bulk and are absent in the outside layer, suggesting that various degrees of degradation occurred inside and outside the silicone piece when burning. The silicone outer layer was completely combusted, attributable to direct contact with oxygen, whereas the inside part degraded into short chains of polysiloxane due to oxygen deficiency. The pulverized silicone combustion product (powder for short) is a mixture of the outside layer and inside bulk.

The powder was distributed on the bottom of a disc mould, then a PP disc piece was inserted and the mould was pressed under 10 MPa pressure for 30 min at 180 °C. As shown in Figure 4.13(a), the composite surface was flat, just like the morphology of the densely packed powder; at a higher magnification it can be seen that the nanoparticles with diameter of 100–200 nm stack together (Figure 4.13(b)), forming a robust rough structure which remained after being smashed and pressed. The water repellency of the coating is highlighted in Figure 4.13(c); the water droplets exhibit spherical shapes on the powder/PP composite surface. Upon immersion in water, the surface acted like a silver mirror when viewed at a glancing angle (Figure 4.13(d)), and the surface was completely dry after being removed from the water without any mass change before and after immersion. The bright, reflective surface visible in the water reveals that an air layer exists between water and the superhydrophobic surface. The porous rough morphology of the surface traps air and thus establishes surface solid–liquid–air interfaces, leading to excellent superhydrophobicity with a water contact angle of 160° (Figure 4.13(e)) and a sliding or roll-off angle of 3.5° (Figure 4.13(f))(Figure 4.14).

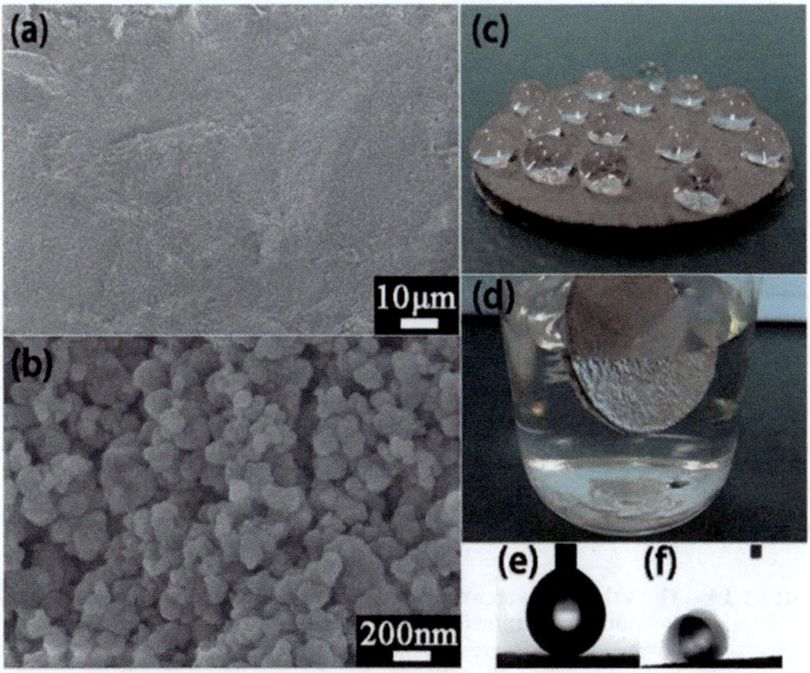

Figure 4.13 SEM images of the superhydrophobic powder/PP composite surface at (a) low and (b) high magnifications; optical images of (c) water droplets on different locations of the composite surface displaying spherical shapes, (d) mirror-like phenomenon appearing when powder/PP composite was immersed in water, (e) water contact angle of 160° and (f) sliding angle of 3.5° of the powder/PP composite surface. Reproduced from ref. 24 with permission from the Royal Society of Chemistry.

Pharmaceutical grade cyclic siloxane polymers (cyclomethicones) have been also reported in the preparation of superhydrophobic polymer composites.[25] These polymers can be used to modify reactive thermosets such as polyurethanes. For instance, nanostructured polyurethane/organoclay composite films were fabricated by dispersing moisture-curable polyurethanes and fatty amine/aminosilane surface-modified montmorillonite clay (organoclay) in cyclomethicone-in-water emulsions. Cyclomethicone Pickering emulsions were made by emulsifying decamethylcyclopentasiloxane (D5), dodecamethylcyclohexasiloxane (D6), and aminofunctional siloxane polymers with water using montmorillonite particles as emulsion stabilizers. Polyurethane and organoclay dispersed emulsions were spray-coated on aluminium surfaces. Upon thermosetting, water-repellent self-cleaning coatings were obtained with measured static water contact angles exceeding 155° and low contact angle hysteresis (<8°). Electron microscopy images of the coating surfaces revealed formation of self-similar hierarchical micro/nanoscale surface structures, as seen in Figure 4.15. The surface

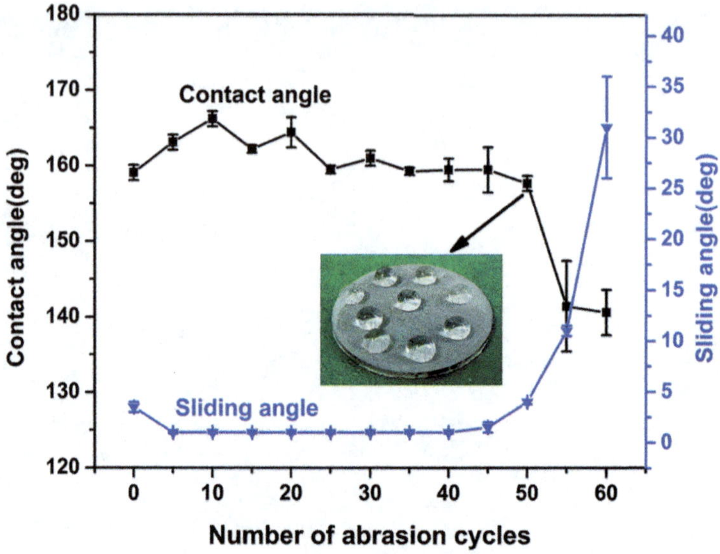

Figure 4.14 The values of water contact angle and sliding angle as a function of the number of abrasion cycles for the PP composite surface. The inset corresponds to water droplets depositing on the powder/PP composite surface after 50 cycles of a sandpaper abrasion test where the surface was dragged in one direction at a speed of 3 cm s^{-1}, and abrasion length of 28 cm, corresponding to a pressure of 5.7 kPa. Reproduced from ref. 24 with permission from the Royal Society of Chemistry.

morphology and the coating adhesion strength to aluminium substrates were found to be sensitive to the relative amounts of dispersed polyurethane and organoclay in the emulsions. The degree of superhydrophobicity was analysed using static water contact angles as well as contact angle hysteresis measurements.

POSS resins are a class of silicon polymers with the empirical formula $RSiO_{1.5}$, where R denotes hydrogen or an alkyl, alkylene, aryl, or arylene group. Among POSS resins, solution-processable hydrogen silsesquioxane (HSQ) or methylsilsesquioxane (MSQ) resins generally find applications as low-dielectric materials for integrated circuit devices and are also used as an alternative route to produce silica coatings. For high-temperature applications, POSS resins may have greater advantages over their organic polymer counterparts due to their inherently higher bond strength. Silsesquioxanes can be produced with a variety of pendant chemical groups (*e.g.* alkyl, aryl), thus allowing appreciable solubility in organic solvents for solution processing; MSQ is expected to be hydrophobic due to the presence of methyl groups. Further enhancement of film hydrophobicity may be attained by introducing hydrophobic inorganic nanoparticles, such as hydrophobic fumed silica (HFS, *i.e.* silica modified by silanes or siloxanes), into the MSQ film. These filler particles not only lower the surface energy of the films, but also affect surface texture, two factors known to significantly influence hydrophobicity,[26] as discussed in earlier examples.

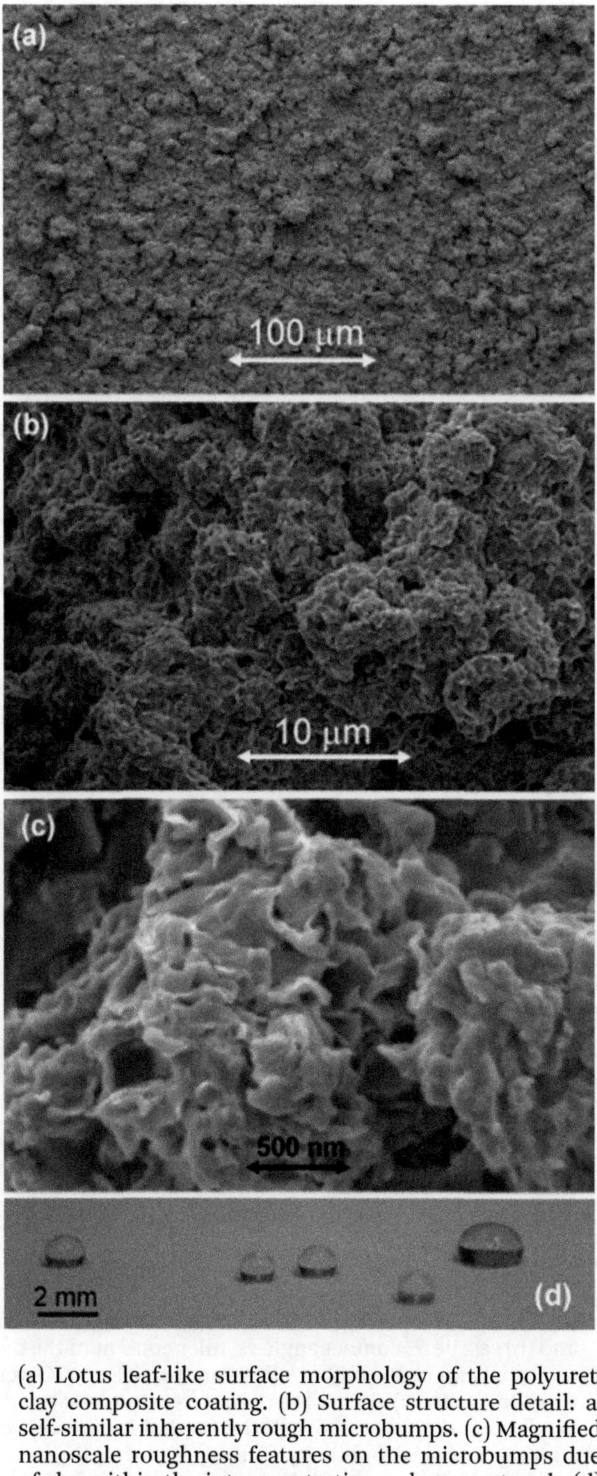

Figure 4.15 (a) Lotus leaf-like surface morphology of the polyurethane/organoclay composite coating. (b) Surface structure detail: appearance of self-similar inherently rough microbumps. (c) Magnified image of the nanoscale roughness features on the microbumps due to assembly of clay within the interpenetrating polymer network. (d) Non-wetting water droplets placed on the surface of the coating. Reproduced from *Applied Surface Science*, **257**, I. S. Bayer *et al.*, Fabrication of superhydrophobic polyurethane/organoclay nano-structured composites from cyclomethicone-in-water emulsions, 823–826. Copyright (2010) with permission from Elsevier.[25]

As shown in Figure 4.16(a), the contact angle hysteresis is 8° at a HFS/MSQ mass ratio of 1.5, which designates a self-cleaning surface (droplets roll off the surface under only a slight substrate tilt). A further increase in hydrophobic filler concentration results in a further decrease of contact angle hysteresis. Figure 4.16(b) shows sessile contact angle values (θ^*) for the previously characterized coatings, before and after flame treatment. For all cases where a hydrophobic silica nanoparticle filler is incorporated into MSQ, the coatings achieve a zero-valued or immeasurably small apparent contact angle after flame treatment, thus indicating superhydrophilicity. High temperature such as flame treatment renders these nanocomposites superhydrophilic. Heat treatment thus allows patterning superhydrophilic regions within the superhydrophobic matrix, as shown in Figure 4.17. In this case, a CO_2 laser was used to pattern the surfaces.

Figure 4.17(a–d) shows examples of surface tension confined microchannels with and without deposited water. Figure 4.17(e and f) show SEM micrographs of the channels with increasing magnification. It is obvious that laser treatment has removed much of the coating material, thus creating a physical channel. Figure 4.17(g) shows a high-magnification micrograph of the patterned area where the laser treatment caused cellular morphology. The laser treatment actually produced its own unique surface texture, facilitating superhydrophilicity. Figure 4.17(f) shows coating areas that have been completely stripped by the laser processing; these areas form apparently bare islands with the remaining coated sections percolating, thus allowing superhydrophilicity to persist. Further optimization of spray

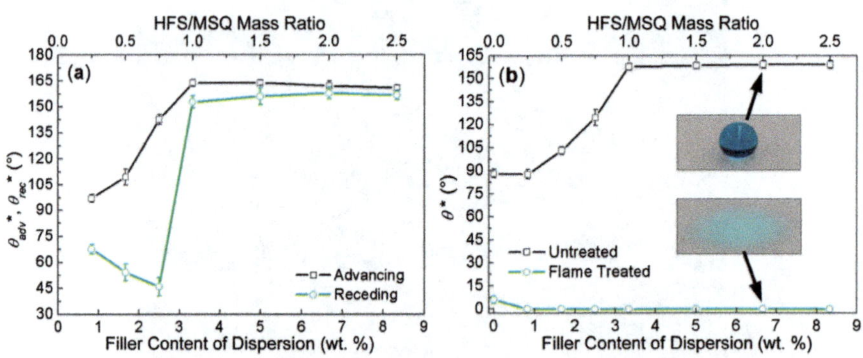

Figure 4.16 (a) Apparent advancing (θ^*adv) and receding (θ^*rec) contact angles, and (b) sessile θ^* contact angle *vs.* filler content of the dispersion (bottom axis) and hydrophobic silica nanoparticles/MSQ weight ratio (top axis). Insets in (b) show blue-dyed water droplets: top demonstrating beading on the untreated superhydrophobic surface; bottom demonstrating superhydrophilicity (fully spread droplet) attained after the coating was flame-treated. Reproduced from ref. 26 with permission from the Royal Society of Chemistry.

processing (*e.g.*, increased coating thickness) should reduce the likelihood of such island-like feature formation. An important property of surface tension confined microchannels is the average flow velocity of water in them. In the case of a 7.2 mm × 0.18 mm channel formed by this procedure, the average water flow velocity was 2.5 mm s^{-1}, which is comparable to speeds attained by others in similar sized channels. Such thermally stable (at temperatures <400 °C) POSS superhydrophobic nanocomposite coatings offer promising applications for enhanced heat transfer applications. Water velocities of 2 mm s^{-1} in true microchannels offer promise for fabricating surface tension confined microchannels. Advantages also include inherent thermal stability over organic-based coatings, as well as scalability to large-area applications.

Superhydrophobic surfaces based on ZnO–PDMS nanocomposite coatings were also demonstrated by a simple wet chemical route.[27] The authors synthesized ZnO nanopowders with average particle size of 14 nm by a low-temperature solution combustion method. The as-formed ZnO coating demonstrated hydrophobic wetting behaviour with water contact angle ~108°, however on modification with polydimethylsiloxane (PDMS), it became superhydrophobic with measured contact and sliding angles for water of 155° and <5° respectively. Advantages of such coatings could be use of cheap and fluorine-free raw materials, environmentally benign solvents, and feasibility for applying on large areas of different substrates. The surface morphology of

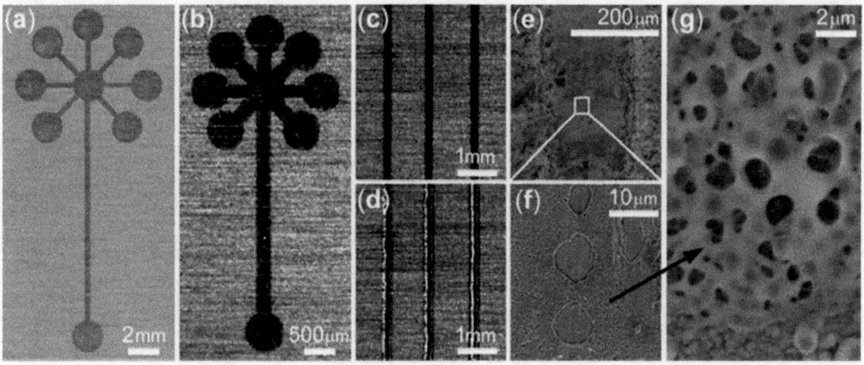

Figure 4.17 (a) Photographic and (b) optical microscope images of superhydrophilic patterns (dark) on a superhydrophobic MSQ–HFS coating; (c) hydrophilic, laser-patterned lines (dark), and (d) surface tension confined channel showing wetting of the lines in (c) by water through capillary action. (e) and (f) SEM images of a surface tension confined microchannel with increasing magnification; (g) SEM image of a laser-patterned area revealing porosity. For 0.18 mm wide surface tension confined microchannels, the water propagation velocity was measured to be 2.5 mm s^{-1}. Reproduced from ref. 26 with permission from the Royal Society of Chemistry.

as-formed ZnO nanopowders coated on aluminium substrates indicate many agglomerates with an irregular morphology. The particles are connected to each other to make large network systems consisting of hierarchical structure based on ZnO nanoparticles with an average diameter in the range of 100–150 nm, as seen in Figure 4.18.

At ambient conditions, ZnO is a hydrophobic material with water contact angle of ~110° on flat surfaces and has good chemical stability. However, when blended with PDMS, it was transformed into a superhydrophobic state with a water contact angle of ~155°, and water droplets roll off (SA < 5°). The modification of blending with PDMS was done in such a way that the original ZnO surface texture had not been modified but rather the textures were rendered hydrophobic by the presence of PDMS.

Similarly, recent work[28] has demonstrated a simple dipping process for the preparation of superhydrophobic coatings based on titanium dioxide nanowires combined with PDMS. The superhydrophobic coatings turned into hydrophilic ones after UV irradiation. It was found that water droplets clean away as they roll over the surface. Furthermore, the coating surfaces demonstrated resistance to solvents. When immersed in various solvents the coatings did not dissolve, and maintained their superhydrophobicity after the solvent-swollen surfaces dried up.

The self-cleaning function of these superhydrophobic coating surfaces was demonstrated with graphite powder as contaminant. The self-cleaning process is shown in Figure 4.19. A sparse layer of contaminant powder was

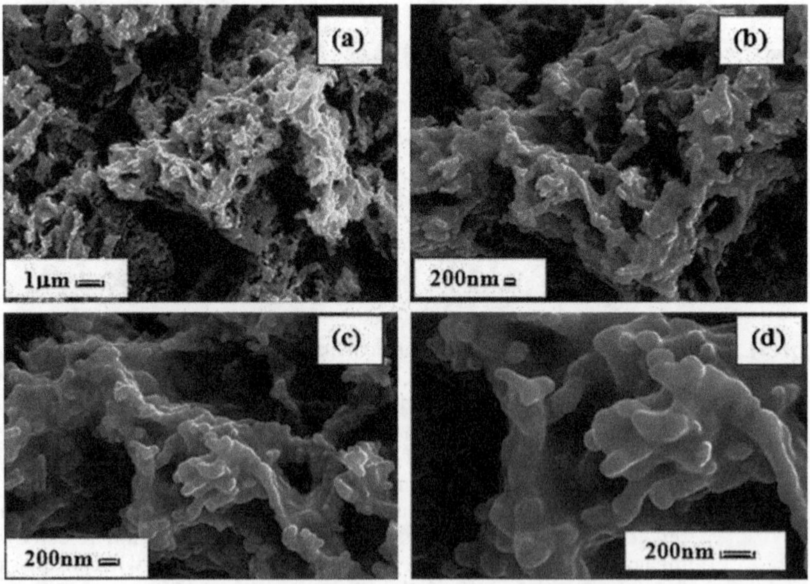

Figure 4.18 SEM micrographs of superhydrophobic ZnO–PDMS coating with different magnifications. Images (a) to (d) display surface texture with higher magnification. Reproduced from ref. 27 with permission of Springer.

sprinkled on the surface and then a water droplet (10 μL) was placed on the contaminated surface. Once the contaminant powders contacted the water droplet, they were immediately adsorbed on the surface of the water droplet. After several sliding process, the water drop adsorbed the contaminant completely (Figure 4.20).

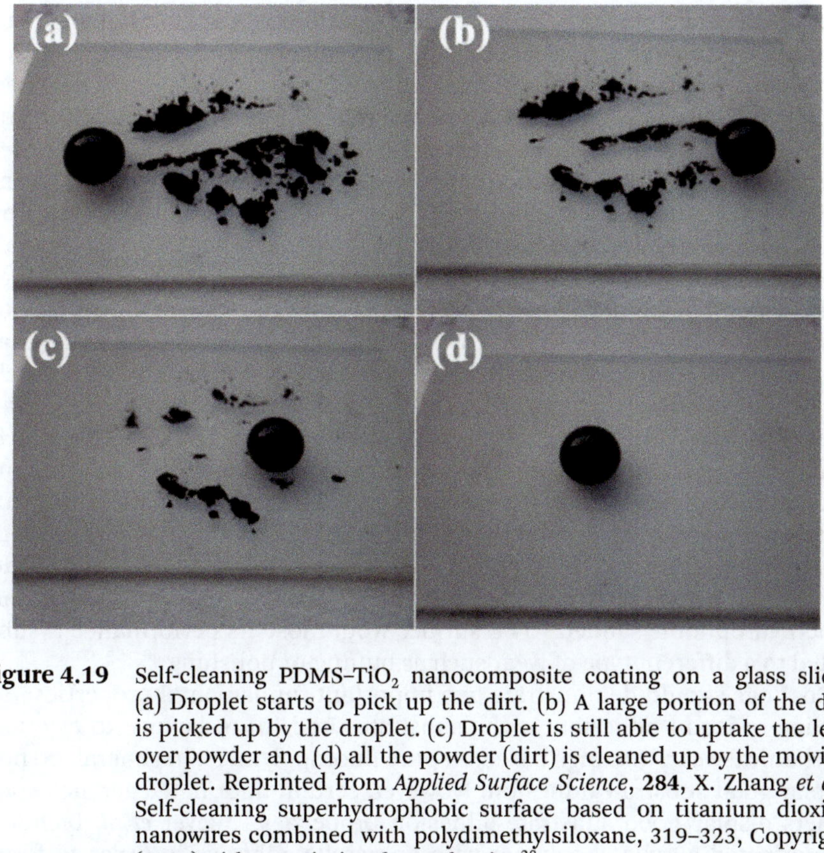

Figure 4.19 Self-cleaning PDMS–TiO_2 nanocomposite coating on a glass slide. (a) Droplet starts to pick up the dirt. (b) A large portion of the dirt is picked up by the droplet. (c) Droplet is still able to uptake the left-over powder and (d) all the powder (dirt) is cleaned up by the moving droplet. Reprinted from *Applied Surface Science*, **284**, X. Zhang *et al.*, Self-cleaning superhydrophobic surface based on titanium dioxide nanowires combined with polydimethylsiloxane, 319–323, Copyright (2013) with permission from Elsevier.[28]

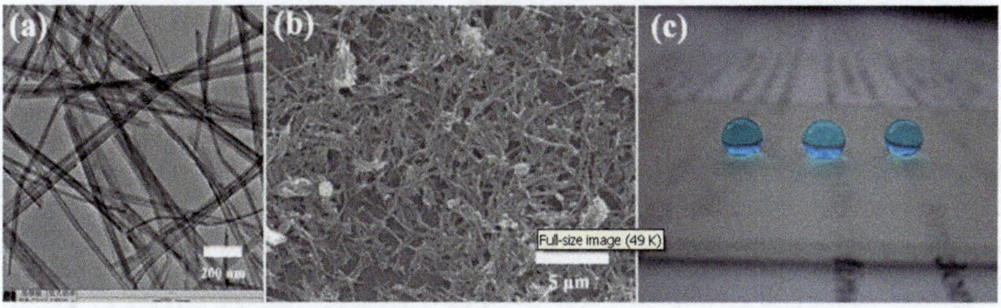

Figure 4.20 (a) TEM image of the TiO_2 nanowires. (b) SEM images of the as-prepared superhydrophobic coating surface. (c) Photographs of water droplets on the superhydrophobic surface. Reprinted from *Applied Surface Science*, **284**, X. Zhang *et al.*, Self-cleaning superhydrophobic surface based on titanium dioxide nanowires combined with polydimethylsiloxane, 319–323. Copyright (2013) with permission from Elsevier.[28]

The morphology of the TiO_2 nanowires, their composite with PDMS, and the resultant water-repellent coatings are shown in Figure 4.18.

4.2.3 Wear Abrasion Resistant Liquid-Repellent Polymer Composites

One of the most challenging aspects of liquid-repellent coatings or treatments is the resistance to wear abrasion. By resistance we mean the ability of the coating to maintain liquid-repellent properties during or after repeated abrasion. This term sometimes is referred to as "mechanical durability", but throughout this chapter we use the term "wear abrasion resistance" in order to address robustness of liquid-repellent polymer (nano)composites. A general review of the mechanical durability of superhydrophobic materials by Verho *et al.* addressed the state of the art for the early 2010s.[29] In general, there are two approaches to creating a non-wetting surface that is wear abrasion resistant: (a) limiting material removal so as to retain superhydrophobicity under abrasion for as long as possible, and (b) developing a material that maintains superhydrophobicity as it wears away. For the latter type, such performance for surfaces under a *single* wear condition is described here as "wear similarity". A simple example of wear similarity is sanded PTFE, which can be rendered superhydrophobic by using fine grit sandpaper, so that continued sanding retains superhydrophobicity until the PTFE material is completely worn away. There are other examples of wear similarity with respect to manual sanding, but these do not demonstrate wear similarity over a wide range of abrasion conditions typical of most applications. For example, the superhydrophobic sanded PTFE surface would lose its performance if subjected to a different type of wear such as buffing or polishing.[30]

Coatings should demonstrate two important fundamental properties: (a) peeling-resistant adhesion to the substrates and (b) resistance to external mechanical wear. Coatings based on pure fluoropolymers, in general, do not display good substrate adhesion. However, certain fluorinated acrylic copolymers display good substrate adhesion properties.[31] Bayer *et al.* blended a fluorinated acrylic copolymer with an acrylic adhesive in order to form polymer–clay nanocomposites demonstrating some resistance against wear abrasion by sandpaper rubbing.[32] The morphology of the superhydrophobic surfaces after sandpaper abrasion is shown in Figure 4.21.

Gentle surface polishing was performed by using a 3M 1000 grit aluminium oxide sandpaper mounted on a rotating disc applying ~0.06 kg cm^{-2} downward force. To quantify the effect of sandpaper polishing on superhydrophobicity of the composites, up to 40 water contact angle and contact angle hysteresis measurements were conducted on various locations on a 10 × 10 cm^2 coated aluminium foil before and after sanding. In addition, coatings made using an industrial-grade anaerobic adhesive (3M Scotch-Weld 3495) were also tested for performance comparisons. On average, static water contact angles after sanding declined from 159° to 154° and hysteresis increased from 4° to 10° while still maintaining self-cleaning superhydrophobicity.[32]

A recent study[33] reports superhydrophobic conductive graphite nanoplatelet (GNP)/vapour-grown carbon fibre (VGCF)/polypropylene (PP) composite coatings with mechanical durability by a hot-pressing method (see Figure 4.22). The as-prepared GNP/VGCF/PP composite coatings showed water contact angle >150° and SA <5°. The superhydrophobicity was improved with the increase of VGCF content in the hybrid GNP and VGCF fillers. The more VGCFs were added in the GNP/VGCF/PP composite coating, the higher was

Figure 4.21 SEM image of surface morphology of fluoromethacrylic latex/bioadhesive organoclay nanocomposites after sanding with 1000 grit Al_2O_3 paper. The arrows indicate wear marks caused by sanding the surface. The leftover debris on the surface after sanding is also shown marked by the circles. Reproduced from ref. 32. Copyright (2009) The Japan Society of Applied Physics.

Figure 4.22 Optical image of GNP/VGCF/PP (GNP : VGCF = 4 : 1) composite coating surfaces prepared with different hybrid filler masses and infiltration times (10, 20, 30, 40, 60, and 100 min, respectively). Reprinted from *Composites Science and Technology*, **117**, L. Shen *et al.*, Facile fabrication of superhydrophobic conductive graphite nanoplatlet/vapor-grown carbon fiber/poly propylene composite coatings, 39–45. Copyright (2015) with permission from Elsevier.[33]

the surface porosity. Compared to the GNP/PP and VGCF/PP composite coatings, the GNP and VGCF hybrid fillers exhibited more remarkable synergistic effect on the electrical conductivity of the GNP/VGCF/PP composite coatings. The GNP/VGCF/PP composite coating with GNP : VGCF = 2 : 1 displayed a sheet resistance of 1 Ω/sq.

The authors performed a mechanical abrasion test by rubbing the as-prepared superhydrophobic surfaces against the print paper. A pressure of 10 kPa (1 N acted on the specimen with an area of 1 cm^2) was applied to the superhydrophobic composite coating surfaces, and the surface was dragged in one direction with a speed and abrasion length of 5 cm s^{-1} and 20 cm, respectively. The surface morphology of the print paper is shown in Figure 4.23. The print paper has a roughness of dozens of micrometres, which could produce an obvious wear while rubbing the composite coating against it, resulting in a loss of VGCF and GNP on the composite coating surface. After one cycle of the abrasion test, the microstructure of the VGCF/PP composite coating disappeared and its surface became smooth with few micropores, showing a loss of superhydrophobicity (Figure 4.23(b1 and b2)). However, the GNP/VGCF/PP (GNP : VGCF = 2 : 1) composite coating still maintained superhydrophobicity

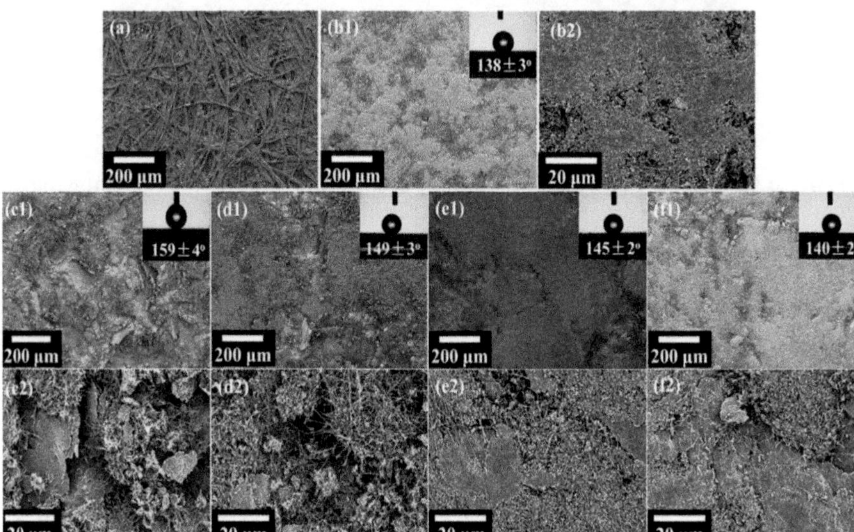

Figure 4.23 (a) Surface morphologies of the print paper used as an abrasive. (b1) SEM images of VGCF/PP composite coating after 1 cycle abrasion. (c1, d1, e1, f1) SEM images of GNP/VGCF/PP (2 : 1) composite coatings after 1, 2, 4, and 8 cycles abrasion, respectively. The inset images correspond to the WCAs of each composite coating after the abrasion test. (b2, c2, d2, e2, f2) SEM images at high magnification corresponding to b1, c1, d1, e1, and f1, respectively. Reprinted from *Composites Science and Technology*, **117**, L. Shen *et al.*, Facile fabrication of superhydrophobic conductive graphite nanoplatlet/vapor-grown carbon fiber/poly propylene composite coatings, 39–45. Copyright (2015) with permission from Elsevier.[33]

because the rough microstructure formed by GNP and VGCF was tougher than VGCF and preserved its high porosity after abrasion for <2 cycles (Figure 4.23(c1–d2)). After >4 cycles of the abrasion test, the surface morphologies became smooth and only a small fraction of the microstructure was preserved (Figure 4.23(e1–f2)), forming water contact angles (WCAs) <150°. By comparison, it is notable that the incorporation of GNPs into the VGCF/PP composite coating remarkably improved the mechanical durability.

Figure 4.24 shows the WCAs of different composite coatings after the abrasion test. Before the test, all of the composite coatings possessed WCA >150°. An obvious decrease of WCA was observed for the VGCF/PP composite coating comparing to the GNP/VGCF/PP composite coatings. After two abrasion cycles, the GNP/VGCF/PP composite coatings with a hybrid filler mass ratio of GNP : VGCF = 2 : 1 kept their superhydrophobicity, however the other two kinds of GNP/VGCF/PP composite coatings lost their superhydrophobicity. After the abrasion test, the rough microstructure of the GNP/VGCF/PP (2 : 1) composite coating was mostly restored and the composite coating retained superhydrophobicity, but not for the VGCF/PP composite coating. When the superhydrophobic surface is mechanically damaged with a loss of superhydrophobicity, it can be easily repaired with adhesive tape. Moreover, superhydrophobicity of the oil-fouled composite surface can be regenerated by wetting the surface with alcohol and subsequently burning the alcohol off.[33]

Alternatively, to create wear abrasion resistant structured surfaces with superhydrophobicity, self-cleaning, and low drag, carbon nanotube (CNT) polymer composite structures were produced by replication of a micropatterned silicon surface using an epoxy resin and deposition of the CNT

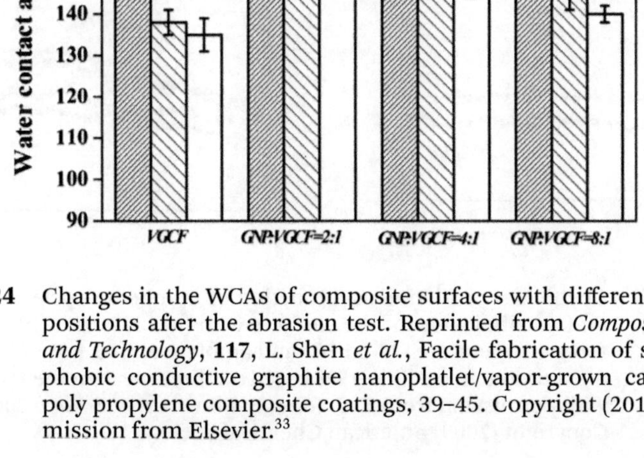

Figure 4.24 Changes in the WCAs of composite surfaces with different filler compositions after the abrasion test. Reprinted from *Composites Science and Technology*, **117**, L. Shen *et al.*, Facile fabrication of superhydrophobic conductive graphite nanoplatlet/vapor-grown carbon fiber/poly propylene composite coatings, 39–45. Copyright (2015) with permission from Elsevier.[33]

composite using a spray method.[34] The hierarchical structure created by the CNTs showed a high static contact angle of 170° and a low contact angle hysteresis of 2°. As a benchmark, structures created using lotus wax were used to compare wear abrasion characteristics of CNT composite structures. Wear and friction studies were performed using AFM and a ball-on-flat tribometer. It was found that superhydrophobic CNT composite structures showed good stability of wetting properties not only from long-term exposure to water but also high water pressure. From wear and friction studies, it was found that the lotus wax nanostructures can be easily damaged at even a small load. However, in this case, the CNT composite structure at the surface showed high mechanical strength and wear resistance due to uniform distribution of the CNTs in the epoxy matrix and the strong bonding maintained between the CNTs and the epoxy.

To investigate the durability of structured surfaces at a high load, conventional ball-on-flat tribometer experiments were conducted for the surfaces with CNTs. Figure 4.25(a) shows the coefficient of friction as a function of number of cycles for the nanoscale and hierarchical structures with CNTs. The data are reproducible within ±5% based on three measurements. The coefficients of friction on both nanoscale and hierarchical structures with CNTs first increased slightly for 20 cycles. Such a trend can be due to the elastic bending or buckling of CNTs by contact with the sapphire ball during the beginning of the scan, resulting in an increase of the contact area. During the entire experiment, the coefficient of friction value of the nanoscale and hierarchical structures with CNTs changed minimally, which indicates that the CNTs were not being worn after 100 cycles. As such, it was observed that there is no or low wear on nanoscale and hierarchical structures after wear tests. No or low wear on the CNT composite structure may possibly be due to the significant increase in the mechanical strength and wear resistance mainly due to the uniform distribution and strong bonding of CNTs on flat epoxy resin and microstructure. The elastic bending or buckling exhibited by

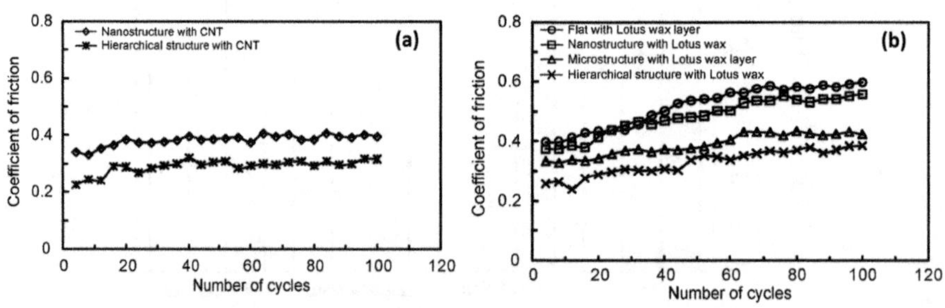

Figure 4.25 Coefficient of friction as a function of number of cycles using a ball-on-flat tribometer for the surfaces with (a) CNTs and (b) lotus wax at room temperature (22 ± 1 °C) and ambient air (45 ± 5% RH). The data are reproducible within ±5% based on three measurements. Reprinted with permission from Y. C. Jung, B. Bhushan. *ACS Nano* 2009, **3**, 4155. Copyright (2009) American Chemical Society.[34]

CNTs makes them exceedingly tough materials and may absorb some of the force at contact, acting as a compliant spring moderating the impact of the ball on the surface.

4.2.4 Environmentally Friendly Processes and Materials for Liquid-Repellent Polymer Composites

Polymers obtained from renewable resources will be more and more important in the future as oil resources are depleted. Plant-based polymers are increasingly being studied in many different applications because of their non-toxic and biodegradation properties. As it is well known, water and liquid repellency requires presence of CH_3 and CF_3 groups at the solid surface–air interfaces, along with the right kind of hierarchical surface roughness. By nature, most biopolymers (excluding natural waxes) are not hydrophobic as they are made up of polysaccharide building blocks, or polyester structures. However, this does not mean that it is impossible to produce liquid-repellent surfaces from biopolymers. After all, the lotus leaf surface is a composite made up of waxes embedded in a cellulosic matrix as a result of millions of years of natural evolution. In liquid-repellent technologies, "mimicking" mostly refers to replicating the natural surface nanostructures and functionalizing them with synthetic chemicals such as fluorinated silanes. However, recently there have been attempts to produce liquid-repellent materials from biopolymers with minimal use of fluorinated compounds or with fluorinated compounds that will not degrade into perfluorooctanoic acid. The aim of this section is to present the latest literature related to liquid-repellent materials utilizing bio-based resources and discuss their relevant shortcomings and future directions. It is hoped that this review will encourage readers to increase their research efforts towards gradually reducing and eventually eliminating the use of hazardous fluorinated compounds in designing liquid-repellent materials.

Earlier work by Bayer *et al.*[35] demonstrated a simple technique to fabricate rubber-toughened biopolymer/organoclay nanocomposite coatings with highly water-repellent surface wetting characteristics and strong adhesion to metal surfaces. The technique combines the principles of phase inversion and atomization of multicomponent polymer/organoclay suspensions containing a biolubricant as the non-solvent. The biolubricant was a blend of cyclomethicone/dimethiconol oil with fruit kernel oils. The ternary system of cellulose nitrate/solvent/biolubricant was blended with rubber dispersed in organoclay nanofluids. Natural, synthetic, and fluoroacrylic latex rubbers were used for the purpose. Self-cleaning superhydrophobic coatings were obtained from synthetic and fluoroacrylic rubbers whereas formulations containing natural rubber resulted in sticky superhydrophobic coatings (see Figure 4.26).

These coatings can be sprayed on to large-area substrates such as the one shown in Figure 4.27, where an aluminium foil is coated with the biopolymer–clay composite and water droplets clean away dirt particles, in this case carbon black.

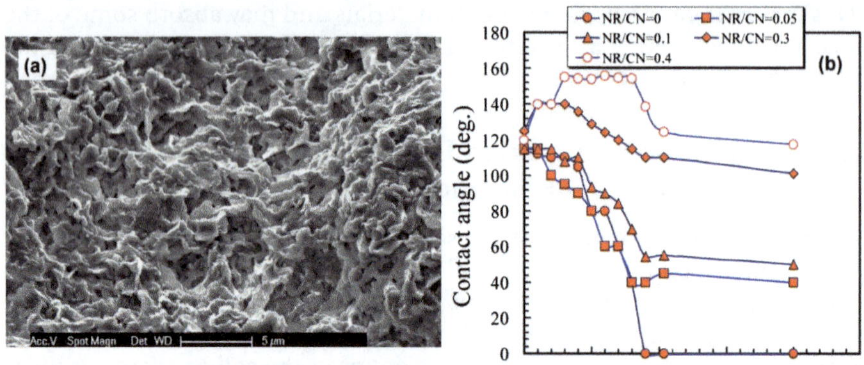

Figure 4.26 (a) SEM morphology of biopolymer (cellulose nitrate)/natural rubber (NR)/organoclay nanocomposite coating and (b) wetting characteristics of the nanocomposite as a function of biopolymer/rubber ratio and added organoclay concentration. No vulcanites were used for the NR latex. Reprinted with permission from I. S. Bayer, A. Steele, P. Martorana, E. Loth, S. J. Robinson, D. Stevenson. *Appl. Phys. Lett.* 2009, **95**, 063702. Copyright (2009), AIP Publishing LLC.[35]

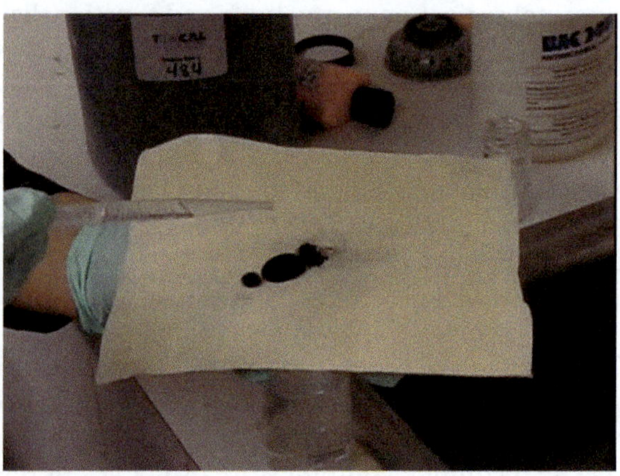

Figure 4.27 A superhydrophobic cellulose nitrate/natural rubber polymer blend matrix containing hydrophobic organoclay particles spray painted on an aluminium foil. Carbon black powder is used as dirt particles to demonstrate the self-cleaning ability. Reprinted with permission from I. S. Bayer, A. Steele, P. Martorana, E. Loth, S. J. Robinson, D. Stevenson. *Appl. Phys. Lett.* 2009, **95**, 063702. Copyright (2009), AIP Publishing LLC.[35] The original video is available at https://www.youtube.com/watch?v=djgiRPDdH00.

A recent study[36] used a commercial thermoplastic starch composite, Materi-Bi®, hydrophobic fumed silica (HMFS) nanoparticles, and lycopodium spores to form water- and oil-repellent composites. It was found that both HMFS and Materi-Bi® had excellent colloidal stability and solubility in organic solvents and they can be spray-coated onto various surfaces such as paper, metals, and semiconductors. By varying the concentration of biopolymer to HMFS, the wetting properties of the nanocomposites could be varied. Superhydrophobic nanocomposites displayed raspberry-like surface roughness (Figure 4.28) with WCA exceeding 160° with very low water droplet roll-off angles (SA ~ 1°). On the other hand, composites of Materi-Bi® and lycopodium spores displayed sticky superhydrophobicity (rose petal effect). Superhydrophobic nanocomposites were found to withstand thermal ageing at 250 °C without loss of properties. Due to the resultant micromorphology of nanocomposites, certain coatings were rendered superoleophobic by functionalizing with a dilute fluoroacrylic polymer (with C6 chemistry) solution in acetone. Oil droplet contact angles reached 166° with droplet roll-off angles of ~15°.

The effect of thin fluoroacrylic polymer coating can be seen in Figure 4.29.[36] All the surfaces with 50% HMFS loading and above retain their excellent water-repellent properties, as shown in Figure 4.29. However, as shown in the table in Figure 4.29, relatively high water roll-off substrate angles were observed, ranging from 17° to 29.3° without showing a specific trend. This is attributed to the potential partial loss of surface morphology due to the presence of an extra polymer layer covering the original submicrometre topological features. In terms of oleophobicity, the coatings of pure Mater Bi®

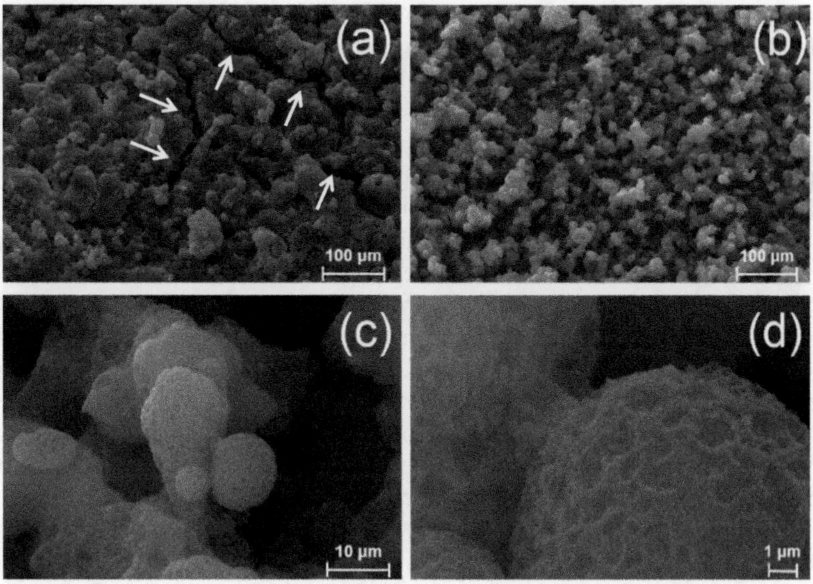

Figure 4.28 Low-magnification SEM images of the nanocomposite loaded with 60 wt% hydrophobic fumed silica (HMFS) nanoparticles before (a) and after (b) thermal treatment. (c,d) Higher-magnification SEM images indicate that the nanoscale roughness due to nanoparticles is still present in the nanocomposite and is responsible for the superhydrophobicity. Reprinted from ref. 36 with permission from the Royal Society of Chemistry.

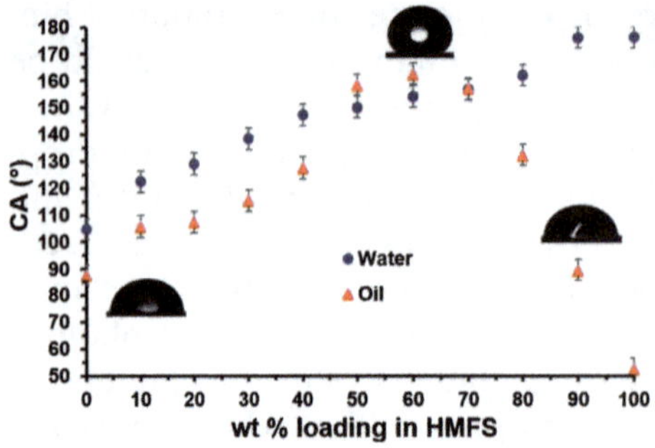

wt% loading in HMFS	RA (°) water	RA (°) oil
50	29.5	pinned
60	17.0	14.7
70	23.1	pinned
80	28.7	pinned
90	29.3	pinned
100	24.2	pinned

Figure 4.29 Water and oil droplet contact angles as a function of hydrophobic silica nanoparticle loading. The table lists water and oil droplet roll off angles as a function of hydrophobic silica nanoparticle loading. All the droplets were pinned when nanoparticle concentration in the polymer matrix was below 50 wt%. Reprinted from ref. 36 with permission from the Royal Society of Chemistry.

with the fluoroacrylic cover layer exhibit an oil contact angle (OCA) of 88°. The addition of HMFS gradually increases the OCA due to its contribution to the surface roughness. An increasing trend in the OCA was observed with a maximum value corresponding to the sample with 60% HMFS loading (OCA = 163°). In particular, the condition for superoleophobicity (OCA > 150°) is satisfied for loadings of HMFS between 50% and 70%. However, even if these OCAs are really high, only the coating with 60% HMFS demonstrates oil droplet roll off at 14.7° substrate tilt angles. In all the other concentrations, the oil droplets remain pinned on the surface. At this particular nanocomposite composition, the fluoropolymer-coated surface texture maintains both the nanoscale and microscale roughness features, as shown in Figure 4.28. Each micron-sized bump has very slight surface roughness which resembles raspberry-like dual-scale structured surfaces (Figure 4.28(d)).

Yohe *et al.*[37] prepared 3D superhydrophobic materials from biocompatible building blocks (Figure 4.30), where air acts as a barrier component in a porous electrospun mesh to control the rate at which a drug is released. Specifically, they fabricated electrospun meshes from poly(ε-caprolactone), a biopolyester, containing poly(glycerol monostearate-*co*-ε-caprolactone) as a hydrophobic polymer dopant. The resulting meshes had a high apparent contact angle, which dictates the rate at which water penetrates into the porous network and displaces entrapped air. The surface wettability was controlled by hydration, which caused a model bioactive agent to penetrate into the pores displacing air. Once the meshes were in contact with an aqueous medium the bioactive agent was slowly released from the pores. It was possible to produce porous electrospun meshes with higher surface area to release drugs more slowly than control non-porous constructs. The drug-loaded meshes were efficacious against cancer cells *in vitro* for more than 60 days, thus demonstrating their applicability for long-term drug delivery, as schematically depicted in Figure 4.31.

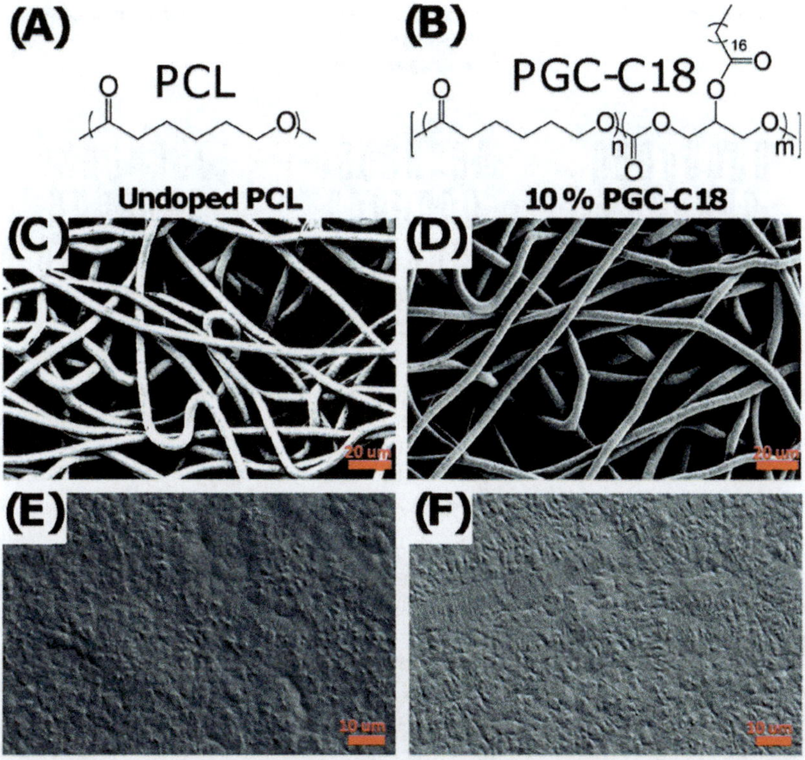

Figure 4.30 (A) Polycaprolactone (PCL) was used as the base polymer for fabrication of electrospun meshes and molten electrospun meshes. (B) Poly(glycerol monostearate-*co*-ε-caprolactone) PGC-C18 was used as the hydrophobic dopant in electrospun PCL meshes to decrease their wettability. (C) Electrospun PCL mesh with an average fibre size of 7.7 ± 1.2 μm. (D) 10% PGC-C18-doped electrospun PCL mesh with an average fibre size of 7.2 ± 1.4 μm. (E) A melted PCL mesh. (F) A melted 10% PGC-C18-doped electrospun PCL mesh. Reprinted with permission from S. T. Yohe, Y. L. Colson, M. W. Grinstaff. *J. Am. Chem Soc.* 2012, **134**, 2016. Copyright (2012) American Chemical Society.[37]

Very recently, a facile, environmentally friendly (water-based) and a cost-effective method to produce large-scale superhydrophobic and superoleophobic polymer–silica nanocomposites by using hydrophilic silica nanoparticles was described.[14] The spraying process reported there requires maintaining the substrate at elevated temperatures in order to improve the quality of the coatings and eliminate dewetting effects due to water's high surface tension and high boiling point. The wetting properties were controlled by changing the polymer/nanosilica ratio in solution. WCAs and OCAs > 159° were achieved, while droplet roll-off angles were as low as 3° and 2° for oil and water, respectively. These environmentally friendly coatings were also able to repel a wide variety of other liquids (see Figure 4.32). In addition, the effect of particle size and morphology was studied and was shown to

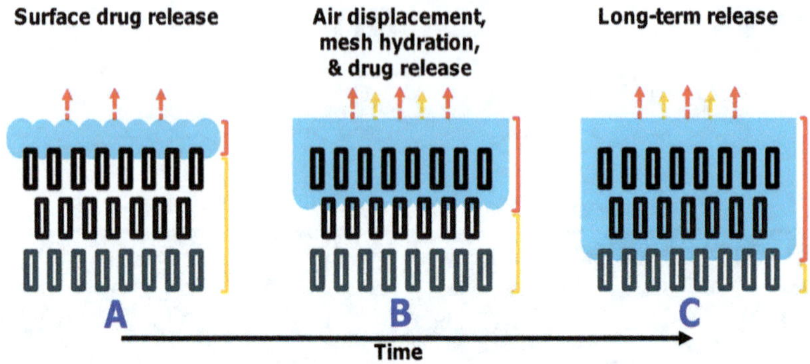

Figure 4.31 Infusion of polymer nanofibre network with water allowing controlled and slow rate drug release. Infusion progresses in time from A to C. Reprinted with permission from S. T. Yohe, Y. L. Colson, M. W. Grinstaff. *J. Am. Chem Soc.* 2012, **134**, 2016. Copyright (2012) American Chemical Society.[37]

Liquid	Contact Angle (°)	Roll Off Angle (°)
Water	159	2
Oil	159	3
Ethylene Glycol	162	3
Hexadecane	146	8
Vinegar	157	5
Coffee	159	4
Milk	149	7
Hot Sauce	143	6
Orange Juice	159	4

Figure 4.32 Left: photograph of various liquids on the nanocomposite surface. The table on the right displays static contact angles as well as droplet roll-off angles for every liquid tested. Surface tensions: water, 72.8 mN m^{-1}; mineral oil, 28.6 mN m^{-1}; ethylene glycol 47.7 mN m^{-1}; hexadecane, 27.5 mN m^{-1}; milk, 41 mN m^{-1}; orange juice, 32 mN m^{-1}; vinegar, 36 mN m^{-1}; hot sauce, 33 mN m^{-1} (values at 20 °C). Reprinted from ref. 14 with permission from the Royal Society of Chemistry.

influence the surface roughness characteristics, wetting, and the resultant wear abrasion resistance of the coatings. The authors also analysed the influence of the chemical composition and the surface texture on the wetting properties based on measurements from white light interferometry, electron microscopy, and X-ray photoelectron spectroscopy. When the substrates were treated with an adhesive primer, the wear abrasion resistance of the coatings was enhanced. The effect of silica nanoparticle dispersion in the coatings on wear abrasion was also presented. Certain nanocomposites, for instance, were found to exhibit good abrasion resistance by retaining their water repellency for up to 60 abrasion cycles under 20.5 kPa applied pressure.

4.3 Conclusions

Although there are many highly effective, facile and rapid methods of producing liquid-repellent materials and surfaces, polymer composite coatings are probably the most suitable candidates for scaling-up and commercialization. The key aspect of polymer composite coatings that can repel water or other low surface tension liquids is the ability to formulate them with the right kind of polymeric binder and nanoscale filler. Super-liquid-phobic polymer composites can easily benefit from all the technologies and recent developments in the paint and coatings industry. In the last few years, important high-performance criteria required for paints and industrial coatings such as scratch resistance, strong adhesion, resistance to oxidation, and wear abrasion have been also requested for liquid-repellent polymer composites. It is indeed a very challenging task to formulate super-liquid-repellent polymer composite paints that can perform similarly to an aerospace coating, for instance. However, considerable progress has been made and many researchers have developed and published impressive liquid-repellent materials that can withstand many cycles of wear abrasion, exposure to acidic and basic media, as well as resistance to water jet or rain erosion. The second most important aspect is to be able to produce super-liquid-repellent coatings or paints by using green materials and environmentally friendly processes. Exciting developments and materials have recently been reported in the literature on this issue as well. Much more fundamental and applied research will be needed to synthesize and design super-liquid-repellent polymer composite coatings based on renewable biomaterials. Within the limited space of this chapter, it has been impossible to present and discuss all this recent work in detail; however, the author believes that the chapter will catalyse further interest in addressing important issues such as commercialization, mechanical and chemical durability, and robustness as well as environmentally benign superhydrophobic and superoleophobic materials.

References

1. B. N. Sahoo and B. Kandasubramanian, *RSC Adv.*, 2014, **4**, 22053.
2. Z. Guo, W. Liu and B.-L. Su, *J. Colloid Interface Sci.*, 2011, **353**, 335.
3. B. Wang, Y. Zhang, L. Shi, J. Li and Z. Guo, *J. Mater. Chem.*, 2012, **22**, 20112.
4. A. Marmur, *Langmuir*, 2004, **20**, 3517.
5. A. Marmur, *Biofouling*, 2006, **22**, 107.
6. C. K. Söz, E. Yilgör and I. Yilgör, *Prog. Org. Coat.*, 2015, **84**, 143.
7. S. Yang, L. Wang, C.-F. Wang, L. Chen and S. Chen, *Langmuir*, 2010, **26**, 18454.
8. P. Van Der Wal and U. Steiner, *Soft Matter*, 2007, **3**, 426.
9. Y. Tang, J. Yang, L. Yin, B. Chen, H. Tang, C. Liu and C. Li, *Colloids Surf., A*, 2014, **459**, 261.
10. W. Wu, X. Wang, X. Liu and F. Zhou, *ACS Appl. Mater. Interfaces*, 2009, **1**, 1656.

11. L. Cao, A. K. Jones, V. K. Sikka, J. Wu and D. Gao, *Langmuir*, 2009, **25**, 12444.
12. C.-H. Xue, Z.-D. Zhang, J. Zhang and S.-T. Jia, *J. Mater. Chem. A*, 2014, **2**, 15001.
13. N. Kudo and Y. Kawashima, *J. Toxicol. Sci.*, 2003, **28**, 49.
14. A. Milionis, K. Dang, M. Prato, E. Loth and I. S. Bayer, *J. Mater. Chem. A*, 2015, **3**, 12880.
15. I. S. Bayer, A. Steele and E. Loth, *Chem. Eng. J.*, 2013, **221**, 522.
16. C. E. Cansoy and U. Cengiz, *Colloids Surf., A*, 2014, **441**, 695.
17. A. Steele, I. Bayer and E. Loth, *Nano Lett.*, 2008, **9**, 501.
18. S. G. Lee, D. S. Ham, D. Y. Lee, H. Bong and K. Cho, *Langmuir*, 2013, **29**, 15051.
19. T. M. Schutzius, I. S. Bayer, M. K. Tiwari and C. M. Megaridis, *Ind. Eng. Chem. Res.*, 2011, **50**, 11117.
20. J. D. Lichtenhan, *Comments Inorg. Chem.*, 1995, **17**, 115.
21. V. A. Ganesh, A. S. Nair, H. K. Raut, T. T. Y. Tan, C. He, S. Ramakrishna and J. Xu, *J. Mater. Chem.*, 2012, **22**, 18479.
22. G. Kickelbick, Silsesquioxanes, in *Functional Molecular Silicon Compounds I*, Springer International Publishing, 2014, pp. 1–28.
23. I. S. Bayer, V. Caramia, D. Fragouli, F. Spano, R. Cingolani and A. Athanassiou, *J. Mater. Chem.*, 2012, **22**, 2057.
24. L. Shen, W. Qiu, B. Liu and Q. Guo, *RSC Adv.*, 2014, **4**, 56262.
25. I. S. Bayer, A. Steele, P. J. Martorana and E. Loth, *Appl. Surf. Sci.*, 2010, **257**, 826.
26. T. M. Schutzius, I. S. Bayer, G. M. Jursich, A. Das and C. M. Megaridis, *Nanoscale*, 2012, **4**, 5385.
27. M.-G. Jeong, H. O. Seo, K.-D. Kim, D. H. Kim, Y. D. Kim and D. C. Lim, *J. Mater. Sci.*, 2012, **47**, 5196.
28. X. Zhang, Y. Guo, Z. Zhang and P. Zhang, *Appl. Surf. Sci.*, 2013, **284**, 319.
29. T. Verho, C. Bower, P. Andrew, S. Franssila, O. Ikkala and R. H. A. Ras, *Adv. Mater.*, 2011, **23**, 673.
30. A. Steele, A. Davis, J. Kim, E. Loth and I. S. Bayer, *ACS Appl. Mater. Interfaces*, 2015, 7, 12695.
31. V. C. Malshe and N. S. Sangaj, *Prog. Org. Coat.*, 2005, **53**, 207.
32. I. S. Bayer, A. Brown, A. Steele and E. Loth, *Appl. Phys. Express*, 2009, **2**, 125003.
33. L. Shen, W. Qiu, W. Wang, G. Xiao and Q. Guo, *Compos. Sci. Technol.*, 2015, **117**, 39.
34. Y. C. Jung and B. Bhushan, *ACS Nano*, 2009, **3**, 4155.
35. I. S. Bayer, A. Steele, P. Martorana, E. Loth, S. J. Robinson and D. Stevenson, *Appl. Phys. Lett.*, 2009, **95**, 063702.
36. A. Milionis, R. Ruffilli and I. S. Bayer, *RSC Adv.*, 2014, **4**, 34395.
37. S. T. Yohe, Y. L. Colson and M. W. Grinstaff, *J. Am. Chem. Soc.*, 2012, **134**, 2016.

CHAPTER 5

Etching Techniques for Superhydrophobic Surface Fabrication

SAMI FRANSSILA[a]

[a]Department of Materials Science and Engineering and Micronova Center for Nanofabrication, Aalto University, Espoo, Finland
*E-mail: sami.franssila@aalto.fi

5.1 Introduction

Single-scale random nanoroughness is very easy to produce by etching. Plasma etching of silicon, silica, and polymers; acidic etching of metals; electrochemical etching of metals and silicon; and metal-assisted wet chemical etching silicon all result in random nanoroughness. Regular nanoscale features are much more difficult to make. The integrated circuits industry produces 20–30 nm structures routinely, but at high cost.[1] This is because nanoscale lithography is expensive, especially when the writable area is large, which is usually the requirement in fluidic experiments. Electron beam lithography is limited in patternable area, or writing speed. Interference lithography equipment is expensive, but in-house built systems are available although they usually have limited writable area.[2] Because most structures for superhydrophobic (SHB) surfaces are regular dot/pillar or ridge arrays, interference lithography would be a good candidate. Nanoimprint

RSC Soft Matter No. 5
Non-wettable Surfaces: Theory, Preparation, and Applications
Edited by Robin H. A. Ras and Abraham Marmur

Published by the Royal Society of Chemistry, www.rsc.org

lithography suffers from thin resists which do not support high aspect ratio etching. This can be solved by hard masks, which add a little complexity to the process. There are only a few publications on SHB surfaces that have lithographically defined nanoscale structures.[3,4]

Microscale roughness is usually produced by lithography and etching. Microlithography is orders of magnitude cheaper than nanolithography, and large areas pose no problems.[5] Etching comes in two major forms: anisotropic and isotropic. Anisotropic etching allows tighter dimensional control, and freely variable pattern densities. Isotropic etching is limited in etched depth. Also, high aspect ratio structures can only be produced by anisotropic etching. Pattern density freedom and high aspect ratios are essential to optimize the solid fraction of SHB surfaces, and to make the structures high enough to prevent sagging. Typical structure dimensions for silicon pillar widths are 1–20 μm with heights from 5–50 μm, with aspect ratios from ~3:1 to 10:1.

Dual-scale roughness is usually achieved by making the microscale structures first, and adding nanoscale roughness all over the microroughness. But there are also applications where microscale pillars with smooth tops and nanoscale sidewall roughness are beneficial, *e.g.* in maximizing slip flow.[6]

5.2 Plasma Etching

5.2.1 Basics

Plasma etch process parameters are power, pressure, flow rate(s), and temperature, once the etch gases have been fixed. The major plasma process outcomes are rate, selectivity (to mask and to underlying layer), and sidewall angle. Additionally, there are general process criteria such as uniformity, wafer-to-wafer reproducibility, and throughput.

We use the terms *plasma etching* and *reactive ion etching (RIE)* interchangeably. Deep reactive ion etching (DRIE) is a variant of RIE: it is characterized by separation of ion generation and acceleration, enabling high etch rates and high selectivities. A high rate is related to high plasma density generated by a powerful (kW) radiofrequency (RF) source for ion generation and high selectivity is due to small ion bombardment because the wafer electrode is biased by another RF source at very low power, typically only a few watts. For recent reviews of RIE and DRIE, see *e.g.* ref. 7–10.

Silicon reacts spontaneously with fluorine (F_2), forming volatile compounds like SiF_4. Because of this reactivity, isotropic etch profile is the norm for fluorine etching of silicon, and much research has been spent on developing anisotropic processes. One route is to use chlorine (Cl_2) or bromine (Br_2) which do not spontaneously etch silicon, but require ion assistance. This is the option chosen by the integrated circuits industry. However, lower reactivity equals lower etch rates (only ~10% of fluorine rates), and the microelectromechanical systems (MEMS) community with deep etch requirements uses SF_6-based etching almost exclusively because it is a safe-to-handle gas which provides lots of fluorine. The solution is sidewall passivation, which prevents lateral etching. The passivation typically forms on all surfaces, vertical ion bombardment removes it from horizontal surfaces, and directional etching results.

The Bosch process (aka time-modulated or cyclic process), is the mainstream DRIE process. It depends on repetition of isotropic SF_6-based etching cycle with a C_4F_8-passivation cycle[7] as shown in Figure 5.1. The resulting sidewall profile is macroscopically vertical but undulating at microscale due to the isotropic nature of the SF_6–silicon reaction. A SEM micrograph of Bosch sidewall undulation is also shown in Figure 5.1.

Bosch process variables include the SF_6 etching cycle time (from a few seconds to 10 s), C_4F_8 cycle time (a few seconds), SF_6 flow (~100–200 sccm (standard cubic centimetres per minute)), C_4F_8 flow (~50–100 sccm), pressure (tens of mTorr), inductively coupled plasma (ICP) power (500 W to 5 kW), and bias power (a few watts). Etch rates vary from a few micrometres per minute to tens of micrometres per minute.

Good thermal contact between the wafer and the electrode is essential to maintain constant temperature and etch rate. Electrostatic or mechanical clamping and helium backside cooling are used to stabilize temperature. A resist mask works well for shallow and medium etch depths. Rates can be tens of micrometres per minute, but optimization of other parameters is usually achieved at the expense of the rate, and in many cases 5 μm min^{-1} or 2 μm min^{-1} might actually be reached.

The cryogenic DRIE process uses temperatures of −120 °C, and good wafer-to-chuck thermal contact is mandatory. Cryoetching is based on continuous SF_6/O_2 flows and continuous replenishing of a SiF_xO_y passivation layer. Even though cryogenic temperature slows spontaneous chemical reactions, the passivation layer is needed for anisotropic profiles. The passivation layer forms when temperatures are below −100 °C. It evaporates when the wafer heats up to room temperature, and no special cleaning is required after

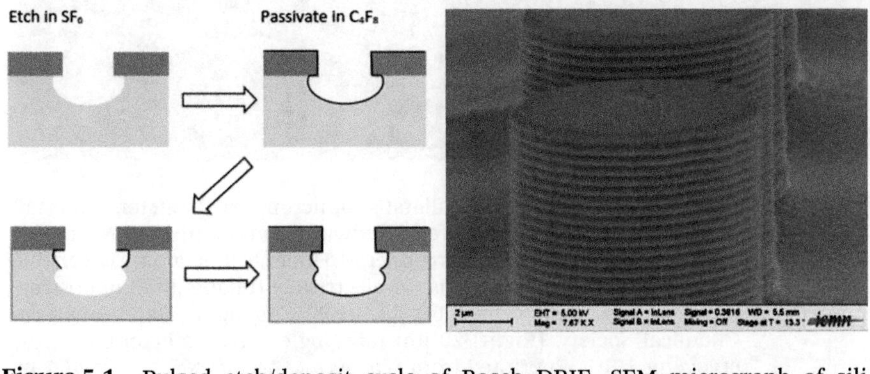

Figure 5.1 Pulsed etch/deposit cycle of Bosch DRIE. SEM micrograph of silicon micropillars show the undulation arising from pulsed etching.[11] Reprinted from *J. Coll. Interf. Sci.*, T. P. N. Nguyen *et al.*, Micro- and nanostructured silicon-based superomniphobic surfaces, 280–288, Copyright (2014) with permission of Elsevier.

cryoetching. Because the process is continuous, the sidewalls are smooth. One example can be seen in Figure 5.2. Sidewall angle can be tailored by process temperature (and other parameters). Etch rates of the cryogenic process are usually smaller than with Bosch, but up to 7 μm min^{-1} has been demonstrated.[13] A special concern is the cracking of photoresists at cryogenic temperatures. This limits the choice of lithographic process somewhat: either thin (<1 μm) or thick resists (*e.g.* >10 μm SU-8) have to be used, because many standard resists will crack.[16]

Cryogenic DRIE process parameters—gas flows, pressure, RF power[18]—are similar to those of the Bosch process, with an additional key parameter, the chuck temperature. Cryo processes tend to be slower than Bosch processes, and rates are typically 2–5 μm min^{-1}.

Resolution, or linewidth, is determined by the lithography process. Plasma etching is generally always available for all linewidths that a lithography process can produce. One key issue is the resist selectivity when high aspect ratio structures need to be produced. In order to minimize linewidth, resist thickness has to be scaled down, and such thin resists are not good for plasma etch masking. Typical resist:silicon selectivities in DRIE are of the order of 100 : 1, so electron beam lithography resists of 100 nm thickness will limit the structure heights to *c.* 10 μm. Standard optical resists for i-line (365 nm) and g-line (405 nm) exposure are roughly 1 μm thick, and suffice for almost all interesting structure heights.

A simple solution to resist selectivity is to use a hard mask. Oxide and metal thin films offer extreme selectivities (60 000 : 1 has been demonstrated),[16] and layers as thin as 1 nm can act as etch masks.[16] Because the hard mask can

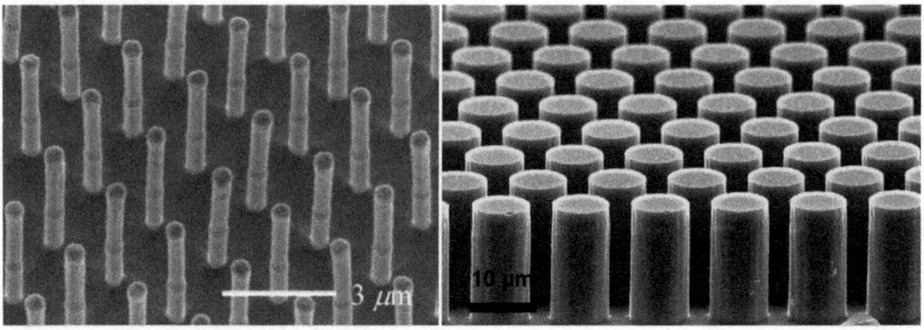

Figure 5.2 (Left) Simple DRIE etched pillars: submicrometer diameter, 5 μm tall polysilicon pillars, from ref. 3, reprinted with permission from T. N. Krupenkin, J. A. Taylor, E. N. Wang, P. Kolodner, M. Hodes, T. R. Salamon: Reversible wetting–dewetting transitions on electrically tunable SHB nanostructured surfaces, *Langmuir* 2007, **23**, 9128. Copyright (2007) American Chemical Society. (Right) 20 μm tall single crystal silicon etched by cryogenic DRIE. Reprinted from *Sensors and Actuators B: Chemical*, **132**, L. Sainiemi *et al.*, Fabrication and fluidic characterization of silicon micropillar array electrospray ionization chip, 380–387, Copyright 2008 with permission from Elsevier.[14]

be extremely thin, resist selectivity in etching the hard mask is not an issue. If, however, the hard mask will also serve as a structural element in the final device, *e.g.* as an overhang, additional factors need to be considered. Hard mask thickness is then probably determined by the device requirements, and overhang length, film thickness, and film stress are coupled. Hard mask materials are, by definition, difficult to etch, and thick hard masks are therefore a bit problematic.

Isotropic etch profile is easier to achieve than anisotropic. Due to the reactivity of fluorine, silicon is easily undercut. The typical isotropic silicon etch is SF_6/O_2.[17] Oxygen is added to scavenge SF_x* radicals and prevent recombination back to SF_6, thereby increasing F* concentration. By changing the SF_6/O_2 ratio, a wide variety of etch rates and profiles are possible, as shown in Figure 5.3.

Isotropic etching is essential in making overhang structures. It is possible to make overhang profiles using isotropic etching alone, but most often overhangs are made by a combination of anisotropic DRIE and an isotropic step. The limitation of the "isotropic only" approach is the limited pattern density: only small solid fraction designs are possible. The isotropic etch process can produce sharp pillars, which is beneficial for a "fakir bed" effect, but not necessarily for mechanical strength. Various isotropically etched pillar surfaces are shown in Figure 5.3. Surface (2) displays contact angle of 160° for *n*-octane ($\gamma = 21.6$ mN m^{-1}).

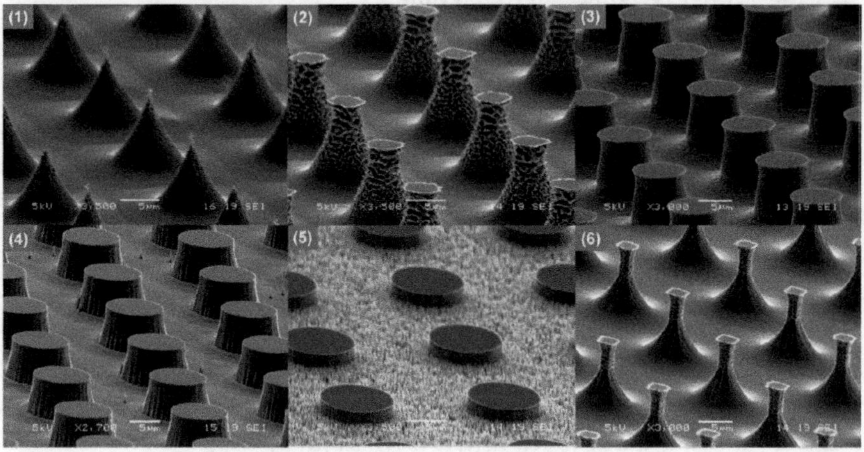

Figure 5.3 The degree of isotropy in pillar etching can be controlled by changing plasma etch parameters, in this case by varying the oxygen flow and keeping constant SF_6 flow. The O_2 flow was: (1) 17.5 sccm, (2) 20 sccm, (3) 25 sccm, (4) 3 sccm, and (5) 50 sccm for 5 min. Sample (6) was processed in two steps: 25 sccm for 2.5 min, followed by 15 sccm for 2.5 min. In each SEM image the scale bar represents 5 µm. Reproduced from ref. 17 © IOP Publishing. Reproduced with permission. All rights reserved.

5.2.2 Limitations in Plasma Etching

Etch rate is not uniquely defined but instead depends on a number of factors. Maximum rates reported refer to cases where etchable area is very small, typically in the range of 1% only, *i.e.* etching of holes. The rates for large etchable fractions, typical of pillar-like SHB surfaces, can be smaller by a factor of 2 easily, and if other optimization criteria need to be considered, such as sidewall angle and sidewall smoothness, maybe only 10% of maximum available rate. Instead of the 50 $\mu m\ min^{-1}$ which is achievable with a 20 kW RF power source, the practical etch rate for sparse pillars (*e.g.* 5% solid fraction) might be 5 $\mu m\ min^{-1}$, and much less if a smaller ICP power source is used. This loading effect, or area-dependent etch rate, is a general phenomenon in all chemical reactors and it depends on the balance between supply and consumption of reactants. Loading may lead to unacceptable etch results if the wafer contains test patterns with highly variable patterns densities: it may be that a small solid fraction area (= high etchable area) will be completely unacceptable, and this large non-uniformity also radiates to neighbouring areas of the wafer, possibly rendering them unacceptable too.

The loading effect also affects sidewall angle. It is typical that in dense structures sidewalls are closer to the ideal vertical, while in sparse structures severe negative angle sloping is encountered, up to the point that pillars are detached from the substrate. Positive angle sloping, which is beneficial for optical matching, can be achieved by process parameter tuning, but that will also affect other responses like etch rate.

5.2.2.1 Mushroom/Overhang/T-Profile

In order to achieve oleophobicity, overhang profiles (aka mushroom pillars or T-pillars) are employed. These are easily achieved by undercutting a hard mask, such as SiO_2,[18,19] as shown in Figure 5.4. The oxide cap, which also serves as a hard mask, has been etched by anisotropic plasma etching. This is preferred, since a wet-etched hard mask with a sloped profile will be consumed during the following etch steps. The main limitation of isotropic undercut etching is the poor dimensional control: the undercut rate is usually not well defined, and needs to be determined case by case.

5.2.2.2 Serif-T/Double Re-Entrant Structures

The omnirepellent serif-T profile,[20] also known as a double re-entrant profile,[21] depends on a T-shape with a vertical serif, a nanoscale downward protrusion, as shown in Figure 5.5. The process described in ref. 23 is based on polymer replication, while the process in ref. 24 depends on plasma etching. The latter utilizes four anisotropic RIE processes (two for oxide and two for silicon), plus an isotropic silicon etch step. Thermal SiO_2 (1 μm thick) is used for the hard mask and it also forms the overhang. The first RIE step is etching of this oxide hard mask anisotropically in $CHF_3/C_4F_8/Ar$ plasma.

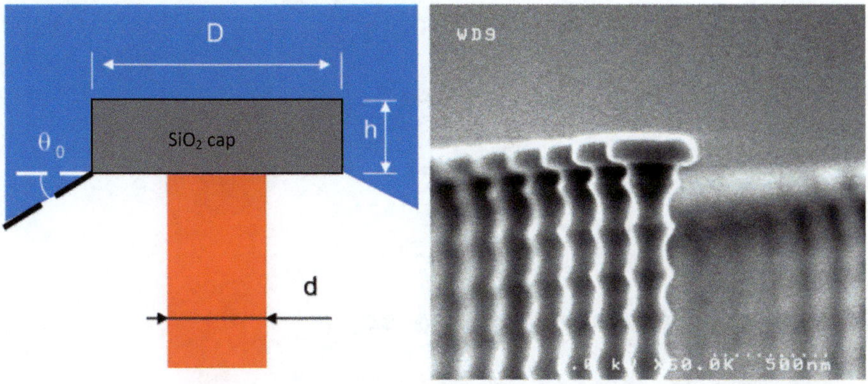

Figure 5.4 Overhang profile created by anisotropic SiO_2 etch, anisotropic silicon DRIE with controlled undercut. Reprinted with permission from A. Ahuja, J. A. Taylor, V. Lifton, A. A. Sidorenko, T. R. Salamon, E. J. Lobaton, P. Kolodner, T. N. Krupenkin: Nanonails: a simple geometric approach to electrically tunable superlyophobic surfaces, *Langmuir* 2008, **24**, 9.[19] Copyright (2008) American Chemical Society.

The first silicon anisotropic etch step is to create a recess which matches the designed serif height (vertical overhang), 1.5 μm in this case. This step is slightly isotropic (as seen in the third step): this undercut compensates for the fact that thermal oxide will used in the following step. The photoresist is removed, and a second thermal oxidation step is used to cover the exposed silicon with 300 nm thick oxide. The third RIE step etches away all the SiO_2 from horizontal surfaces, leaving 300 nm wide spacers (vertical overhangs). The fourth anisotropic step is the silicon DRIE step to a depth about half that of the pitch (*e.g.* 50 μm). The fifth and final etching step isotropically undercuts the hard mask oxide, revealing the serifs. Tungsten-coated and parylene-coated versions were made by coating the silicon/SiO_2 pillars with the respective materials.

5.2.3 DRIE for Shapes Other than Pillars

One key performance factor where DRIE excels over wet etching is shape freedom: whatever patterns lithography can produce, DRIE can etch. Shapes studied include microbricks and honeycombs (Figure 5.6). The work described in ref. 3 and 22 concentrated on developing electrically tuneable wetting, and the honeycombs and microbricks behaved similarly to pillar surfaces in terms of wettability switching. The special requirement for these structures is electric isolation: a thin SiO_2 layer was deposited before thick polysilicon for isolation. The capability to deposit such thick polysilicon layers is beyond most laboratories. Radial silicon ridges have been utilized for droplet positioning.[22] Geometrical gradient drives the droplet to the centre of the spoked wheel (Figure 5.7).

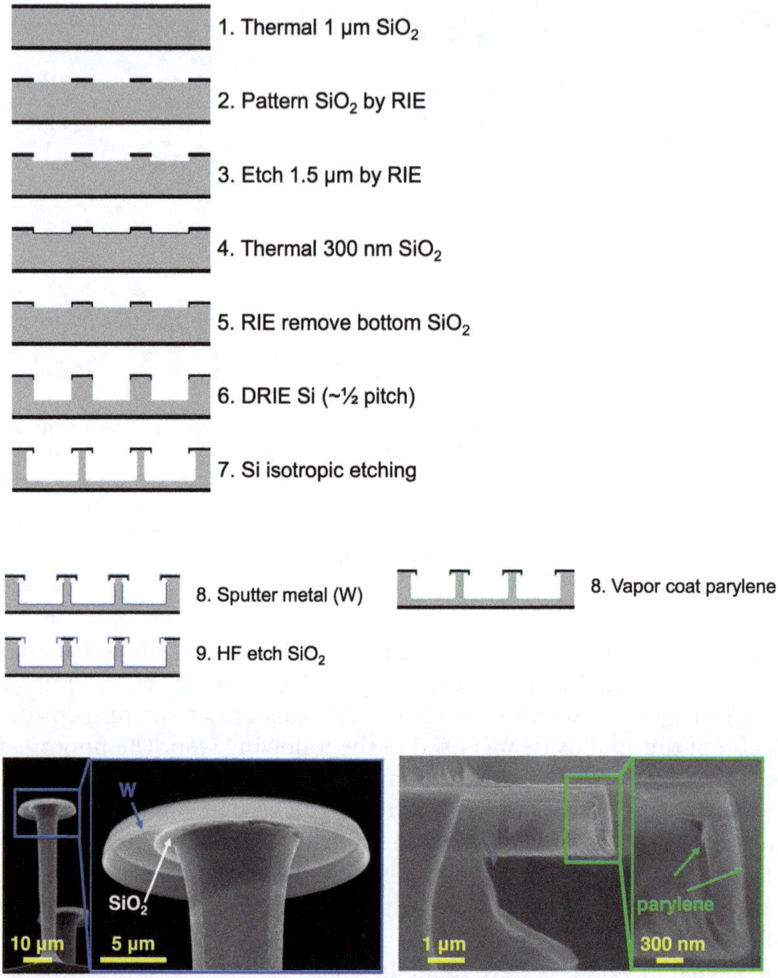

Figure 5.5 Process flow to fabricate serif-T/double re-entrant posts. SEM of tungsten-coated and parylene-coated versions. From ref. 21 with permission from The American Association for the Advancement of Science.

5.2.4 Nanoroughness by Non-Masked Plasma Etching

Plasma etching of a blank silicon wafer without any mask patterns produces nanoscale structure under suitable conditions. Two main mechanisms are responsible: overpassivation and micromasking from non-etchable residues and redeposits.

Black silicon (also known as silicon nanograss) is a prototypical material produced by overpassivation.[23] C_4F_8 produces a Teflon-like passivation film in the Bosch process. The process must be finely balanced so that the following etch step will be able to remove this passivation film from horizontal surfaces, but retain it on the sidewalls. If this is not the case, passivation film will start accumulating on horizontal surfaces, and these deposits will act as local etch masks. The resulting surface consists of nanocones and spikes hundreds of

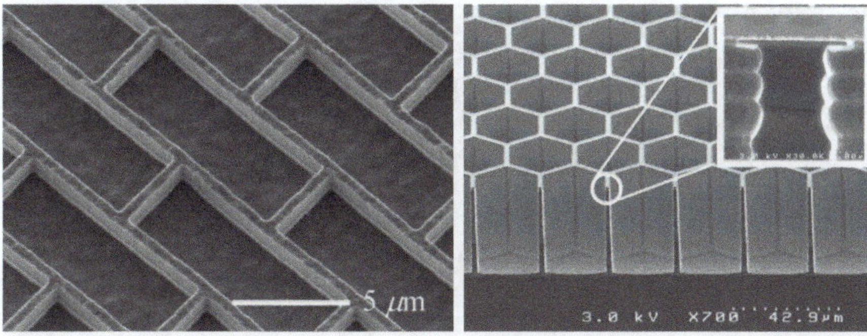

Figure 5.6 Microbrick geometry in 7 μm high polysilicon, with permission from T. N. Krupenkin, J. A. Taylor, E. N. Wang, P. Kolodner, M. Hodes and T. R. Solomon, Reversible wetting–dewetting transitions on electrically tunable superhydrophobic nanostructured surfaces, *Langmuir*, 2007, **23**, 9128. Copyright (2007) American Chemical Society; honeycomb 40 μm high, with permission from A. Ahuja, J. A. Taylor, V. Lifton, A. A. Sidorenko, T. R. Salamon, E. J. Lobaton, P. Kolodner, T. N. Krupenkin: Nanonails: a simple geometric approach to electrically tunable super-lyophobic surfaces, *Langmuir* 2008, **24**, 9.[19] Copyright (2008) American Chemical Society.

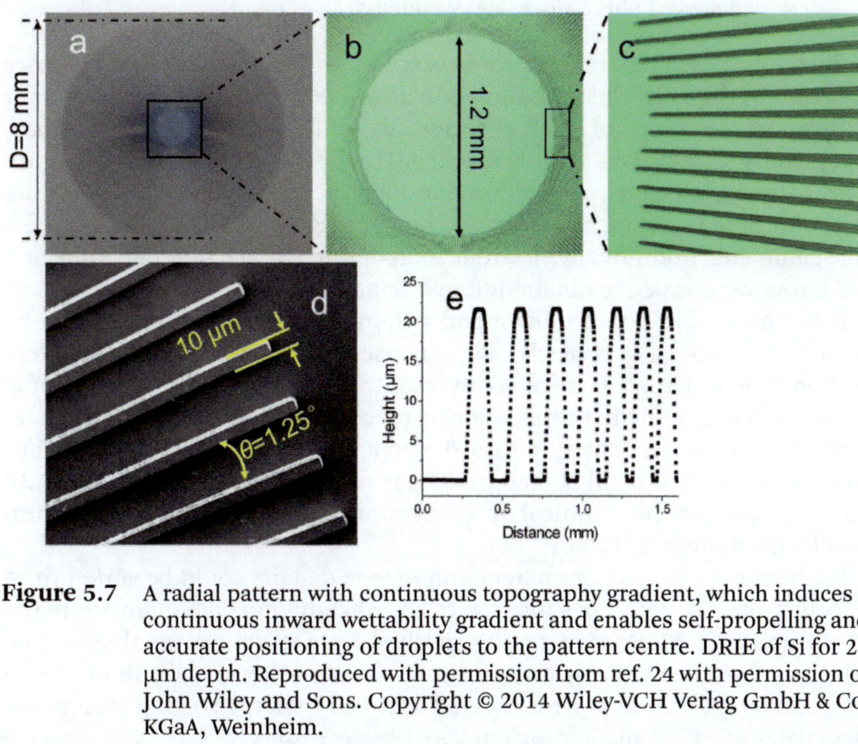

Figure 5.7 A radial pattern with continuous topography gradient, which induces a continuous inward wettability gradient and enables self-propelling and accurate positioning of droplets to the pattern centre. DRIE of Si for 22 μm depth. Reproduced with permission from ref. 24 with permission of John Wiley and Sons. Copyright © 2014 Wiley-VCH Verlag GmbH & Co. KGaA, Weinheim.

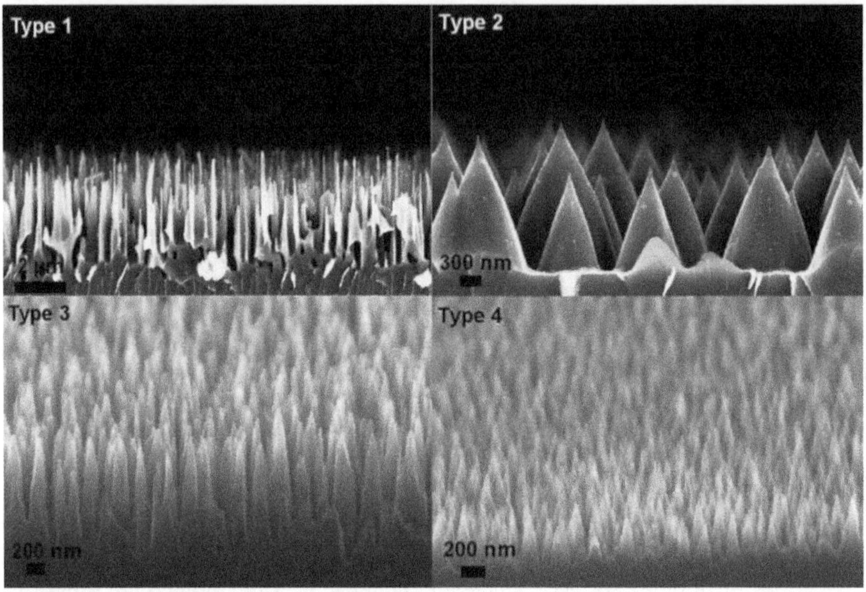

Figure 5.8 Black silicon (silicon nanograss) formed under different non-masked cryogenic plasma etching conditions. Reprinted from ref. 25 with permission from John Wiley and Sons. Copyright © 2011 Wiley-VCH Verlag GmbH & Co. KGaA, Weinheim.

nanometres in size. Its reflectance is very low and it is visually black (hence the name). Originally the formation of black silicon was used as a marker for anisotropy: conditions of DRIE were adjusted until black silicon was formed, and then passivation was slightly reduced. This operating point was a compromise between adequate passivation and high etch rate. A similar procedure works for SF_6/O_2 passivation in cryoetching, using oxygen flow as the adjustable parameter. Additionally electrode temperature, SF_6/O_2 flow rate ratio, and coil power were varied to obtain different nanograss types (Figure 5.8).

Both sparse and dense pillars, and pyramidal shapes and spikes can be formed.[23–25] Especially if black silicon is used as a master for polymer replication, sidewall angle control is important. In Figure 5.8, the type 2 profile is best suited for polymer replication. A type 2 surface exhibits a 170° water contact angle when coated with a fluoropolymer and its polymer replica 160°. In ref. 29 two blanket etch processes were combined: cryogenic DRIE was used to generate conical micropyramids, and Bosch DRIE to form needle-like nanograss (Figure 5.9).

By changing plasma parameters, microcone density could be varied from 86 cones mm^{-2} to 355 cones mm^{-2} with cone height and base diameter in the 20–30 µm range. Microcone–nanograss dual-scale structure was shown to be robust against mechanical damage, especially at the low density limit. When C_4F_8-based passivation polymer served as the low energy coating, water contact angles of >170° and sliding angles <1° were observed.

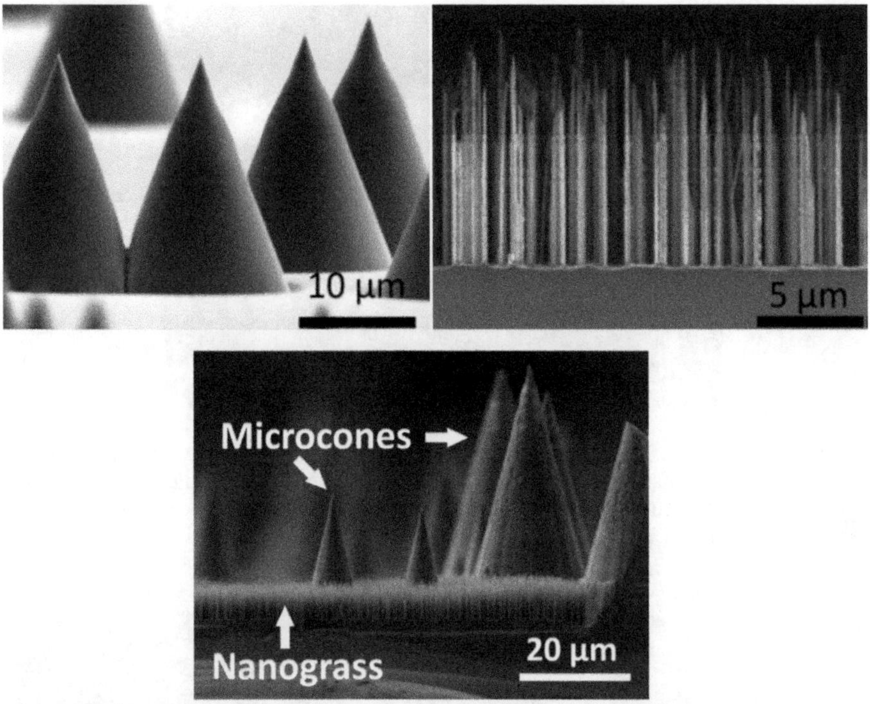

Figure 5.9 Microcones formed by cryogenic plasma etching (top left) and nanograss formed by Bosch process (top right); combined micro/nano rough surface (bottom). Reprinted with permission from V. Kondrashov, J. Rühe: Microcones and nanograss: toward mechanically robust SHB surfaces, *Langmuir* 2014, **30**, 4342.[26] Copyright (2014) American Chemical Society.

5.3 Silicon Anisotropic Wet Etching

Silicon anisotropic wet etching relies on etch rate differences between silicon crystal planes. Crystal planes ⟨110⟩ and ⟨100⟩ are fast etching, and the ⟨111⟩ plane is slow etching, with rates of the order of 1 µm min^{-1} *vs.* 10 nm min^{-1}. Etching at 80 °C in strong alkaline etchants like KOH or TMAH (tetramethyl ammonium hydroxide) means that there are materials limitations: photoresists and most non-noble metals will be attacked. Common materials that survive alkaline etching include silicon dioxide, silicon nitride, diamond-like carbon (DLC), chromium, and gold. For crystal plane dependent etching of silicon see for instance ref. 1 and 27.

Because of crystal plane limitations only fairly simple shapes like pillars and pyramids are possible. In Figure 5.10 random pyramidal shapes are shown.[29] The process is copied from the solar cell industry where similar pyramidal etching is used to reduce reflectance. Regular, lithographically defined pyramids are shown in Figure 5.11.[21] The first etching is TMAH-based

Figure 5.10 Non-lithographically KOH etched ⟨100⟩ silicon. Reprinted with permission from: Y. Liu, Y. Xiu, D. W. Hess, C. P. Wong: Silicon surface structure-controlled oleophobicity, *Langmuir* 2010, **26**, 8908.[28] Copyright (2010) American Chemical Society.

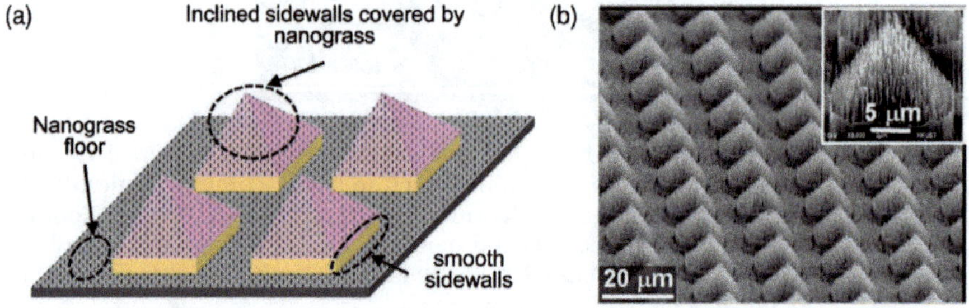

Figure 5.11 Schematic drawing and scanning electron microscopy (SEM) of nanograssed micropyramid arrays. (a) Schematic drawing showing different surface roughness on the smooth sidewalls of the pyramids, nanograssed sidewalls, and nanograssed floor. (b) SEM image showing the as-fabricated nanograssed micropyramid arrays on a Si wafer. From ref. 29 with permission from John Wiley and Sons. Copyright © 2011 Wiley-VCH Verlag GmbH & Co. KGaA, Weinheim.

anisotropic wet etching to create the pyramids. DRIE for black silicon formation is done next. This DRIE step also results in smooth sidewalls because nanograss formation takes some time and during that period silicon is consumed. Static contact angle (with perfluorooctyl trichlorosilane (PFOS) surface treatment) was 160°.

Undercut etching can be done by anisotropic wet etching, in KOH for example. As shown in Figure 5.12, undercut exhibits crystal plane anisotropy. After etching the oxide caps (by RIE), silicon stem (by DRIE), and undercut by KOH, the whole structure was coated by 100 nm thick DLC layer, as shown in Figure 5.12. This hydrophilic coating resulted in contact angles of

Figure 5.12 T-shaped oxide caps and silicon stem are all covered by DLC, resulting in contacts angle of *c.* 160° without further fluoropolymer coatings. Reprinted with permission from. Wang, F. Liu, H. Chen, D. Chen: SHB behavior achieved from hydrophilic surfaces, *Appl. Phys. Lett.* 2009, **95**, 084104.[30] Copyright 2009, AIP Publishing LLC.

c. 160° depending on details of structure sizes and spacings.[30] Similar undercutting could also be obtained by isotropic plasma etching, XeF_2 dry etching, or isotropic wet etching in HF–HNO_3.

5.3.1 Silicon Nanostructures by Metal-Assisted Wet Etching

Metal-assisted chemical etching (MaCE), also known as metal-catalysed etching or galvanic displacement reaction, is a simple way of producing nanowires and nanopores in silicon.[12,31] Noble metals catalyse silicon oxidation, and hydrofluoric etchant removes the formed oxide.

The silicon etching reaction in MaCE is not fully understood; see ref. 32 for a recent review. Dissolution in the tetravalent state directly as shown below is one possibility:

$$Si + 4H^+ + 4HF \rightarrow SiF_4 + 4H^+$$

$$SiF_4 + 2HF \rightarrow H_2SiF_6$$

It is also possible that direct dissolution of silicon takes place in the divalent state:

$$Si + 4HF_2^- \rightarrow SiF_2^- + 2HF + H_2(g) + 2e^-$$

A mixed divalent/tetravalent reaction for the dissolution of silicon in MaCE has the overall reaction:

$$Si + n/2H_2O_2 + 6HF \rightarrow nH_2O + H_2SiF_6 + (4 - n)/2H_2(g)$$

There are five main implementations of MaCE nanowire/nanopore formation:

- lithographic patterning of noble metal films
- nanobead/nanosphere/colloidal lithography
- metal nanoparticle deposition on silicon
- noble-metal-containing etchant
- thin metal film dewetting.

The first two result in regular or almost regular patterns, while the latter three result in random patterns.

Patterning of noble metal films include all the standard techniques: optical lithography or interference lithography are typical. Metal film thicknesses are in the tens of nanometres. Colloidal lithography (nanoparticle self-assembly/nanobead lithography) typically uses polystyrene nanoparticles (100–1000 nm) as a self-assembled mask pattern (Figure 5.13). Oxygen plasma etching can be used to decrease particle size, and to increase pillar spacing. The nanoparticles then act as a lift-off mask for noble metal deposition.

Metal nanoparticle deposition has been explored in various ways: direct noble metal nanoparticle deposition from solution, or island growth in

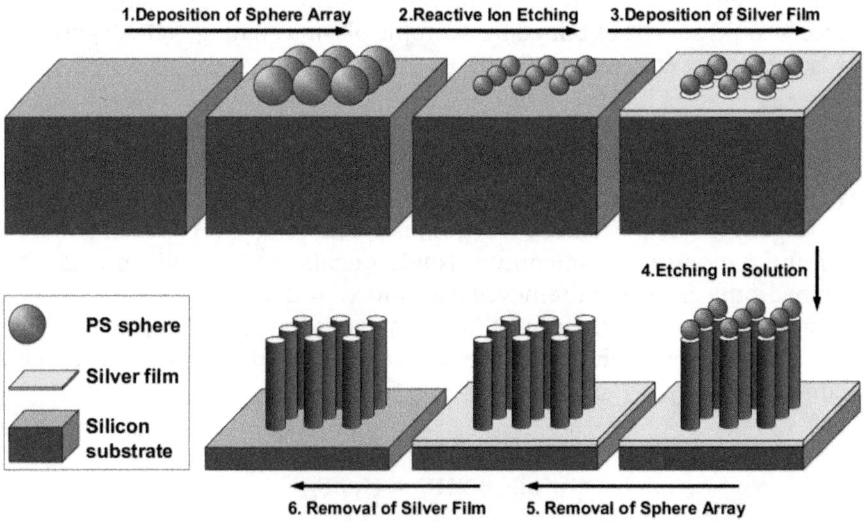

Figure 5.13 Fabrication of quasiregular silicon nanopillars by nanosphere lithography and MaCE: (1) polystyrene (PS) bead lithography; (2) isotropic RIE oxygen plasma narrowing; (3) silver evaporation; (4) MaCE; (5) acetone removal of polymer spheres; (6) aqua regia etching of silver. From ref. 31 with permission from John Wiley and Sons. Copyright © 2007 Wiley-VCH Verlag GmbH & Co. KGaA, Weinheim.

early-stage atomic layer deposition (ALD) of platinum. Silver is the most widely used noble metal but gold and platinum can also be used, with minor differences in etch rate and other process responses. $AgNO_3$–HF is the main noble-metal-containing etchant.

Very thin metal films (5–10 nm) dewet upon thermal annealing, forming islands in the size range of tens of nanometres. Metal minimizes its surface energy by ball formation if this is favourable compared with metal–substrate interfacial energy. These metal islands can then act as etch masks. Annealing of silver films leads to balling-up and formation of nanoislands which can act as seeds for metal-catalysed etching.[33]

At room temperature HF–H_2O_2–H_2O etches silicon at a rate of *c.* 0.3–0.5 µm min^{-1}. Depending on the lithographic method, high aspect ratios of 20 : 1 are possible. Silver and gold are removed in boiling aqua regia, 3 : 1 (v/v) HCl–HNO_3 for 15 min. Drying of the nanowire array is critical because capillary forces can easily lead to the collapse of nanowires. Collapsed nanowire bundles, however, have been shown to display omniphobicity without fluoropolymer coatings.[34] This results from overhang structures formed in the collapse.

The rate is reduced upon longer etching times and deeper etching is not usually an option. Most of the early papers on MaCE were about nanowire fabrication and reduced-dimensionality systems to be used as high surface area or quantum effect sensors in chemical and photonic systems.[34] The use of MaCE for micrometre patterning is rare.[35] With fluoropolymer coating, MaCE nanowire arrays display water contact angles of 150–160° depending on the solid fraction.

5.4 Combined Processes

DRIE combined with MaCE produces nanostructured micropillars (Figure 5.14). If the DRIE mask material is not removed before MaCE, the pillar tops will remain featureless, but the sidewalls will be nanostructured. A SEM micrograph of resulting nanoporous pillar is shown in Figure 5.15. Pore size is typically in the 10–100 nm range.

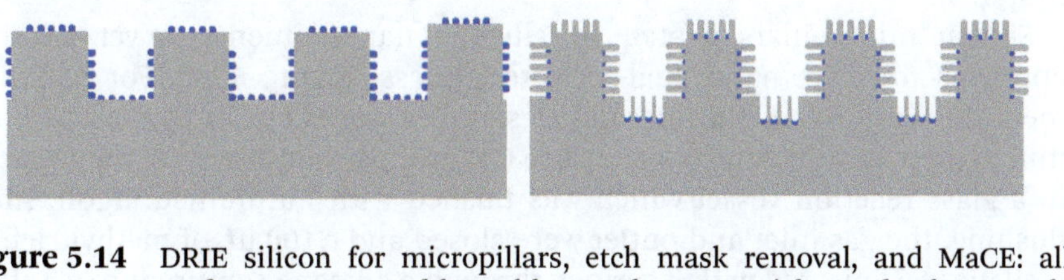

Figure 5.14 DRIE silicon for micropillars, etch mask removal, and MaCE: all surfaces are covered by noble metal nanoparticles and subsequently nanopores. Reprinted from *Journal of Colloid and Interface Science*, **364**, Y. He *et al.*, SHB silicon surfaces with micro-nano hierarchical structures *via* DRIE and galvanic etching, 219–229, Copyright 2011 with permission from Elsevier.[12]

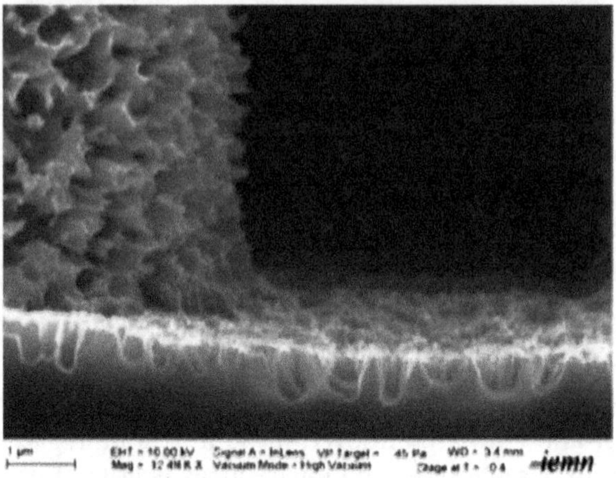

Figure 5.15 Bosch DRIE silicon micropillars and a detail of pillar foot after MaCE to create nanorough surface. Reprinted from *Journal of Colloid and Interface Science*, **416**, T. Phuong *et al.*, Micro- and nanostructured silicon-based superomniphobic surfaces, 280–288, Copyright 2014 with permission from Elsevier.[11]

A combination of DRIE and black silicon formation is a common route to dual-scale roughness. Two main variants exist: formation of micropillars and nanograss formation on pillar tops.[36] This approach leads to water contact angles of 160° without polymer coating. The other process variant produces nanoroughness on top of pillars and also on the bottom surface.[37] These two cases can both be replicated into polymers, as discussed below.

A combination of anisotropic DRIE and isotropic RIE has been used to create sharp microneedles,[38] as shown in Figure 5.16. After DRIE of cylindrical micropillars, an isotropic step sharpens and finally cuts the pillars. One problematic aspect of this process is that the mask oxide plus some silicon is cut off, and has to be etched away in subsequent steps. The process was further continued by deposition of ZnO nanowires on silicon microtips. Such surfaces displayed 160° contact angles and <5° contact angle hysteresis.

Silicon micropillars coated by silicone nanofilaments is yet another approach to micro/nano dual-scale roughness (Figure 5.17). For nanofilament deposition, silicon micropillar samples were kept in O_2 plasma for 5 min to increase the density of surface OH groups. Samples were then placed in a glass reaction vessel which was flushed with humidified argon. After flushing, the gas inlet and outlet were closed and *c.* 100 μL of methyltrichlorosilane was injected with a syringe through a silicone septum into a Teflon cup inside the vessel. After 14 h the samples were taken out and rinsed with deionized water. The resulting dual-scale structures showed a "nano-Cassie" state: even though the spaces between micropillars were filled with water, air remained in between the nanofilaments. This enabled reversal back to the "micro-Cassie" state.

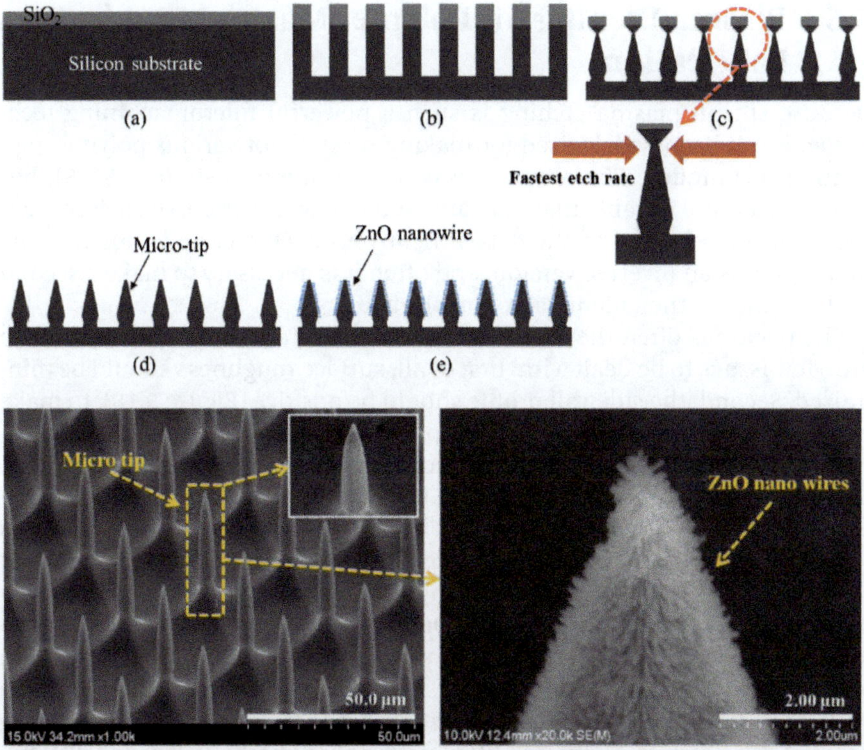

Figure 5.16 Fabrication process for sharp silicon pillars: (a) deposition of SiO_2 etch mask, (b) patterning of the SiO_2 mask and anisotropic etching of the silicon substrate using the DRIE process, (c) isotropic etching of the silicon pillar with the RIE process to form sharp micro-tip end, (d) removal of the etch mask, and (e) decoration with ZnO nanowires. SEM micrographs show silicon pillars before and after ZnO nanorod formation. Reprinted from *Current Applied Physics*, **14**, SHB properties of a hierarchical structure using a silicon micro-tip array decorated with ZnO nanowires, 665–671, Copyright 2014 with permission from Elsevier.[38]

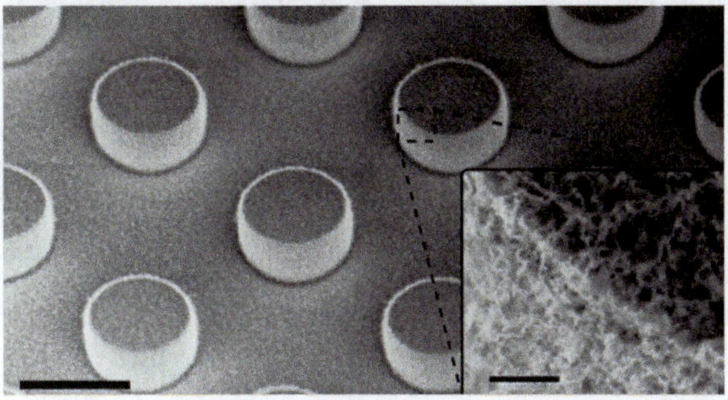

Figure 5.17 Silicon micropillars coated with silicone nanofilaments. Scale bar is 10 µm in the main figure and 500 nm in the inset. From ref. 15, Copyright the National Academy of Sciences.

5.5 Plasma Etching for Polymer Master Mould Fabrication

Because silicon plasma etching is such a powerful micromachining technique, it has been widely used for making masters for various polymer replication techniques. The alternative is to use photoresists (*e.g.* SU-8), but silicon is a more durable material, and allows a wide variety of surface coatings that are essential in the detachment phase. One issue is polarity: the first replica is an inverted version, and often it is necessary to make a second replica which is then identical to the original.

The release is often the most difficult step in a replication process. There are a few issues to be dealt with: first of all, surface roughness should be minimized. Second, the sidewall profile should be positive (Figure 5.18), to make release easy. Even though polymer replication cannot reproduce the sharpness of the original silicon, the water contact angles for the replicas are 160°.

Replication of overhang structures is difficult. There are reports of successful release[39] but in most cases this is limited to elastomers only. Softness (elasticity) maybe beneficial for mechanical robustness, but it might also lead to micro- and nanostructures attaching to themselves, especially when very soft and sticky elastomers are used. One solution to release overhangs is to sacrifice the whole substrate wafer, known as a dissolved wafer process in the MEMS community. An example of this is shown in Figure 5.19. Anisotropic DRIE and isotropic etch of silicon were combined to produce inverted mushrooms.[40] The etched cavity was filled with poly(dimethyl siloxane) (PDMS), and the substrate silicon wafer was completely removed in KOH (10% by weight, 60 °C).

Replication of nanoscale roughness from silicon masters is somewhat more demanding than microstructure replication. Filling of nanostructures with polymers is not straightforward, and for example standard Sylgard 184 PDMS does not fill nanoscale structures with high fidelity. Other polymers,

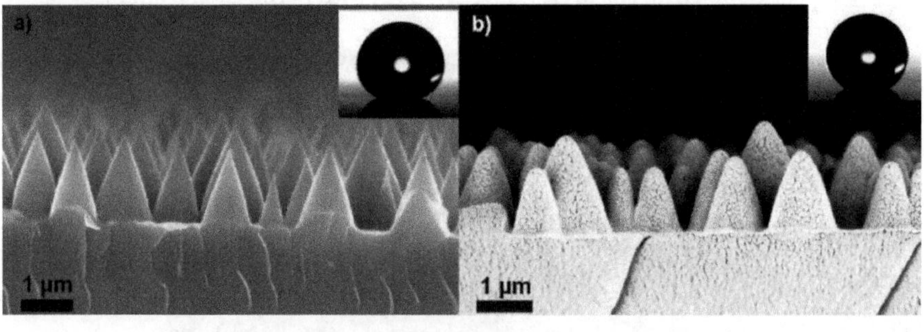

Figure 5.18 (a) Black silicon master with water contact angle (WCA) of 170° (b) replicated in hybrid polymer Ormocomp with WCA of 160° (intermediate mould not shown). Reprinted from ref. 25 with permission from John Wiley and Sons. Copyright © 2011 Wiley-VCH Verlag GmbH & Co. KGaA, Weinheim.

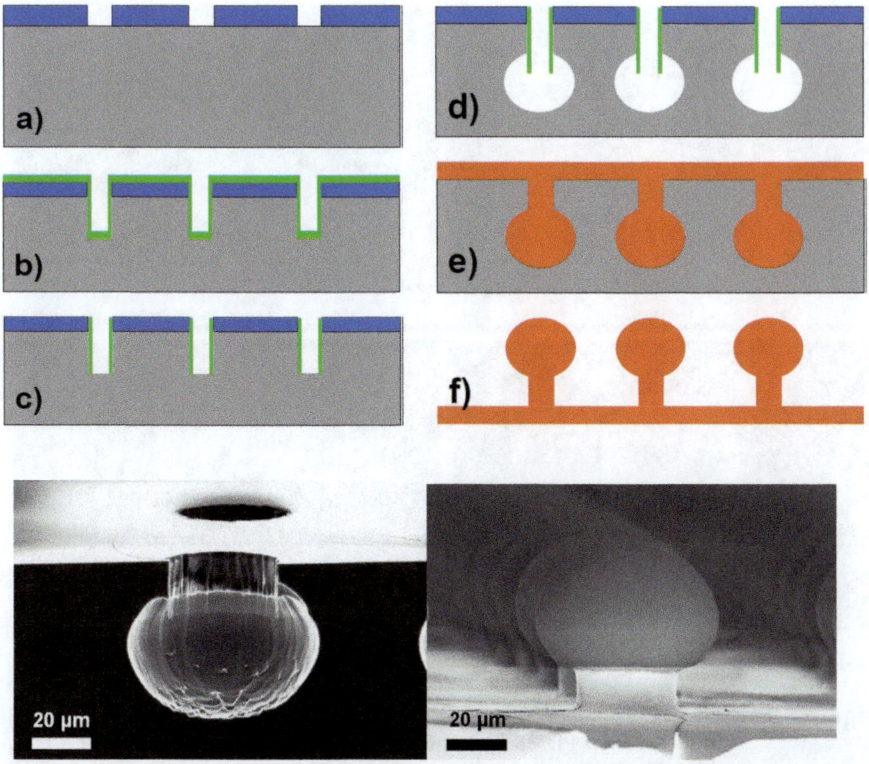

Figure 5.19 Fabrication of PDMS mushrooms by sacrificial wafer process: (a) hard mask; (b) silicon DRIE and spacer deposition; (c) spacer etch; (d) isotropic etch; (e) spacer removal and polymer casting; (f) silicon wafer sacrifice. In SEM micrographs the silicon master and the PDMS replica are shown. From ref. 40.

like hard PDMS (h-PDMS), are more suitable.[41] An example of SHB surface replication from a dual-scale roughness silicon master is shown in Figure 5.20. It displays a contact angle of 167°. An intermediate h-PDMS/PDMS composite master is used to emboss spincoated and cured fluoropolymer CYTOP™.

5.6 Glass Plasma Etching

Plasma etching of SiO_2 films, and of glass (SiO_2 mixed with metal oxides) and fused silica wafers (pure crystalline SiO_2) suffers from the need for considerable ion bombardment because of the strength of the Si–O bond. Etch selectivity of 5 : 1 translates to a 100 μm thick mask for through-etching of a 500 μm glass wafer. In the extreme case a silicon wafer, 500 μm thick, was etched through by DRIE, and then used as an etch mask for deep glass etching, to ensure that even though the mask is consumed during etching, it is not completely gone.[42]

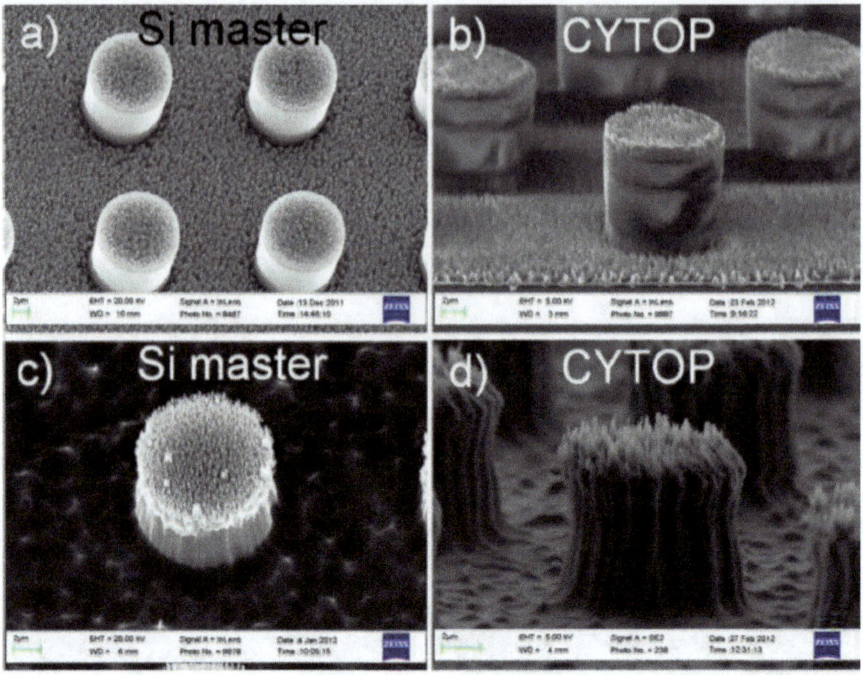

Figure 5.20 (a) Silicon master (DRIE micropillars, black silicon nanoroughness) replicated *via* PDMS intermediate mould into fluoropolymer CYTOP. (b) CYTOP replica. (c) Si master with non-rough bottom. (d) CYTOP replica of (c). Contact angle of the CYTOP is 167°. Reprinted from *Colloids and Surfaces A: Physicochemical and Engineering Aspects*, **434**, V. Jokinen *et al.*, Durable SHBity in embossed CYTOP fluoropolymer micro and nanostructures, 207–212, Copyright 2013 with permission from Elsevier.[41]

Mask erosion leads to profile sloping and it is very difficult to achieve vertical walls for deeper structures. Glasses contain metals in addition to SiO_2, which means that fluorine plasmas do not produce volatile products, and metal fluoride residues result in micromasking and rough surfaces.

Metals masked by dewetting have been used as etch masks for glass etching. As an example, a copper/glass combination can be used.[43] Sputtered copper, 4–6 nm thick, was annealed at 750 °C in rapid thermal annealing (RTA) for *c.*1–2 min in a nitrogen atmosphere. Subsequently the glass was plasma etched in CHF_3/Ar plasma, and the metal mask removed by acidic etching. Using glass as the material resulted in a favourable combination of SHB properties (water contact angle 156°, oil contact angle 116°) and low omnidirectional reflectivity because of subwavelength feature sizes. Durability tests showed that 5000 mechanical wipes did not adversely affect these properties. As a limitation, the mask is very thin, and pillar heights were limited to 600 nm.

The poor mask selectivity has been utilized in making shaped pillars.[44] By carefully controlling the erosion rate of the mask, a gently sloping oxide

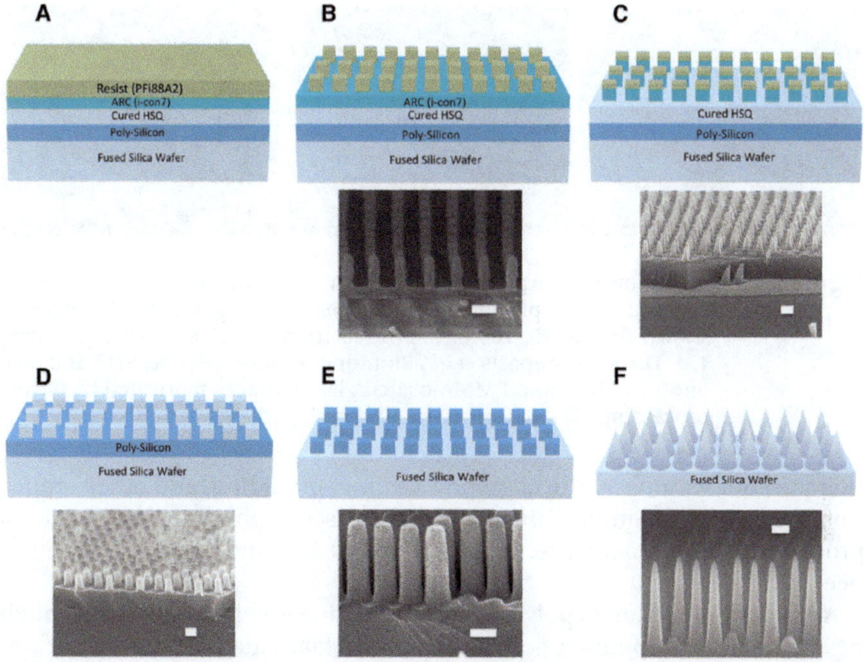

Figure 5.21 Stages of the silica SHB fabrication process. (A) Deposition of multiple coating layers. (B) Development of photoresist pattern. Subsequent etching steps of antireflective coating layer (C), cured HSQ layer (D), polysilicon layer (E), and fused silica wafer (F). All of the white scale bars on the micrographs represent 200 nm. Reproduced from K.-C. Park, H. J. Choi, R. E. Cohen, G. H. McKinley, G. Barbastathis: Nanotextured silica surfaces with robust SHBity and omnidirectional broadband supertransmissivity, *ACS Nano*, 2012, 6, 3789.[44] Copyright (2012) American Chemical Society.

profile has been created (Figure 5.21). This is beneficial for SHB properties because breakthrough pressure increases towards the bottom, and also because the structure acts as an index-matching anti-reflection coating. Contact angles of 160° and contact angle hysteresis <3° were demonstrated.

5.7 Polymer Plasma Etching

Plasma etching of polymers in DRIE-like processes is possible but etch rates are typically only *c.*10% of silicon etch rates.[7] Various etch chemistries, for instance O_2, O_2/Ar, O_2/CF_4, O_2/SF_6 and O_2/C_4F_8, have been used. The O_2/C_4F_8 process is an analogue of the Bosch silicon etching process: the oxygen step etches polymer and the C_4F_8 step passivates the sidewalls.[45]

Hard masks of metals and oxides are used. There are no known processes where the microscale roughness of a SHB surface has been produced by polymer plasma etching. The reader is referred to the polymer processing

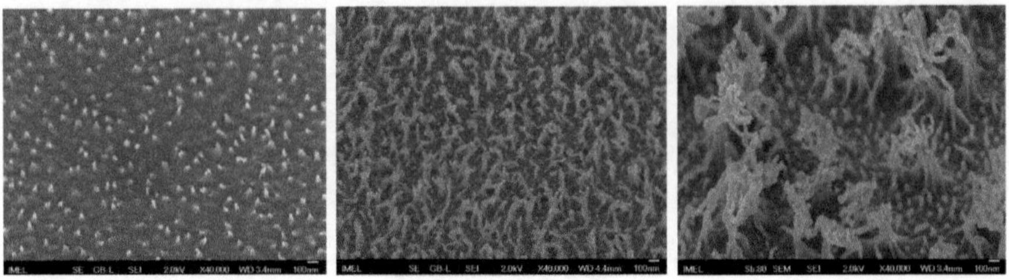

Figure 5.22 Evolution of morphology on PMMA substrates etched and nanotextured in oxygen plasma as a function of etching time SEM images are shown for 30, 90, 180 s. Reprinted from *Microelectronic Engineering*, **121**, D. Kontziampasis *et al.*, Biomimetic, antireflective, SHB and oleophobic PMMA and PMMA-coated glass surfaces fabricated by plasma processing, 33–38, Copyright 2014 with permission of Elsevier.[48]

literature for various soft lithography, casting, moulding, and other replication processes.[46,47] Simple lithographic processes in photopolymers can also produce the required microscale geometry, but low surface energy coating is needed.

A typical application of polymer plasma etching is to introduce nanoroughness as the last step into a polymer structure fabricated in some other way. With etch rates of *c.* 0.5–1 μm min^{-1} an etch step of a 1–2 min produces random nanograss with high aspect ratios.[48] Etching conditions for poly(methyl methacrylate) (PMMA) in Figure 5.22 were: power 1900 W, bias power 250 W (bias voltage −100 V), electrode temperature 15 °C, gas pressure 0.75 Pa, oxygen gas flow 100 sccm.

Polyurethane of 70D hardness was plasma etched in Ar/SF_6 (30 sccm/10 sccm), 50 mTorr, 150 W to create nanoroughness.[49] The treatment resulted in a water contact angle of 168°. In the same study Teflon (PTFE) was etched to achieve a contact angle of 161°.

5.8 Plasma Etcher as a Deposition Tool

Plasma etchers are almost identical to plasma enhanced chemical vapour deposition (PECVD) reactors. Whether etching or deposition takes place depends on the details of plasma chemistry, as already discussed in connection with Bosch-process passivation film. By suitable process parameter selection C_4F_8 and CHF_3 source gases can be used to deposit $(CF_2)_n$ film, *i.e.* Teflon-like material. These materials are usually characterized by contact angle measurements to be Teflon-like (flat surface contact angles ~120°), but most often bond identification, *e.g.* by X-ray photoelectron spectroscopy (XPS), is not carried out. Because the possibility of depositing these Teflon-like passivation layers is available to everybody with a plasma etcher, they are often used as low surface energy coatings for SHB surfaces.[50] When C_4F_8 is used, the reaction depends on process details: high energy (high RF power to the bottom/wafer electrode) leads to etching of SiO_2, while low (or zero) power to the bottom electrode leads to a deposition process (this is similar to the Bosch passivation cycle, which is run with zero or low power, *e.g.* 3 W). Depending on coil power, the deposition rate of polymer is in the range of hundreds of nanometres per minute. Refractive index at 632 nm was measured to be 1.389 ± 0.007 for a 50 mT/800 W deposition. This is somewhat above the values for Teflon (1.35–1.38).

Local fluoropolymer coatings can be obtained by both additive and subtractive methods. Global fluoropolymer deposition followed by lithography, etching, and resist stripping leads to SHB/SHL patterns with micrometre dimensional control.[27] Simple resist stripping in acetone can be used because it does not attack plasma-deposited fluoropolymer.

Alternatively, a shadow mask can be applied, and fluoropolymer deposited through a stencil. Because the contact between the shadow mask and the wafer is not intimate, some deposition takes place under the mask, leading to loss of lateral definition. This approach is best suited for applications that only require coarse patterns, in the hundreds of micrometres range.

Compared with fluorinated self-assembled monolayers (SAMs), the plasma deposition process is quick and easy and the plasma films are much thicker. Conformality cannot match surface-controlled deposition of SAMs, and because of plasma, there is no guarantee of stoichiometry. Thickness is advantageous with regards to abrasive wear; on the other hand, adhesion is not necessarily as good as with covalently bonded SAMs.

The C_4F_8-passivation cycle renders all surfaces with a Teflon-like fluoropolymer coating, and this is quick way of making a surface hydrophobic. However, if hydrophilic surfaces are desired, it is essential to remove all remains of this passivation polymer. It is not soluble in acetone, and can leave veils, thin hydrophobic sidewalls, that prevent liquid spreading (Figure 5.23).[51] Oxygen plasma ashing or ozone resist removal is needed. Contact angle difference between the samples with oxygen plasma *vs.* acetone photoresist removal was 50°.

5.9 Conclusions

Etching techniques are superior in the control of feature sizes, and that has made them important in making model surfaces for basic studies of SHB and oleophobicity. Equally important is the ability to make any shape whatsoever. If a resist pattern has been made, plasma etching is able to delineate that shape into the underlying material. The limitations come from the cost of lithographic patterning, especially at the nanoscale. Some of the silicon structures can be replicated into polymers, alleviating the cost issues of silicon microfabrication.

Etching of silicon is limited by the wafer sizes available. Most the reported research has been done on small wafers, usually 100 mm diameter, and often the usable area has been much smaller than that. The same size limitations apply to glass wafers. Polymer plasma etching approaches can be extended to larger sizes when random, non-lithographic, patterns are made. Compared

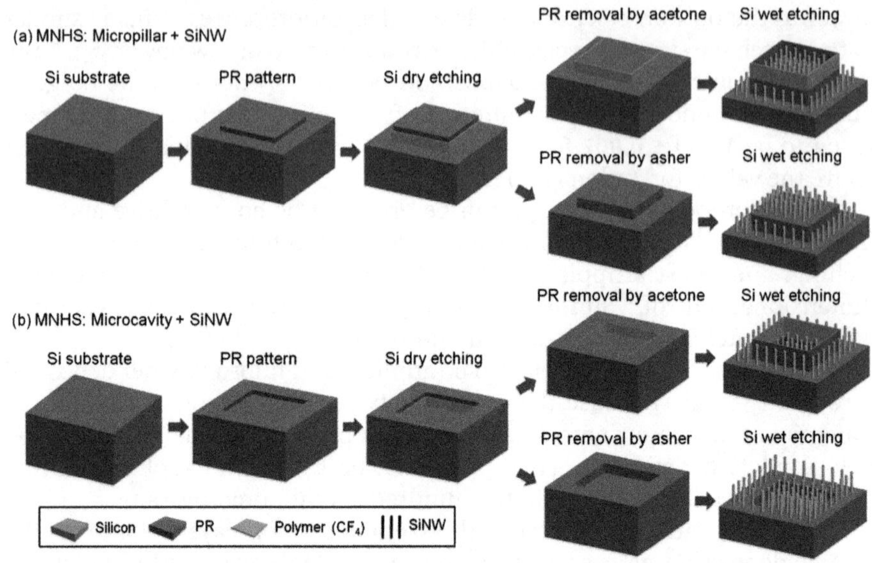

Figure 5.23 (a) Elevated and (b) recessed pillar arrays made by micro/nano hybrid structures (MNHS/silicon nanowires Si NW) formed by a combined plasma etching and MaCE. Photoresist (PR) removal by acetone leaves the sidewalls protected by Bosch passivation polymer while PR removal by oxygen plasma asher removes it. From ref. 51 with permission of Springer. Published under the CC-BY licence.

with nanoparticle deposition or fibre-based approaches, plasma etching of the most accurate geometries will remain a research tool. Wet etching of silicon is more easily extended in size, and batch fabrication of multiple wafers simultaneously helps to reduce costs.

Plasma etching (and MaCE) enable fabrication of SHB, and even oleophobic, surfaces without (fluoro)polymer coatings. This has important ramifications for the durability and robustness of such surfaces because thin coatings are prone to mechanical damage as well as chemical, thermal, and UV degradation.

References

1. S. Franssila, *Introduction to Microfabrication*, John Wiley & Sons, 2nd edn, 2010.
2. W. K. Choi, T. H. Liew, M. K. Dawood, H. I. Smith, C. V. Thompson and M. H. Hong, Synthesis of silicon nanowires and nanofin arrays using interference lithography and catalytic etching, *Nano Lett.*, 2008, **8**, 3799.
3. T. N. Krupenkin, J. A. Taylor, E. N. Wang, P. Kolodner, M. Hodes and T. R. Solomon, Reversible wetting–dewetting transitions on electrically tunable superhydrophobic nanostructured surfaces, *Langmuir*, 2007, **23**, 9128.

4. R. Hensel, A. Finn, R. Helbig, H.-G. Braun, C. Neinhuis, W.-J. Fischer and C. Werner, Biologically inspired omniphobic surfaces by reverse imprint lithography, *Adv. Mater.*, 2013, **26**, 2029.
5. S. Franssila and S. Tuomikoski, "MEMS lithography", in *"Handbook of silicon based MEMS materials and technologies"*, ed. M. Tilli, T. Motooka, V-M. Airaksinen, S. Franssila, M. Paulasto-Krockel and V. Lindroos, Elsevier, 2nd edn, 2015.
6. C. Lee and C. J. Kim, Maximizing the giant liquid slip on superhydrophobic microstructures by nanostructuring their sidewalls, *Langmuir*, 2009, **25**, 12812.
7. F. Laermer, S. Franssila, L. Sainiemi and K. Kolari: "DRIE", in *"Handbook of silicon based MEMS materials and technologies"*, ed. M. Tilli, T. Motooka, V-M. Airaksinen, S. Franssila, M. Paulasto-Krockel and V. Lindroos, Elsevier, 2nd edn, 2015.
8. S. Franssila and L. Sainiemi, "RIE", in *"Encyclopedia of Micro- and nanofluidics"*, ed. Dongqing Li, Springer, 2nd edn, 2008.
9. K. R. Williams, K. Gupta and M. Wasilik, Etch rates for micromachining processes, *J. Microelectromech. Syst.*, 2003, **12**, 761.
10. A. Wu, B. Kumar and S. Pamarthy, High aspect ratio silicon etch: a review, *J. Appl. Phys.*, 2010, **108**, 051101.
11. T. P. N. Nguyen, R. Boukherroub, V. Thomy and Y. Coffinier, Micro-and nanostructured silicon-based superomniphobic surfaces, *J. Colloid Interface Sci.*, 2014, **416**, 280.
12. Y. He, C. Jiang, H. Yin, J. Chen and W. Yuan, Superhydrophobic silicon surfaces with micro–nano hierarchical structures *via* deep reactive ion etching and galvanic etching, *J. Colloid Interface Sci.*, 2011, **364**, 219.
13. L. Sainiemi and S. Franssila, Mask material effects in cryogenic DRIE, *J. Vac. Sci. Technol.*, 2007, **B25**, 801.
14. L. Sainiemi, T. Nissilä, V. Jokinen, T. Sikanen, T. Kotiaho, R. Kostiainen, R. A. Ketola and S. Franssila, Fabrication and characterization of silicon micropillar array electrospray ionization chip, *Sens. Actuators, B*, 2008, **132**, 380.
15. T. Verho, L. Sainiemi, V. P. Jokinen, J. T. Korhonen, C. Bower, K. Franze, S. Franssila, P. Andrew, O. Ikkala and R. H. A. Ras, Reversible Switching between Superhydrophobic States on a Hierarchically Structured Surface, *Proc. Natl. Acad. Sci. U. S. A.*, 2012, **109**, 10210.
16. K. Grigoras, S. Franssila and V.-M. Airaksinen, Investigation of sub-nm ALD aluminum oxide films by plasma assisted etch-through, *Thin Solid Films*, 2008, **516**, 5552.
17. A. Susarrey-Arce, A. G. Marin, S. Schlautmann, L. Lefferts, J. G. E. Gardeniers and A. van Houselt, One-step sculpting of silicon microstructures from pillars to needles for water and oil repelling surfaces, *J. Micromech. Microeng.*, 2013, **23**, 025004.
18. A. Tuteja, W. Choi, M. Ma, J. M. Mabry, S. A. Mazzella, G. C. Rutledge, G. H. McKinley and R. E. Cohen, Designing superoleophobic surfaces, *Science*, 2007, **318**, 1618.

19. A. Ahuja, J. A. Taylor, V. Lifton, A. A. Sidorenko, T. R. Salamon, E. J. Lobaton, P. Kolodner and T. N. Krupenkin, Nanonails: a simple geometric approach to electrically tunable superlyophobic surfaces, *Langmuir*, 2008, **24**, 9.
20. R. Hensel, R. Helbig, S. Aland, H. G. Braun, A. Voigt, C. Neinhaus and C. Werner, Wetting resistance at its topographical limit: the benefit of mushroom and serif-T structures, *Langmuir*, 2013, **29**, 1100.
21. T. Liu and C. J. Kim, Turning a surface superrepellent even to completely wetting liquids, *Science*, 2014, **346**, 1096.
22. J. Li, X. Tian, A. Pyymaki Perros, S. Franssila and V. Jokinen, Self-propelling and positioning of droplets using continuous topography gradient surface, *Adv. Mater. Interfaces*, 2014, **1**, 1400001.
23. H. V. Jansen, M. J. de Boer, S. Unnikrishnan, M. C. Louwerse and M. C. Elwenspoek, Black silicon method X: a review on high speed and selective plasma etching of silicon with profile control: an in-depth comparison between Bosch and cryostat DRIE processes as a roadmap to next generation equipment, *J. Micromech. Microeng.*, 2009, **19**, 033001.
24. V. Jokinen, L. Sainiemi and S. Franssila, Complex droplets on chemically modified silicon nanograss, *Adv. Mater.*, 2008, **20**, 3453.
25. L. Sainiemi, V. Jokinen, A. Shah, M. Shpak, S. Aura, P. Suvanto and S. Franssila, Non-reflecting silicon and polymer surfaces by plasma etching and replication, *Adv. Mater.*, 2011, **23**, 122.
26. V. Kondrashov and J. Rühe, Microcones and nanograss: toward mechanically robust superhydrophobic surfaces, *Langmuir*, 2014, **30**, 4342.
27. I. Zubel, E. Viinikka and M. Gosalves, Wet etching of silicon, in *"Handbook of silicon based MEMS materials and technologies"*, ed. M. Tilli, T. Motooka, V-M. Airaksinen, S. Franssila, M. Paulasto-Krockel and V. Lindroos, Elsevier, 2nd edn, 2015.
28. Y. Liu, Y. Xiu, D. W. Hess and C. P. Wong, Silicon surface structure-controlled oleophobicity, *Langmuir*, 2010, **26**, 8908.
29. X. Chen, J. Wu, R. Ma, M. Hua, N. Koratkar, S. Yao and Z. Wang, Nanograssed micropyramidal architectures for continuos dropwise condensation, *Adv. Funct. Mater.*, 2011, **21**, 4617.
30. J. Wang, F. Liu, H. Chen and D. Chen, Superhydrophobic behavior achieved from hydrophilic surfaces, *Appl. Phys. Lett.*, 2009, **95**, 084104.
31. Z. Huang, H. Fang and J. Zhu, Fabrication of silicon nanowire arrays with controlled diameter, length, and density, *Adv. Mater.*, 2007, **19**, 744.
32. Z. Huang, N. Geyer, P. Werner, J. de Boor and U. Gösele, Metal-Assisted Chemical Etching of Silicon: A Review, *Adv. Mater.*, 2011, **23**, 285.
33. B. Dou, R. Jia, H. Li, C. Chen, Y. Meng, W. Ding, X. Liu, T. Ye and Y. Wang, Maskless fabrication of selectively sized silicon nanostructures for solar cell application, *J. Vac. Sci. Technol.*, 2012, **B 30**, 041401.
34. S. Hoshian, V. Jokinen, V. Somerkivi, A. Lokanathan and S. Franssila, Robust superhydrophobic silicon without a low surface-energy hydrophobic coating, *ACS Appl. Mater. Interfaces*, 2015, **7**, 941.

35. M. Zahedinejad, S. D. Farimani, M. Khaje, H. Mehrara, A. Erfanian and F. Zeinali, Deep and vertical silicon bulk micromachining using metal assisted chemical etching, *J. Micromech. Microeng.*, 2013, **23**, 055015.
36. C. K. Kang, S. M. Lee, I. D. Jung, P. G. Jung, S. J. Hwang and J. S. Ko, The fabrication of patternable silicon nanotips using deep reactive ion etching, *J. Micromech. Microeng.*, 2008, **18**, 075007.
37. G. Sun, T. Gao, X. Zhao and H. Zhang, Fabrication of micro/nano dual-scale structures by improved deep reactive ion etching, *J. Micromech. Microeng.*, 2010, **20**, 075028.
38. Y.-M. Shin, S.-K. Lee, S. Jang and J.-H. Park, Superhydrophobic properties of a hierarchical structure using a silicon micro-tip array decorated with ZnO nanowires, *Curr. Appl. Phys.*, 2014, **14**, 665.
39. C. N. LaFratta, T. Baldacchini, R. A. Farrer, J. T. Fourkas, M. C. Teich, B. E. A. Saleh and M. J. Naughton, Replication of two-photon-polymerized structures with extremely high aspect ratios and large overhangs, *J. Phys. Chem.*, 2004, **108**, 11356.
40. L. Sainiemi, V. Jokinen, S. Aura and S. Franssila, Transparent, flexible and superhydrophobic material based on three-dimensional mushroom-shaped PDMS microstructures, *35th International Conference on Micro & Nano Engineering (MNE2009)*, Ghent, Belgium, 2009.
41. V. Jokinen, P. Suvanto, A. R. Garapaty, J. Lyytinen, J. Koskinen and S. Franssila, Durable superhydrophobicity in embossed CYTOP fluoropolymer micro and nanostructures, *Colloids Surf., A*, 2013, **434**, 207.
42. K. Kolari, V. Saarela and S. Franssila, Deep plasma etching of glass for fluidic devices with different mask materials, *J. Micromech. Microeng.*, 2008, **18**, 064010.
43. D. Infante, K. W. Koch, P. Mazumder, L. Tian, A. Carrilero, D. Tulli, D. Baker and V. Pruneri, Durable, superhydrophobic, antireflection, and low haze glass surfaces using scalable metal dewetting nanostructuring, *Nano Res.*, 2013, **6**, 429.
44. K.-C. Park, H. J. Choi, R. E. Cohen, G. H. McKinley and G. Barbastathis, Nanotextured silica surfaces with robust superhydrophobicity and omnidirectional broadband supertransmissivity, *ACS Nano*, 2012, **6**, 3789.
45. E. Meng, P.-Y. Li and Y. C. Tai, Plasma removal of Parylene C, *J. Micromech. Microeng.*, 2008, **18**, 045004.
46. H. Becker and C. Gärtner, Polymer microfabrication technologies for microfluidic systems, *Anal. Bioanal. Chem.*, 2008, **390**, 89.
47. D. B. Weibel, W. R. DiLuzio and G. M. Whitesides, Microfabrication meets microbiology, *Nat. Rev. Microbiol.*, 2007, **5**, 209.
48. D. Kontziampasis, G. Boulousis, A. Smyrnakis, K. Ellinas, A. Tserepi and E. Gogolides, Biomimetic, antireflective, superhydrophobic and oleophobic PMMA and PMMA-coated glass surfaces fabricated by plasma processing, *Microelectron. Eng.*, 2014, **121**, 33.

49. Y. Xiu, Y. Liu, D. W. Hess and C. P. Wong, Mechanically robust superhydrophobicity on hierarchically structured Si surfaces, *Nanotechnol.*, 2010, **21**, 155705.
50. K. Kolari and A. Hokkanen, Tunable hydrophilicity on a hydrophobic fluorocarbon polymer coating on silicon, *J. Vac. Sci. Technol.*, 2006, **A 24**, 1005.
51. B. S. Kim, S. Shin, S. J. Shin, K. M. Kim and H. H. Cho, Micro-nano hybrid structures with manipulated wettability using a two-step silicon etching on a large area, *Nanoscale Res. Lett.*, 2011, **6**, 333.

CHAPTER 6

Design Principles for Robust Superoleophobicity and Superhydrophobicity

KOCK-YEE LAW*[a] AND HONG ZHAO[b]

[a]Research and Innovative Solutions, Penfield, NY 14526, USA; [b]Department of Mechanical and Nuclear Engineering, Virginia Commonwealth University, Richmond, VA 23284, USA
*E-mail: lawkockyee@gmail.com

6.1 Introduction

The lotus effect describes the super-repellency of water droplets on lotus leaves, characterized by the extremely large water contact angle (WCA) of >150°. Water droplets bead up on the leaf surface and roll off at very small tilt angles.[1,2] Dust and dirt adhere to the water droplet as it rolls off, resulting in the famous self-cleaning effect. This phenomenon turns out to be very general in nature: many plants,[3–8] waterfowl,[7] and insects[8–10] also display similar properties. The lotus leaf has a multi-scale surface structure comprising micrometre-size clusters of wax tubules randomly distributed on the ~10–20 μm papillae, with the entire leaf surface carpeted with a layer of waxy nanohairs.[1,6,11]

Owing to the fascinating self-cleaning effect, superhydrophobic surfaces have been exploited for many potential applications, such as self-cleaning windows and textiles; oil- and soil-resistant clothing; anti-smudge surfaces

RSC Soft Matter No. 5
Non-wettable Surfaces: Theory, Preparation, and Applications
Edited by Robin H. A. Ras and Abraham Marmur

Published by the Royal Society of Chemistry, www.rsc.org

for smart phones and displays; anti-icing coatings for power lines, roofs, and aircraft; corrosion-resistant coatings for bridges and other metal structures; drag reduction in ships; gas and fuel transportation; and microfluidic devices. Numerous artificial superhydrophobic surfaces/coatings have been reported to date and the subject has been reviewed frequently in the recent literature (within the last 3 years).[12–17]

In contrast, surfaces that repel hydrocarbon oils (oleophobic) are less well known. Hydrocarbon oils are low surface tension, non-polar liquids and they wet most surfaces. Having a surface that can repel hydrocarbon oil well is extremely rare. In this chapter, we define a surface as superoleophobic when it exhibits a hexadecane contact angle of >150° along with a small sliding angle (~10°). Admittedly, contact angle 150° is arbitrary and appears to be just a convenient consensus by the scientific community.[18] Many terminologies have been used in the literature to describe super-repellency with non-aqueous solvents, such as superoleophobicity, superomniphobicity, superamphiphobicity, superlyophobicity. To justify the use of these different terminologies, a variety of solvents, including hexadecane (also dodecane, decane, and octane), alcohols (methanol, ethylene glycol, glycerol), paraffin oil, rapeseed oil, methylene iodide and so on, have been used as probing liquids. Since it is well known that wettability increases as the surface tension of the wetting liquid decreases, we feel that defining surfaces with hexadecane contact angle >150° and sliding angle ~10° superoleophobic is reasonable. The surface tensions of most other non-aqueous solvents are higher than that of hexadecane. Surfaces that are superoleophobic are expected to be super-repelling against other non-aqueous solvents too.

Since the first report fabrication of superoleophobic surfaces prepared by electrospinning and photolithography,[19] research in this field has been very active. Several reviews have already appeared.[20–24] Despite the intense effort, adoption of self-cleaning technology has lagged. Part of the cause can be attributed to the insufficient identification of functional requirements required for a self-cleaning surface. Most researchers have focused on three basic design parameters: large contact angle, small sliding angle, and low hysteresis. These parameters are good for leaves and feathers, which in nature only have to deal with water. Any mechanical damage to the leaf/feather can be regrown or replaced. Synthetic superhydrophobic surfaces, on the other hand, will unavoidably be exposed to volatile organics in urban and industrial environments. They will also be subjected to handling during manufacturing and normal use, where some degree of abrasion may occur.

Most superhydrophobic surfaces are superoleophilic. For instance, droplets of hexadecane are shown to wet and spread both on lotus leaves and on feathers.[25,26] Hence, the designed self-cleaning function will be degraded once the surface is contaminated by organic pollutants. Similarly, damage to superhydrophobicity has also been reported as a result of mechanical abrasion.[27–29] While working in industry, we recognized these shortfalls at the very beginning. We[30] were given the opportunity to examine the competitive advantage that a superoleophobic surface may provide to printing in general.

In addition to the usual desirable features of large hexadecane contact angle and small sliding angle, we always had our eyes on mechanical robustness, wetting stability under pressure, and large-area manufacturing issues.

In this chapter, the progression of the activity is summarized. Using a model system, we first study the basic design parameters that lead to superoleophobicity. This is followed by a systematic investigation of the design space and the trade-off between hysteresis and wettability. The robustness of superoleophobic surfaces is examined from the viewpoints of mechanical abrasion and wetting stability under external pressure. The principle and the latitude space for designing a superoleophobic surface, *i.e.* robustness to pressure (wetting and mechanical) is summarized. The chapter concludes with a discussion of issues and challenges related to large-area, large-scale manufacturing of superoleophobic and superhydrophobic surfaces. Underwater superoleophobicity is not within the scope of this chapter.

6.2 Study of a Model Superoleophobic Surface

6.2.1 Fabrication and Characterization of a Model Textured Surface

By mimicking the surface structure found in nature, many artificial superhydrophobic surfaces have been made with a combination of micro/nanoscale roughness and a hydrophobic coating. However, an oleophobic coating with hexadecane contact angle >90° is not known. Realizing this challenge, we launched a systematic study to elucidate the basic parameters for superoleophobicity. Figure 6.1 summarizes the surface texturing and chemical modification procedure for the fabrication of the model surface used in this investigation. The pillar array texture was made by the conventional photolithographic technique and the resulting textured surface was chemically modified with a fluorinated self-assembled monolayer obtained by molecular vapour deposition of tridecafluoro-1,1,2,2-tetrahydrooctyltrichlorosilane (FOTS) onto the textured surface. Details of the processes have been published elsewhere.[30]

Figure 6.2(a) shows a SEM micrograph of the FOTS pillar array surface, comprised of a ~3 μm diameter pillar array ~7.8 μm in height with a pitch of ~6 μm.

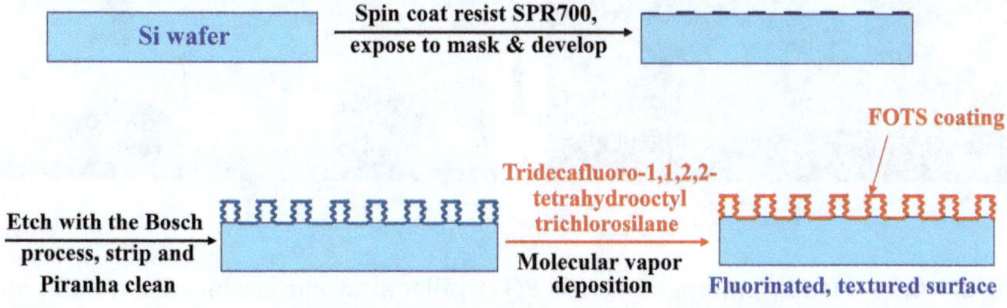

Figure 6.1 Fabrication of a fluorinated pillar array surface on silicon wafer. Adapted with permission from H. Zhao, K. Y. Law, and V. Sambhy, *Langmuir* 2011, 27, 5927–5935. Copyright (2011) American Chemical Society.[30]

The high-magnification SEM micrograph (insert) reveals that the sidewall in each pillar is not smooth; rather, it consists of a ~300 nm wavy structure from top to bottom, attributable to the Bosch etching process. The surface property of the FOTS pillar array surface was studied by static, advancing, and receding contact angle measurements. The static contact angle data with water and hexadecane (~5 μL droplets) are given in Figure 6.2(b). The water and hexadecane contact angles for the FOTS pillar array surface are at 156° and 158°, respectively, and are significantly higher than those of the controlled smooth surfaces, which are 107° and 73°, respectively. The results suggest that the large contact angles observed for the FOTS pillar array surface are the result of both surface texturing and fluorination. The sliding angles (~10 μL droplets) for the FOTS pillar array surface are found to be ~10° with both water and hexadecane. The high contact angles coupled with the low sliding angles lead us to conclude that the FOTS pillar array surface is both superhydrophobic and superoleophobic with low contact angle hysteresis.

6.2.2 Basic Design Parameters for Superoleophobicity

The use of lithographic and e-beam procedures and hydrophobic coatings to create micro/nano textured surfaces is not new and has been reported in the literature.[31–33] In all of these studies, the sidewalls of all the pillar structures are smooth, and only superhydrophobicity was reported, not superoleophobicity. Recently, Tuteja and coworkers[19] reported the fabrication of electrospun mats that exhibited superoleophobicity. The mat consists of nanofibres made from F-POSS (1*H*,1*H*,2*H*,2*H*-heptadecafluorodecyl polyhedral oligomeric silsesquioxane) and PMMA (poly(methyl methacrylate)) blends. The

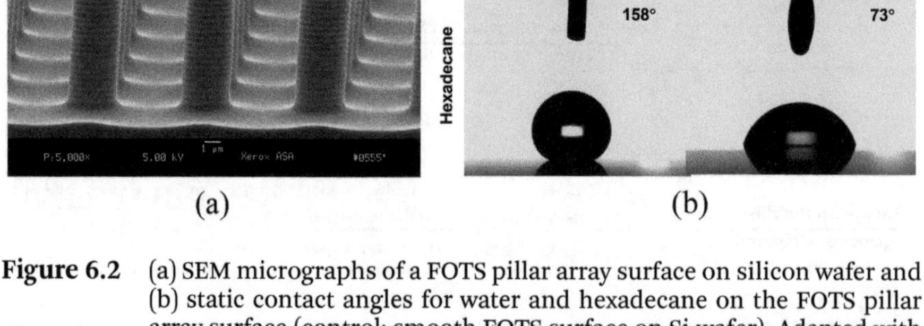

Figure 6.2 (a) SEM micrographs of a FOTS pillar array surface on silicon wafer and (b) static contact angles for water and hexadecane on the FOTS pillar array surface (control: smooth FOTS surface on Si-wafer). Adapted with permission from H. Zhao, K. Y. Law, and V. Sambhy, *Langmuir* 2011, 27, 5927–5935. Copyright (2011) American Chemical Society.[30]

flat surface of the same material is oleophilic with a hexadecane advancing contact angle of ~80°. To elucidate the mechanism for the superoleophobicity of the electrospun mat, these authors created the so-called "micro-hoodoo" structure on a silicon wafer using photolithography and surface fluorination. These authors concluded that the re-entrant geometry in the micro-hoodoo structure is critical to achieving superoleophobicity and the electrospun mat has a similar geometry at the liquid–solid interface. The conclusion agrees with the theoretical prediction by Marmur[34] and is substantiated by additional experimental and modelling studies.[25,26,35]

As mentioned earlier, the sidewall in each pillar consists of a ~300 nm wavy structure (Figure 6.2(a)). In view of the reports by Tuteja and coworkers, we suggest that the superoleophobicity observed in the model FOTS pillar array surface may be due either to the re-entrant structure at the top of the wavy sidewall, or to the entire wavy structure. To differentiate these possibilities, FOTS pillar array surfaces with (a) a smooth straight sidewall and (b) a straight sidewall with an overhang structure were fabricated. Figure 6.3(a and b) show SEM micrographs of these FOTS pillar array surfaces and the contact angle data of water and hexadecane on these two surfaces.

Both pillar array surfaces are superhydrophobic, with WCA and sliding angle ~152° and ~12°, respectively. Interesting and different results are obtained with hexadecane. On the FOTS pillar array surface with a straight sidewall, the hexadecane contact angle is about 120° and the droplet does not slide even at a 90° tilting angle (Figure 6.3(a)). The result indicates that the hexadecane droplet is pinned on the FOTS pillar array surface. On the

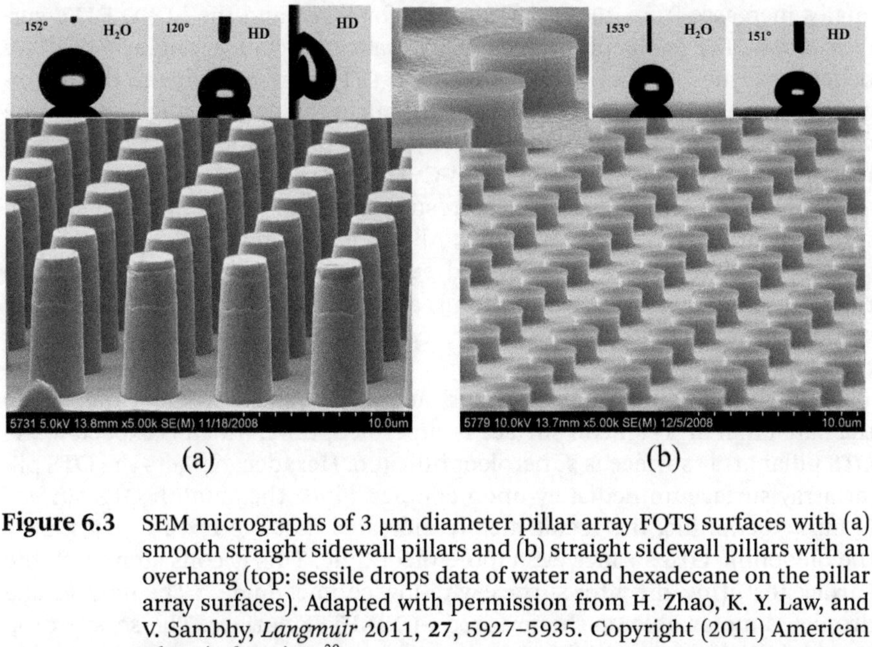

Figure 6.3 SEM micrographs of 3 μm diameter pillar array FOTS surfaces with (a) smooth straight sidewall pillars and (b) straight sidewall pillars with an overhang (top: sessile drops data of water and hexadecane on the pillar array surfaces). Adapted with permission from H. Zhao, K. Y. Law, and V. Sambhy, *Langmuir* 2011, **27**, 5927–5935. Copyright (2011) American Chemical Society.[30]

other hand, for the FOTS pillar array surface with an overhang structure (Figure 6.3(b)), a hexadecane contact angle of 151° and sliding angle of ~10° are observed. The similarity in surface properties between the model surface with the wavy sidewall and the pillar surface with the overhang leads to the conclusion that the re-entrant structure at the top of the wavy sidewall is a key contributor to the superoleophobic property exhibited by the model FOTS pillar array surface. This conclusion is not only in agreement with those reported by Tuteja and coworkers,[19,25,26,35] but also consistent with other observations reported by Fujii *et al.*,[36] Cao *et al.*,[37] Ahuja *et al.*[38] and Kumar and coworkers,[39] who all pointed to the importance of the re-entrant or overhang structure in achieving surface superoleophobicity.

To investigate the importance of surface chemistry for superoleophobicity, we prepared several pillar array surfaces with the texture as shown in Figure 6.2, but applied a different surface coating to the textured surface. The results are summarized in Table 6.1. Compared to the control (bare silicon wafer), all three surface coatings (on silicon wafer) are hydrophobic with water contact angles ranging from 107° to 139°. While the WCA for the OTS coating (a self-assembled monolayer from octadecyltrichlorosilane) is 109° and is in agreement with the literature value,[40] the WCA for the i-CVD (initiated chemical vapour deposition) poly(tetrafluorotheylene) (PTFE) coating on the silicon wafer is significantly higher than that of a smooth PTFE film, which is typically ~112–117°,[41,42] and is attributable to the fibrous texture of the surface due to the CVD deposition process.[43] As for the oleophobicity, there is a significant difference among the four surfaces. The bare silicon is superoleophilic. The hexadecane drop wets and spreads on the surface. The hexadecane contact angles increases from 40° for OTS to 73° for FOTS and the i-CVD PTFE surface. Since PTFE is oleophilic and has a hexadecane contact angle of 48°,[44] we believe that the 73° observed for the i-CVD PTFE surface is due to the nanoroughness on the surface which tends to inflate the contact angle value. The data in Table 6.1 therefore indicate that the oleophobicity of the four surfaces increases from bare silicon to OTS to i-CVD PTFE to FOTS.

For the pillar array surfaces, the results in Table 6.1 show that the bare textured surface is superhydrophilic. All three hydrophobic coatings impart superhydrophobicity to the textured surfaces. Their water contact angles range from 152° to 157° with sliding angles varying from 8° to 12°. The results are consistent with other textured or pillar array surfaces reported in the literature.[31–33]

On the other hand, different results are obtained with hexadecane. While the bare pillar array silicon surface is superoleophilic, which is expected, the OTS pillar array surface is superoleophilic too. Hexadecane wets the OTS pillar array surface immediately upon contact. Since the smooth OTS surface is fairly oleophilic, the result merely indicates that the texture just makes the oleophilic OTS surface even more oleophilic. This is consistent with the surface texturing effect for surface with low contact angle.[45] The hexadecane droplet is metastable on the textured i-CVD PTFE surface. The sessile drop has an initial contact angle of ~150°. Upon slight vibration on the bench, or after several minutes sitting, the hexadecane drop transitions to a stable sessile drop with a hexadecane contact angle of ~116° on the goniometer. Although the oleophobicity of this textured surface is quite high, the hexadecane drop shows extremely high adhesion towards the i-CVD PTFE pillar array surface. It does not slide and is actually stuck on the textured surface.

Only the FOTS pillar array surface is superoleophobic, with a hexadecane contact angle of 158° and a sliding angle of 10°. Since FOTS is the most oleophobic of the four surfaces tested in Table 6.1, the result re-confirms that one of the drivers for superoleophobicity is oleophobicity of the surface coating. The overall results indicate that the basic parameters for superoleophobicity are: chemistry (hexadecane contact angle of the coating), roughness, and re-entrant geometry, which are graphically depicted in Figure 6.4.

Table 6.1 Effect of surface chemistry on the surface properties of smooth and textured surfaces.[h]

		Water		Hexadecane	
Surface	Coating	CA[b]	SA[c]	CA[b]	SA[c]
Si-wafer	None	12.3°	*d*	2.6°	*e*
Textured[a] Si-wafer	None	Super wetting	—	Super wetting	—
Smooth OTS	OTS	109°	13°	40°	8°
Textured[a] OTS	OTS	157°	8°	Super wetting	0°
Smooth i-CVD PTFE	i-CVD PTFE	139°	*f*	73°	*f*
Textured[a] i-CVD PTFE	i-CVD PTFE	152°	10°	150° → 116°[g]	*f*
Smooth FOTS	FOTS	107°	14°	73°	9°
Textured[a] FOTS	FOTS	156°	10°	158°	10°

[a]Pillar diameter ~2.7 μm, height ~7 μm, pitch 6 μm, with a wavy sidewall.
[b]Contact angle, measurement error ~2°.
[c]Sliding angle, measurement error ~2°.
[d]Drop starts to flow at 13°.
[e]Drop starts to flow at 5°.
[f]Drop does not slide and is pinned on the surface even at 90° tilting.
[g]Drop is unstable and transitions to 116° during the measurement.
[h]Adapted from table in ref. 30, Copyright 2011 The American Chemical Society.

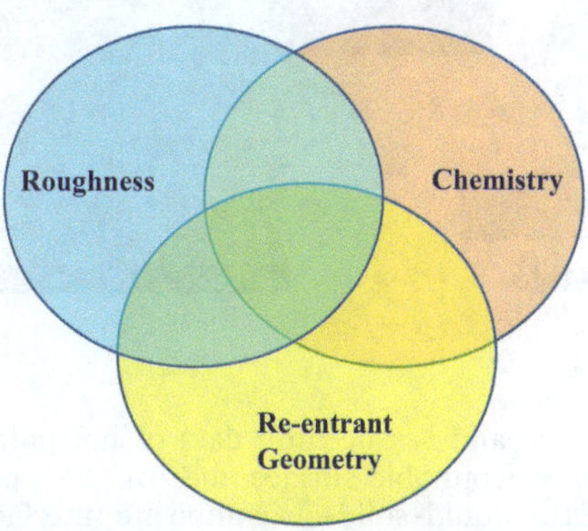

Figure 6.4 Graphic representation of basic design parameters for superoleophobicity.

6.2.3 Composite Liquid–Solid–Air Interface and Pinning Location

Figure 6.5 depicts the two states commonly used to describe the composite liquid–solid–air interface between liquid droplets and rough surfaces.[45,46] In the Wenzel state, the liquid fully wets the rough surface, whereas in the Cassie–Baxter state air pockets are trapped at the liquid–solid interface.

Figure 6.6(a) shows a sessile droplet of hot polyethylene wax on the model pillar array surface at 105 °C. The contact angle and the sliding angle of the droplet were measured as 155° and 33°, respectively, consistent with the hexadecane data. When the wax drop was cooled to room temperature, it was carefully detached from the textured surface and the geometry of the composite interface was examined by SEM microscopy. Figure 6.6(b) shows the SEM micrograph of the liquid–solid–air composite interface. The result indicates that the wax surface is "flat" with holes corresponding to the location of the pillars. From the thickness of the "rim", one can estimate the penetration

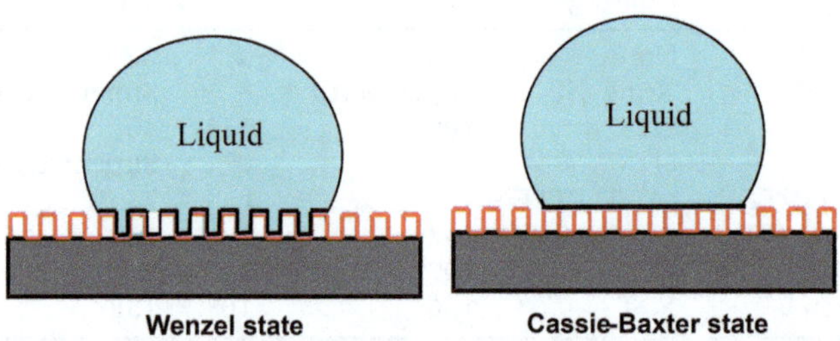

Figure 6.5 Phenomenal states of liquid droplets on rough surfaces.

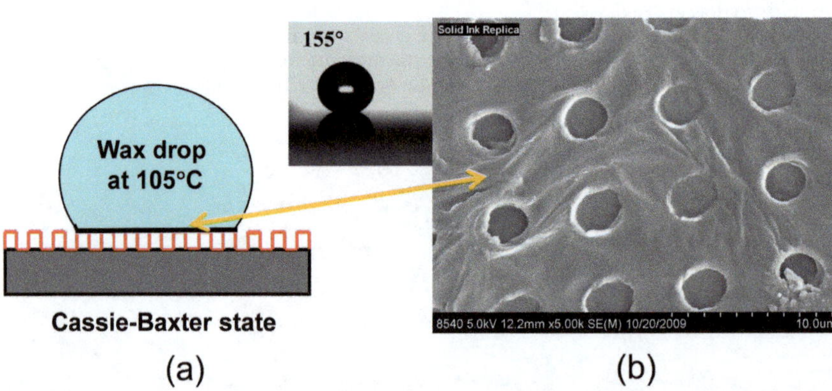

Figure 6.6 (a) Schematic and sessile drop data of hot polyethylene wax on the model superoleophobic surface and (b) SEM micrograph of the wax replica of the liquid–solid–air composite interface. Adapted with permission from H. Zhao, K. Y. Law, and V. Sambhy, *Langmuir* 2011, **27**, 5927–5935. Copyright (2011) American Chemical Society.[30]

depth of the molten wax droplet into the void space between the pillars as <0.5 μm. Since the height of the pillar is ~7 μm, this result positively reveals that the molten wax droplet is indeed sitting on air at the interface of the superoleophobic surface. However, the composite interface is not perfectly flat; the molten wax appears to penetrate into the void space between the pillars. Although we have not imaged the composite interface for water on the FOTS pillar array surface, we believe that the water droplet is in the Cassie–Baxter state too, based on the contact angle and sliding angle data.

To gain insight on the pinning location of water and hexadecane in the pillar array surface, a series of 3 μm T-shaped pillar array surfaces (straight sidewall with overhang thickness varying from 95 to 1030 nm) were fabricated.[47] Representative SEM micrographs are shown in Figure 6.7(a), along with the sessile drop data with water, hexadecane, and octane (Figure 6.7(b)). Surface Evolver[48] simulations were carried out for water and hexadecane to aid visualization of the pinning location and liquid sagging at the interface (Figure 6.7(c)). For water, which has a contact angle of 107° on FOTS, the simulations show that it wets the top of the pillar surface and pins at the edge of the overhang. On

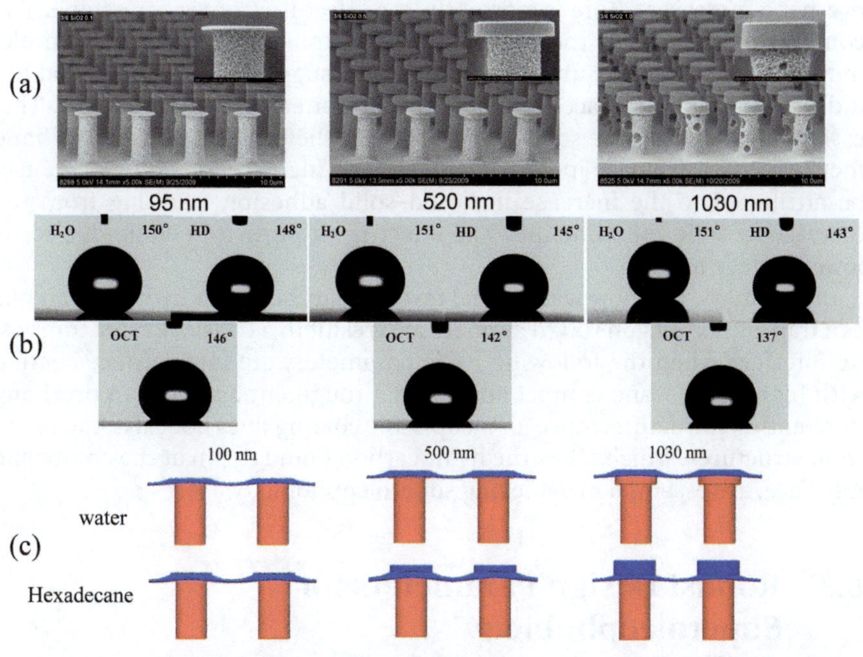

Figure 6.7 (a) SEM micrographs of T-shaped pillar array surfaces with overhang thickness of 95, 520, and 1030 nm. (b) Water, hexadecane, and octane sessile drop data of T-shaped pillar array surfaces in (a). (c) Surface Evolver simulation of the wetting of T-shape pillar array surfaces by water and hexadecane (black/blue: liquid interface, grey/red: pillar). Adapted with permission from H. Zhao, C. K. Park, and K. Y. Law, *Langmuir* 2012, **28**, 14925–14934. Copyright (2012) American Chemical Society.[47]

the other hand, the simulation shows that hexadecane not only wets the top surface, it penetrates into the sidewall and pins at the lower edge of the overhang. The difference in pinning location, attributable to the lower hexadecane contact angle of 73° on FOTS, is clearly seen when compared the structure of the liquid–solid–air composite interfaces between water and hexadecane for array surfaces with overhang thicker than 500 nm. We believe that octane also pins at a similar location on the wavy pillar array surfaces.

The advancing and receding contact angle data for all the T-shape pillar array surfaces are plotted in Figure 6.8(a and b). The overall results indicate that all the T-shape pillar array surfaces are superhydrophobic with advancing angles of ~160°, receding angles of ~140°, and sliding angles of ~10°. The surface property is independent of the thickness of the overhang and is consistent with the results obtained from Surface Evolver simulations. Since water droplets only wet the pillar surface and pin at the edge of the overhang, the absence of any overhang thickness effect is rational.

On the contrary, the thickness of the overhang is shown to profoundly influence the wetting properties of hexadecane and octane. Advancing and receding angles decrease and sliding angle increases as the thickness of the overhang increases. More importantly, the effect is stronger for octane. The contact angle data and the Surface Evolver simulation results are complementary. For instance, Surface Evolver results suggest that hydrocarbon liquids wet the pillar surface and pin at the lower edge of the overhang. This essentially increases the solid area fraction as the thickness of the overhang increases. The decrease in receding angle and increase in sliding angle can be attributed to the increase in liquid–solid adhesion resulting from the increase in solid area fraction. The effect is stronger for octane due to its lower surface tension.

In summary, we show with model textured surfaces that hydrocarbon oils can trap air pockets on rough surfaces and result in a Cassie–Baxter composite interface when the following basic parameters are met: surface coating with high hexadecane contact angle and a rough surface with an overhang or re-entrant structure. Since an oleophobic coating does not exist, the re-entrant structures, which allow the hydrocarbon liquid to pin at the composite interface, are essential in achieving superoleophobicity.

6.3 Robust Design Parameters for Superoleophobicity

Robustness has multiple meanings as one works towards commercialization. A surface can be defined as robust if it can maintain superoleophobicity during its intended functional life. Depending on the application or the market it serves, the surface may have to be mechanically strong to handle abrasion during manufacturing and normal use. The surface may also have to maintain its superoleophobic state when the liquid droplet is experiencing pressure at the composite interface. In other words, a robust

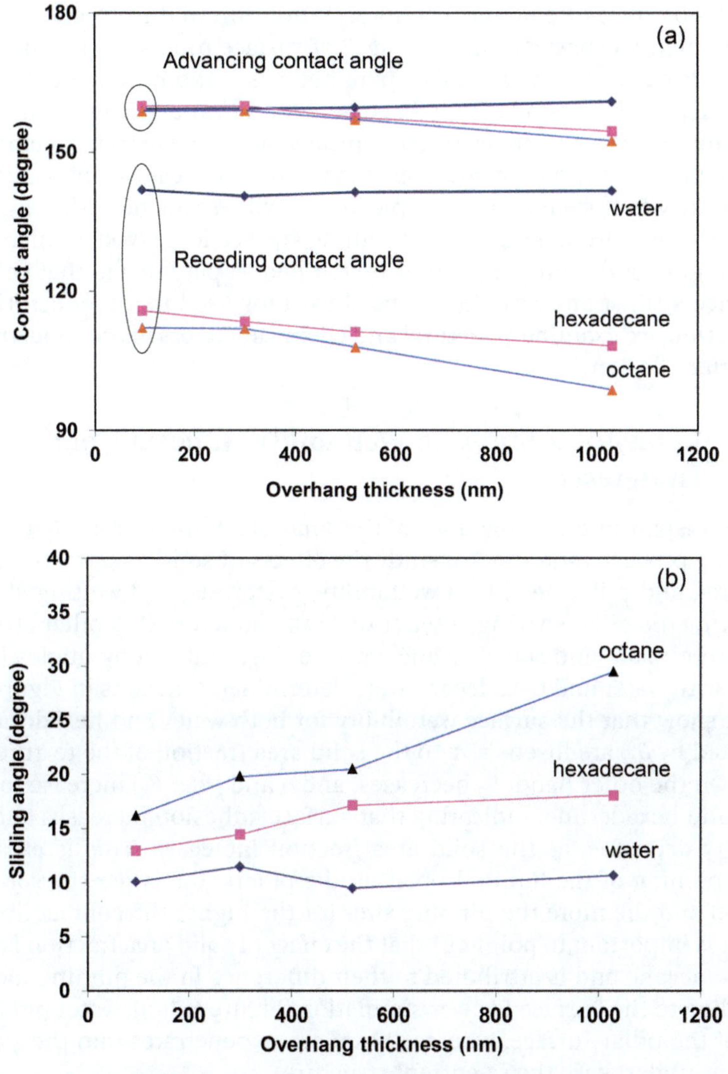

Figure 6.8 Plots of (a) advancing and receding contact angle and (b) sliding angle as a function of the thickness of the overhang for T-shaped pillar array surfaces. Reprinted with permission from H. Zhao, C. K. Park, and K. Y. Law, *Langmuir* 2012, **28**, 14925–14934. Copyright (2012) American Chemical Society.[47]

superoleophobic surface should have specifications in abrasion resistance and boundary of wetting stability, in addition to the usual large contact angle and low hysteresis. The specifications will depend on the intended end use.

Another issue is the robustness of the design parameters. In manufacturing, variations induced by the different processing steps are unavoidable. It is important to point out that factors influencing the stability or movement of the contact line are at the molecular level, whereas contact angles are macroscopic measurements. The information inferred from contact angle measurements is basically an "average" of surface properties for the sessile droplet. What if a certain design parameter is unknowingly defined near some catastrophic boundary? In that case, process variation during manufacturing may drive some areas microscopically out of the design specification. This microscopic variation could lead to microscopic catastrophic failure. It is very likely that such a microscopic failure will escape detection by a contact angle measurement unless a scanning surface force type of microscope is used. Robust design parameters are defined as parameters that will have tolerance against any variations caused by steps used in manufacturing. In this section, we examine structural and chemical factors in the robustness of the surface design.

6.3.1 Robustness Study on Wettability, Adhesion, and Hysteresis

To gain insight into the tolerance of the superoleophobic property to manufacturing process variation, we study the effects of solid area fraction, pillar diameter, and pillar height on wettability, hysteresis, and wetting stability. By varying the pillar spacing, a series of 3 μm diameter FOTS pillar array surfaces were made and their advancing, receding, and sliding angles (θ_A, θ_R, and α) with water and hexadecane were determined.[47] Results in Figure 6.9(a and b) show that the surface wettability for both water and hexadecane, as indicated by θ_A, are insensitive to the solid area fraction of the textured surfaces. On the other hand, θ_R decreases, and α and $(\theta_A - \theta_R)$ increase for both water and hexadecane, indicating that surface adhesion increases and drop mobility decreases as the solid area fraction increases. This is attributed to the pinning of the liquid droplet on the pillars, the larger the solid area fraction and the more the pinning sites (or the higher the contact line density). It is important to point out that the effect of solid area fraction is larger for hexadecane and is attributed to their difference in the pinning location. According to the Surface Evolver simulation (Figure 6.7(c)), water pins at the edge of the pillar surface, whereas hexadecane penetrates into the sidewall and pins underneath the re-entrant structure.

Also included in Figure 6.9(a and b) are results from pillar array surfaces with 1 and 5 μm diameter pillars (represented by data points X and O, respectively). These data points are completely compatible with the results of the 3 μm pillar array surfaces, indicating that surface adhesion and drop mobility are governed primarily by the density of the contact lines, not the geometry of the texture. In any event, the results in Figure 6.9(a and b) show clearly that a slight variation of pillar size due to variation in manufacturing will have little effect on wettability, but may negatively affect surface adhesion and drop mobility. Fortunately, these are not catastrophic failure. A robust design here will just involve a proper centreline specification of the design parameters.

The effect of pillar height on the superoleophobic properties was studied using 3 μm pillar array surfaces with a centre-to-centre spacing of 12 μm.

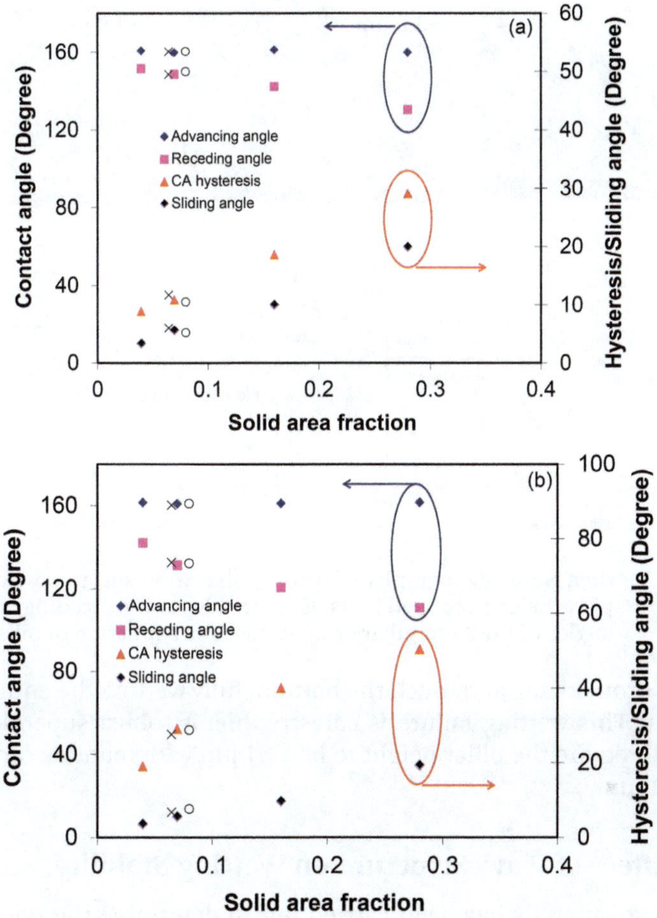

Figure 6.9 Plots of advancing and receding contact angle data of (a) water and (b) hexadecane *versus* solid area fraction for 3 μm pillar array surfaces. (Insert: advancing and receding contact angle data from a 1 μm pillar array (X) and a 5 μm pillar array (O) surface). Adapted with permission from H. Zhao, C. K. Park, and K. Y. Law, Langmuir 2012, 28, 14925–14934., Copyright (2012) American Chemical Society.[47]

This pillar spacing is wider than the surface shown in Figure 6.2 and represents a more stressful case for liquid sagging. The pillar height was controlled by the number of Bosch etching cycles and varied from ~8 to ~0.8 μm. Figure 6.10 shows the SEM micrographs of three representative surfaces along with the plots of contact angle data against the pillar height for both water and hexadecane. The results show that the pillar array surfaces are in their Cassie–Baxter state with both water and hexadecane at pillar heights taller than 1 μm. For surfaces with a pillar height of 0.8 μm, both water and hexadecane droplets are unstable. Water and hexadecane are shown to wet the entire surface upon standing in ambient conditions. This indicates that

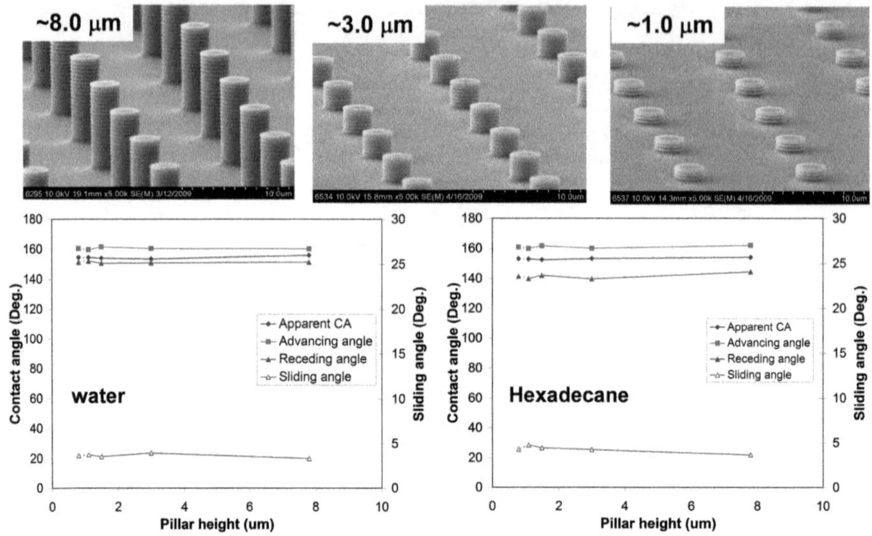

Figure 6.10 (*top*) SEM micrographs of 3 μm pillar array surfaces with different pillar height. (*bottom*) Plots of static, advancing, receding, and sliding angles of the 3 μm pillar array surfaces as a function of pillar height.

the liquid droplets sag and touch the bottom, fully wetting the entire surface as a result. This wetting failure is catastrophic. A robust superoleophobic design will require the pillar height to be ≥1.1 μm with manufacturing variation < ±0.1 μm.

6.3.2 Effect of Wavy Structure on Wetting Stability

Gibbs energy analysis has been carried out to determine the possible pinning locations and stability of the wetting states when liquid wets the FOTS pillar array surfaces.[49] Figure 6.11(a) shows dimensions of the straight sidewall pillar structure (r = 1.3 μm, d = 3 μm, H = 9 μm measured from SEM images), Figure 6.11(b) shows pining of the three-phase contact line (TCL) at the edge of the pillar surface, and Figure 6.11(c) depicts the advance of the TCL along the sidewall. We made two assumptions: (1) the local contact angle could differ from θ_{rec} to θ_{adv} when pinned, as shown in Figure 6.11(b). Here θ_{rec} and θ_{adv} refer to the receding and advancing contact angles of the smooth surface, not the contact angles on the textured surfaces. (2) The TCL moves upon external energy inputs, *e.g.* external pressure, after depinning. The TCL moves downward and starts sagging with a constant $\delta\theta = \theta_{adv} - \pi/2$.

The process of liquid sagging into the straight sidewall follows two steps: (1) TCL is at the edge of the pillar surface, with $h = 0$ while $\delta\theta$ varies from θ_{rec} to θ_{adv}; and (2) TCL advances along the sidewall with varying h and a constant $\delta\theta = \theta_{adv} - \pi/2$. Recently this assumption has been validated by laser scanning confocal measurement of the contact line.[50] For each step, Gibbs energy is formulated and calculated based on the straight sidewall pillar geometry,

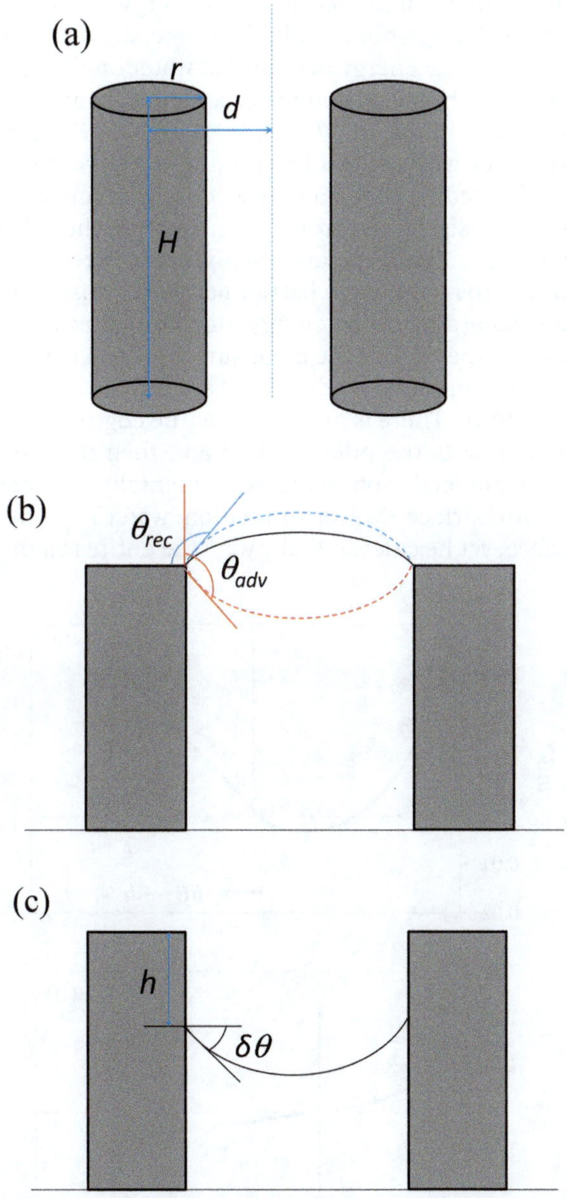

Figure 6.11 (a) Dimensions of straight sidewall pillar structure, radius of the cylinder r, half centre-to-centre distance d, pillar height H, and sagging height h. (b) TCL pinned at the edge of the pillar surface, the local contact angle changes from θ_{rec} to θ_{adv}. (c) TCL moves downward for a sagging height h with a constant local contact angle equal to θ_{adv}. Reprinted from figures in ref. 49, Copyright 2014 ICE Publishing.

local contact angle at the TCL, and advancing/receding contact angles on the smooth FOTS surface. In this work, Gibbs energy is normalized with the Gibbs free energy of a droplet in its spherical state, denoted as G^*.

Figure 6.12(a) shows the energy diagram for water on the straight sidewall pillar array surface at the measured water apparent contact angle of 152°. There exist two local minima, at $\delta\theta = 0°$, $h = 0$ and $h = H$, corresponding to the Cassie–Baxter state with contact line pinning at the edge of the pillar surface and the fully wetted Wenzel state, respectively. G^* increases with h and when $\delta\theta > 0°$, which resists water from advancing into the sidewall. In other word, on the straight sidewall pillars, even without a re-entrant structure, water still needs to cross an energy barrier before it transits to the fully wetted Wenzel state. According to the energy diagram, water is in a metastable state when pinned at the edge of the pillar surface with a flat liquid–air interface. However, for hexadecane, G^* decreases sharply as $\delta\theta$ and h increase, as shown in Figure 6.12(b). There is no pinning at the edge of the pillar surface. Instead, hexadecane wets the pillar surface and then the straight sidewall spontaneously. This indeed is observed experimentally (Figure 6.3(a)) and is also consistent with Surface Evolver simulation: water is pinned at the edge of the pillar surface, yet hexadecane fully wets the entire rough surface.

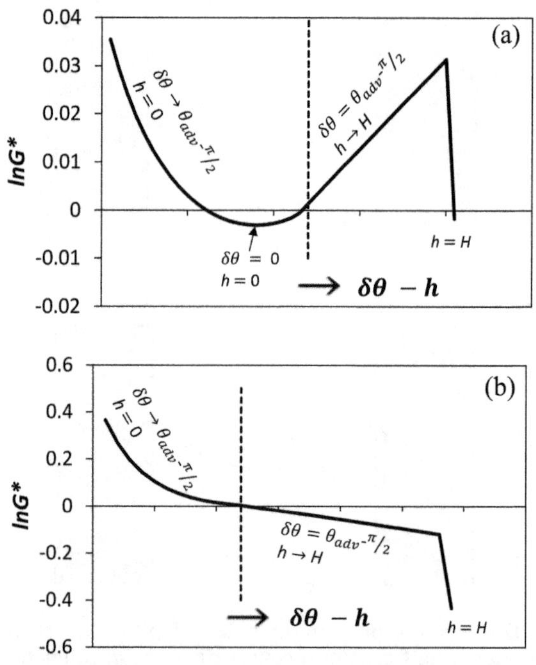

Figure 6.12 Energy diagram of the straight sidewall pillar array surface for (a) water and (b) hexadecane: ln G^* evolvement with sagging angle $\delta\theta$ and sagging height h at the measured liquid apparent contact angle. The x-axis represents the two steps of TCL movement illustrated in Figure 6.11. Adapted from figures in ref. 49. Copyright 2014 ICE Publishing.

Figure 6.13(a) shows the dimension of the wavy sidewall pillar structure (r = 1.35 µm, d = 3 µm, H_1 = 7 µm, wave structure of H_2 = 300 nm, and φ = 45° measured from SEM images). For water, the droplet is expected to be pinned at the edge of the pillar surface. However, water may still penetrate into the sidewall structure under external pressure. Due to the reduced surface tension, hexadecane will fully wet the pillar surface and penetrate into the wavy sidewall structure. To determine how the TCL advances on the wavy sidewall, one can visualize the wetting of the wave structure as two half-units. Figure 6.13(b and c) represent the upper and lower half of the wave, respectively. The process of wetting into the wavy sidewall structure can be visualized in four steps: (1) TCL is at the edge of the pillar surface with h = 0 and varying $\delta\theta$; (2) TCL advances along the upper half of the unit cell (re-entrant structure) with a constant $\delta\theta = \theta_{adv} - \varphi$ and varying h; (3) TCL is at the concave location with $h = H_2/2$ and varying $\delta\theta$, and (4) TCL advances along the lower half of the wave with a constant $\delta\theta = \theta_{adv} + \varphi - \pi$ and varying h. For each step, Gibbs energy is formulated and calculated, based on the geometry of the wavy sidewall, local contact angle at the TCL, and advancing/receding contact angles on the smooth FOTS surface.

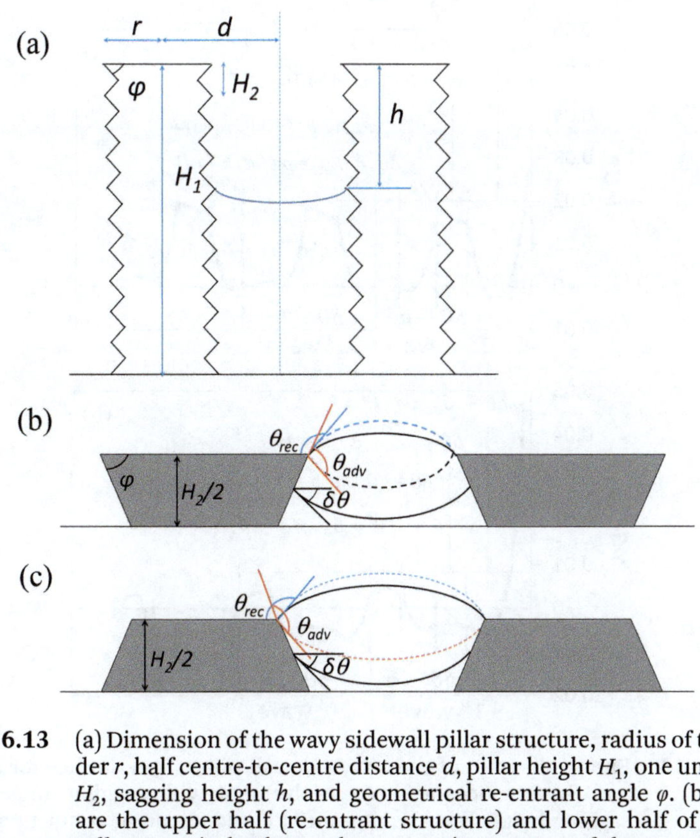

Figure 6.13 (a) Dimension of the wavy sidewall pillar structure, radius of the cylinder r, half centre-to-centre distance d, pillar height H_1, one unit height H_2, sagging height h, and geometrical re-entrant angle φ. (b) and (c) are the upper half (re-entrant structure) and lower half of the unit cell, respectively, that make up a unit structure of the wavy sidewall. Reprinted from figures in ref. 49, Copyright 2014 ICE Publishing.

The energy diagram for water advancing along the wavy sidewall structure is shown in Figure 6.14(a) at the measured water apparent contact angle of 156°. The Gibbs energy is calculated when water wets the wavy sidewall following the four steps described above. In addition to the edge of the pillar surface, every protruding corner with $h = nH_2$, and concave site with $h = (n + 1/2)H_2$, is a possible pinning site for water with a local energy minimum. Similar to the straight sidewall, water is pinned at the edge of the pillar surface with a local energy minimum. Increases of h and $\delta\theta$ (Figure 6.13(b), upper half unit) increase G^*, indicating that pinning at the edge results in a metastable Cassie–Baxter interface (Figure 6.14(a)). An external energy input (*e.g.* external pressure) is needed to de-pin the TCL from the edge of the pillar surface. The TCL will advance and may be pinned at the concave position with $h = H_2/2$. It is interesting to note that the re-entrant structure provides little stability to the water TCL as the energy barrier at the concave position is relative small. The likely stable metastable location with an energy minimum will be at the next protrusion corner at $h = H_2$. As shown in Figure 6.14(a), the energy barriers created at the edge of the pillar and the protrusion corners

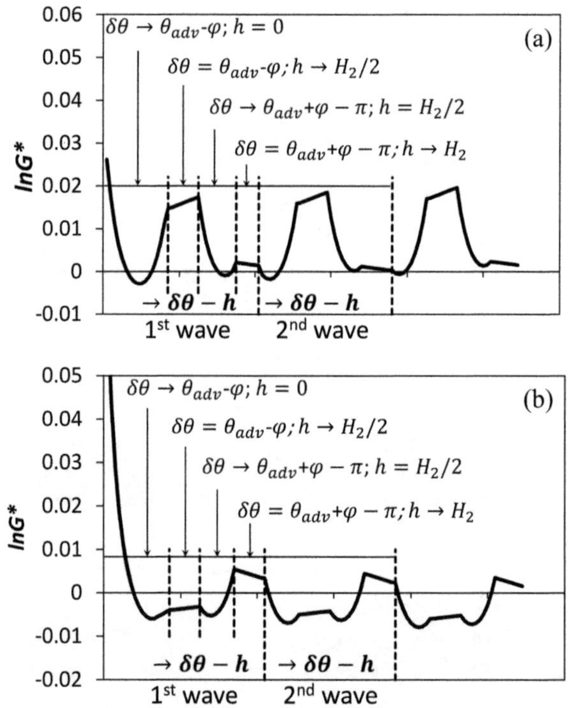

Figure 6.14 Energy diagram of the wavy sidewall pillar array surface for (a) water and (b) hexadecane: $\ln G^*$ evolvement with sagging angle $\delta\theta$ and sagging height h at the measured liquid apparent contact angle. The x-axis represents the four steps of TCL movement illustrated in Figure 6.13. Adapted from figures in ref. 49, Copyright 2014 ICE Publishing.

in the sidewall are comparable. Therefore, the protrusion corners ($h = n\text{H}_2$) along the wavy sidewall will not offer any additional metastability to the composite interface. The wavy sidewall, however, offers an insurance policy for protecting the Cassie–Baxter state. For example, during manufacturing, what if the re-entrant structure at the top is off specification? The pillar array surface may still survive without collapsing to the Wenzel state due to additional layers of protection by the protrusion corners.

The energy diagram for hexadecane is similar to that of water, shown in Figure 6.14(b) at the measured hexadecane apparent contact angle of 158°. In addition to the edge of the pillar surface, two pinning sites at both the protruding and concave corners are energy local minima. Due to the shallow energy barrier, hexadecane is not pinned at the edge of the pillar surface (at $h = 0$) under most wetting conditions. Instead, the contact line continues to advance through the upper half of the unit cell (Figure 6.13(b)) and is pinned at the next energy minimum, the concave position. Further advance of the TCL to the lower half of the unit cell (Figure 6.13(c)), from $\text{H}_2/2$ to H_2, requires additional energy input. This creates an energy barrier and results in stabilization of the metastable Cassie–Baxter state for hexadecane. In other words, results from the Gibbs free energy analysis, which is in agreement with experimental data and Surface Evolver simulation, re-confirm that hexadecane pins beneath the re-entrant structure of the wavy sidewall. Analogous to water, the wavy sidewall also offers a multi-level of protection for the Cassie–Baxter state due to the multiple pinning locations along the sidewall.

6.3.3 Effect of Re-Entrant Geometry on Wetting Stability

The Gibbs energy analysis can also be used to examine the effect of re-entrant geometry on wetting stability. We are particularly interested in investigating the re-entrant angle of the wavy sidewall on the wetting stability against the external energy input (*e.g.* pressure) using the Gibbs energy analysis. From the discussion above, re-entrant structure (or $\varphi < 90°$) in the textured surface allows for the formation of a composite interface (Cassie–Baxter metastable state) and results in extremely high apparent contact angles even if the surface coating material has a contact angle <90°. Figure 6.15 plots the Gibbs energy of an advancing hexadecane droplet on the wavy sidewall with various re-entrant angles, assuming the same unit cell height. The energy barriers $\Delta \ln G^*$ between the different steps on the wavy sidewall structure are compared for various re-entrant angles φ. The results indicate that a smaller φ (sharper corner) would lead to a higher energy barrier, increasing the stability of the metastable state and a more robust superoleophobicity. Specifically, the increase of energy barrier will increase the resistance of the Cassie–Baxter state to become fully wetted. A ~5 times improvement in $\Delta \ln G^*$ is obtained (inset in Figure 6.15) when the re-entrant angle changes from 60° to 30°. Recently, Liu *et al.*[51] created micro pillars with negative re-entrant angles which enabled superoleophobicity on the bare silicon oxide surface. This textured surface can even repel fluorinated solvents with

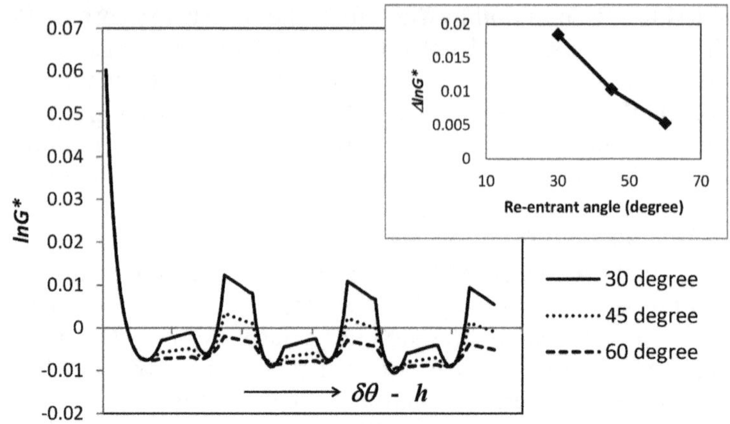

Figure 6.15 Plot of $\ln G^*$ *vs.* $\delta\theta - h$ for the wavy sidewall pillar array surface. The x-axis represents the four steps of TCL movement illustrated in Figure 6.13. The three curves show the effect of re-entrant angle on the energetics. *Inset*: plot of energy barrier ($\Delta\ln G^*$) *vs.* re-entrant angle on the wavy sidewall structure. Reprinted from figures in ref. 49, Copyright 2014 ICE Publishing.

surface tension as low as 10 mN m^{-1}. This work not only demonstrates that re-entrant structure is an essential enabler for superoleophobicity, it also shows the profound effect of the re-entrant angle (φ) on the metastability: the smaller the re-entrant angle the higher the energy barrier between the Cassie–Baxter and Wenzel state and the higher the wetting stability.

6.3.4 Effect of Breakthrough Pressure on Superoleophobicity

The effect of pressure on the wetting stability is studied using Surface Evolver simulation on 3 μm diameter T-shape pillar array surfaces. Results in Figure 6.16(a) shows that as the pressure across the liquid–air interface increases, liquid sagging increases. The breakthrough pressure is defined as the critical pressure when the liquid droplet transitions from the Cassie–Baxter composite state to the fully wetted Wenzel state. There are two modes of wetting failure as the pressure continues to increase. If the pillar height (H) is short, pressure can induce excessive sagging of the hexadecane drop at the liquid–air interface, which may lead to wetting of the bottom of the pillar structure and the fully wetted Wenzel state (Figure 6.16(b)). This is designated as H^* failure.[26] Increase in external pressure can also increase the capillary pressure, which de-pins the contact line and results in wetting of the entire pillar structure (Figure 6.16(c)). This is designated as T^* failure.[26] Between these two types of failure, H^* can be avoided by making sure that the pillar height is taller than the critical height (H_C). In that case, T^* will become the major wetting failure when the breakthrough pressure is reached.

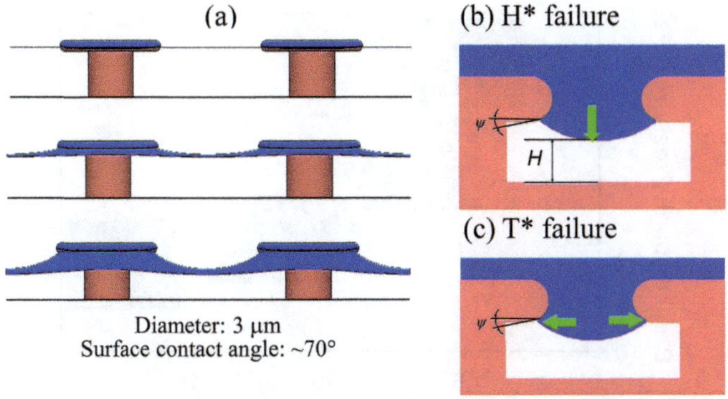

Figure 6.16 Surface Evolver simulation: (a) wetting of a FOTS pillar array surface with an overhang structure by hexadecane. (b) H^* failure, sagging induced wetting failure due to short pillar height. (c) T^* failure, wetting failure due to pressure-induced depinning of the contact line (grey/red: pillar, black/blue: hexadecane interface). Adapted with permission figures in H. Zhao, C. K. Park, and K. Y. Law, *Langmuir* 2012, **28**, 14925–14934., Copyright (2012) American Chemical Society.[47]

The robustness of T^* against the basic design parameters for superoleophobicity was studied using Surface Evolver simulation for a series of 3 μm diameter pillar array surfaces with varying spacing. The structural parameter to describe the spacing is D^*. D^* is defined as $(W + D)^2/W^2$ where $2W$ is the diameter of the pillar and $2D$ is the separation distance between two pillars. A plot of the breakthrough pressure with hexadecane against D^* is shown in Figure 6.17(a). The result indicates that, for the 3 μm FOTS pillar array surfaces, the hexadecane breakthrough pressure increases from ~1.3 to ~12.4 kPa, when the separation between pillars decreases from 9 to 1.5 μm. Breakthrough pressure can increase further if D^* continues to decrease.

Figure 6.17(b) plots the breakthrough pressure as a function of the smooth surface contact angle for pillar array surfaces with different pillar diameters. The D^* value is kept at 4 ($W = D$) in the plot. As expected, breakthrough pressure increases as the smooth surface contact angle increases. More importantly, the results also show that there is a strong pillar size effect on the breakthrough pressure. For a given D^*, the smaller the pillar diameter, the higher the breakthrough pressure against wetting. Therefore, the key design parameters for robustness against wetting breakthrough pressure are: smooth surface contact angle, pillar diameter, and pillar spacing, assuming the pillar height is $>H_C$. The overall results in Figure 6.17(a and b) suggest that superoleophobic surface with wetting breakthrough pressure at ~70 kPa is attainable with a 0.5 μm diameter pillar array surface at a pillar separation of 0.25 μm. It is interesting to note that the breakthrough pressure for the lotus leaf was estimate to be ~20 kPa[52] and the pressure generated from a heavy rain storm can reach 100 kPa.[53] The present work illustrates

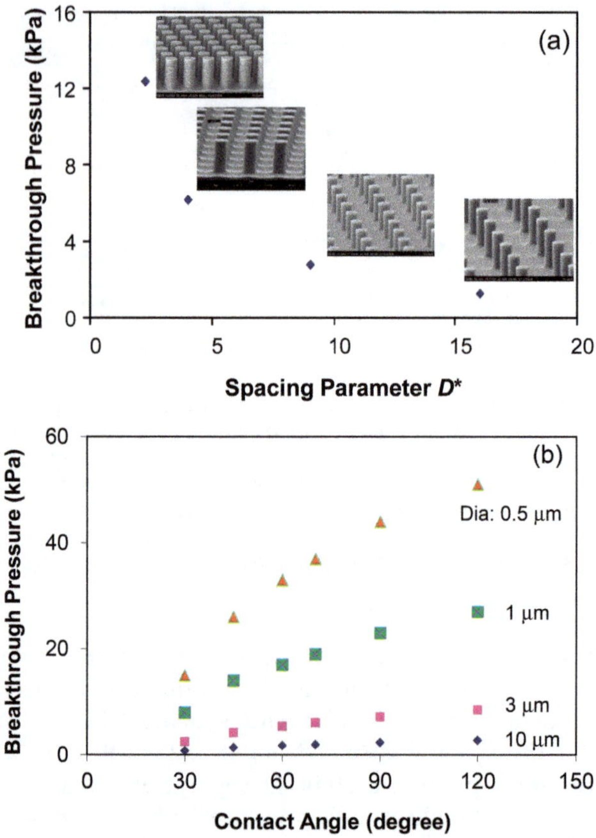

Figure 6.17 Plot of breakthrough pressure as a function of (a) spacing parameter D^* for FOTS pillar array surfaces (hexadecane case) and (b) static contact angle for pillar array surfaces with different pillar diameters at $D^* = 4$. Reprinted with permission from H. Zhao, C. K. Park, and K. Y. Law, *Langmuir* 2012, **28**, 14925–14934., Copyright (2012) American Chemical Society.[47]

that making a superoleophobic surface with wetting resistance comparable to surfaces found in nature is definitely possible.

6.3.5 Mechanical Robustness Against Abrasion

For the pillar array surface, key mechanical failures would be pillar bending and buckling. Figure 6.18 shows the mechanical model used to assess the robustness of the pillar array surface. Assuming an arbitrary force F_{global} is applied to area A on the pillar array surface with dimensions indicated in Figure 6.18, two dimensionless design parameters, bending parameter S^* and buckling parameter N^*, can be developed according to the classic solid mechanics bending theory of a circular cross-section beam with one fixed

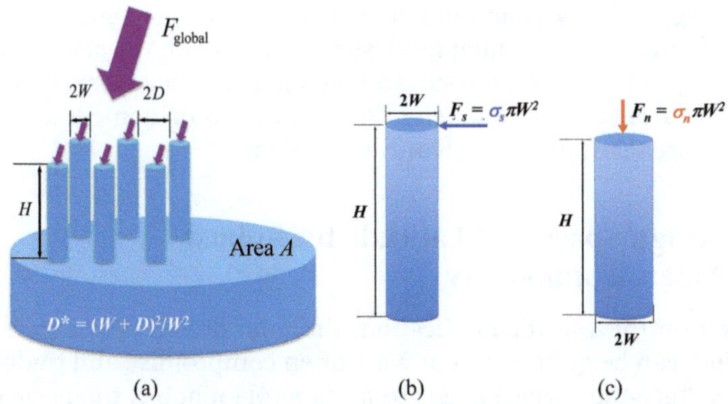

Figure 6.18 (a) Model textured surface used to model the deformation of the pillar array surface against external force F_{global}. (b) Bending failure of a single pillar. (c) Buckling failure of a single pillar. Reprinted with permission from H. Zhao, C. K. Park, and K. Y. Law, *Langmuir* 2012, **28**, 14925–14934. Copyright (2012) American Chemical Society.[47]

end and one free end.[54] S^* and N^* are the ratios of the maximum theoretical stress the pillar can withstand to the actual stress exerted to the pillar structure under shear and normal stress condition, respectively. They are given by eqn (6.1) and (6.2), respectively.

$$S^* = \frac{\pi}{16}\left(\frac{E\gamma}{r_0}\right)^{\frac{1}{2}}\left(\frac{F_{s,global}}{A}\right)^{-1}(D^*)^{-1}\left(\frac{H}{W}\right)^{-1} \tag{6.1}$$

$$N^* = \frac{\pi^3 E}{64}\left(\frac{F_{n,global}}{A}\right)^{-1}(D^*)^{-1}\left(\frac{H}{W}\right)^{-2} \tag{6.2}$$

where E is the Young's modulus of the pillar material, γ is the surface energy of the material, r_0 is the distance between neighbouring atoms, F_s is the shear force generated from F, and F_n is the normal force generated by F. $D^* = (W + D)^2/W^2$, W, D, and H are structural parameters as shown in Figure 6.18.

According to the definition, the pillar structure will survive without failure when $S^* > 1$ and the higher the S^* value the more robust the pillar structure is. The pillar structure will have a bending failure when $S^* < 1$. When $S^* = 1$, the critical stress can be calculated by applying all the related properties of silicon and the geometrical parameters of the pillar array surface (*e.g.* for the surface in Figure 6.2, $2W = 3$ μm, $2D = 3$ μm, $D^* = 4$, and $H = 7.8$ μm). From eqn (6.1), $F_{s,global}/A$ is calculated to be ~1.7×10^5 kPa. Similarly, $F_{n,global}/A$ is found to be ~8.5×10^5 kPa according to eqn (6.2) when $N^* = 1$. Since a smaller $F_{s,global}/A$ value is obtained from eqn (6.1), the comparison suggests that the likely mechanical failure for the pillar array surface is pillar bending. Indeed, bending failure has been reported for a poly(dimethylsiloxane) (PDMS) pillar array surface.[55,56] Initial assessment of the bending failure of the pillar array surface in Figure 6.2 by a nano-indenter also suggests that bending is the primary failure mode for the pillar array surface. Although the above mechanical modelling is for pillar array surfaces, geometrical considerations suggest that rough surfaces with low height-to-width ratio or dome-like structure would be more resistant to mechanical abrasion.

6.3.6 Design Space and Latitude for Robust Superoleophobicity

Depending on the specific application, the final design parameters for an end product can be quite different. Very often compromise and trade-off are necessary. Since the surface needs to be superoleophobic, the basic parameters the surface must have are high hexadecane contact angle and some sort of micro/nano rough texture with a re-entrant structure at the liquid–solid interface. Three different application specific superoleophobic surface design spaces are shown in Figure 6.19 and discussed below. These spaces are then used as a springboard for discussions of design latitude and trade-off areas for technologically more challenging surface applications.

- *Low-adhesion and non-sticky superoleophobic surface*. For indoor, maintenance-free, always clean surfaces, such as the glass in museum display cabinets or jewellery stores, there is less concern about breakthrough pressure or mechanical abrasion. The surfaces need to be non-sticky and repel grease and oily materials with low adhesion against dust or dirt. The preferred design is to have an optimally large spacing parameter D^* (equivalent to small solid area fraction) to render low adhesion and low contact angle hysteresis. The spacing should be smaller than the sizes of the dust and dirt but large enough to deliver the low-hysteresis feature.

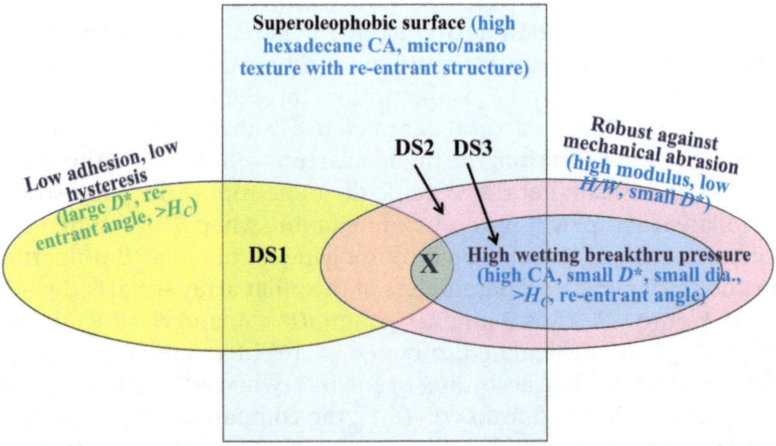

Figure 6.19 Graphic representation of design spaces for superoleophobic surface with different characteristics.

The height of the features is not critical as long as it is $>H_C$, while meeting the not too stringent mechanical abrasion resistance requirements for normal manufacturing handling and irregular touching during use. This design space is designated as *DS1* in Figure 6.19.

- *High mechanical abrasion resistance superoleophobic surface*. For applications such as outdoor self-cleaning windows and other surfaces, superoleophobicity is primarily used to protect the surface from organic and oily pollutants, which may degrade the self-cleaning function. Because of the re-entrant structure on the surface, the wetting breakthrough pressure against water during rain will be superior to most superhydrophobic surface designs. On the other hand, the surface may be abraded by dust and dirt particles during wind shear as well as human touch. Key design parameters for mechanical abrasion resistance include the selection of a material with high Young's modulus, small spacing parameter D^*, and a low aspect ratio of H/W, where H and W are the height and width of the rough feature in the surface. This design space is designated as *DS2* in Figure 6.19.
- *Superoleophobic surface with high wetting breakthrough pressure*. For use in oil pipelines, a surface may need to be superoleophobic and remain in the Cassie–Baxter state under a certain external pressure during the intended use. If the superoleophobic state is compromised under pressure and the surface becomes fully wet, friction and drag will increase and the surface will become non-functional. Key design parameters for high wetting breakthrough pressure include smaller D^* (equivalent to larger solid area fraction), smaller roughness feature size, and height $>H_C$. An additional way to further enhance the breakthrough pressure is through fine tuning of the re-entrant angle at the liquid–solid interface: the sharper the re-entrant angle, the higher the resistance to being wetted. This is a set of very demanding design parameters and is designated as *DS3* in Figure 6.19.

In reality, most applications demand more than *DS1*, *DS2*, or *DS3* alone. It is not unusual for an application to require very low hysteresis as well as very strong mechanical abrasion resistance. In that case, the design space would be the overlap area between *DS1* and *DS2*. The space shrinks as the design requirement becomes more stringent. In fact, there are applications that demand requirements from all three design spaces. For instance, for an anti-smudge surface for a smartphone or display will need to repel oil and grease from human hands, while being mechanically strong enough to withstand the constant pressure and abrasion from typing. At the same time, the surface will need to have sufficient breakthrough pressure to resist being fully wetted when the surface is typed on by greasy fingers. As a result, the design space is very limited; the area is labelled as X in Figure 6.19. It is important to note that this limited space will shrink further, or may even disappear if the requirement for the wetting pressure is too high to be designed for.

6.4 Discussion of Robust Design Parameters for Superhydrophobicity

Inspired by the lotus effect, research activities on superhydrophobicity have grown exponentially since the 1997 *Planta* article[1] and the subject has been reviewed several times recently.[12–17] As a result, basic design parameters for superhydrophobicity are well documented. Here we are only concerned with factors that have direct effects on the robustness of superhydrophobic surfaces.

6.4.1 Re-Entrant and Overhang Structures

Like superoleophobicity, superhydrophobicity is also an interplay of surface chemistry and surface roughness. Because water has a much higher surface tension than oils, surface design for superhydrophobicity is less demanding. A number of less widely known reports have shown that superhydrophobicity can be attained with moderately hydrophilic materials and a re-entrant structure or an overhang.[34,57–59] Using photolithography, Cao *et al.*[57] fabricated a pillar array surface with an overhang on silicon wafer (Figure 6.20(a)). Since no extra coating is used, the surface of the textured surface is basically hydrogen-terminated silicon, which has a contact angle of 74°. The water contact angles of the textured surfaces range between 150° and 160° at solid area fractions smaller than 0.07. Similarly, Wang *et al.*[58] created superhydrophobic surfaces with T-shaped pillars on silicon wafer followed by coating the textured surfaces with a ~100 nm thickness of diamond-like carbon (DLC) film (Figure 6.20(b)). The WCA of a smooth DLC film is ~72° and the textured surfaces exhibit superhydrophobic-like contact angles of ~160°.

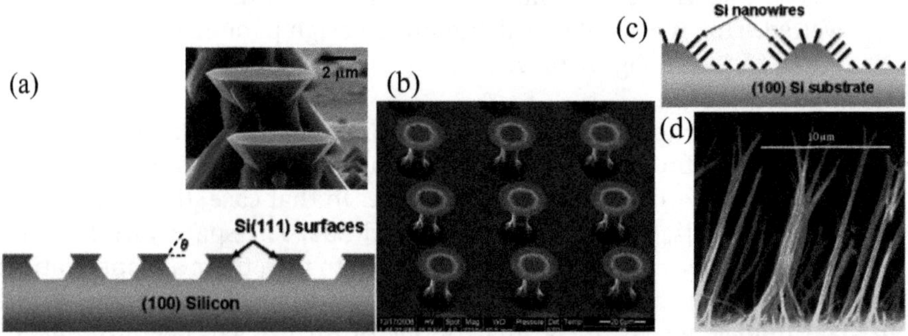

Figure 6.20 Mosaic of superhydrophobic surfaces with (a) overhang, (b) re-entrant structure, (c) silicon nanowires, and (d) PVA nanofibres. Figures (a) and (c) reproduced with permission from L. Cao, H. H. Hu, and D. Gao, *Langmuir* 2007, **23**, 4310–4314. Copyright (2007) American Chemical Society.[57] Figure (b) Reprinted with permission from J. Wang, F. Liu, H. Chen, and D. Chen, *Appl. Phys. Lett.* 2009, **95**, 084104. Copyright (2009), AIP Publishing LLC. Figure (d) Reprinted from ref. 59 with permission from John Wiley and Sons. Copyright © 2002 Wiley-VCH Verlag GmbH & Co. KGaA, Weinheim.

Superhydrophobic surfaces can also be fabricated with hydrophilic materials at very low solid area fraction. For example, Cao *et al.* also fabricated a rough surface decorated with silicon nanowires (Figure 6.20(c)).[57] Again, native silicon is hydrophilic but a very large contact angle (~160°) was obtained. Similarly, Feng and co-workers[59] were able to grow a poly(vinyl alcohol) (PVA) nanofibre forest (Figure 6.20(d)) using an aluminium oxide template and observed an apparent contact angle of 171°. The PVA polymer used is hydrophilic with a WCA of ~72°. The common feature of the latter two surfaces is that they both have a very low solid area fraction. Even though water is pinned at the tip of the nanowire or fibre, the close proximity of the pinning location, coupled with the high water surface tension, enables the formation of air pockets and thus superhydrophobicity. The design rules for superhydrophobicity are similar to the rules for superoleophobicity: chemistry, roughness, and re-entrant or overhang structure. The above studies show that although meeting two out of the three requirements will be sufficient for superhydrophobicity, having all three requirements would be beneficial for robustness, particularly the wetting breakthrough pressure.

6.4.2 Hierarchical, Multi-Scale Roughness

One of the misconceptions in the literature of superhydrophobicity is that the wax on a lotus leaf is hydrophobic. Although the chemical structure of the plant wax has not been fully characterized, Cheng and coworkers showed that the waxy material on the lotus leaf is moderately hydrophilic with a WCA of 74°.[11] The hydrophilic nature of the leaf surface is supported by experimental observation, where water was shown to condense onto the leaf surface at high humidity.[60,61] Careful examination of the high-magnification SEM of the lotus leaf indicates that the entire leaf surface, including the 10–20 μm papillae, is covered with waxy tubules (Figure 6.21). It is likely that when water wets the hierarchical structure of the leaf surface, a re-entrant structure is established due to the tubular nature of the interface. This further reduces the solid area fraction of the interface and creates mini air pockets within the micrometre-size air pockets from the 10 μm papillae. As consistently mentioned in the literature, the multi-scale roughness structure has contributed, not only to the large contact angle, but also low hysteresis and high wetting stability.[11,52,62]

The effect of multi-scale roughness on superhydrophobicity have been experimentally and analytically studied by many research groups.[63–70] Thermodynamic free energy analysis has been analysed on textured surfaces with multiple scale roughness at different wetting states.[66,67] On a microscale-only textured surface, the condensed droplet is shown to exist in the Wenzel state.[66] On the other hand, the condensed droplet on a properly designed micro/nano dual-scale superhydrophobic surface can transition to the Cassie–Baxter state with external stimulation. Three-scale roughness is found to be more robust to wetting than mono- or dual-scale roughness.[67] The secondary and ternary roughness play a significant role in preventing water from penetrating into troughs of the rough structures. Su *et al.*[68] further investigated both

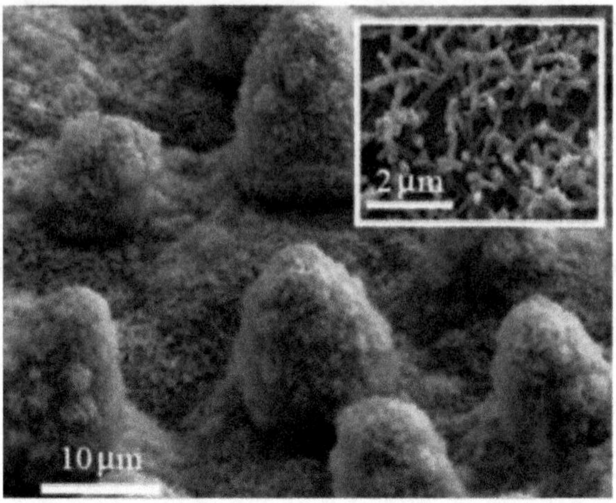

Figure 6.21 SEM micrographs of lotus leaf. Reprinted from figures in *Progress in Materials Science*, **54**, K. Koch *et al.*, Multifunctional surface structures of plants: an inspiration for biomimetics, 137–178, Copyright (2009) with permission from Elsevier.[6]

analytically and numerically that the two-level hierarchy roughness stabilizes the superhydrophobic non-wetting state, which also allows the higher-level structures to restore the Cassie–Baxter state after the impact of rainfall. The combination of multi-scale roughness and hydrophobic coating facilitates the formation of multi-scale air pockets, significantly reducing the wetted area at the liquid–solid interface. This lowers the contact angle hysteresis and surface adhesion and results in enhancement of the self-cleaning performance.[69] This conclusion is supported by Gibbs free energy analysis, where the energy barriers during contact line receding from three- (dual)-scale roughness exhibit three (two) levels of fluctuations, splitting a large receding energy barrier into many small ones, and hence decreasing the receding energy barrier.[67]

From the mechanical robustness point of view, multi-scale roughness can reduce the height of the roughness features while still being able to keep water in the Cassie–Baxter state.[70,71] Compared to the fragile nanoscale roughness, microscale roughness is better equipped to handle wear and abrasion. In addition, even if some features on the surfaces are worn and flattened, the multi-scale nature of the unworn portion will still be able to maintain the Cassie–Baxter state, while single-scale nanoroughness will be easily worn off and single-scale microroughness will likely lose the Cassie–Baxter wetting state after being partially abraded.

6.4.3 Design Parameters for Robust Superhydrophobicity

Remember that robustness in this chapter means high wetting breakthrough pressure, strong abrasion resistance, and longevity in large contact angle and low hysteresis. Due to the high surface tension of water, more tools become

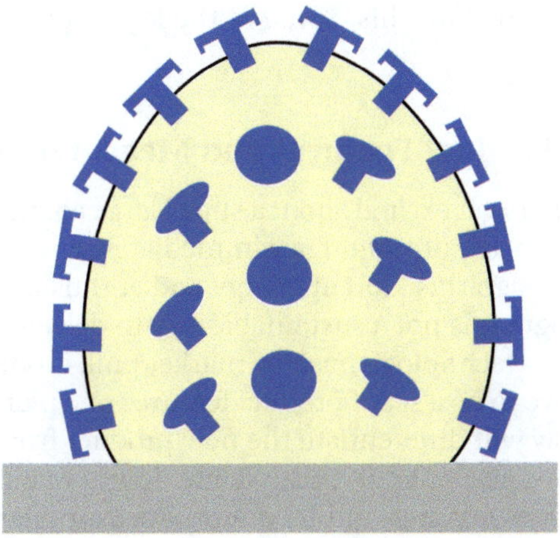

Figure 6.22 Graphic representation of a robust superhydrophobic surface.

available. These tools include the availability of hydrophobic materials and the ability to create multi-level air pockets in the hierarchical surface texture in addition to the re-entrant structure. Although the design space map for robust superhydrophobicity is similar to that of superoleophobicity (Figure 6.19), having a hydrophobic material will not only increase the latitude space, it has a direct consequence in delivering a better performance. For instance, the breakthrough pressure with water is always higher than that with hexadecane for a given surface structure (Figure 6.17). Using Gibbs free energy analysis,[49] we can compare the difference in breakthrough pressure for straight sidewall pillar array surfaces (~3 μm diameter with a 6 μm pitch) with and without the re-entrant structure. The analysis reveals that the breakthrough pressure with the re-entrant structure pillar array is 18 times higher.

Designing a robust low-adhesion and low-hysteresis superoleophobic surface with high abrasion resistance (*e.g.* the overlap area between *DS1* and *DS2* in Figure 6.19) presents a dilemma. However, with the hierarchical surface structure, the design space just opens up tremendously. In the hierarchical surface, the micrometre-size feature is responsible for the mechanical property and the nanoscale features will deliver the low-adhesion and low-hysteresis property. If the desire is to have superhydrophobicity with low hysteresis, low adhesion, high abrasion resistance, and high wetting breakthrough pressure (area *X* in Figure 6.19), a dome-like hierarchical surface decorated with nano-mushrooms will have a better chance of success (Figure 6.22).

6.5 Summary and Remarks

Ever since the publication of the inspirational *Planta* article in 1997,[1] research into various aspects of the lotus effect and superhydrophobicity has been non-stop.[72,73] Admittedly, there are a lot of hypes and the implementation of the self-cleaning technology has lagged. As described in the introduction, there are several reasons for this. Some of the key issues are summarized in the following.

6.5.1 Gaps in Product Features and Measurements

Very often, researchers get excited about a super-large contact angle or how the liquid drops are bouncing around. They immediately envision applications or even products. Although this kind of energy and enthusiasm are needed for a technology campaign, it is not a sustainable way to deliver a product. Regardless of whether it is a technology push or market pull, if one wants to develop a product, there have to be a set of product features one plans to deliver. These features presumably will differentiate the new product from competing alternatives in the marketplace. What most people do not realize is that there are gaps between product features and laboratory measurements. The key question to ask is: have we identified all the parameters and measurements to quantify the intended product features in the lab? This is not a simple question and is often overlooked after an exciting scientific discovery.

There are many steps in translating product features to lab measurements. For instance, once product features are identified or agreed on, one has to understand the physical processes or chemical interactions that enable these features. After all the physical processes and chemical interactions have been identified, a set of enablers must be developed to make it happen. These enablers include for instance material properties and ranges of operating conditions and parameters. From there, the types of measurements to monitor and quantify the material properties and device performance will be identified. Material specifications and product functional parameters will then be developed and nominal performance will be released. As you can see, it is a long way from lab measurements to product features. Insufficient understanding of the product delivery process has partially contributed to the lagging development of self-cleaning technology.

6.5.2 Compromises and Trade-Off

For product delivery, the tendency is to aim at a perfect design that will meet all facets of the requirements and 100% of the market space. For instance in self-cleaning technology, the ideal surface should have high water and oil repellency, low adhesion, non-stick, and high abrasion resistance along with high liquid breakthrough pressure. This translates to a very small design space, such as the space *X* in Figure 6.19. On the other hand, not all applications require this set of demanding requirements. If the requirement on wetting breakthrough pressure is relaxed, the design space immediately expands from *X* to the overlap area between *DS1* and *DS2*. Although this may seem like a compromise, it is actually a smart way of not overdesigning a product that no one will pay money for.

What happens if the properties in space *X* seem essential to a product design? It is possible that by putting all the design parameters together one may end up cornered into a tight design space that is very difficult to manufacture. One approach to get around this design dilemma was reported by Zhao and Law in 2012.[74] Figure 6.23(a) shows the SEM micrograph of a groove-structured surface fabricated by the Bosch etching process followed by surface fluorination with FOTS. The surface comprises 3 μm wide parallel grooves (~4 μm height) separating by a 6 μm pitch. The SEM micrograph clearly shows the wavy sidewall structure due to the Bosch etching process. This groove surface was found to exhibit directional wetting properties with both water and hexadecane. Details of the wetting study and the evidence for hydrocarbon fluid forming the Cassie–Baxter state on the groove-structured surface have been published.[74] The contact angles of the hexadecane sessile droplet taken in directions parallel and orthogonal to the grooves are given in Figure 6.23(b). In the direction orthogonal to the groove, superhydrophobic- and superoleophobic-like contact angles were observed. However, both hysteresis and sliding angle are large. On the other hand, smaller contact angles are obtained in the parallel direction and the hysteresis and sliding angle are unexpectedly small. Table 6.2 compares the advancing and receding contact angle data and mechanical properties of three pillar array FOTS surfaces with the groove FOTS surface. The model superoleophobic surface with SEM micrograph shown in Figure 6.2 is the reference surface (row 1). With hexadecane, this surface is shown to exhibits a hysteresis of 40°, a sliding angle of 10°, and a wetting breakthrough pressure of 12.4 kPa. Increasing the pitch of the pillar (row 2) increases the D^* value, leading to a twofold improvement in both hysteresis and sliding angle. Unfortunately, there is a significant setback in the wetting breakthrough pressure. Reduction of the pillar height from row 2 to row 3 improves the mechanical

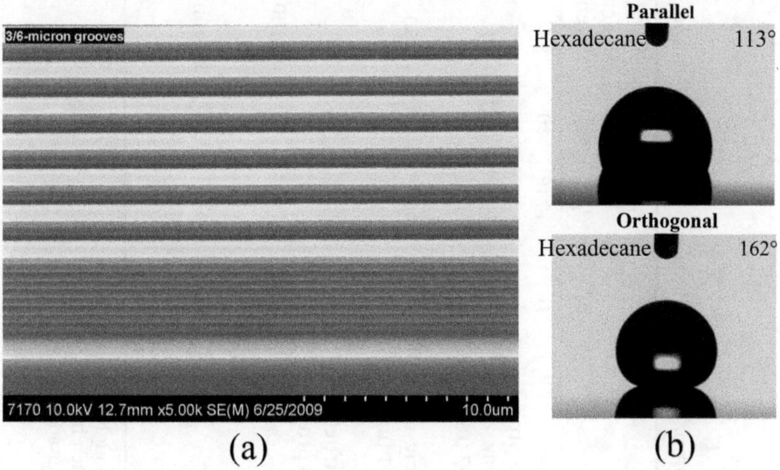

Figure 6.23 (a) SEM micrograph of a FOTS groove surface with wavy sidewall. (b) Hexadecane sessile drop data in parallel and orthogonal direction relative to the groove structure. Adapted with permission from H. Zhao and K. Y. Law, *Langmuir* 2012, **28**, 11812.[74] Copyright (2012) American Chemical Society.

Table 6.2 Comparison of surface and mechanical properties between pillar array and groove superoleophobic surfaces.

Surface texture	$\theta_A{}^c$	$\theta_R{}^c$	α^c	Relative mechanical strength	D^*	Breakthrough pressure	Remarks
FOTS pillar array (3 μm/6 μm/7.8 μm)[a]	161°	121°	10°	×1	2.25	12.4 kPa	Reference surface
FOTS pillar array (3 μm/12 μm/~7 μm)[a]	162°	145°	3.7°	×1	16	1.3 kPa	Hysteresis: 2× improvement Slide angle: ~2× reduction Breakthrough pressure: 10x worsening
FOTS pillar array (3 μm/12 μm/~1 μm)[a]	160°	140°	4.8°	×50	16	1.3 kPa	Hysteresis: 2× improvement Slide angle: 2× reduction Breakthrough pressure: 10x worsening
FOTS groove structure (3 μm/6 μm/~4 μm)[b]							
Parallel	119°	102°	4°	×100	16	~12.4 kPa	Hysteresis: >2× improvement Slide angle: 2× reduction Breakthrough pressure: the same
Orthogonal	164°	98°	34°	×100	16	~12.4 kPa	Hysteresis: ~2× worsening Slide angle: ~4× worsening Breakthrough pressure: the same

[a] Diameter/pitch/height.
[b] Groove width/pitch/height.
[c] Measured with hexadecane.

property significantly, but the low wetting breakthrough pressure remains as the weak link of the surface design. With the groove surface (row 4), the abrasion resistance is definitely better and the wetting breakthrough is expected to be slightly higher than the reference. In the orthogonal direction, although large contact angles are obtained, there are serious drawbacks with the hysteresis and sliding angle performance. In contrast, improvements in hysteresis and sliding angle are obtained in the parallel direction despite the smaller contact angle. The overall results suggest that the groove surface should exhibit a self-cleaning property in the parallel direction. Since drop sliding is driven by gravity, the only constraint the groove surface imposes on the application is the requirement to have the grooves align with the gravitational pull when incorporated into a self-cleaning device. This exercise illustrates that one can improve hysteresis, sliding angle, and mechanical properties by simply converting a pillar array design to a groove design. The slight reduction in contact angle and the need for device alignment in product design seem a small price to pay.

6.5.3 Challenges in Manufacturing

Numerous artificial superhydrophobic and superoelophobic surfaces have been reported in the literature. They exist in many different forms and shapes, with the rough structures ranging from being totally random to regularly patterned. They can be made from a bottom-up or top-down approach as well as by moulding, embossing and nano-imprinting. A summary of the different fabrication methods can be found in recent reviews.[21,22] Here we are concerned with generic issues relating to large-scale, large-area manufacturing.

6.5.3.1 Process Variations and Latitude

Process variations are unavoidable during manufacturing and they can cause variations in surface performance. Variations can be chemical or physical in nature. For example, a slight variation in coating concentration can lead to a slight change in surface property or film thickness. Slight dimensional/geometrical variation can be caused by tools or operator during coating, lithography or moulding. This kind of variation can be reduced through better engineering practices. Any negative impact from these variations can also be controlled by choosing a surface design (including material, fabrication process, and surface property achieved) that is robust and less sensitive to process variation.

Manufacturers prefer processes with wide latitude, more forgiving in performance when small changes in material concentration or process parameters occur. An example to illustrate the complex relationship between filler material concentration, surface repellency, and mechanical abrasion resistance can be found in a recent study by Campos and coworkers, who reported the fabrication of a series of super liquid-repellent surfaces with varying FF-silica (fluoroalkyl-funcationalized silica)[75] concentration in a fluoroelastomer by the spray-coating technique.[76] The contact angles and sliding angles of the surfaces as a function of FF-silica loading were studied with four solvents, including water and hexadecane. The results show that superhydrophobicity (WCA >150°, sliding angle <20°) is observed for surfaces with FF-silica loading at ≥20%. On the other hand, superoleophobicity (hexadecane contact angle >150°, sliding angle ~20°) is only achieved at 80% FF-silica loading. Scratch resistance tests were performed on these surfaces, revealing that FF-silica particles start to come off the surface when rubbed at loading >30%. By overlaying the liquid repellency and scratch resistance performance as a function of FF-silica loading, it becomes clear that there is only a 10% window (20–30%) in which to fabricate a scratch-resistant, superhydrophobic surface from this material package; however, the window to make a scratch-resistant superoleophobic surface is non-existent.

6.5.3.2 *Manufacturing Defects*

Defects are a fact of life in manufacturing technology. The way to cope with them is to reduce the rate of occurrence to background level. Presumably, the level of tolerance will depend on the type of defect and the targeted application. Very little has been known about the effects of defects on superhydrophobicity or superoleophobicity till recently. Fang and Amirfazli[77] compared the anti-icing performance of two chemically identical superhydrophobic surfaces prepared by coating a Teflon-AF solution onto a rough aluminium substrate, which itself is superhydrophilic. Two methods were employed to apply the Teflon coating, spray-coating and dip-coating. The water advancing contact angles for both surfaces are the same at 151°, whereas the spray-on sample is found to exhibit a smaller receding angle (138° *vs.* 148°) and a larger hysteresis. An anti-icing experiment showed that the dip-coated superhydrophobic surface displays the best delay freeze behaviour. The performance of the spray-coated superhydrophobic surface is comparable to that of the superhydrophilic surface, the worst among the four surfaces tested. The result is attributable to pinholes in the Teflon coating due to the spray-coating process. The pinhole defects lead to the formation of microscopic superhydrophilic areas, which become nucleation sites for the freezing process. More work is needed to understand the roles of defects on the performance of superhydrophobic and superoleophobic surfaces.

6.5.4 Concluding Remarks

The surface community has come a long way in terms of developing and understanding the design rules for creating superhydrophobic and superoleophobic surfaces. Significant knowledge of wetting fundamentals has also been gained in the last decade.[78] There has been increased attention to addressing the robustness issues. Studies of manufacturing issues and their relationship to surface performance, materials design, and fabrication processes are starting to appear. It is now about 20 years since the report of the lotus effect. With the continuous market pull and the combined knowledge of the community, commercialized superoleophobic surfaces may be just a few years away from reality!

Acknowledgements

The authors thank Dr Kyoo-Chul Park (Harvard) for the help in preparing some of the figures in this work and Professor Eric Loth (University of Virginia) for helpful comments on the manuscript.

References

1. W. Barthlott and C. Neinhuis, *Planta*, 1997, **202**, 1.
2. C. Neinhuis and W. Barthlott, *Ann. Bot.*, 1997, **79**, 667.
3. W. Barthlott, T. Schimmel, S. Wiersch, K. Koch, M. Brede, M. Barczewski, S. Walheim, A. Weis, A. Kaltenmaier, A. Leder and H. F. Bohn, *Adv. Mater.*, 2010, **22**, 2325.
4. K. Koch and W. Barthlott, *Philos. Trans. R. Soc., A*, 2009, **367**, 1487.
5. K. Koch, H. F. Bohn and W. Barthlott, *Langmuir*, 2009, **25**, 14116.
6. K. Koch, B. Bhushan and W. Barthlott, *Prog. Mater. Sci.*, 2009, **54**, 137.
7. L. Zheng, X. Wu, L. Zeng and D. Wu, *Chin. Sci. Bull.*, 2004, **49**, 1779.
8. K. Liu, X. Yao and L. Jiang, *Chem. Soc. Rev.*, 2010, **39**, 3240.
9. Y. M. Zheng, X. F. Gao and L. Jiang, *Soft Matter*, 2007, **3**, 178.
10. X. Gao and L. Jiang, *Nature*, 2004, **432**, 36.
11. Y. T. Cheng, D. E. Rodak, C. A. Wong and C. A. Hayden, *Nanotechnology*, 2006, **17**, 1359.
12. K. Liu and L. Jiang, *Annu. Rev. Mater. Res.*, 2012, **42**, 231.
13. Y. Y. Yan, N. Gao and W. Barthlott, *Adv. Colloid Interface Sci.*, 2012, **169**, 80.
14. I. Sas, R. E. Gorga, J. A. Joines and K. A. Thoney, *J. Polym. Sci., Part B: Polym. Phys.*, 2012, **50**, 824.
15. S. Nishimoto and B. Bhushan, *RSC Adv.*, 2013, **3**, 671.
16. E. Celia, T. Darmanin, E. T. de Givenchy, S. Amigoni and F. Guittard, *J. Colloid Interface Sci.*, 2013, **402**, 1.
17. J. Drelich and A. Marmur, *Surf. Innovations*, 2013, **2**, 211.
18. P. Roach, N. J. Shirtcliffe and M. I. Newton, *Soft Matter*, 2008, **4**, 224.
19. A. Tuteja, W. Choi, M. Ma, J. M. Mabry, S. A. Mazzella, G. C. Rutledge, G. H. McKinley and R. E. Cohen, *Science*, 2007, **318**, 1818.
20. A. K. Kota, J. M. Mabry and A. Tuteja, *Surf. Innovations*, 2013, **1**, 77.
21. K. Liu, Y. Tian and L. Jiang, *Prog. Mater. Sci.*, 2013, **58**, 503.
22. K. Y. Law, in *Dekker Encyclopedia of Nanoscience and Nanotechnology*, ed. S. Lyshevski, CRC Press, New York, 3rd edn, 2014, p. 4657.
23. H. Bellanger, T. Darmanin, E. T. deGivenchy and F. Guittard, *Chem. Rev.*, 2014, **114**, 2694.
24. Z. Chu and S. Seeger, *Chem. Soc. Rev.*, 2014, **43**, 2784.
25. A. Tuteja, W. Choi, G. H. McKinley, R. E. Cohen and M. F. Rubner, *MRS Bull.*, 2008, **33**, 752.

26. A. Tuteja, W. Choi, J. M. Mabry, G. H. McKinley and R. E. Cohen, *Proc. Natl. Acad. Sci.*, 2008, **105**, 18200.
27. Y. Xiu, Y. Liu, D. W. Hess and P. C. Wong, *Nanotechnology*, 2010, **21**, 155705.
28. A. Steele, R. K. Nayak, A. Davis, M. C. Gupta and E. Loth, *J. Micromech. Microeng.*, 2013, **23**, 115012.
29. Y. H. Yeong, A. Davis, A. Steele, E. Loth and I. S. Bayer, *Surf. Innovations*, 2014, **2**, 70.
30. H. Zhao, K. Y. Law and V. Sambhy, *Langmuir*, 2011, **27**, 5927.
31. D. Oner and T. J. McCarthy, *Langmuir*, 2000, **16**, 7777.
32. R. Furstner, W. Barthlott, C. Neinhuis and P. Walzel, *Langmuir*, 2005, **21**, 956.
33. E. Martines, K. Seunarine, H. Morgan, N. Gadegaard, C. D. W. Wilkinson and M. O. Riehle, *Nano Lett.*, 2005, **5**, 2097.
34. A. Marmur, *Langmuir*, 2008, **24**, 7573.
35. W. Choi, A. Tuteja, S. Chhatre, J. M. Mabry, R. E. Cohen and G. H. McKinley, *Adv. Mater.*, 2009, **21**, 2190.
36. T. Fujii, Y. Aoki and H. Habazaki, *Langmuir*, 2011, **27**, 11752.
37. L. Cao, T. P. Price, M. Weiss and D. Gao, *Langmuir*, 2008, **24**, 1640.
38. A. Ahuja, J. A. Taylor, V. Lifton, A. A. Sidorenko, T. R. Salamon, E. J. Lobaton, P. Kolodner and T. N. Krupenkin, *Langmuir*, 2008, **24**, 9.
39. R. T. R. Kumar, K. B. Mogensen and P. Boggild, *J. Phys. Chem. C*, 2010, **114**, 2936.
40. N. Tillman, A. Ulman, J. S. Schildkraut and T. L. Penner, *J. Am. Chem. Soc.*, 1988, **110**, 6136.
41. H. Murase and T. Fujibayashi, *Prog. Org. Coat.*, 1997, **31**, 97.
42. B. Samuel, H. Zhao and K. Y. Law, *J. Phys. Chem. C*, 2011, **115**, 14852.
43. W. E. Tenhaeff and K. K. Gleason, *Adv. Funct. Mater.*, 2008, **18**, 979.
44. S. Lee, J. S. Park and T. R. Lee, *Langmuir*, 2008, **24**, 4817.
45. R. N. Wenzel, *Ind. Eng. Chem.*, 1936, **28**, 988.
46. A. B. D. Cassie and S. Baxter, *Trans. Faraday Soc.*, 1944, **40**, 546.
47. H. Zhao, C. K. Park and K. Y. Law, *Langmuir*, 2012, **28**, 14925.
48. K. A. Brakke, The Surface Evolver, *Exp. Math.*, 1992, **1**, 141.
49. J. L. Zhang and H. Zhao, *Surf. Innovations*, 2014, **2**, 103.
50. P. Lv, Y. Xue, Y. Shi, L. Lin and H. Duan, *Phys. Rev. Lett.*, 2014, **112**, 196101.
51. T. Liu and C. J. Kim, *Science*, 2014, **346**, 1096.
52. C. W. Extrand, *Langmuir*, 2011, **27**, 6920.
53. M. A. Nearing, J. M. Bradford and R. D. Holtz, *Soil Sci. Soc. Am. J.*, 1987, **51**, 1302.
54. L. S. Srinath, *Advanced Mechanics of Solids*, Tata McGraw-Hill Education, 3rd edn, 2009.
55. Y. Zhang, C. W. Lo, A. Taylor and S. Yang, *Langmuir*, 2006, **22**, 8595.
56. C. Greiner, A. Campo and E. Arzt, *Langmuir*, 2007, **23**, 3495.
57. L. Cao, H. H. Hu and D. Gao, *Langmuir*, 2007, **23**, 4310.
58. J. Wang, F. Liu, H. Chen and D. Chen, *Appl. Phys. Lett.*, 2009, **95**, 084104.

59. L. Feng, Y. Song, J. Zhai, B. Liu, J. Xu, L. Jiang and D. Zhu, *Angew. Chem., Int. Ed.*, 2003, **42**, 800.
60. Y. T. Cheng and D. E. Rodak, *Appl. Phys. Lett.*, 2005, **86**, 144101.
61. J. B. Boreyko and C. H. Chen, *Phys. Rev. Lett.*, 2009, **103**, 174502.
62. L. Gao and T. J. McCarthy, *Langmuir*, 2006, **22**, 2966.
63. B. Bhushan, Y. C. Jung and K. Koch, *Philos. Trans. R. Soc., A*, 2009, **367**, 1631.
64. W. Li and A. Amirfazli, *Soft Matter*, 2008, **4**, 462.
65. D. Ge, L. Yang, Y. Zhang, Y. Rahmawan and S. Yang, *Part. Part. Syst. Charact.*, 2014, **31**, 763.
66. T. Liu, W. Sun, X. X. Sun and H. Ai, *Langmuir*, 2010, **26**, 14835.
67. H. H. Liu, H. Y. Zhang and W. Li, *Langmuir*, 2011, **27**, 6260.
68. Y. Su, B. Ji, K. Zhang, H. Hao, Y. Huang and K. Hwang, *Langmuir*, 2010, **26**, 4984.
69. E. J. Ensikat, P. Ditsche-Kuru, C. Neinhuis and W. Barthlott, *Beilstein J. Nanotechnol.*, 2011, **2**, 152.
70. E. Bittoun and A. Marmur, *Langmuir*, 2012, **28**, 13933.
71. T. Verho, C. Bower, P. Andrew, S. Franssila, O. Ikkala and R. H. A. Ras, *Adv. Mater.*, 2011, **23**, 673.
72. L. C. Gao, T. J. McCarthy and X. Zhang, *Langmuir*, 2009, **25**, 14100.
73. K. Y. Law, *Surf. Innovations*, 2013, **1**, 57.
74. H. Zhao and K. Y. Law, *Langmuir*, 2012, **28**, 11812.
75. R. Campos, A. J. Guenthner, S. Haddad and J. M. Mabry, *Langmuir*, 2011, **27**, 10206.
76. R. Campos, A. J. Guenthner, A. J. Meuler, A. Tuteja, R. E. Cohen, G. H. McKinley, S. Haddad and J. M. Mabry, *Langmuir*, 2012, **28**, 9834.
77. G. Fang and A. Amirfazli, *Surf. Innovations*, 2014, **2**, 94.
78. K. Y. Law and H. Zhao, *Surface Wetting. Characterization, Contact Angle, and Fundamentals*, Springer, New York, 2015.

CHAPTER 7

Patterned Superhydrophobic Surfaces

ERICA UEDA[a] AND PAVEL A. LEVKIN*[a,b]

[a]Institute of Toxicology and Genetics, Karlsruhe Institute of Technology, Postfach 3640, 76021 Karlsruhe, Germany; [b]Department of Applied Physical Chemistry, Heidelberg University, Postfach 10 57 60, 69047 Heidelberg, Germany
*E-mail: levkin@kit.edu

7.1 Introduction

In this chapter, we review the fabrication and application of patterned superhydrophobic surfaces developed in recent years. We focus on how the properties and functionalities arising from patterns of superhydrophobicity combined with other surface properties, such as superhydrophilicity, on a substrate can be utilized for a range of diverse and interesting applications. The typical methods for creating uniform superhydrophobic surfaces are not always applicable for creating surfaces with patterns of wettability due to the complexity arising from the need to spatially impart different chemistry or morphology in specific locations on the surface.

The development of surfaces with patterns of wettability is being actively explored and various practical applications have already been realized, often through inspiration from nature, with the potential for many more. Nature has designed surfaces with patterns of varying wettability that are important, for example, for the survival of insects and plants. Desert beetles have

RSC Soft Matter No. 5
Non-wettable Surfaces: Theory, Preparation, and Applications
Edited by Robin H. A. Ras and Abraham Marmur

Published by the Royal Society of Chemistry, www.rsc.org

both wettable and non-wettable regions on their backs. They collect water from fog by nucleation on the non-waxy hydrophilic peaks until the water droplet grows to a critical size and then rolls down the waxy hydrophobic bumps.[1] Many groups have fabricated patterned superhydrophobic surfaces to try to mimic the beetle's ability to capture water from humid air.[2–6] The carnivorous *Nepenthes* pitcher plant also has regions of hydrophilicity and hydrophobicity to help it capture its prey.[7] The unique slippery, liquid, and self-restoring nature of the inside surface of the *Nepenthes* pitcher plant has inspired a relatively new class of surfaces termed slippery liquid-infused porous surfaces (SLIPS) that have already been developed for diverse applications such as anti-biofouling and anti-icing.[8]

These examples of wettable, non-wettable, and patterned surfaces that occur in nature can inspire new surface designs for real-world applications. In this chapter, we focus on some of the practical advantages that arise from the difference in wettability between wettable and non-wettable regions patterned on a surface: (A) wettability patterns can form surface tension-confined microchannels; (B) superhydrophobic regions in a Cassie–Baxter state can control bioadhesion on surfaces; (C) discontinuous dewetting can passively dispense aqueous solutions into wettable regions surrounded by a non-wettable background; (D) the shape and positioning of liquid droplets, particles, or microchips can be easily controlled; and (E) droplets of liquid can be efficiently collected by directing the flow of droplets. Recent methods for creating surfaces with patterns of wettability and their specific applications are discussed.

7.2 Fabrication of Surfaces with Patterned Wettability

A variety of methods are available to fabricate or tune the chemistry and morphology of surfaces to produce various wettability characteristics. However, creating surfaces patterned with combinations of extreme wetting properties, such as superhydrophilicity and superhydrophobicity, that are robust, stable, and relatively easy to fabricate is still challenging and being actively explored. In this section, we present some of the methods used to create surfaces with patterned wettability.

7.2.1 UV Light Irradiation

Takai and coworkers fabricated patterned surfaces by exposing superhydrophobic surfaces to UV light to transform the treated regions to superhydrophilic.[9] Superhydrophobic films were deposited on glass plates or Si wafers by microwave plasma-enhanced chemical vapour deposition (CVD) of a trimethylmethoxysilane and Ar gas mixture. Then, UV light with a wavelength of 172 nm was irradiated on the substrate for 30 min through a photomask to decompose the methyl groups to create superhydrophilic regions.

The process resulted in an irregular surface topography composed of granular particles and nanoscale pores on the order of a few hundred nanometers in diameter, which contributed to the superhydrophobicity. The static (θ_{st}), advancing (θ_{adv}), and receding (θ_{rec}) water contact angles of the superhydrophobic surface were 155°, 157°, and 153°, respectively. The θ_{st} of the superhydrophilic surface was 0°.

7.2.2 Phase Separation and UVO Irradiation

Mano and coworkers used a phase separation method to transform smooth hydrophobic polystyrene (PS) surfaces to rough superhydrophobic surfaces, followed by UV/ozone (UVO) irradiation to create superhydrophilic patterns on the surface.[10–13] The detailed procedure described here is from Oliveira *et al.*[11] A solution of PS (70 mg ml^{-1}) in tetrahydrofuran (THF) was prepared, and then ethanol (100% v/v) was added to the PS/THF solution at a ratio of 1.35 : 2 (v/v). A few drops of this mixture were applied to smooth PS surfaces of 0.25 mm thickness for 5 s, after which the excess mixture was removed and the substrate immersed in ethanol. The substrates were dried at room temperature, and the resulting random nano- and microstructures created an average surface roughness of 13 μm and transformed the surface to superhydrophobic. The rough PS surfaces had a θ_{st} of 151°. To create superhydrophilic–superhydrophobic patterned surfaces, the rough superhydrophobic PS surfaces were modified by UVO irradiation through a hollowed mask for 18 min to create superhydrophilic regions with a θ_{st} of 0°.

7.2.3 Hydrophilic–Superhydrophobic Black Silicon Patterned Surfaces

Chang and coworkers used single-side-polished silicon wafers as substrates to fabricate patterned hydrophilic–superhydrophobic surfaces.[14] First, 700 nm of silicon dioxide (silica) hard mask material was deposited onto the silicon wafer using plasma-enhanced CVD, and then standard photolithography with positive tone resist was used to spin and pattern a photoresist (AZ5214E) on top of the silica layer. The photoresist served as a mask for the hydrophilic sites while the unmasked silica was etched in buffered hydrofluoric acid, and then the photoresist was removed in an acetone bath. Next, black silicon was formed by the method of cryogenic inductively coupled plasma (SF_6/O_2) reactive ion etching and consisted of a random array of vertical nanospikes that contribute to the antireflective properties of black silicon.[15] Lastly, a thin layer (~50 nm) of a low surface energy fluoropolymer was deposited using CHF_3 in a reactive ion etcher, and was then lifted off from the hydrophilic sites in buffered hydrofluoric acid. This method resulted in hydrophilic sites patterned on a superhydrophobic black silicon surface.

7.2.4 UV-Initiated Free Radical Polymerization and Photografting

Polymer substrates are advantageous because of the diversity in chemical composition, surface and bulk properties, and processing techniques that are possible.[16,17] Photoinitiated polymerization and grafting allows precise control over where porous polymers, or monoliths, are formed and where surface modification takes place by irradiating specific locations with UV or visible light.

Methods based on UV-initiated free radical polymerization and surface grafting have been used to create superhydrophilic and superhydrophobic porous polymer films using different monomers and porogen ratios to control the bulk chemistry, morphology, and porosity.[18–27] The polymerization mixtures contained monovinyl and divinyl monomers, initiator, and a mixture of porogens that were required to generate porosity and surface roughness. The porous structures and globules introduced both micro- and nanoscale roughness to the material and could be easily tuned, without changing the chemistry, by changing the porogens while keeping the monomer content the same in the polymerization mixture.

Levkin *et al.* introduced a method to produce superhydrophobic porous polymer films based on UV- and thermo-initiated polymerization of alkyl methacrylates.[18] Han *et al.* and Zahner *et al.* polymerized a thin film of microporous (1–4 µm pore size) superhydrophobic or nanoporous (100–200 nm pore size) hydrophobic butyl methacrylate crosslinked with ethylene dimethacrylate (BMA-EDMA) and then modified it with a hydrophilic monomer by UV-initiated photografting through a photomask to create superhydrophilic micropatterns.[19,20] Positively charged, negatively charged, or neutral functionality could be introduced into the superhydrophilic micropatterns.[20] It is important to note that photografting occurred through the whole thickness of the porous polymer matrix resulting in the formation of three-dimensional, superhydrophilic, surface tension-confined microchannels.

Auad *et al.* used the attributes of this BMA-EDMA polymer film (the porous structure and the whole-thickness modification) to develop a simple method to rapidly create multiple superhydrophilic–superhydrophobic patterned substrates from a single template.[24] Each time adhesive tape was pressed onto the surface of a 125 µm thick BMA-EDMA polymer film and then peeled off, a thin layer of the patterned polymer was transferred to the tape and could be used as a patterned substrate. This method allowed up to 12 copies to be produced from one polymer film, thereby saving significant time and expense in creating patterned substrates.

As an alternative approach to making superhydrophilic–superhydrophobic patterned polymer surfaces, Geyer *et al.*, Ueda *et al.*, and Efremov *et al.* first prepared 12.5 µm thin, nanoporous, superhydrophilic poly(2-hydroxyethyl methacrylate-*co*-ethylene dimethacrylate) (HEMA-EDMA) polymer films by UV-initiated free radical polymerization, which were then modified with 2,2,3,3,3-pentafluoropropyl methacrylate (PFPMA) by UV-initiated photografting through a quartz photomask to create superhydrophobic micropatterns with defined geometries (Figure 7.1).[21–23,25,27] This method is fast, flexible with respect to the monomers that can be used, and enables large areas to be patterned at once.

7.2.5 Surface Patterning *Via* Thiol-yne Click Chemistry

Photoinitiated click reactions have also been actively investigated for creating patterned surfaces due to their excellent spatial and temporal control over photochemical processes. Thiol-yne reactions are particularly advantageous because they can proceed efficiently and rapidly at room temperature and in the presence of oxygen or water, they do not require expensive or toxic catalysts, and they are compatible with a wide range of functional groups.

Patton and coworkers used thiol-yne chemistry in conjunction with UV lithography to create hydrophilic–hydrophobic patterned surfaces.[28] Poly(propargyl methacrylate) brushes with "yne" functionalities were

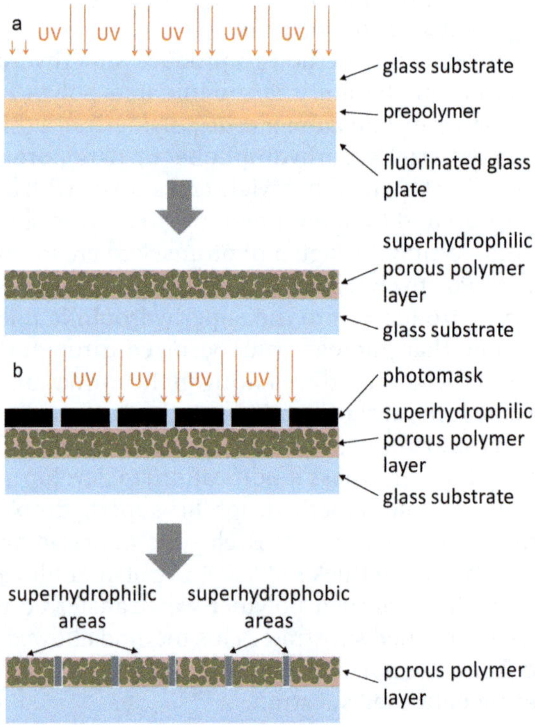

Figure 7.1 (a) Schematic of the fabrication of a superhydrophilic porous polymer film on a glass substrate by UV-initiated free radical polymerization. (b) Schematic of the fabrication of a superhydrophobic grid-like pattern on the superhydrophilic surface by UV-initiated photografting. Adapted with permission from John Wiley and Sons ref. 21. Copyright © 2011 Wiley-VCH Verlag GmbH & Co. KGaA, Weinheim.

produced *via* surface-initiated photopolymerization and subsequently functionalized with commercially available thiols.

Feng *et al.* also demonstrated the use of UV-induced sequential thiol-yne click chemistry, but as an extremely fast and initiator-free approach to create superhydrophilic–superhydrophobic micropatterns (Figure 7.2).[29] Since the thiol-yne reaction could also be performed at room temperature in water, this method was able to produce surfaces patterned with peptides as well as a variety of reactive functional groups containing a terminal thiol (*e.g.* OH, NH_2, or COOH). First, a 12.5 μm thin, porous (50% porosity, 80–250 nm pores) polymer layer of poly(2-hydroxyethyl methacrylate-*co*-ethylene dimethacrylate) (HEMA-EDMA) was prepared on a glass substrate.[21] Second, the HEMA-EDMA layer was modified with 4-pentynoic acid through a standard esterification procedure to create an intermediate, reactive alkyne surface. The resulting porous polymer bearing alkyne groups was then functionalized *via* thiol-yne click reactions initiated by irradiation with 260 nm UV light (12 mW cm^{-2}) at room temperature to transform the surface to either superhydrophobic or superhydrophilic, depending on whether hydrophobic or hydrophilic thiols were used. The reaction proceeded extremely fast, requiring only 0.5 s of UV irradiation in the presence of an initiator (2,2-dimethoxy-2-phenylacetophenone) and only 5 s without any initiator; no reaction occurred without UV light. Functionalization of the alkyne surface with cysteamine transformed the hydrophobic alkyne polymer (θ_{st} = 124°) into a superhydrophilic surface (θ_{st} = 4.4°), whereas modification with 1-dodecanethiol or 1*H*,1*H*,2*H*,2*H*-perfluorodecanethiol resulted in a superhydrophobic surface with θ_{adv}, θ_{st}, and θ_{rec} measured to be 171°, 169°, and 162° or 173°, 170°, and 164°, respectively. The porous structure of the HEMA-EDMA polymer layer resulted in a rough surface, which was proved to be an important feature for fabricating the superhydrophilic or superhydrophobic surfaces.

To create a surface with patterned wettability, the reactive alkyne surface was first modified with 5% (v/v) 1*H*,1*H*,2*H*,2*H*-perfluorodecanethiol in acetone in specific areas by irradiation with UV light through a photomask. After rinsing the substrate with acetone, the remaining non-irradiated, unmodified, reactive alkyne groups were subject to a thiol-yne reaction with 15 wt% cysteamine hydrochloride in an ethanol–water solution (1 : 1) without the need for a photomask during UV irradiation. This resulted in a surface patterned with both superhydrophilic and superhydrophobic properties, and pattern sizes as small as 10 μm could be produced.

Simply substituting the thiols with those of other functionalities during the sequential thiol-yne reactions can produce surfaces patterned with different chemistries. Since functionalization of the alkyne surface could be performed without an initiator in either apolar or polar solvents, including water, this allowed compatibility of the method with thiol-containing biomolecules, such as proteins or peptides. This was demonstrated by patterning a peptide containing a terminal cysteine residue (fluorescein-β-Ala-GGGGC) on the reactive alkyne-functionalized surface.

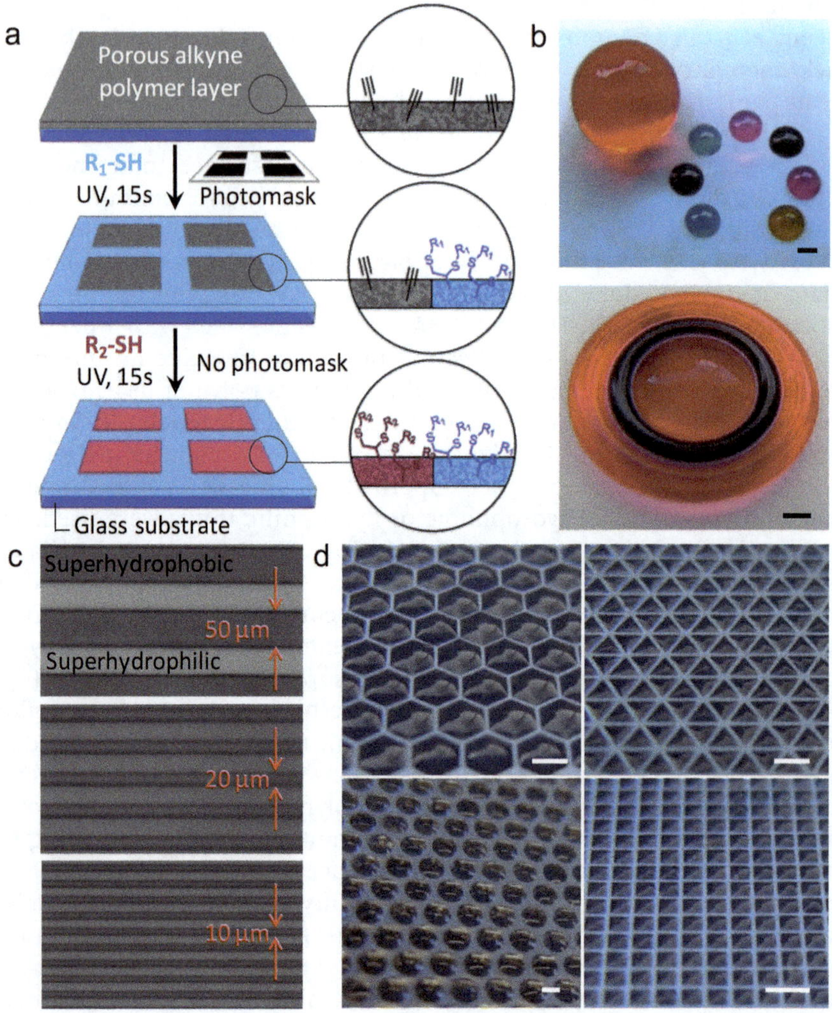

Figure 7.2 Fabrication of superhydrophilic–superhydrophobic patterns *via* thiol-yne photo-click reactions. (a) Schematic representation of the thiol-yne photo-click reaction for creating superhydrophobic–superhydrophilic micropatterns using an alkyne-modified porous polymer layer as a substrate. Optical images of (b) superhydrophilic–superhydrophobic patterns filled with dye–water solutions; superhydrophobic gap between the two rings is 100 μm. (c) Superhydrophilic regions (light areas) separated by superhydrophobic gaps (dark areas) of different widths. (d) Droplet-Microarrays formed by dipping the superhydrophobic–superhydrophilic arrays with different geometries into water. Wetted parts become transparent (dark). Scale bars are 1 mm. Reproduced with permission from John Wiley and Sons ref. 29 Copyright © 2014 Wiley-VCH Verlag GmbH & Co. KGaA, Weinheim.

7.2.6 Surface Functionalization *Via* Thiol-ene Reaction

Li *et al.* introduced a surface modification method based on creating a superhydrophobic surface with reactive vinyl groups functionalized with either molecules bearing thiol groups through a UV-triggered thiol-ene reaction or with molecules bearing disulfide groups through a UV-triggered disulfide-ene reaction.[30] They hypothesized that disulfides could react with alkenes in a similar way that thiols react with alkenes under UV light since sulfenyl radicals can be produced from disulfides upon UV irradiation. First, trichlorovinylsilane was polycondensed on a glass substrate to create thin, transparent, porous silicone nanofilaments (~30–50 nm in diameter) bearing reactive vinyl groups that formed a photoactive, inscribable, non-wettable, and transparent surface (PAINTS). Although no fluoro-containing functionalities were present, the PAINTS was superhydrophobic with θ_{st} = 166° and a water contact angle hysteresis of ~2°. High surface roughness of the silicone filaments as well as the porosity of the bulk nanofilament film probably contributed to the superhydrophobicity. This method allowed PAINTS to be easily fabricated on 3D glass objects of complex shapes, such as the inside of a glass vial and the convex side of a watch glass, without compromising their transparency.

Both the thiol-ene and disulfide-ene reactions were used to modify PAINTS to create superhydrophobic–hydrophilic patterned surfaces. To demonstrate modification using the thiol-ene reaction, a PAINTS-coated glass slide was wetted with a 10% (v/v) cysteamine in ethanol solution and irradiated with 260 nm UV light (~9 mW cm^{-2}) for 15 s. This transformed the superhydrophobic PAINTS into a highly hydrophilic surface possessing a θ_{st} of ~6°. If the UV irradiation was done through a photomask, only the irradiated regions on the PAINTS became highly hydrophilic. The non-irradiated regions still possessed reactive vinyl groups that were then modified through another thiol-ene reaction using 1*H*,1*H*,2*H*,2*H*-perfluorodecanethiol to create a superhydrophobic–hydrophilic patterned surface.

For modification through the disulfide-ene reaction, the PAINTS was wetted with a 10% (v/v) 3,3-dithiodipropionic acid in ethanol solution and irradiated with 260 nm UV light for 3 min. Again, the superhydrophobic PAINTS was transformed into a highly hydrophilic surface with a θ_{st} of ~5.1°. If a photomask covered the PAINTS during UV irradiation, then a highly hydrophilic micropattern was formed on the superhydrophobic PAINTS.

7.2.7 Surface Functionalization *Via* UV-Induced Tetrazole–Thiol Reaction

Feng *et al.* introduced a versatile UV-induced, tetrazole–thiol reaction that can be used for rapid catalyst-free polymer–polymer conjugation, efficient surface functionalization and patterning, and the functionalization of (bio) molecules bearing periphery thiol groups.[31] The method is based on 1,3-dipolar nucleophilic addition of thiols to tetrazoles, which when induced by

UV light allows the reaction to proceed rapidly at room temperature without a catalyst, with high yields, and in both polar protic and aprotic solvents. Superhydrophobic–hydrophilic micropatterns were created using this method by sequentially modifying a tetrazole-functionalized porous polymer surface with hydrophobic and hydrophilic thiols.

First, a 12.5 μm thin, hydrophilic porous polymer film composed of poly(2-hydroxyethyl methacrylate-*co*-ethylene dimethacrylate) (HEMA-EDMA) was created on a glass substrate.[21] Second, esterification of the hydroxyl groups on the HEMA-EDMA surface by 4-(2-phenyl-2*H*-tetrazol-5-yl) benzoic acid was carried out to transform the hydrophilic HEMA-EDMA surface (θ_{st} = 5°) into a hydrophobic tetrazole surface (θ_{st} = 115°). Then, sequential modifications of the tetrazole surface through UV-induced tetrazole–thiol reactions were used to create patterns of wettability on the surface. The surface was site-selectively modified with a 20% (v/v) 1*H*,1*H*,2*H*,2*H*-perfluorodecanethiol in ethyl acetate solution by irradiation with 260 nm UV light (5 mW cm^{-2}) through a photomask for 2 min, and then subsequently modified with a 20 wt% cysteamine hydrochloride in 1 : 1 ethanol–water solution under UV irradiation without a photomask. Regions of the surface modified with 1*H*,1*H*,2*H*,2*H*-perfluorodecanethiol exhibited superhydrophobicity with θ_{st}, θ_{adv}, and θ_{rec} as high as 167°, 170°, and 161°, respectively, whereas regions modified with cysteamine hydrochloride transformed the hydrophobic tetrazole surface to hydrophilic (θ_{st} = 22°). Patterns with feature sizes as small as 10 μm were feasible using this method.

7.2.8 Surface Modification Through Polydopamine

In recent years, a novel, relatively simple, and versatile method for surface modification inspired by the adhesive ability of mussels has been actively researched and developed since it was first introduced by Messersmith, Lee and coworkers.[32] Small molecules containing catecholamine functional groups, such as dopamine, are used as structural mimics of 3,4-dihydroxy-L-phenylalanine, a critical molecule found in adhesive proteins produced by mussels, and *in situ* oxidative polymerization of dopamine into a thin layer of polydopamine (PDA) is used to coat and subsequently immobilize molecules on surfaces.[33,34] PDA can be coated onto a wide variety of substrates such as ceramics, glass, metals, oxides, polymers, and silica. Surfaces can be functionalized in one step by simply coating the surface with a mixture of dopamine and the molecules to be immobilized at alkaline pH.[35–38] Polymerization of dopamine can also be controlled by exposure to UV light, even in acidic and neutral conditions, which also allows micropatterns of polydopamine to be created.[39] Further insight into applications of PDA is provided in several in-depth reviews.[40,41]

Lee and coworkers used oxidative self-polymerization of dopamine to transform superhydrophobic to hydrophilic surfaces, and created patterned surfaces by partially exposing the surface to a dopamine solution for 18 h through micromoulded capillaries.[42] The superhydrophobic surfaces were created by coating anodic aluminium oxide (AAO) membranes with fluorosilane by gas-phase deposition. When the superhydrophobic AAO surfaces were immersed in a dopamine solution for 18 h, the surfaces changed from superhydrophobic to hydrophilic with a decrease in θ_{st} from 158.5 ± 2.8° to 37.3 ± 2.6°. To fabricate superhydrophobic–hydrophilic patterned surfaces, an alkaline dopamine solution (2 mg ml^{-1}) was injected into the microchannels and incubated for 18 h to create hydrophilic line patterns 50 μm wide.

Wang and coworkers introduced a mask-free method for creating well-defined, superhydrophilic micropatterns on a superhydrophobic surface based on the use of a piezoelectric-based inkjet printer to dispense picolitre drops of dopamine solution directly onto the superhydrophobic surface, followed by *in situ* polymerization of dopamine to PDA.[43] The dopamine solution was optimized to achieve a Wenzel wetting state to maximize interaction between the dopamine and rough superhydrophobic surface, while also having a high contact angle to precisely control the deposition of the droplet of dopamine solution on the superhydrophobic surface. In addition, to allow enough time for the oxidative self-polymerization of dopamine to take place, the surface tension as well as the vapour pressure of the aqueous dopamine droplets was reduced by adding water-miscible solvents with low surface tension (*e.g.* ethanol) or low vapour pressure (*e.g.* ethylene glycol) to induce a transition from a Cassie to a Wenzel wetting state and to prolong the time available for polymerization before evaporation of the droplet.

The superhydrophobic substrates were fabricated by spin-coating silica nanoparticles and PS granules (1–2 mm, MW 350 000) onto precleaned glass slides, calcination to fuse the silica nanoparticles together, and then coating with a semifluorinated silane of 1*H*,1*H*,2*H*,2*H*-perfluorooctyltriethoxysilane by CVD. The superhydrophobic surface exhibited θ_{st} of approximately 157° and a sliding angle of <1°. To create the superhydrophilic micropatterns on the superhydrophobic surface, the printer cartridge was filled with a freshly prepared dopamine solution (5.0 mg ml^{-1}) in a mixture of water, ethanol, and ethylene glycol (1:1:1, v/v/v), and then the inkjet printer dispensed 10 pL droplets of the dopamine solution in predesigned patterns onto the superhydrophobic substrate at room temperature. The substrate was then stored in a sealed chamber at 50 °C for 36 h to allow the dopamine to polymerize. The thickness of the PDA coating was ~45 nm and exhibited a θ_{st} of 0°. The smallest achievable feature size of the printed pattern was ~50 μm. The PDA patterns were stable for at least 1 year under ambient conditions, and could withstand flushing with water as well as with organic solvents such as acetone and ethanol.

7.2.9 Superomniphobic–Superomniphilic Patterned Surfaces

Tuteja and coworkers developed superomniphobic–superomniphilic patterned surfaces that were compatible with aqueous as well as non-aqueous, low surface tension liquids such as oil and alcohol.[44] Superomniphobic surfaces are characterized as both superhydrophobic and superoleophobic, and similarly superomniphilic surfaces are characterized as both superhydrophilic and superoleophilic. Superomniphobic surfaces were fabricated by electrospinning solutions of 50 wt% 1*H*,1*H*,2*H*,2*H*-heptadecafluorodecyl polyhedral oligomeric silsequioxane (fluorodecyl POSS) and poly(methyl methacrylate) (PMMA). Superomniphobicity was achieved due to the highly porous, re-entrant, bead morphology of the electrospun surfaces and the low surface energy of the fluorodecyl POSS–PMMA blend. With both water and heptane, the surfaces exhibited $\theta_{adv} = 162°$ and low contact angle hysteresis $\Delta\theta = 2°$. Superomniphilic surfaces ($\theta_{adv} = \theta_{rec} \approx 0°$ for water and heptane) were obtained by exposing the superomniphobic surfaces to O_2 plasma, which most likely led to oxygen enrichment and simultaneous degradation of the fluorinated end groups in fluorodecyl POSS. Exposing the superomniphobic surface to O_2 plasma treatment through a photomask resulted in patterned surfaces with superomniphilic regions on a superomniphobic background.

7.2.10 Amine-Reactive Modification of Superhydrophobic Polymers

Lynn and coworkers presented a rapid and simple modification method based on printing an amine-containing "ink" on a superhydrophobic polymer layer to form superhydrophilic patterns.[45] The amine groups in the ink covalently attached to the amine-reactive azlactone groups of the superhydrophobic polymer, which was fabricated by layer-by-layer assembly of poly(ethyleneimine) (PEI) and poly(2-vinyl-4,4-dimethylazlactone) (PVDMA). Azlactone groups can rapidly react with primary amines through ring–opening reactions in the absence of catalysts and without the generation of by-products. The superhydrophobic PEI/PVDMA multilayers were modified through the whole thickness of the multilayer by D-glucamine (hydrophilic amine) and subsequently by n-decylamine (hydrophobic amine). First, a poly(dimethylsiloxane) (PDMS) stamp was used to contact print superhydrophilic glucamine spots (~300 μm^2) onto the PEI/PVDMA substrates, and then the substrates were immersed in a decylamine solution for 1 h to render the non-reacted azlactone groups superhydrophobic and non-wettable. Superhydrophilic patterns could also be directly drawn on the surface using a homemade glucamine-ink pen or a glucamine-soaked thread.

7.2.11 Patterns of Reversible Wettability

The methods described so far for creating surfaces with patterns of wettability are basically irreversible. Creating patterned surfaces that can easily revert back to their unmodified state is useful for applications constantly requiring many templates or changes in design patterns, such as lithography.

Fujishima and coworkers made superhydrophilic–superhydrophobic patterned substrates using a mask-free method, which were ultimately used for offset printing.[46] A rough, anodized Al plate was coated with TiO_2, modified with octadecylphosphonic acid (ODP) self-assembled monolayers (SAMs) ($\theta_{st} \approx 154°$), and patterned with UV light-resistant ink by inkjet printing, which acted as a photomask. The surface was irradiated with UV light to photocatalytically decompose the ODP SAMs not covered by ink and render the regions superhydrophilic. After the ink was removed by washing with water, a superhydrophilic–superhydrophobic patterned substrate was obtained. The whole substrate could be rendered superhydrophilic again by simply irradiating the surface with UV light for 5 h. The whole process could be repeated again from the ODP modification to generate another patterned surface.

Chi and coworkers also demonstrated that patterns of wettability on a superhydrophobic surface could be alternately generated and then erased when using an alcohol-based ink.[47] Additionally, the superhydrophobic regions could be reversibly adjusted between a sticky or sliding behaviour of droplets on the surface. An electrochemical anodizing process was used to fabricate a surface composed of vertically aligned and uniformly distributed TiO_2 nanotubes of 100 nm outer diameter and 400 nm length. Water on the TiO_2 nanotube array (TNA) film rapidly spread and wetted the surface, but after modification with 1*H*,1*H*,2*H*,2*H*-perfluorooctyltriethoxysilane the surface became superhydrophobic ($\theta_{st} = 160°$) with water droplets exhibiting a spherical shape and low adhesion to the surface. A pen or inkjet printer was used to dispense a smooth layer of the waterproof, alcohol-based ink (surface tension in the range of 20–32 mN m^{-1}) on the surface of the superhydrophobic TNA. The smoothness of the hydrophilic ink layer reduced the surface roughness and increased the solid–liquid contact area, resulting in enhanced water adhesion and hydrophilicity ($\theta_{st} \approx 82°$) on the inked regions.

Superhydrophobic–superhydrophilic patterns were fabricated by taking advantage of the photocatalytic property of TiO_2 and the ability of the ink to act as a photomask to shield the underlying surface from UV light. First, a hydrophilic ink pattern was drawn with a pen on the superhydrophobic TNA film. Then the surface was irradiated with UV light, during which the fluoroalkyl groups not protected by ink were decomposed by the TiO_2 film and resulted in a wettability change from superhydrophobic to superhydrophilic. After removing the ink layer by rinsing the surface with a methanol solution, the superhydrophobic TNA surface had superhydrophilic micropatterns. Larger and complex ink patterns can be achieved by using an inkjet printer to dispense ink on the superhydrophobic TNA surface.

To adjust the superhydrophobic surface from a sliding Cassie–Baxter state to a sticky Wenzel state, a micro-sized ink layer was created on the superhydrophobic TNA surface. Droplets were pinned on ink dots on the superhydrophobic TNA surface even at a tilt angle of 90° or 180°. To revert back to the sliding state, the alcohol-based ink was removed from the surface by simply washing the surface in a methanol solution for 30 s. Water droplets rolled off the recovered superhydrophobic TNA surface when the substrate was tilted >10°. This quick and reversible adhesion switching on the superhydrophobic TNA surface could be repeated many times by printing then erasing the ink.

Li *et al.* presented a rapid and simple method of printing an "ink", a phospholipid in ethanol solution, onto a thin, microporous, superhydrophobic polymer layer to create superhydrophilic spots within a superhydrophobic background.[48] The ink was simply dispensed on the surface and then dried under ambient conditions. The method is compatible with various printing technologies such as contact printers, dip-pen nanolithography, and inkjet printers to make high-density arrays or intricate patterns of superhydrophilic spots. Chemicals that were added to the solution of phospholipid ink printed onto the superhydrophobic surface resulted in arrays of superhydrophilic spots prefilled with chemicals. This allows easy multiplexing and patterning of deposited substances.[49]

Liu and coworkers used an atmospheric-pressure plasma jet (APPJ) to create superhydrophilic patterns on various superhydrophobic metal surfaces by decreasing the hydrophobic fluorine-containing functional groups and increasing the hydrophilic oxygen-containing functional groups on the APPJ-treated regions.[50] The APPJ (~4 mm in diameter) was generated by bare electrode discharge without the need for expensive vacuum equipment. The superhydrophobic substrates were fabricated by first electrochemically treating polished metal plates (*e.g.* aluminium, copper, titanium, or zinc) to create roughness on the surface, and then immersing the substrates in a 1 wt% fluoroalkylsilane ($C_6F_{13}C_2H_4Si(OCH_2CH_3)_3$, FAS) in ethanol solution to fluorinate the surface and lower the surface free energy. Masks were used to selectively expose the superhydrophobic metal substrates to the APPJ to create superhydrophilic patterns such that the θ_{st} decreased from 159° to <5° after 45 s of exposure. APPJ treatment did not significantly affect the surface morphology. The superhydrophobicity of the surface could be recovered simply by immersing the substrate in the FAS–ethanol solution again for 2 min, and this reversible wettability transition could be repeated for at least five cycles.

7.3 Applications of Patterned Superhydrophobic Surfaces

In this section, we present a diverse range of applications that take advantage of the unique wetting properties of surfaces with patterned wettability. In general, these patterned surfaces are designed with wetting and non-wetting regions to control the deposition, flow, or position of liquids on surfaces, or alternatively to create bioadhesive and non-bioadhesive regions.[51]

7.3.1 Open Microfluidic Channels

Microfluidic devices are able to manipulate minute volumes of samples and thus can be used to carry out miniaturized chemical reactions and biological assays. However, external components such as pumps and valves are usually needed and add complexity to the platform. As an alternative to these microfluidic devices, the use of surface tension-confined microfluidic

(STCM) devices based on the principle of surface tension to passively control the movement of liquids, instead of pumps, is being actively developed for a variety of applications.[52] STCM devices transport small volumes of liquid along tracks that exhibit a higher surface energy compared to the background region through the use of surface tension gradients, capillary force, gravitational force, and pressure differences, without the need for pumps or physical walls along the tracks. For example, aqueous solutions can be transported along high surface energy, hydrophilic 2D channels while the surrounding hydrophobic regions act as virtual walls that contain the liquid along the hydrophilic path.

Fréchet and coworkers photopatterned a superhydrophilic channel in a superhydrophobic, porous polymer layer and used the channel to separate peptides of different isoelectric point and hydrophobicity by 2D thin layer chromatography.[19] The separated peptides were identified by desorption electrospray ionization mass spectroscopy directly from the polymer surface, which was possible due to the open nature of the device. There is great potential for open microfluidic devices used in conjunction with a variety of detector systems in the field of miniaturized separation and diagnostics.

Hancock *et al.* used surfaces with patterns of wettability to passively generate gradients of chemicals or particles.[53,54] Hydrophobic boundaries were created on a hydrophilic glass slide simply by using tape to mask the desired hydrophilic pattern and applying a hydrophobic spray to the background. When a solution was dispensed at one end of a hydrophilic stripe or more complex shape, the difference in curvature pressure generated a flow and created a concentration gradient by convection along the hydrophilic region as the hydrophobic boundaries contained the flow. Hancock *et al.* used the hydrophilic–hydrophobic patterns to also create gradients of solutes, cells, and microspheres encapsulated in 3D hydrogels by photocrosslinking prepolymeric solutions in the hydrophilic stripes.[55,56] Lin and coworkers also created 3D hydrogels using complex, geometric hydrophilic–hydrophobic patterns, which were subsequently used as moulds to form PDMS channels.[57]

Efremov *et al.* used surfaces patterned with arrays of hydrophilic spots surrounded by hydrophobic barriers to quickly generate various customized liquid structures on the same substrate without the need for complex equipment.[27] The method, termed "digital liquid patterning", was inspired by the working principle of a digital scoreboard. Similar to lighting up specific bulbs to form a pattern on the scoreboard, specific hydrophilic spots were filled with aqueous solutions to form a pattern of drops among the array of hydrophilic spots. The individual drops in adjacent hydrophilic spots could also be made to coalesce across the hydrophobic barrier using a pipette tip to create continuous liquid patterns or channels. This method was also used to create hydrogel structures as well as patterns and gradients of silica microparticles and cells within the surface tension-confined liquid channels. An advantage of this maskless method is that the desired liquid pattern can be changed on demand simply by manual pipetting.

Lee and coworkers used STCM channels to create a gravity-driven microfluidic device to move and mix droplets.[58] The 2D channels were formed by micropatterning lines of hydrophilic polydopamine (PDA) (60 μm wide, 200–300 nm thick) on nanostructured, superhydrophobic, AAO surfaces. Gravity caused water droplets to move along the patterned PDA microlines when the patterned surface was tilted downwards by 5°. To promote the mixing of two droplets, Y-shaped PDA microlines with a square micropatch (200 μm wide) of PDA at the intersection were patterned on the superhydrophobic AAO substrate. The first water droplet (10 μl) rolling down one arm of the Y-pattern was captured on the micropatch due to surface tension, but after the second water droplet (10 μl) was released down the other arm of the Y-pattern and mixed with the first droplet, the coalesced droplets continued to move downwards. The balance between the gravitational force and the surface tension applied to the droplets on the PDA micropatch was enough to capture the first droplet, but not enough to hold both droplets after mixing. The capability of this mixing device was further demonstrated by using it to synthesize gold nanoparticles. Thus, open STCM devices have proved to be a relatively simple tool to control the movement of droplets on surfaces and can be utilized as chemical microreactors.

7.3.2 Cell Patterning and Cell Microarrays

In this section, we discuss how patterns of wettability can be used to control the adhesion of proteins, cells, bacteria, and microorganisms on surfaces. Often, a non-wettable background is used to confine solutions within the wettable regions on a surface. Similarly, a background exhibiting a superhydrophobic Cassie–Baxter state is able to confine cells within the hydrophilic regions due to the inability of cells to adhere firmly to the surface because of the limited cell–surface contact area and reduced protein adsorption.

Superhydrophobic surfaces exhibiting the Cassie–Baxter state effectively trap air in the surface asperities, preventing the penetration of aqueous solutions.[59,60] Cell–surface interactions are influenced by proteins present on the surface and those that adsorb onto the surface from the culture medium, followed by deposition of the cell's own extracellular matrix (ECM). Since only a small fraction of the superhydrophobic surface contacts the culture medium, this likely reduces the number of sites available for protein adsorption and deposition of the cell's own ECM as well as subsequent focal adhesions.[61] Even at these limited points of contact between the culture medium and the superhydrophobic surface, adsorbed proteins such as fibronectin can demonstrate alterations of the domain conformations involved in cell adhesion, and reorganization or exchange of the adsorbed proteins with those deposited by the cells can be significantly inhibited on superhydrophobic surfaces.[61,62] These events would further discourage cell adhesion on superhydrophobic surfaces.

Takai and coworkers demonstrated that the design of superhydrophilic–superhydrophobic patterns can influence cell adhesion and morphology, as well as ECM production.[9] Mouse embryonic fibroblast (3T3) cells spontaneously recognized and migrated towards the superhydrophilic regions after being seeded on the patterned surface, and immediately adhered there and grew to confluence after 24 h. Maintenance of the cells within an array of superhydrophilic spots depended on the distance between the spots. At inter-spot distances of 150 μm, confluent cells in the superhydrophilic spots began to make contact with cells in adjacent spots. At longer inter-spot distances of 400 μm, no physical cell–cell contact occurred between cells in adjacent spots. In contrast, cells required 24–72 h after seeding to adhere to the superhydrophobic background. This difference in cell adhesion behaviour was attributed to the difference in protein adsorption on the superhydrophilic *versus* superhydrophobic regions. Since proteins adsorbed less efficiently on the superhydrophobic regions, the cells needed time to produce their own ECM to form a protein layer suitable for adhesion.

Mano and coworkers studied cell adhesion and proliferation on rough *versus* smooth hydrophobic and hydrophilic PS surfaces.[11] Human primary osteosarcoma cells (SaOs-2) adhered more on the smooth hydrophilic and rough superhydrophilic PS surfaces, and did not significantly attach or proliferate on either the smooth hydrophobic or rough superhydrophobic PS surfaces even after 6 days of culture. In contrast, mouse lung fibroblast cells (L929) reached confluence on both hydrophobic surfaces after 6 days of culture, which was attributed to the ability of fibroblasts to proliferate quickly, and specifically to the greater ability of L929 cells (compared to other cell types) to proliferate even in unfavourable culture conditions. Enhanced initial cell adhesion on the rough superhydrophilic *versus* smooth hydrophilic surface was attributed to the surface roughness, which increased the surface area available for cell–surface contacts and promoted more exposure of oxygen-rich chemical groups on the surface that can bind cells and support high cell adhesion. When superhydrophobic substrates patterned with superhydrophilic squares (1 mm side length) were immersed and cultured in medium containing SaOs-2 cells for 6 days, the superhydrophilic regions became densely populated with cells while only a few cells occupied the superhydrophobic background.

Further examples are available where superhydrophobicity is used to confine cells within (super)hydrophilic patterns on a surface. Chi and coworkers found that 3T3 cells cultured on a superhydrophobic TiO_2 nanotube array surface patterned with hydrophilic ink regions preferentially immobilized on the ink-covered hydrophilic regions and were confined in the ink patterns by the surrounding superhydrophobic region.[47] Feng *et al.* used UV-induced, sequential thiol-yne click chemistry to create superhydrophilic–superhydrophobic micropatterns.[29] Human cervical adenocarcinoma cells (HeLa-GFP) adhered well to the superhydrophilic microspots, and less than 1% of the cells occupied the superhydrophobic regions separating the microspots after 2 days of culture. Li *et al.* created superhydrophobic–hydrophilic micropatterns on a surface composed of a thin layer of silicone nanofilaments.[30] They demonstrated cell patterning on these surfaces and reported that 94% of the human embryonic kidney cells (HEK 293) cultured on the surface after 24 h were confined within the hydrophilic squares and only 6% of the cells occupied the superhydrophobic barriers. Feng *et al.* created superhydrophobic–hydrophilic micropatterns by sequentially modifying a tetrazole-functionalized porous polymer surface with hydrophobic and hydrophilic thiols.[31] Rat mammary carcinoma cells (MTly) cultured on the patterned surfaces adhered well to the hydrophilic areas, but the superhydrophobic barriers demonstrated efficient cell repellency.

Efremov *et al.* cultured multiple cell types separately in microreservoirs formed within hydrophilic patterns on a superhydrophobic, nanoporous polymer substrate.[23] They demonstrated that hydrophilic–superhydrophobic patterned surfaces facilitated cell patterning as well as the study of cell–cell signalling processes *in vitro*. In this case, two different cell populations were initially seeded in adjacent microreservoirs and subsequently cultured in shared medium after adhesion of the cells to the surface, and then the cross-talk between the two cell populations *via* Wnt signalling molecules was monitored and analysed.

The unique properties arising from patterns of superhydrophilicity and superhydrophobicity on a surface also offer advantages in the field of cell microarrays. Cell microarrays have become a valuable tool for high-throughput screening due to their miniaturized format and the economical use of precious cells, reagents, and consumables compared with conventional microtitre plates. For example, reverse cell transfection microarrays have been developed and implemented for genetic screening to discover the role of genes in cellular processes or to identify drug targets.[63–72] Typically, oligonucleotides and transfection reagents are mixed with ECM components that promote cell adhesion, and then printed onto a surface. Cells are then seeded over the whole surface to ultimately form clusters of transfected cells within a background of non-transfected cells, or the background is passivated such that the cells mainly adhere to the spots of interest. Instead of passivating the background with surface coatings such as poly(ethylene glycol) (PEG) or albumin to prevent cell adhesion, superhydrophobicity can be used to control the patterning of liquids and cells on microarray surfaces.[73–78]

Superhydrophilic–superhydrophobic patterned surfaces offer several advantages over the conventional non-patterned glass slides that are often used for cell microarrays: higher spot density due to confinement by the superhydrophobic barriers of both the printed solutions and cells within the superhydrophilic spots; and low variation in the spot size and resulting concentration of the printed samples on the surface (independent of the properties of the aqueous solution or surface) due to confinement of the printed solutions by the superhydrophobic barriers. Typical layouts for cell microarrays have spot diameters in the range of 200–500 μm printed at a pitch of 500–1500 μm.[66,77,79–81]

Geyer *et al.* fabricated densely packed, superhydrophilic squares (335 μm side length) separated by narrow (60 μm wide), superhydrophobic barriers on a nanoporous polymer substrate to create a cell microarray with a sixfold higher spot density compared with a non-patterned glass slide.[21] Aqueous solutions dispensed in individual superhydrophilic squares completely wetted the squares, but were completely contained by the watertight superhydrophobic barriers. MTly, HEK 293, and mouse hepatoma (Hepa) cells all preferentially adhered and grew to confluence on the superhydrophilic squares. The superhydrophobic barriers prevented cell proliferation on the barriers and cell migration between adjacent spots, which was attributed to reduced cell–surface contact due to air being trapped inside and on the surface of the rough, nanoporous, superhydrophobic barriers exhibiting the Cassie–Baxter state. This notion was further demonstrated by showing that almost no cell pattern formed when the superhydrophobic barriers were transitioned from the dry Cassie–Baxter to the wetted Wenzel state. The patterned substrates were used for reverse cell transfection experiments by dispensing a plasmid DNA transfection mixture onto the superhydrophilic squares before cell seeding. In principle, superhydrophilic–superhydrophobic patterned substrates can be used for cell screening with a variety of bioactive molecules.

Although surface chemistry can be used to control cell adhesion, its effectiveness can be highly dependent on the cell type and the composition of the culture medium as the original chemical pattern is obscured over time with adsorbed proteins. Mano and coworkers found that less protein was adsorbed on rough superhydrophobic surfaces, independent of the underlying surface chemistry, and attributed this to the Cassie–Baxter effect.[82] This also resulted in a lower affinity of cells to the rough superhydrophobic surfaces. Thus, superhydrophobicity can be used as a more general approach for controlling protein and cell adhesion since liquid–surface interactions are minimized, and using superhydrophobic barriers to confine cells can be an interesting and effective alternative to using physical barriers for applications such as cell patterning, cell microarrays, and lab-on-a-chip or diagnostic devices.

7.3.3 Cell or Chemical Screening in Arrays of Liquid or Hydrogel Droplets

Cell microarray technology is usually limited to screening adherent cells in a 2D format, and, depending on the diffusion rate of the components of the printed solutions or the factors secreted by the cells, neighbouring spots can cross-contaminate each other by diffusion of chemicals through the shared culture medium. Cell screening platforms are being developed to address these issues, such as the use of surfaces with patterned wettability to encapsulate chemicals and cells in individual droplets or hydrogels arrayed on the surface. These droplet arrays enable screening of adherent as well as non-adherent cells with chemical stimuli in a 2D or 3D microenvironment.

Mano and coworkers deposited droplets of cell suspension in hydrophilic spots patterned on a rough, superhydrophobic PS substrate and cultured the cells in the isolated droplets.[12] Aqueous solutions were confined within the square (super)hydrophilic spots (1 mm side length) due to the large difference in surface tension compared with the superhydrophobic background. Hydrophilic spots were individually coated with different concentrations of human serum albumin and human plasma fibronectin, and then a 10 μl droplet of cell suspension was manually dispensed onto each hydrophilic spot and cultured for 4 h. In general, more cells were detected on the spots with higher amounts of human plasma fibronectin, which is cell adhesive due to the presence of integrin binding domains, whereas albumin is a passivating protein.

Mano and coworkers also performed a combinatorial screen of the chemical composition and cytocompatibility of 3D hydrogels in an array format on hydrophilic–superhydrophobic patterned substrates.[13] The hydrophilic squares (2 mm side length) were separated by 0.5 mm on a superhydrophobic PS surface. Chitosan, collagen, or hyaluronic acid was combined with alginate at different ratios to create 24 combinations of alginate-based polymers. The pre-polymer solutions were mixed with cells and then dispensed as droplets on the hydrophilic squares using a micropipette. $CaCl_2$ was added to the droplets to ionically crosslink the polymers to form hydrogels, and then the substrate was immersed in culture medium. Fibroblasts (L929) or pre-osteoblasts (MC3T3-E1) were cultured in the hydrogels for 24 h, and both non-destructive (live/dead cell staining and image analysis) and destructive (proliferation assay and dsDNA quantification) methods were used to analyse the cytocompatibility of the hydrogels. This platform demonstrates the potential of surfaces with patterned wettability to create hydrogel arrays to study cell–material or cell–molecule interactions in 3D microenvironments.

Arrays of droplets or hydrogels are typically formed by manual pipetting, but this considerably limits the scale of spots that can be made in practice, and should be replaced by a technology more suitable for high-throughput screening. Chatelain and coworkers developed the "DropChip" microarray to perform multiplexed cell-based assays in isolated droplets in a high-throughput manner.[83] The DropChip featured up to 100 nl droplets on hydrophilic spots of bare glass (500 μm diameter, 1 mm spot pitch) surrounded by a hydrophobic perfluoro-octylsilane background. A piezoelectric-controlled spotting device was used to precisely dispense droplets of cells, drugs, and nucleic acids onto the DropChip. The spots can be assayed in parallel simply by dipping the slide in solutions such as those for fixing or staining cells.

Ueda *et al.* used the concept of discontinuous dewetting on a superhydrophilic–superhydrophobic patterned surface to create high-density arrays of aqueous droplets or hydrogels encapsulating cells, without the need to manually pipette each droplet or use automated equipment.[22] The simple method relies on the large difference in wettability between the superhydrophilic and superhydrophobic regions to passively dispense thousands of isolated droplets in the superhydrophilic spots, referred to as a Droplet-Microarray. During discontinuous dewetting, the trailing edge of an aqueous solution moving along the patterned surface becomes pinned at the borders between the superhydrophilic and superhydrophobic regions due to a sudden and extreme change in θ_{rec}, until the liquid film thins and finally breaks to form a droplet in each superhydrophilic spot.[84–87] The volume of the individually formed droplets ranged from 700 pl to 3 μl depending on the size and geometry of the superhydrophilic spots as well as on the surface tension of the solution. The Droplet-Microarray can be an alternative to growing suspension cells in microwells or in a hanging drop. Furthermore, discontinuous dewetting was used to create crosslinked maleimide–polyvinyl alcohol hydrogel micropads encapsulating cells that were immobilized within the superhydrophilic spots. After the array of hydrogels is formed, the substrate can be immersed in shared culture medium or droplets of medium can be formed in each superhydrophilic spot to isolate each hydrogel. When hydrogels encapsulating cells were incubated individually on superhydrophilic spots pre-printed with increasing amounts of doxorubicin, a cancer drug, a concentration-dependent effect on cell viability was observed. This demonstrated that chemicals printed and dried in the superhydrophilic spots were able to diffuse into the isolated hydrogels without cross-contaminating the adjacent spots. The patterned substrates could also be compatible with other published methods for forming hydrogels, such as dispensing a hydrogel precursor solution that crosslinks upon UV irradiation or forming alginate hydrogels by ionic crosslinking with the addition of calcium chloride.[13,88,89]

To address the challenge of simultaneously adding substances to each individual droplet on a Droplet-Microarray, Popova *et al.* developed a "Droplet-Array (DA) sandwich chip" that involved sandwiching a Droplet-Microarray with a substrate printed with substances in a matching array format.[90] The sandwich method is applicable for high-throughput, cell-based screenings with the possibility of (1) one-step seeding of adherent or non-adherent cells in individual droplets and culturing for at least 24 h, (2) simultaneous addition of substances into individual droplets without cross-contamination, (3) chemically treating or transfecting cells in droplets, and (4) low consumption of reagents and cells. The DA slide consisted of an array of superhydrophilic, nanoporous 2-hydroxyethyl methacrylate-*co*-ethylene dimethacrylate (HEMA-EDMA) spots (1 mm diameter) on a superhydrophobic background (Figure 7.3). Discontinuous dewetting was used to form an array of isolated microdroplets (60 nl) encapsulating cells by sliding a cell suspension down a tilted patterned substrate. A library microarray (LMA) slide was prepared by printing drugs or transfection mixtures onto a bare glass slide using a non-contact, ultralow volume dispenser in an array format corresponding to the pattern of the DA slide. Simultaneous addition of substances into each individual droplet was achieved by precisely aligning and sandwiching the DA slide with the LMA slide. The printed chemicals and transfection mixtures dissolved and diffused into the individual droplets, after which the LMA was removed and the DA was cultured in medium.

When HeLa and HEK293 cells were cultured in droplets for 24 h, both cell types showed a typical spread morphology with viability rates of 96% and 98%, respectively. The sandwiching approach was used for the parallel addition of different amounts of the drug doxorubicin printed onto a LMA

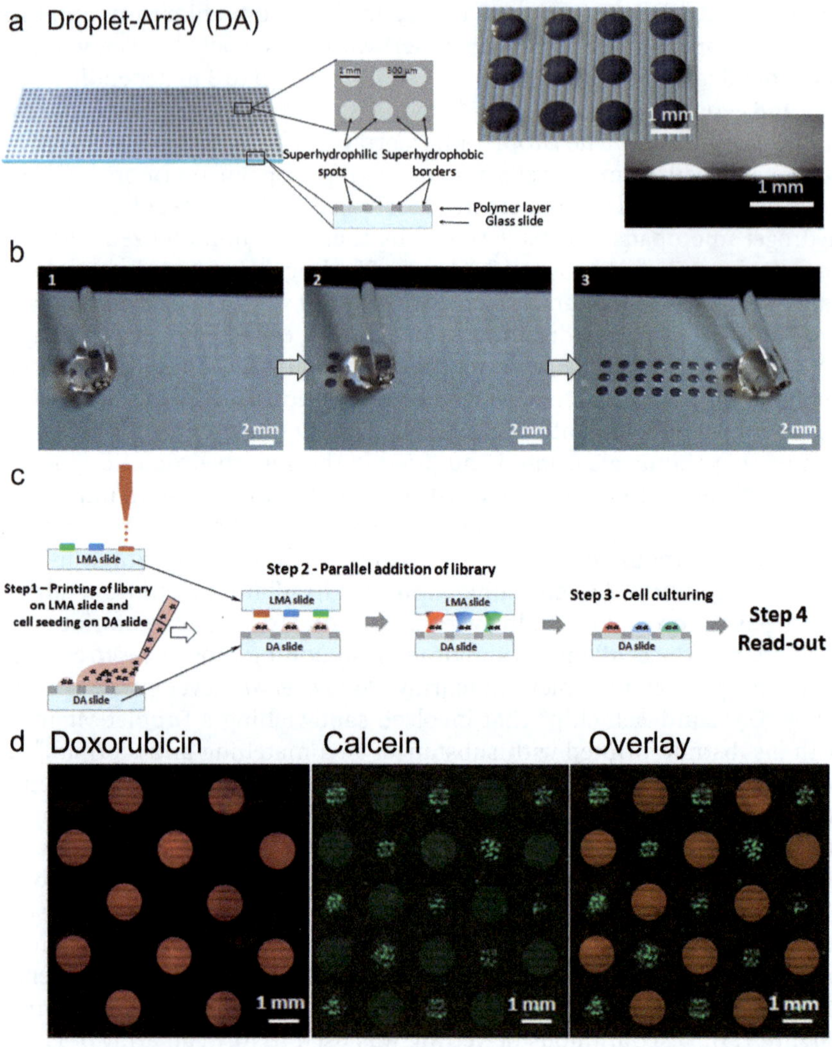

Figure 7.3 (a) Schematic of a Droplet–Array (DA) slide and images of droplets formed on a superhydrophobic–superhydrophilic pattern. (b) Snapshots of the process of discontinuous dewetting leading to the formation of an array of microdroplets. (c) Schematic of the workflow for cell-based screenings using the DA sandwich chip. Library microarray (LMA) slide is prepared by printing substances of interest on a glass slide (Step 1). DA slide is prepared by seeding cells using discontinuous dewetting (Step 1). For parallel addition of a library into individual droplets, the LMA slide is aligned and sandwiched with the DA slide containing cells (Step 2). After the substances are transferred into the droplets, the LMA slide is removed and the DA slide is placed into a cell culture incubator (Step 3). (d) Fluorescence microscopy images of a DA slide containing HeLa cells 18 h after treatment with doxorubicin. Doxorubicin shows red fluorescence (left). Calcein-stained HeLa cells on the same DA slide (middle). Overlay of the images (right). Adapted with permission from John Wiley and Sons ref. 90. Copyright © 2015 Wiley-VCH Verlag GmbH & Co. KGaA, Weinheim.

hydrophobic glass slide to a DA slide with microdroplets containing live HeLa cells. The cells in droplets sandwiched with the spots on a LMA printed with 25 ng of doxorubicin had few live cells, as indicated by Calcein staining at 18 h of culture post sandwiching (Figure 7.3). The sandwich method was also used to perform parallel gene overexpression in HEK293 cells seeded on a DA slide by sandwiching with a LMA slide arrayed with dried transfection mixtures. HEK293 cells showed a transfection efficiency of ~20%.

So far, substrates with patterns of wettability have shown potential for performing more sophisticated cell screenings than 2D cell screening platforms. Although small pattern sizes allow high-density arrays to be achieved, the limitations of working with very small droplet volumes for culturing cells are preventing droplet evaporation and nutrient starvation, especially during long culture times. Another important issue to address is how to exchange the solution in droplets, especially for those containing non-adherent cells, for processes such as medium exchange or performing assays. In principle, hydrogels can be functionalized or incorporated with ECM proteins, signalling molecules, or chemicals to enable sophisticated cell screenings in a 3D microenvironment. Culturing cells in 3D microenvironments, such as hydrogels, *versus* 2D cell culture gives the opportunity to study cells in an environment that more accurately resembles the *in vivo* situation.[88,91] In addition, the physical properties of hydrogels such as stiffness and degradability can be tuned.[92] Even plasmids can be incorporated into the hydrogels for transfection in 3D.[93] High-throughput screening of cells in 3D systems is an important tool that can provide valuable information, but the methods are still being established.

Arrays of droplets on hydrophilic–hydrophobic patterned surfaces can also be used for the synthesis and quantitative analysis of chemicals in a high-throughput manner. Balakirev and coworkers used nanodroplets positioned on a micropatterned surface to synthesize and profile new enzyme inhibitors of the NS3/4A serine protease of the hepatitis C virus (HCV), which plays an essential role in the maturation and immune evasion of HCV.[94] Nanodroplets were printed onto hydrophilic spots (500 μm diameter) patterned on a hydrophobic surface. Potent inhibitors of the NS3/4A protease were identified within a set of 200 hydrazides and 20 100 distinct dihydrazones synthesized inside the nanodroplets.

Balakirev and coworkers also used droplets on the hydrophilic–hydrophobic patterned surfaces for the synthesis and screening of 1600 unique fluorophores with a drug-like scaffold that could act as bioimaging probes.[95] The crucial role of the amidine structure for fluorescence was discovered. These examples demonstrate that droplets on patterned surfaces can be an extremely useful tool for creating combinatorial libraries and miniaturizing drug or molecular discovery efforts.

Tuteja and coworkers used superomniphobic–superomniphilic patterned surfaces such that arrays of droplets with both aqueous or non-aqueous solutions could be formed.[44] When the patterned surface was dipped in heptane, the extreme wettability contrast of the patterned surface caused heptane to selectively wet the superomniphilic regions and resulted in the self-assembly of heptane droplets within these regions. These droplets of organic liquids have the potential to serve as microchannels or microreactors for liquid-phase reactions in the future.

Similarly, Feng *et al.* used discontinuous dewetting on patterned surfaces to create high-density arrays of microdroplets of organic liquids with surface tensions as low as 18.4 mN m^{-1}.[96] Low surface tension organic liquids wetted and formed droplets in the cysteamine micropatterns fabricated on a chloro(dimethyl)vinylsilane-coated flat glass surface, but did not wet the surrounding *1H,1H,2H,2H*-perfluorodecanethiol-modified areas. This approach was further used for the formation of homogeneous arrays of hydrophobic nanoparticles, polymer micropads of controlled shapes, and polymer microlens arrays. To demonstrate the potential for performing miniaturized and parallel high-throughput chemical reactions in organic solvents without multiple pipetting steps, a library microarray (LMA) slide printed with two different dyes was sandwiched on top of a slide with an array of 1-butanol microdroplets. This led to the dissolution of the chemicals in the individual microdroplets without cross-contamination between adjacent droplets. In addition, water–oil interfaces in microdroplets were formed by sandwiching an oil microdroplet array with an aqueous microdroplet array formed on a superhydrophobic–superhydrophilic pattern, which could enable miniaturized parallel liquid–liquid microextractions or heterophasic organic reactions.

7.3.4 Positioning or Sorting Particles

Fan and Stebe evaporated particle suspensions on lyophobic surfaces patterned with lyophilic regions to position and sort particles by size.[97] As the solution evaporates, the contact line recedes and fills the lyophilic features with discrete droplets by way of discontinuous dewetting. Sorting occurs by creating droplets on the lyophilic features with dimensions larger than the particles of interest and smaller than those to be excluded. For example, 210 nm amidine-functionalized PS particles could be separated from a suspension also containing 810 nm particles by evaporating the suspension on a surface patterned with 1 μm-wide stripes of wettable COOH-terminated SAMs surrounded by non-wettable CH_3-terminated SAMs on a gold-coated surface. In another example, a solution containing 900 nm and 5.46 μm PS particles functionalized with streptavidin and bound to biotinylated dye biotin-4-fluorescein or Alexa-fluor 594 biocytin, respectively, was evaporated on a surface patterned with 5 μm and 25 μm square lyophilic patches. All of the lyophilic spots contained the smaller 900 nm particles that fluoresced green, while only the 25 μm squares contained the larger 5.46 μm particles and fluoresced red. Particles could also be deposited and sorted by dip-coating the patterned surface in a solution. This method is applicable for sorting any particles that can be suspended in a solvent and the ability to create patterned lyophilic and lyophobic regions, and thus maybe be useful for creating arrays for biosensors and in microphotonics.

Vermant and coworkers used a Langmuir–Blodgett deposition technique to assemble micrometre-sized, sulfate-modified PS particles (2.9 μm diameter, 9.7 μC cm^{-2} surface charge density) from an oil–water interface onto a hydrophobic silicon substrate patterned with hydrophilic regions.[98] Decane–water was used as the fluid–fluid interface to enhance dipolar repulsion of colloids to form a highly ordered and stable monolayer at the interface during Langmuir–Blodgett deposition, which resulted in planar colloidal clusters on the hydrophilic regions where the cluster size depended on the pattern size. Particles were deposited onto square hydrophilic spots (side length = 7.5 μm or 5 μm, spacing = 5 μm or 6 μm, respectively) using the Langmuir–Blodgett technique and resulted in an average cluster size of 12.74 and 6.25 particles per cluster, respectively, with low dispersity in size. Improving self-assembly of the colloids into more complex forms by increasing the geometric complexity of the hydrophilic pattern should be further studied.

Tuteja and coworkers used superomniphobic–superomniphilic patterned surfaces for the wettability-driven self-assembly of microparticles and polymers from both aqueous solutions and liquids with low surface tension (*e.g.* heptane).[44] When the patterned surface was dipped in an aqueous or organic solution, the extreme wettability contrast of the patterned surface caused the solution to selectively wet the superomniphilic spots and resulted in the self-assembly of droplets within these spots. Dispersions of UV fluorescent green microspheres in water or UV fluorescent red microspheres in heptane became confined within the superomniphilic regions when sprayed onto patterned surfaces, demonstrating the compatibility of the surfaces with both high and low surface tension liquids. Such precise control over the site-selective, self-assembly of particles and films can be useful for fabricating electronic and optical devices, patterning cells, or forming well-defined thin films and nanostructures.

Patterning nanorods, nanowires, and nanotubes with good control of their assembly and alignments is also of great interest as building blocks in nanoelectronics and photonics.[99] Again, a general approach is to use surfaces with patterns of contrasting wettability to selectively deposit and align the nanomaterials based on the interactions between the solvent and the patterned surface.[51]

Bao and coworkers controlled the assembly and alignment of palladium nanorods (250 nm diameter, 6 μm length) by depositing palladium nanorod suspensions on surfaces patterned with hydrophobic (1-hexadecanethiol-functionalized gold, $\theta_{st} \approx 98°$) and hydrophilic (bare gold, $\theta_{st} \approx 64°$) regions.[100] After the nanorods were allowed to settle for 10 min, the excess suspension was removed from the substrate using a pipette resulting in small, discrete droplets of the nanorod suspension on the hydrophilic regions. Nanorods could be aligned in hydrophilic regions with widths less than the length of the nanorods by dip-coating the patterned substrate in a nanorod suspension and evaporating the droplets. Using superhydrophilic–superhydrophobic patterned surfaces instead would result in a higher contrast in wettability between the two regions and could allow finer pattern designs and more precise arrangement of the nanorods, as well as more options in terms of suspension medium and particle properties. Zhang and coworkers implemented a similar approach by using SiO_2 substrates with patterns of wettability to grow or deposit arrays of organic nanowires.[101]

Bao and coworkers demonstrated selective deposition of organic semiconducting crystals onto and between source and drain electrodes to create arrays of single-crystal organic field-effect transistors.[102] Despite their high performance, integrating single crystals into electronic devices is challenging to mass-produce because the single crystals are usually handpicked and placed onto the device substrate, and thus nearly impossible to fabricate devices using submicrometre-sized organic crystals. The method presented by Bao and coworkers was based on solvent wetting and dewetting to assemble organic semiconducting crystals of 5,5′-bis(4-*tert*-butylphenyl)-2,2′-bithiophene (tPTTPt) directly onto transistor source–drain electrodes. A suspension of tPTTPt was allowed to settle on the hydrophilic electrodes for 10 min and then the excess solution was removed from the hydrophobic dielectric layer with a pipette, leaving behind small, discrete droplets of tPTTPt suspension on the electrodes due to discontinuous dewetting. Field-effect transistor behaviour was observed with tPTTPt crystals, but further work is needed to improve the crystal quality and optimize the assembly conditions and contact between the dielectric/semiconductor and electrode/semiconductor.

So far, we have presented how surfaces with wettability patterns could be used to create 1D/2D structures. Song and coworkers were able to use hydrophilic–hydrophobic patterned surfaces to create 3D microstructures.[103] They controlled pinning of a droplet containing nanoparticles at the three-phase contact line (TCL) by tuning the surface energy difference at the interface of hydrophilic and hydrophobic regions as well as the properties of the solution, and promoted asymmetric dewetting to manipulate the droplet morphology (Figure 7.4). For liquid droplets containing nanoparticles, the TCL slides along a surface with low surface energy in a centripetal direction, leading to a dome-like deposition. On a surface with high surface energy, TCL pinning usually occurs and results in a ring-like deposition. 3D colloidal crystals were deposited from a microdroplet onto specially designed hydrophilic pinning patterns fabricated by introducing hydrophilic pinning points ($\theta_{st} = 58.3° \pm 1.5°$) onto a homogeneous hydrophobic surface ($\theta_{st} = 109.4° \pm 1.1°$). A monodisperse solution of poly(styrene-methyl methacrylate-acrylic acid) nanoparticles (mean diameter = 210 nm) was printed with a commercial inkjet printer so that each droplet covered three hydrophilic pinning points arranged in a triangle. The nanoparticles moved along with the asymmetrically shrinking TCL of the droplets to form 3D triangle-shaped microcolloidal crystals. When the TCL was pinned on the hydrophilic points, capillary flow from the hydrophobic to the hydrophilic regions drove the nanoparticles to the hydrophilic regions and they settled along the curved surface of the droplet. However, the TCL on the hydrophobic background retracted inward but was restrained by the pinning points and resulted in triangular droplet morphologies. This method can be further developed for creating 3D microstructures

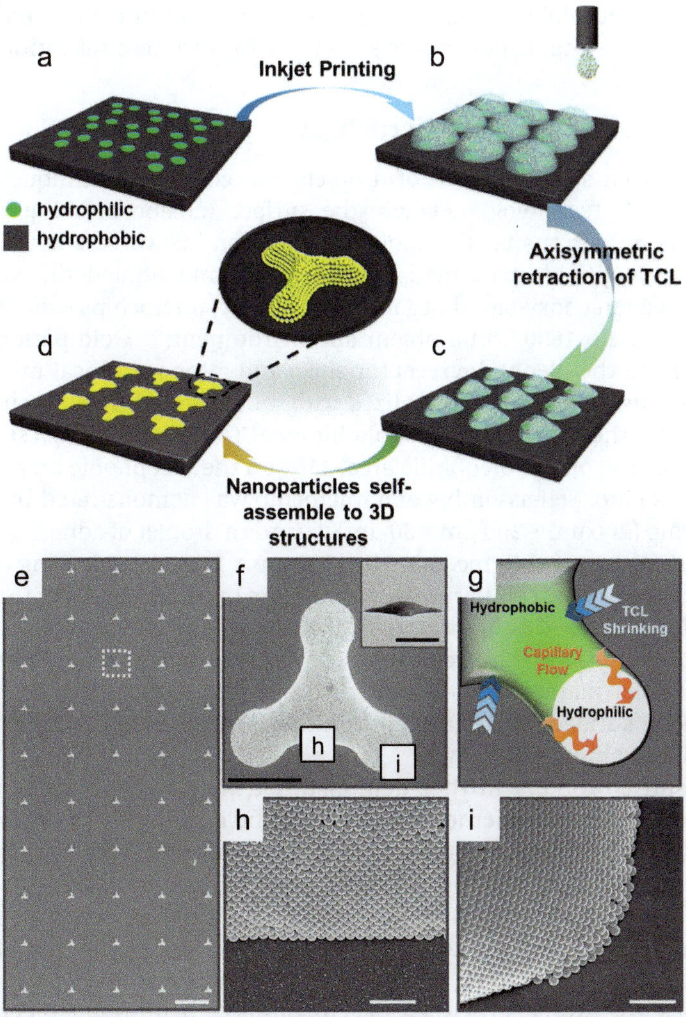

Figure 7.4 (a–d) Schematic of the formation of 3D microcolloidal crystal patterns. (a) A hydrophobic silicon wafer with patterned hydrophilic pinning spots (green) is used as the substrate. (b) Inkjet printing is used to dispense an array of droplets containing nanoparticles at designed locations. (c) An array of triangular droplets is formed by hydrophilic pattern-induced asymmetric dewetting. (d) Arrayed 3D microcolloidal crystals with controllable morphology are achieved. (e–i) Typical morphology of the 3D microcolloidal crystals. (e) SEM image of arrayed microcolloidal crystals. Scale bar is 200 μm. (f) Top-view SEM image of a compactly assembled microcolloidal crystal structure. Inset: Side-view SEM image of the microcolloidal crystal. Scale bars are 20 μm. (g) Schematic of the mechanism of nanoparticle assembly on the hydrophilic region. (h–i) SEM images of the hydrophobic and hydrophilic regions shown in (f). Scale bars are 10 μm. Adapted with permission from John Wiley and Sons ref. 103 Copyright © 2015 Wiley-VCH Verlag GmbH & Co. KGaA, Weinheim.

with other functional materials, such as quantum dots and metal nanoparticles, to achieve suitable defect-free structures for various applications.

7.3.5 Self-Assembly of Microchips

Capillary-driven self-alignment of droplets is a common technique used in microassembly technologies, where the surface tension of a liquid aligns microchips to patterns on the substrate. Zhou and coworkers developed an oleophilic–oleophobic patterned surface and demonstrated the self-alignment of SU-8 (an epoxy-based negative photoresist) microchips using droplets of adhesive (Delo 18507) in ambient air environment.[104] Gold patterns were used to create the oleophilic receptor sites, and a topographical microstructure of porous ormocer functionalized with a fluorinated trichlorosilane was fabricated for the surrounding oleophobic area. The adhesive had a static contact angle of 53° on the oleophilic area, 119° on the oleophobic area, and 47° on the SU-8 chip. Self-assembly of a microchip was demonstrated by moving an SU-8 chip (200 μm × 200 μm × 50 μm) towards a droplet of adhesive (0.5–1.5 nl) dispensed onto a gold receptor site (200 μm × 200 μm), releasing the chip, and then allowing the capillary force to self-align the chip according to the gold pattern (Figure 7.5a–d). Zhou and coworkers later demonstrated self-assembly of microchips, but using hydrophobic receptor sites (silica functionalized with fluoropolymer) patterned on a superhydrophobic surface (nanostructured black silicon surface functionalized with fluoropolymer) and a water droplet as the self-alignment medium.[105]

Zhou and coworkers further demonstrated that capillary self-transport and self-alignment of microchips could still be achieved without requiring any initial overlap between the microchips and the receptor sites.[14] In this case, the surface consisted of hydrophilic silicon receptor sites surrounded by superhydrophobic black silicon coated with a fluoropolymer, and microscopic rain (droplets of 1–5 μm diameter) was used as the medium for capillary self-transport. When a sufficient amount of water droplets rained on the surface, the droplets between the chip and the hydrophilic receptor site grew until a meniscus was established between the chip and the receptor site. This meniscus and the water film on the receptor site coalesced, and the meniscus then reduced its total surface energy by pulling the chip towards and aligning with the hydrophilic receptor site (Figure 7.5e–j). This method of applying microscopic rain on hydrophilic–superhydrophobic patterned surfaces greatly improved the capability, reliability, and error tolerance of the chip self-assembly process.

7.3.6 Lithographic Printing

Lithographic printing uses a chemically patterned surface with anisotropic wetting such that hydrophobic areas accept ink and hydrophilic areas repel ink for reprography. When the plate is introduced to an ink and water mixture, the ink adheres to the hydrophobic graphic areas to be printed while the water cleans the hydrophilic background. This flat plate can be used directly, or it can be offset by transferring the graphic onto a flexible sheet (*e.g.* rubber) to enable much longer and more detailed print runs. Surfaces with sharply contrasting patterns of hydrophobicity and hydrophilicity are crucial for forming high-resolution images when using printing techniques based on wetting and non-wetting regions.

Fujishima and coworkers used superhydrophilic–superhydrophobic patterned substrates for offset printing by wetting the superhydrophilic regions

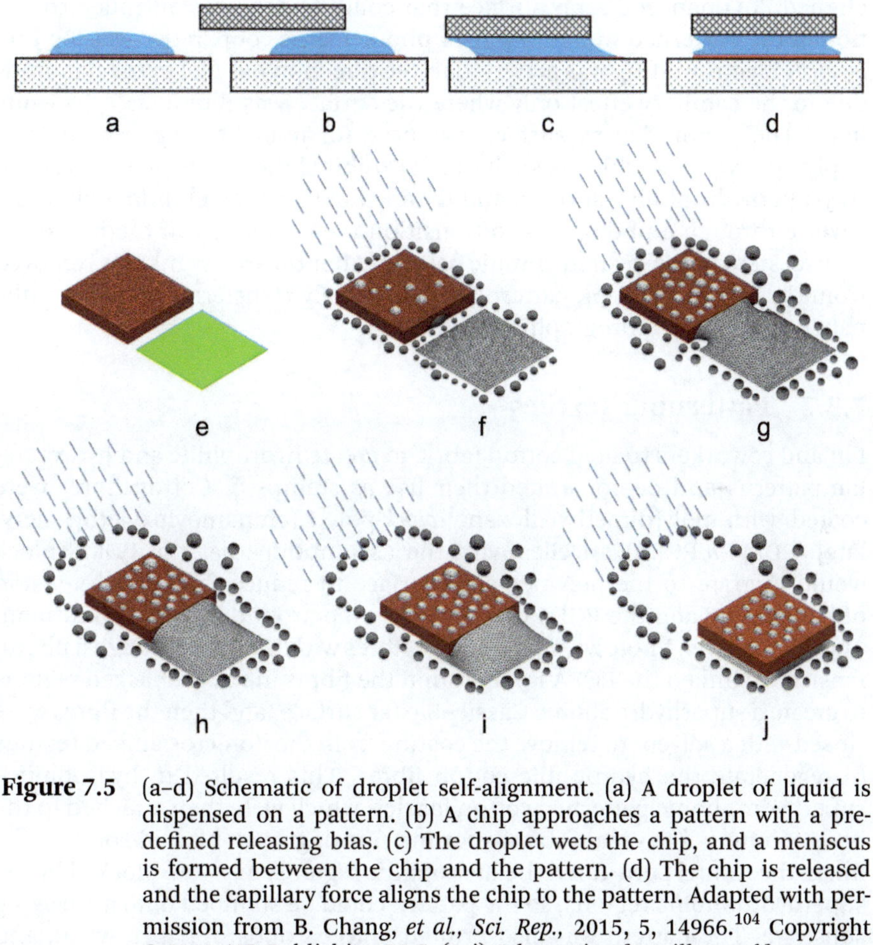

Figure 7.5 (a–d) Schematic of droplet self-alignment. (a) A droplet of liquid is dispensed on a pattern. (b) A chip approaches a pattern with a predefined releasing bias. (c) The droplet wets the chip, and a meniscus is formed between the chip and the pattern. (d) The chip is released and the capillary force aligns the chip to the pattern. Adapted with permission from B. Chang, *et al., Sci. Rep.*, 2015, 5, 14966.[104] Copyright © 2011 AIP Publishing LLC. (e–j) Schematic of capillary self-transport using microscopic rain. (e) A chip (red) is placed outside a receptor site (green) with the same shape and size. (f) Microscopic rain is delivered onto the assembly site and water droplets accumulate on the hydrophilic receptor site and superhydrophobic substrate. (g) A water meniscus is formed between the chip and the receptor site. (h–j) The chip is dragged towards the receptor site and finally aligned with the receptor site. Adapted with permission ref. 14. Copyright © 2015 Nature Publishing Group.

with water and allowing the water-insoluble ink to coat only the superhydrophobic graphic regions.[46] Colour printing at a resolution of 133 lines per inch was achieved. After the print run was complete, the hydrophobic regions could be photocatalytically decomposed by irradiation with UV light to obtain the original superhydrophilic surface for re-patterning with another graphic. A single superhydrophilic–superhydrophobic patterned plate was also able to print 40 000 copies per hour with a web press, demonstrating the robustness of the substrate.

Jiang and coworkers developed a superhydrophobic, photoconductive, aligned ZnO nanorod-array surface that could undergo a controlled transition of the patterned wettability *via* a photoelectric cooperative wetting process.[106] Electrowetting was activated in the direction parallel to the nanorods due to the capillary effect only where the surface was illuminated by white light. The nanorod-array surface was used for liquid reprography by first applying a water-soluble ink such that it exhibited the Cassie–Baxter state on the superhydrophobic surface, and then white light was illuminated on the surface through a photomask to transition the wettability of the ink to the Wenzel state in the desired graphic pattern. After the excess ink was removed from the surface, the ink pattern was then easily transferred to hydrophilic paper to print the reprographic image.

7.3.7 Patterning Textiles

Liu and coworkers treated cotton fabric to create hydrophilic and hydrophobic patterns and demonstrated their use as stamps.[107] Cotton fibres were coated with poly(dimethylsiloxane)-*block*-poly[2-(cinnamoyloxy)ethyl acrylate] (PDMS-*b*-PCEA) micelles with the assumption that the PDMS block would migrate to the polymer–air interface to reduce the surface tension of the coating and the PCEA block would wrap around the fibre to form an underlying layer. Photolysing the cotton fibres with UV light through a photomask crosslinked the PCEA layer around the fibres in the unmasked regions to create a superhydrophobic Cassie–Baxter surface, and then the fibres were rinsed with a solvent to remove the coating from the non-crosslinked regions to regenerate the hydrophilic cotton fibres. This resulted in hydrophilic–superhydrophobic patterned cotton fabrics, which were then attached to the base of a funnel to create stamps for ink printing. Water-based solutions of ink in the funnel permeated the hydrophilic regions, but were blocked by the superhydrophobic regions, so the pattern could be stamped onto a receiving substrate. The pattern was also printed onto cardboard, paper, wood, and aluminium foil.

Pan and coworkers used a different method to create a micropatterned superhydrophobic textile (MST), and demonstrated its potential use for sweat removal.[108] Surface tension-induced pressure gradients produced by extreme differences in wettability as well as capillarity facilitate interfacial microfluidic transport along the fibrous MST. Patterns of hydrophilic yarns were stitched onto a superhydrophobic textile coated with a thin layer of fluoropolymer microparticles to create a MST. High rates of sweat removal along the hydrophilic yarns were achieved while also maintaining dryness and high gas permittivity through the superhydrophobic textile. A micro-stitched pattern with five inlets (1 mm diameter) connected by hydrophilic yarn to one outlet (6 mm diameter) was designed for efficient sweat removal by Laplace pressure gradients generated by the curved surface of droplets. This microfluidic transport mechanism on MST is sustainable even during heavy perspiration conditions in humid weather because sweat is continuously transported to the exterior surface and removed by rolling off the superhydrophobic textile. The applications of this interfacial microfluidic transport by MST can be extended to other needs for biofluidic transport, for example removal of urine or collection and drainage of wound exudates.

7.3.8 Patterning Slippery Lubricant-Infused Porous Surfaces

Aizenberg and coworkers recently introduced SLIPS as a water-immiscible fluid, liquid-repellent, durable, defect-free, self-healing interface formed by locking a perfluorinated lubricant within a porous hydrophobic structure.[8] The principle of SLIPS is different from superhydrophobic Cassie–Baxter surfaces in that a fluid–fluid interface between a lubricant and an immiscible liquid is relied upon liquid repellency due to elimination of pinning points and low sliding angles. Since then, there has been much interest in the development of SLIPS for a wide variety applications ranging from anti-biofouling and anti-marine fouling to anti-icing.[109–112]

Many studies on preventing cell adhesion have relied on trapping an air layer within and on top of porous, superhydrophobic surfaces so that liquid on the surface is in the Cassie–Baxter state. However, the cell-repellent properties can diminish after long culture times as the cells grow to confluence and proteins eventually cover the superhydrophobic surface. Despite a lot of research activity in this field, surfaces possessing a combination of (a) efficient cell repellency, (b) long-term stability, and (c) ability to be patterned are still very rare. Ueda and Levkin introduced a newer concept based on SLIPS for patterning cell-repellent regions, which have proven to be more efficient than conventional cell-repellent coatings such as PEG-functionalized surfaces.[25] They developed a method to selectively and precisely pattern SLIPS (hydrophobic liquid) on porous superhydrophilic–superhydrophobic patterned surfaces to efficiently prevent cell adhesion and create cell microarrays.[25] First, a nanoporous polymer surface with superhydrophilic spots separated by superhydrophobic barriers was wetted with water to form droplets in the superhydrophilic spots by discontinuous dewetting (Figure 7.6).[22] Second, a thin layer of hydrophobic liquid (Krytox® GPL 103) was spread over the surface including over the water droplets, but the hydrophobic liquid only penetrated the superhydrophobic barriers. Third, the hydrophobic liquid layer was washed off the superhydrophilic spots, but it did not wash away from the superhydrophobic barriers and formed a stable micropattern on the polymer surface. The hydrophobic liquid barriers confined adherent cell

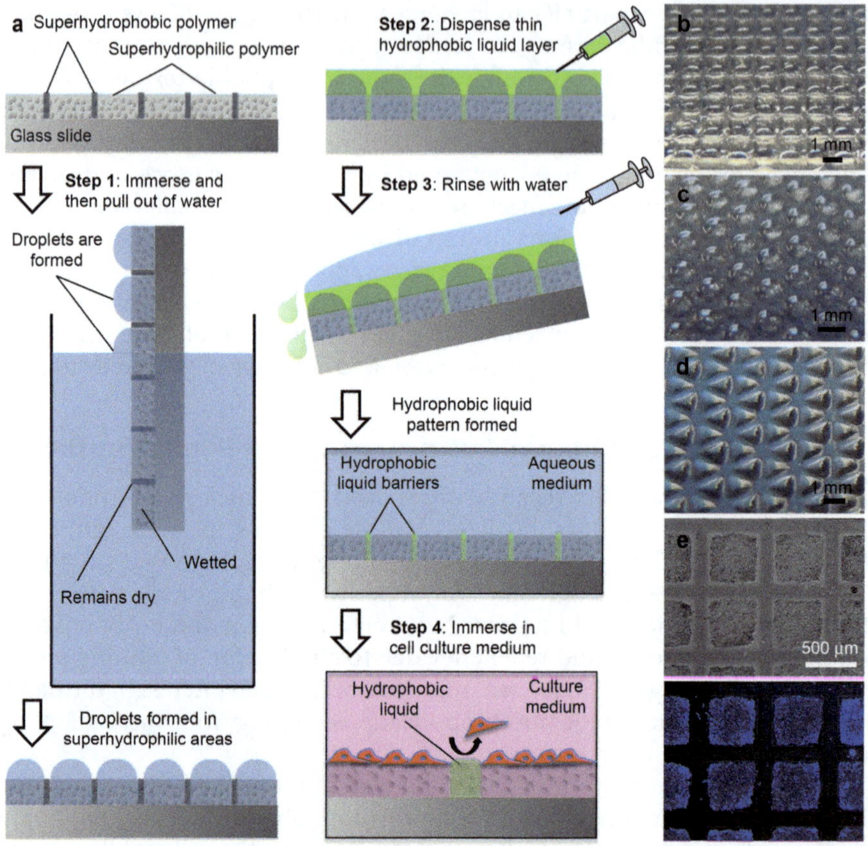

Figure 7.6 (a) Schematic of a superhydrophilic nanoporous polymer surface with superhydrophobic moieties. Step 1: when the superhydrophilic–superhydrophobic patterned substrate is immersed in water, the superhydrophilic areas become easily wetted while the superhydrophobic areas remain dry. When the substrate is pulled out of water, only the superhydrophilic areas remain filled with water and distinct droplets are formed. Step 2: a thin layer of hydrophobic liquid is applied over the water droplets, but infuses only the non-wetted superhydrophobic areas. Step 3: the surface is rinsed with water to wash off the unstable hydrophobic liquid layer covering the water droplets. The hydrophobic liquid infused in the superhydrophobic areas remains stable, resulting in the formation of a hydrophobic liquid pattern. Step 4: cells cultured on hydrophobic liquid patterns adhere to the superhydrophilic areas, but are easily removed from the hydrophobic liquid barriers by weak shear forces. (b–d) Water droplets in superhydrophilic spots separated by 100 µm hydrophobic liquid barriers for different array pattern geometries: (b) 1 mm side length square, (c) 1 mm diameter circle, and (d) 1 mm side length triangle. (e) HEK 293 cells cultured on a hydrophobic liquid micropattern (500 µm side length square, 100 µm barrier). Brightfield and DAPI channel (blue) images are shown. Reproduced with permission from John Wiley and Sons ref. 25. Copyright © 2013 Wiley-VCH Verlag GmbH & Co. KGaA, Weinheim.

types within the superhydrophilic spots, even when the cell population grew to a high density and virtually no cells migrated across the hydrophobic liquid barrier (see Figure 7.6). The few cells that were present on the hydrophobic liquid barriers after 7 days of culture displayed a rounder morphology compared with cells on the superhydrophilic spots and were easily removed by shear forces when gently washing the substrate. This suggested that the cells did not readily adhere and grow on the hydrophobic liquid, but no cytotoxic effects were observed. The slippery, water-immiscible, and self-healing nature of the stable hydrophobic liquid layer formed on the porous surfaces most likely attributed to the cell-resistant behaviour due to the inability of the cells to become anchored to the liquid interface.[8,109] The hydrophobic liquid micropatterns were stable and demonstrated long-term repellent properties against eukaryotic cells, which surpassed that of a PEG-functionalized surface.

Vogel *et al.* also later demonstrated that Krytox® 100 lubricant could be patterned on a substrate and could be used to spatially confine a variety of fluids, including low surface tension liquids or biological fluids.[113]

Lee and coworkers developed a micro-omnifluidic (μ-OF) system based on a phenomenon called microchannel induction that spontaneously occurs when droplets of solvents are applied on omniphilic micropatterned regions of a SLIPS.[114] Photolithography was used to create 2D omniphilic paths on rough fluorinated surfaces, and then the substrates were infused with lubricant. The μ-OF system was compatible with all solvents tested that had a surface tension greater than that of the lubricant used (here, Fluorinert™ FC-70, 17.1 mN m^{-1}) and all solvents that were also able to repel the infused lubricant and settle on the omniphilic micropatterns. Gravity triggered the controlled movement of droplets, which were guided by the induced 2D omniphilic microchannels. The μ-OF system was also designed for mixing droplets using a Y-shaped omniphilic pattern (60 μm line width) with a large omniphilic square patch (200 μm side length) at the Y-junction for droplet mixing. This μ-OF device does not require a pump and is compatible with almost any solvent. Although poly(dimethylsiloxane) (PDMS) microchannels are widely used, one critical issue is the incompatibility of PDMS with organic solvents. Thus, the μ-OF system has potential to be used for microfluidic channels when working with organic solvents or other harsh chemicals.

Manna and Lynn designed SLIPS by infusing silicone oil into nanoporous, hydrophobic polymer multilayers fabricated by reactive or covalent layer-by-layer assembly.[115] The reactivity of the azlactone-functionalized multilayers provided a means to tune the surface and bulk wetting behaviours by treatment with amine-functionalized molecules. The multilayers were chemically patterned with hydrophilic D-glucamine followed by functionalization of the background with 1-decylamine, and then infused with oil. Sticky regions devoid of oil were obtained that could stop the sliding of aqueous liquids, extract samples of liquid, and control the trajectories of sliding droplets. These SLIPS-coated surfaces patterned with sticky patches captured droplets from aqueous dye solutions by dip-coating. If the droplets are large enough compared with the spacing between sticky patches, then the velocity and trajectory of the droplets can be controlled.

Since SLIPS have also demonstrated to be effective against fouling from bacteria, marine organisms, and ice, patterned SLIPS could be useful for applications, for example, in the medical, marine, and transportation industries to design sophisticated anti-fouling surfaces.[109–112]

7.3.9 Fog Collection

Jiang and coworkers combined both wettability and shape gradients for more efficient water collection compared with simple circle-shaped wettability patterns or uniform superhydrophilic or superhydrophobic surfaces.[116] Star-shaped wettability patterns were designed to integrate surface energy gradients and Laplace pressure gradients to quickly drive tiny water droplets towards more wettable regions (Figure 7.7). Reducing the pattern size further improved the water collection efficiency. Highly porous superhydrophilic surfaces (~19.2 μm thick, $\theta_{st} \approx 0°$) were fabricated by spin-coating a TiO_2 slurry onto a bare glass substrate, and then superhydrophobic surfaces ($\theta_{st} > 150°$) could be fabricated by treating the superhydrophilic surface with heptadecafluorodecyl-trimethoxysilane (FAS). Superhydrophilic patterns ($\theta_{st} < 5°$) in the shape of circles or 4-, 5-, 6-, and 8-pointed stars were created by exposure of FAS-modified superhydrophobic surface to UV light (365 nm, ~25 mW cm^{-2}) for 60 min through a photomask to photocatalytically decompose the FAS monolayer in the exposed regions.

The water collection properties from a flow of fog generated by a humidifier (relative humidity >95%) were tested on four kinds of surfaces: uniformly superhydrophilic, uniformly superhydrophobic, circle-patterned, and eight-pointed star-patterned. Droplets spread immediately when captured on the uniformly superhydrophilic surfaces, whereas droplets maintained a spherical shape and frequently coalesced with neighbouring droplets on the uniformly superhydrophobic surfaces. On the circular patterns, droplets were mostly collected and easily coalesced on the outer superhydrophobic region and then were driven inward to the wettable circular region by the gradient in surface energy. Actually capturing the water droplets from the air is a crucial step for efficient water collection, thus an eight-pointed star-shaped pattern was designed to capture tiny water droplets, quickly let them coalesce into larger droplets before they evaporate, and transport them to a reservoir. On the eight-pointed star-shaped patterned surface, droplets were initially captured everywhere on the surface but were then driven by the gradient in surface energy from the outer superhydrophobic region to the superhydrophilic star pattern to form larger droplets. Furthermore, the tips of the star generated a Laplace pressure gradient due to the shape gradient and further enhanced the directional movement of water droplets. Surfaces with the star-shaped pattern were more efficient at collecting water than the circular patterns (2.11 *vs.* 2.78 g cm^{-2} h^{-1}). These experiments indicate that gradients in surface energy

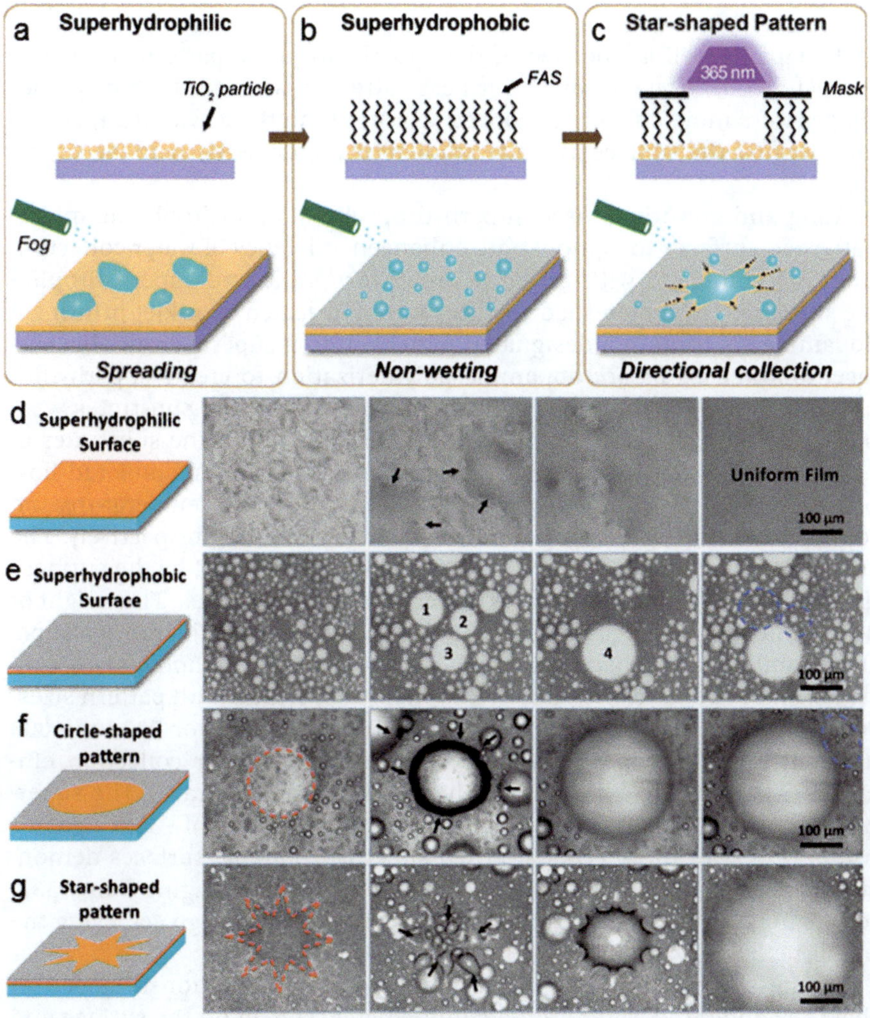

Figure 7.7 Schematic of the fabrication of surfaces with different wettabilities. (a) Superhydrophilic surface composed of TiO_2 nanoparticles, where fog droplets spread. (b) Superhydrophobic surface modified with heptadecafluorodecyl-trimethoxysilane (FAS) showing a non-wetting property to fog droplets. (c) Bioinspired gradient surface with a star-shaped wettability pattern fabricated by illuminating the FAS-modified film through a photomask with UV light. The fog droplets are collected directionally towards the star-shaped region, which is more wettable. (d–g) Water collection from fog on surfaces with various wettability features. (d) On a uniformly superhydrophilic surface, water droplets spread over surface. (e) On a uniformly superhydrophobic surface, individual water droplets coalesce randomly (*e.g.* droplet 1 + 2 + 3 to 4). (f, g) On surfaces with patterns of wettability, tiny water droplets are collected directionally toward the more wettable region (indicated by the arrows). The water collecting processes are continuous because new droplets appear immediately after the previous ones move away, which enhances the fog-collecting efficiency. Adapted with permission from John Wiley and Sons ref. 116. Copyright © 2014 Wiley-VCH Verlag GmbH & Co. KGaA, Weinheim.

and shape as well as the size of the superhydrophilic patterns influence the water collection efficiency. Surfaces patterned with similar but smaller shapes were more efficient at collecting water than those with a larger pattern because the Laplace pressure gradient is highly sensitive to the length scale.

Wang and coworkers used superhydrophilic–superhydrophobic micropatterned surfaces to demonstrate enhanced efficiency of water collection from fog compared with uniform superhydrophilic and superhydrophobic surfaces.[43] The patterned surfaces were fabricated by inkjet printing a dopamine solution in a designated pattern onto a superhydrophobic surface, followed by *in situ* dopamine polymerization to create superhydrophilic regions. For the water collection experiments, the substrates were placed on a thermoelectric cooling module to maintain the substrates at ~4 °C, which is lower than the dew point of 20 °C, and a simulated flow of fog (~10 cm s^{-1}) was generated by a humidifier. The temperature and relative humidity were approximately 22 °C and 90–95%, respectively. The substrates were vertically oriented and the water collected by the surfaces drained by gravity into containers underneath the substrates. The weight of water collected was measured after 1 h for five different substrates: a superhydrophilic glass substrate with $\theta_{st} < 5°$; a superhydrophobic glass substrate; and PDA-patterned superhydrophobic substrates with pattern sizes/separations of either 200 μm/400 μm, 200 μm/1000 μm, or 500 μm/1000 μm. The superhydrophilic surface had the lowest water collection efficiency (~14.9 mg cm^{-2} h^{-1}) among the five different substrates. The superhydrophobic surface reached a water collection efficiency of ~30.0 mg cm^{-2} h^{-1}; however, all three PDA-patterned superhydrophobic surfaces demonstrated even more enhanced water collection efficiency ranging from ~33.2 to 61.8 mg cm^{-2} h^{-1}, with the 500 μm/1000 μm pattern design achieving the highest efficiency.

On the superhydrophilic surface, film-wise condensation occurred in that the condensed water droplets immediately spread on the surface and formed a thin water film. On the superhydrophobic surface, tiny spherical water droplets condensed on the surface and gradually merged into larger droplets until reaching a threshold and rolling off the vertical surface. Self-clearing of the droplets from the superhydrophobic surface allowed continuous nucleation and growth of new droplets, resulting in more efficient collection of water. On the PDA-patterned superhydrophobic surfaces, condensation of tiny water droplets occurred initially on the superhydrophobic regions, but then the droplets preferentially moved towards the PDA-modified superhydrophilic regions due to the difference in wettability and subsequently coalesced into bigger droplets in these regions until reaching a threshold and rolling off the surface. The superhydrophilic–superhydrophobic patterned surfaces were seemingly more efficient at collecting water from fog due to the simultaneous enhancement of droplet nucleation on the superhydrophilic regions along with droplet removal on the superhydrophobic regions.

7.3.10 Heat Transfer During Boiling

Tuteja and coworkers developed superomniphobic–superomniphilic patterned surfaces to improve the heat transfer for more efficient boiling with both aqueous and non-aqueous liquids with low surface tension (*e.g.* oil, alcohol).[44] Surfaces with low surface energy are not easily wetted by boiling liquid and exhibit a high boiling heat transfer coefficient because they facilitate bubble nucleation. Superomniphobic surfaces yield high heat transfer coefficient values even for low surface tension, heat transfer liquids. When the critical heat flux is reached during boiling, the rate of bubble nucleation increases until finally the bubbles coalesce to form a continuous vapour film between the heated surface and the boiling liquid. This vapour film has a high thermal resistance and acts as a barrier to heat transfer. Patterned superomniphobic–superomniphilic surfaces increase the heat transfer coefficient and critical heat flux even more since the superomniphobic regions promote high nucleation rates, while the superomniphilic regions help to prevent the formation of a continuous vapour film. Heptane preferentially condensed within patterned superomniphilic regions and methanol preferentially boiled on patterned superomniphobic regions.

7.4 Conclusions

Research involving surfaces patterned with extreme differences in wettability is actively progressing. A variety of different techniques to produce patterned surfaces have already been developed, but there is still room to improve the robustness, stability, and ease of fabrication and modification of such surfaces. More importantly, novel and practical applications of patterned surfaces are still being developed. In this chapter, we have introduced some of the applications that have been utilized for patterned surfaces: STCM devices for separation or liquid control applications; using superhydrophobic regions in the Cassie–Baxter state to control protein and cell adhesion as well as cell migration; creating ultrahigh-density cell or droplet arrays; controlling the shape and positioning of liquid droplets or microparticles; patterning ink for lithographic printing; patterning hydrophobic lubricants for highly efficient cell repellency; efficient water collection and droplet transport in low moisture conditions; and improved nucleation of bubbles and heat transfer during boiling.

Since the adhesion of molecules and cells was well controlled by patterns of superhydrophobicity in the Cassie–Baxter state, it is an interesting alternative to using physical barriers for applications such as cell patterning, cell screening using microarrays, performing bioassays, controlling the adhesion of biomolecules and cells in complex 2D or 3D architectures, tissue engineering, bioimplants, or performing high-throughput combinatorial chemical screens. Using layers of air trapped on the surface is a more general method for controlling protein and cell adhesion since interactions with the surface are minimized and seem to be mostly independent of the protein structure or chemical composition. In addition, protein and cell repellence would not rely on serum-free or serum-depleted conditions.

Culturing cells in arrays of droplets or hydrogels opens up the possibilities of screening non-adherent cells and cells in 3D microenvironments. In addition to demonstrating the cell-repellent properties of superhydrophobic surfaces, we showed that hydrophobic lubricant surfaces also possess excellent and long-term cell-repellent properties. The concept of hydrophobic lubricant-infused porous surfaces for anti-biofouling applications has recently been introduced, but the mechanism of cell repellency has not been confirmed and should be explored in detail. This information could lead to the better design of non-fouling surfaces. In the near future, surfaces with patterns of wettability will be further implemented to advance the performance and potential of existing or new technologies.

Acknowledgements

The authors are grateful to the European Research Council (ERC Starting Grant, DropCellArray 337077) for the financial support.

References

1. A. R. Parker and C. R. Lawrence, *Nature*, 2001, **414**, 33.
2. L. Zhai, M. C. Berg, F. C. Cebeci, Y. Kim, J. M. Milwid, M. F. Rubner and R. E. Cohen, *Nano Lett.*, 2006, **6**, 1213.
3. R. P. Garrod, L. G. Harris, W. C. E. Schofield, J. McGettrick, L. J. Ward, D. O. H. Teare and J. P. S. Badyal, *Langmuir*, 2007, **23**, 689.
4. C. Dorrer and J. Rühe, *Langmuir*, 2008, **24**, 6154.
5. D. Ishii, H. Yabu and M. Shimomura, *Chem. Mater.*, 2009, **21**, 1799.
6. S. C. Thickett, C. Neto and A. T. Harris, *Adv. Mater.*, 2011, **23**, 3718.
7. C.-P. Hsu, Y.-M. Lin and P.-Y. Chen, *JOM*, 2015, **67**, 744.
8. T.-S. Wong, S. H. Kang, S. K. Y. Tang, E. J. Smythe, B. D. Hatton, A. Grinthal and J. Aizenberg, *Nature*, 2011, **477**, 443.
9. T. Ishizaki, N. Saito and O. Takai, *Langmuir*, 2010, **26**, 8147.
10. N. M. Oliveira, A. I. Neto, W. Song and J. F. Mano, *Appl. Phys. Express*, 2010, **3**, 1.
11. S. M. Oliveira, W. Song, N. M. Alves and J. F. Mano, *Soft Matter*, 2011, **7**, 8932.
12. A. I. Neto, C. A. Custódio, W. Song and J. F. Mano, *Soft Matter*, 2011, **7**, 4147.
13. C. L. Salgado, M. B. Oliveira and J. F. Mano, *Integr. Biol.*, 2012, **4**, 318.
14. B. Chang, A. Shah, Q. Zhou, R. H. A. Ras and K. Hjort, *Sci. Rep.*, 2015, **5**, 14966.
15. L. Sainiemi, V. Jokinen, A. Shah, M. Shpak, S. Aura, P. Suvanto and S. Franssila, *Adv. Mater.*, 2011, **23**, 122.
16. N. M. Alves, I. Pashkuleva, R. L. Reis and J. F. Mano, *Small*, 2010, **6**, 2208.
17. F. Svec, *J. Chromatogr., A*, 2010, **1217**, 902.

18. P. A. Levkin, F. Svec and J. M. J. Fréchet, *Adv. Funct. Mater.*, 2009, **19**, 1993.
19. Y. Han, P. Levkin, I. Abarientos, H. Liu, F. Svec and J. M. J. Fréchet, *Anal. Chem.*, 2010, **82**, 2520.
20. D. Zahner, J. Abagat, F. Svec, J. M. J. Fréchet and P. A. Levkin, *Adv. Mater.*, 2011, **23**, 3030.
21. F. L. Geyer, E. Ueda, U. Liebel, N. Grau and P. A. Levkin, *Angew. Chem., Int. Ed.*, 2011, **50**, 8424.
22. E. Ueda, F. L. Geyer, V. Nedashkivska and P. A. Levkin, *Lab Chip*, 2012, **12**, 5218.
23. A. N. Efremov, E. Stanganello, A. Welle, S. Scholpp and P. A. Levkin, *Biomaterials*, 2012, **34**, 1757.
24. P. Auad, E. Ueda and P. A. Levkin, *ACS Appl. Mater. Interfaces*, 2013, **5**, 8053.
25. E. Ueda and P. A. Levkin, *Adv. Healthcare Mater.*, 2013, **2**, 1425.
26. K. Kreppenhofer, J. Li, R. Segura, L. Popp, M. Rossi, P. Tzvetkova, B. Luy, C. J. Kähler, A. E. Guber and P. A. Levkin, *Langmuir*, 2013, **29**, 3797.
27. A. N. Efremov, M. Grunze and P. A. Levkin, *Adv. Mater. Interfaces*, 2014, **1**, 1.
28. R. M. Hensarling, V. A. Doughty, J. W. Chan and D. L. Patton, *J. Am. Chem. Soc.*, 2009, **131**, 14673.
29. W. Feng, L. Li, E. Ueda, J. Li, S. Heißler, A. Welle, O. Trapp and P. A. Levkin, *Adv. Mater. Interfaces*, 2014, **1**, 1.
30. J. Li, L. Li, X. Du, W. Feng, A. Welle, O. Trapp, M. Grunze, M. Hirtz and P. A. Levkin, *Nano Lett.*, 2015, **15**, 675.
31. W. Feng, L. Li, C. Yang, A. Welle, O. Trapp and P. A. Levkin, *Angew. Chem., Int. Ed.*, 2015, **54**, 8732.
32. H. Lee, S. M. Dellatore, W. M. Miller and P. B. Messersmith, *Science*, 2007, **318**, 426.
33. H. O. Ham, Z. Liu, K. H. A. Lau, H. Lee and P. B. Messersmith, *Angew. Chem., Int. Ed.*, 2011, **50**, 732.
34. B. P. Lee, P. B. Messersmith, J. N. Israelachvili and J. H. Waite, *Annu. Rev. Mater. Res.*, 2011, **41**, 99.
35. K. Chawla, S. Lee, B. P. Lee, J. L. Dalsin, P. B. Messersmith and N. D. Spencer, *J. Biomed. Mater. Res., Part A*, 2009, **90**, 742.
36. S. M. Kang, N. S. Hwang, J. Yeom, S. Y. Park, P. B. Messersmith, I. S. Choi, R. Langer, D. G. Anderson and H. Lee, *Adv. Funct. Mater.*, 2012, **22**, 2949.
37. C. Rodriguez-Emmenegger, C. M. Preuss, B. Yameen, O. Pop-Georgievski, M. Bachmann, J. O. Mueller, M. Bruns, A. S. Goldmann, M. Bastmeyer and C. Barner-Kowollik, *Adv. Mater.*, 2013, **25**, 6123.
38. L. Zhang, H. Yu, N. Zhao, Z.-M. Dang and J. Xu, *J. Appl. Polym. Sci.*, 2014, **131**, 1.
39. X. Du, L. Li, J. Li, C. Yang, N. Frenkel, A. Welle, S. Heissler, A. Nefedov, M. Grunze and P. A. Levkin, *Adv. Mater.*, 2014, **26**, 8029.
40. J. Sedó, J. Saiz-Poseu, F. Busqué and D. Ruiz-Molina, *Adv. Mater.*, 2013, **25**, 653.

41. Y. Liu, K. Ai and L. Lu, *Chem. Rev.*, 2014, **114**, 5057.
42. S. M. Kang, I. You, W. K. Cho, H. K. Shon, T. G. Lee, I. S. Choi, J. M. Karp and H. Lee, *Angew. Chem., Int. Ed.*, 2010, **49**, 9401.
43. L. Zhang, J. Wu, M. N. Hedhili, X. Yang and P. Wang, *J. Mater. Chem. A*, 2015, **3**, 2844.
44. S. P. R. Kobaku, A. K. Kota, D. H. Lee, J. M. Mabry and A. Tuteja, *Angew. Chem., Int. Ed.*, 2012, **51**, 10109.
45. U. Manna, A. H. Broderick and D. M. Lynn, *Adv. Mater.*, 2012, **24**, 4291.
46. S. Nishimoto, A. Kubo, K. Nohara, X. Zhang, N. Taneichi, T. Okui, Z. Liu, K. Nakata, H. Sakai, T. Murakami, M. Abe, T. Komine and A. Fujishima, *Appl. Surf. Sci.*, 2009, **255**, 6221.
47. Y. Lai, F. Pan, C. Xu, H. Fuchs and L. Chi, *Adv. Mater.*, 2013, **25**, 1682.
48. J. S. Li, E. Ueda, A. Nallapaneni, L. X. Li and P. A. Levkin, *Langmuir*, 2012, **28**, 8286.
49. G. Arrabito and B. Pignataro, *Anal. Chem.*, 2012, **84**, 5450.
50. I. P. Parkin, L. Wang, S. Huang, J. Song, Y. Lu, W. Xu, F. Chen and X. Liu, *Micro Nano Lett.*, 2015, **10**, 105.
51. L. Wen, Y. Tian and L. Jiang, *Angew. Chem., Int. Ed.*, 2015, **54**, 3387.
52. I. You, N. Yun and H. Lee, *ChemPhysChem*, 2013, **14**, 471.
53. M. J. Hancock, J. He, J. F. Mano and A. Khademhosseini, *Small*, 2011, **7**, 892.
54. M. J. Hancock, F. Yanagawa, Y. Jang, J. He, N. N. Kachouie, H. Kaji and A. Khademhosseini, *Small*, 2012, **8**, 393.
55. M. J. Hancock, F. Piraino, G. Camci-Unal, M. Rasponi and A. Khademhosseini, *Biomaterials*, 2011, **32**, 6493.
56. F. Piraino, G. Camci-Unal, M. J. Hancock, M. Rasponi and A. Khademhosseini, *Lab Chip*, 2012, **12**, 659.
57. X. Liu, Q. Wang, J. Qin and B. Lin, *Lab Chip*, 2009, **9**, 1200.
58. I. You, S. M. Kang, S. Lee, Y. O. Cho, J. B. Kim, S. B. Lee, Y. S. Nam and H. Lee, *Angew. Chem., Int. Ed.*, 2012, **51**, 6126.
59. A. B. D. Cassie and S. Baxter, *Trans. Faraday Soc.*, 1944, **40**, 546.
60. T. Koishi, K. Yasuoka, S. Fujikawa, T. Ebisuzaki and X. C. Zeng, *Proc. Natl. Acad. Sci.*, 2009, **106**, 8435.
61. J. Ballester-Beltrán, P. Rico, D. Moratal, W. Song, J. F. Mano and M. Salmerón-Sánchez, *Soft Matter*, 2011, **7**, 10803.
62. L. Baugh and V. Vogel, *J. Biomed. Mater. Res., Part A*, 2004, **69**, 525.
63. J. Ziauddin and D. M. Sabatini, *Nature*, 2001, **411**, 107.
64. H. Erfle, B. Neumann, U. Liebel, P. Rogers, M. Held, T. Walter, J. Ellenberg and R. Pepperkok, *Nat. Protoc.*, 2007, **2**, 392.
65. M. Stürzl, A. Konrad, G. Sander, E. Wies, F. Neipel, E. Naschberger, S. Reipschläger, N. Gonin-Laurent, R. E. Horch, U. Kneser, W. Hohenberger, H. Erfle and M. Thurau, *Comb. Chem. High Throughput Screening*, 2008, **11**, 159.
66. J. K. Rantala, R. Mäkelä, A.-R. Aaltola, P. Laasola, J.-P. Mpindi, M. Nees, P. Saviranta and O. Kallioniemi, *BMC Genomics*, 2011, **12**, 1.
67. E. Palmer, *Methods Mol. Biol.*, 2011, **706**, 1.
68. K. Papp, Z. Szittner and J. Prechl, *Cell. Mol. Life Sci.*, 2012, **69**, 2717.

69. K. C. Wood, D. J. Konieczkowski, C. M. Johannessen, J. S. Boehm, P. Tamayo, O. B. Botvinnik, J. P. Mesirov, W. C. Hahn, D. E. Root, L. A. Garraway and D. M. Sabatini, *Sci. Signaling*, 2012, **5**, 1.
70. R. Onuki-Nagasaki, A. Nagasaki, K. Hakamada, T. Q. Uyeda, M. Miyake, J. Miyake and S. Fujita, *BMC Genet.*, 2015, **16**, 1.
71. M. Buchholz, T. Honstein, S. Kirchhoff, R. Kreider, H. Schmidt, B. Sipos and T. M. Gress, *PLoS One*, 2015, **10**, 1.
72. H. M. Botelho, I. Uliyakina, N. T. Awatade, M. C. Proença, C. Tischer, L. Sirianant, K. Kunzelmann, R. Pepperkok and M. D. Amaral, *Sci. Rep.*, 2015, **5**, 1.
73. A. Revzin, R. G. Tompkins and M. Toner, *Langmuir*, 2003, **19**, 9855.
74. H. Ma, D. Li, X. Sheng, B. Zhao and A. Chilkoti, *Langmuir*, 2006, **22**, 3751.
75. A. L. Hook, H. Thissen and N. H. Voelcker, *Biomacromolecules*, 2009, **10**, 573.
76. S. Fujita, R. Onuki-Nagasaki, J. Fukuda, J. Enomoto, S. Yamaguchi and M. Miyake, *Lab Chip*, 2012, **13**, 77.
77. M. R. Carstens, R. C. Fisher, A. P. Acharya, E. A. Butterworth, E. Scott, E. H. Huang and B. G. Keselowsky, *Proc. Natl. Acad. Sci.*, 2015, **112**, 8732.
78. Y. Arima and H. Iwata, *Biomaterials*, 2007, **28**, 3074.
79. S. Mousses, N. J. Caplen, R. Cornelison, D. Weaver, M. Basik, S. Hautaniemi, A. G. Elkahloun, R. A. Lotufo, A. Choudary, E. R. Dougherty, E. Suh and O. Kallioniemi, *Genome Res.*, 2003, **13**, 2341.
80. S. Baghdoyan, Y. Roupioz, A. Pitaval, D. Castel, E. Khomyakova, A. Papine, F. Soussaline and X. Gidrol, *Nucleic Acids Res.*, 2004, **32**, 1.
81. B. Neumann, M. Held, U. Liebel, H. Erfle, P. Rogers, R. Pepperkok and J. Ellenberg, *Nat. Methods*, 2006, **3**, 385.
82. B. N. Lourenço, G. Marchioli, W. Song, R. L. Reis, C. A. van Blitterswijk, M. Karperien, A. van Apeldoorn and J. F. Mano, *Biointerphases*, 2012, **7**, 1.
83. B. Schaack, J. Reboud, S. Combe, B. Fouqué, F. Berger, S. Boccard, O. Filhol-Cochet and F. Chatelain, *NanoBiotechnology*, 2005, **1**, 183.
84. A. J. You, R. J. Jackman, G. M. Whitesides and S. L. Schreiber, *Chem. Biol.*, 1997, **4**, 969.
85. R. J. Jackman, D. C. Duffy, E. Ostuni, N. D. Willmore and G. M. Whitesides, *Anal. Chem.*, 1998, **70**, 2280.
86. J. H. Butler, M. Cronin, K. M. Anderson, G. M. Biddison, F. Chatelain, M. Cummer, D. J. Davi, L. Fisher, A. W. Frauendorf, F. W. Frueh, C. Gjerstad, T. F. Harper, S. D. Kernahan, D. Q. Long, M. Pho, J. A. Walker and T. M. Brennan, *J. Am. Chem. Soc.*, 2001, **123**, 8887.
87. Y. Xia, Y. Yin, Y. Lu and J. McLellan, *Adv. Funct. Mater.*, 2003, **13**, 907.
88. L. Jongpaiboonkit, W. J. King and W. L. Murphy, *Tissue Eng., Part A*, 2009, **15**, 343.
89. J. W. Nichol, S. T. Koshy, H. Bae, C. M. Hwang, S. Yamanlar and A. Khademhosseini, *Biomaterials*, 2010, **31**, 5536.
90. A. A. Popova, S. M. Schillo, K. Demir, E. Ueda, A. Nesterov-Mueller and P. A. Levkin, *Adv. Mater.*, 2015, **27**, 5217.

91. M. W. Tibbitt and K. S. Anseth, *Biotechnol. Bioeng.*, 2009, **103**, 655.
92. C. A. DeForest and K. S. Anseth, *Annu. Rev. Chem. Biomol. Eng.*, 2012, **3**, 421.
93. P. Lei, R. M. Padmashali and S. T. Andreadis, *Biomaterials*, 2009, **30**, 3790.
94. L. Mugherli, O. N. Burchak, L. A. Balakireva, A. Thomas, F. Chatelain and M. Y. Balakirev, *Angew. Chem., Int. Ed.*, 2009, **48**, 7639.
95. O. N. Burchak, L. Mugherli, M. Ostuni, J. J. Lacapère and M. Y. Balakirev, *J. Am. Chem. Soc.*, 2011, **133**, 10058.
96. W. Feng, L. Li, X. Du, A. Welle and P. A. Levkin, *Adv. Mater.*, 2016, **28**, 3202.
97. F. Fan and K. J. Stebe, *Langmuir*, 2005, **21**, 1149.
98. C. L. Wirth, M. De Volder and J. Vermant, *Langmuir*, 2015, **31**, 1632.
99. S.-Y. Zhang, M. D. Regulacio and M.-Y. Han, *Chem. Soc. Rev.*, 2014, **43**, 2301.
100. S. Liu, J. B.-H. Tok, J. Locklin and Z. Bao, *Small*, 2006, **2**, 1448.
101. R.-R. Bao, C.-Y. Zhang, X.-J. Zhang, X.-M. Ou, C.-S. Lee, J.-S. Jie and X.-H. Zhang, *ACS Appl. Mater. Interfaces*, 2013, **5**, 5757.
102. S. Liu, W. M. Wang, S. C. B. Mannsfeld, J. Locklin, P. Erk, M. Gomez, F. Richter and Z. Bao, *Langmuir*, 2007, **23**, 7428.
103. L. Wu, Z. Dong, M. Kuang, Y. Li, F. Li, L. Jiang and Y. Song, *Adv. Funct. Mater.*, 2015, **25**, 2237.
104. B. Chang, V. Sariola, S. Aura, R. H. A. Ras, M. Klonner, H. Lipsanen and Q. Zhou, *Appl. Phys. Lett.*, 2011, **99**, 1.
105. B. Chang, A. Shah, I. Routa, H. Lipsanen and Q. Zhou, *Appl. Phys. Lett.*, 2012, **101**, 1.
106. D. Tian, Q. Chen, F.-Q. Nie, J. Xu, Y. Song and L. Jiang, *Adv. Mater.*, 2009, **21**, 3744.
107. Y. Wang, X. Li, H. Hu, G. Liu and M. Rabnawaz, *J. Mater. Chem. A*, 2014, **2**, 8094.
108. S. Xing, J. Jiang and T. Pan, *Lab Chip*, 2013, **13**, 1937.
109. A. K. Epstein, T.-S. Wong, R. A. Belisle, E. M. Boggs and J. Aizenberg, *Proc. Natl. Acad. Sci.*, 2012, **109**, 13182.
110. J. Li, T. Kleintschek, A. Rieder, Y. Cheng, T. Baumbach, U. Obst, T. Schwartz and P. A. Levkin, *ACS Appl. Mater. Interfaces*, 2013, **5**, 6704.
111. L. Xiao, J. Li, S. Mieszkin, A. Di Fino, A. S. Clare, M. E. Callow, J. A. Callow, M. Grunze, A. Rosenhahn and P. A. Levkin, *ACS Appl. Mater. Interfaces*, 2013, **5**, 10074.
112. P. Kim, T.-S. Wong, J. Alvarenga, M. J. Kreder, W. E. Adorno-Martinez and J. Aizenberg, *ACS Nano*, 2012, **6**, 6569.
113. N. Vogel, R. A. Belisle, B. Hatton, T.-S. Wong and J. Aizenberg, *Nat. Commun.*, 2013, **4**, 1.
114. I. You, T. G. Lee, Y. S. Nam and H. Lee, *ACS Nano*, 2014, **8**, 9016.
115. U. Manna and D. M. Lynn, *Adv. Mater.*, 2015, **27**, 3007.
116. H. Bai, L. Wang, J. Ju, R. Sun, Y. Zheng and L. Jiang, *Adv. Mater.*, 2014, **26**, 5025.

CHAPTER 8

Natural and Artificial Surfaces with Superwettability for Liquid Collection

JIE JU[a], XI YAO[b] AND LEI JIANG*[c]

[a]Biomaterials Innovation Research Center, Division of Engineering in Medicine, Department of Medicine, Brigham and Women's Hospital, Harvard Medical school, Cambridge, MA 02139, USA; [b]School of Engineering and Applied Sciences, Kavli Institute for Bionano Science and Technology, Harvard University, Cambridge, MA 02138, USA; [c]Technical Institute of Physics and Chemistry, Chinese Academy of Sciences, Beijing 100190, China
*E-mail: jianglei@iccas.ac.cn

8.1 Introduction

Surfaces with superwettability have many remarkable prospective applications. Examples include the self-cleaning property of superhydrophobic and low-adhesive surfaces;[1,2] the anti-fogging/icing of superhydrophilic surfaces;[3–5] anti-biofouling surfaces with superhydrophobicity in air as well as underwater superoleophobicity;[6] enhanced heat-conducting surfaces with gradient wettability;[7] and oil/water separating surfaces with different affinities for water and oil,[8,9] the use of which in liquid collection is attracting increasing interest due to the supply pressure on clean water and furnace oil worldwide.

RSC Soft Matter No. 5
Non-wettable Surfaces: Theory, Preparation, and Applications
Edited by Robin H. A. Ras and Abraham Marmur

Published by the Royal Society of Chemistry, www.rsc.org

In this chapter, we concentrate mainly on the liquid collection performance of natural and artificial surfaces characterized by asymmetrically geometrical/structural features and chemical components as well as surfaces with patterned wettability. We first discuss desert beetles, followed by spider silk, then cacti and finally other kinds of surfaces with superwettability suitable for liquid collection. In each section, the order of discussion is from natural to artificial surfaces.

8.2 Liquid Collection on Natural and Artificial Desert Beetles

A shortage of fresh water is a global resource crisis that we face today.[10] Creatures living in drought areas have given us good examples on how to deal with the problem of water shortage. In this section, we describe the fog collection behaviour of a desert beetle, *Physosterna cribripes*, and the development of artificial liquid harvesters inspired by these beetles.

There are two main sources of water from air: humid vapour and fog. Fog consists of numerous tiny water droplets carried by wind. The collection of fog therefore largely involves the collision, interception, and coalescence of water droplets and does not involve a phase change during the process.[11,12] In contrast, the collection of water vapour mainly refers to water condensation on a subcooled surface accompanying a phase change from vapour to liquid.[13–15] These two means of liquid collection on superwettable surfaces, *i.e.* subcooled condensation and fog harvesting at ambient temperature, are described in this section.

8.2.1 Liquid Collection on Natural Desert Beetles

In the Namib desert of south-west Africa, several kinds of darkling beetle have evolved special surface morphology with particular wettability features. As reported by A. R. Parker and C. R. Lawrence in *Nature* in 2001, *Physosterna cribripes* possesses the ability to effectively harness water from the dense fog carried on the early-morning wind.[16] This ability was attributed to a special function of the beetle's elytra (the black wing cases). The elytra have mainly two types of morphology, described as bumps and valleys (Figure 8.1). The bumps were identified as wax-free areas which exhibit hydrophilicity, whereas the valleys show superhydrophobicity because they are textured and covered with hydrophobic wax. As also shown in Figure 8.1, the beetle adopts a characteristic posture facing upwind.

The beetle's structure and wettability arrangements facilitate the capture and transportation of water. Specifically, when the tiny water droplets are brought into contact with the surface of the elytra by a strong wind, they first accumulate on the relatively smooth and hydrophilic surface of the bumps. Since the valleys that surround the bumps are superhydrophobic, the accumulating water grows into large beads on the bumps instead of spreading to the valleys. During the whole process, the beetle inclines itself at such an angle that the growing water droplets eventually slide off the bumps and are

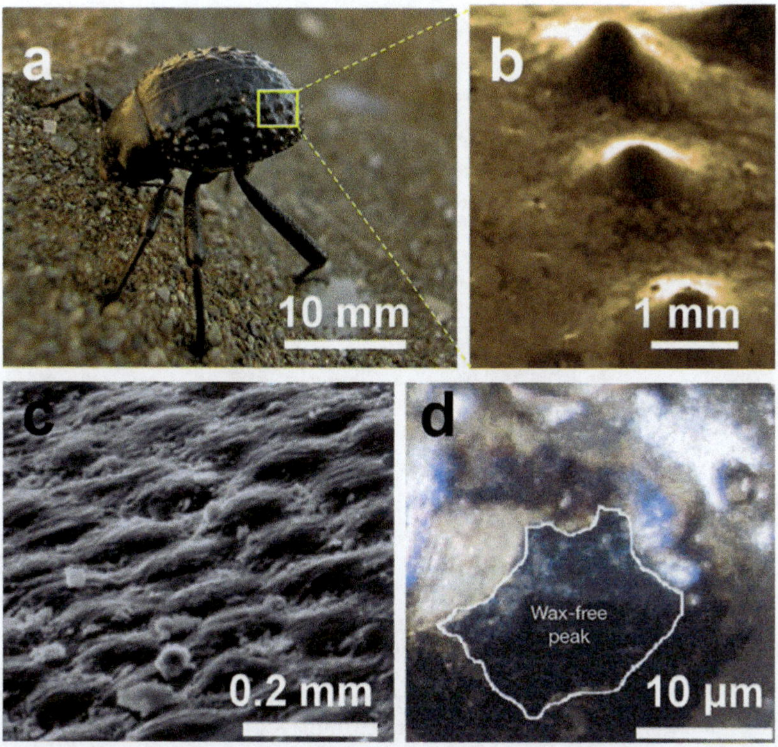

Figure 8.1 Morphology and patterned wettability of the elytra of the desert beetle *Physosterna cribripes*. (a) Typical inclination posture of the beetle for facilitating capture of fog. (b) The elytra are covered with micro- to millimetre-sized bumps. (c) SEM images of the depressed area, indicting densely distributed hemispheres in microscale. (d) The peak of the bump is wax-free according to the dying test (Adapted by permission from Macmillan Publishers Ltd: *Nature*.[16] Copyright (2001)).

directed to its mouth by gravity, leaving the bumps free for the next collection cycle. There are two main concerns relating to this scientific finding:[17] first, the chemical composition of the beetle's back; second, whether this insect actually collects the fog water for survival. Nevertheless, it is true that the water harvesting mechanism proposed by Parker and Lawrence proves effective.

8.2.2 Surfaces with Patterned Wettability Used for Dew Collection *Via* Subcooling Condensation

Inspired by these special hydrophilic/hydrophobic surface arrangements, many fog-collecting systems have been developed. As reported by Xuemei Chen *et al.*, nanograss pyramid arrays were prepared on a silicon nanograss floor by anisotropic wet etching combined with deep reactive ion etching (DRIE).[18] The authors fabricated nanograss pyramids which had smooth cylindrical fences (Figure 8.2a). These smooth fences acted as hydrophilic spots to hold on to water droplets in the vicinity, like the hydrophilic bumps

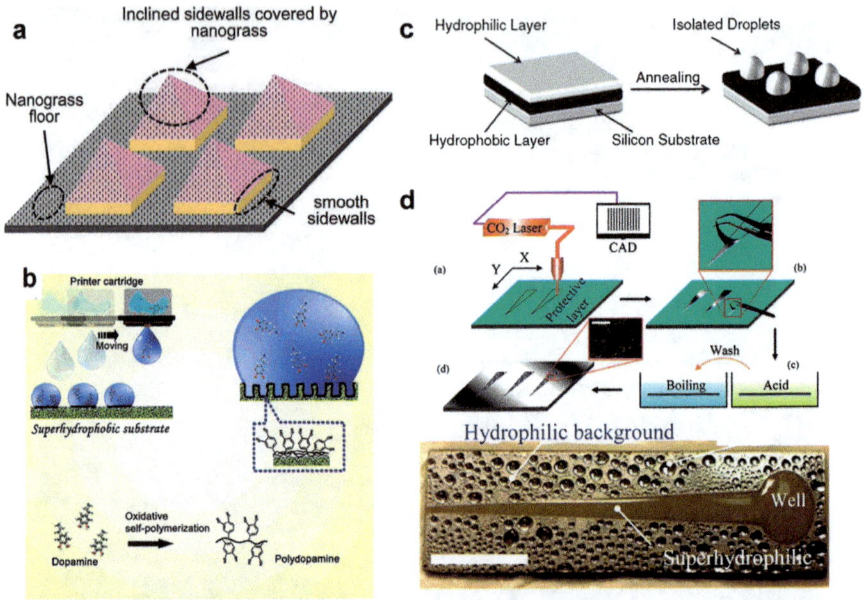

Figure 8.2 Various approaches to fabricating surfaces with patterned wettability for dew collection. (a) Constructing distinct roughness in different areas. (Adapted from ref. 18 with permission from John Wiley and Sons. Copyright © 2011 Wiley-VCH Verlag GmbH & Co. KGaA, Weinheim) (b) Inkjet printing. (Adapted from ref. 19 with permission from the Royal Society of Chemistry) (c) Polymer dewetting. (Adapted from ref. 20 with permission from John Wiley and Sons. Copyright © 2011 Wiley-VCH Verlag GmbH & Co. KGaA, Weinheim) (d) Laser-patterned masking and chemical etching. (Adapted with permission from A. Ghosh, *et al.*, *Langmuir*, 2014, **30**, 13103–13115. Copyright (2014) American Chemical Society).[21]

on the beetle's back. As the water droplet continued to grow in volume, it would become jammed between two adjacent pyramids which would eventually propel the droplet to detach from the fences and roll off. As a result, the surface coverage by water droplets was kept as low as ~25%. Although the authors did not carry out further experiments to measure the heat transfer efficiency, enhanced heat transfer on a sample surface with such a low surface water coverage is guaranteed.

Lianbin Zhang *et al.* used an inkjet printing technique to prepare a hydrophilic pattern on a hydrophobic substrate.[19] As shown in Figure 8.2b, the hydrophilic pattern was prepared by printing droplets of dopamine solution on a superhydrophobic surface. Dopamine is a natural glue extracted from mussels and can self-polymerize. The simple oxidation polymerization and the hydrophilic property of the resultant polydopamine make it a very convenient tool for creating pattered superhydrophilicity on surfaces. During a liquid collection test, the different sample surfaces were subjected to artificial fog at a subcooling temperature. The results showed that the superhydrophilic/superhydrophobic surface had better liquid collection efficiency. Moreover, among samples with different sizes of hydrophilic spots, the one closest in size to the hydrophilic pattern on the desert beetle's back exhibited the highest liquid collection efficiency. This means that the authors have successfully built up an artificial fog harvester, which in turn proves the substantive rationality of its natural counterpart.

Generally, a layer of hydrophilic chemicals tends to dewet on a relatively hydrophobic surface when the temperature is beyond the hydrophilic layer's melting point. As shown in Figure 8.2c, Stuart C. Thickett *et al.* took advantage of this phenomenon and prepared hydrophilic poly(4-vinyl pyridine) domes on a hydrophobic polystyrene (PS) substrate.[20] Different-sized hydrophilic spots can be controlled by applying hydrophilic polymer layers of different thickness. The resulting hydrophilic spots further shape into hemispheres, similar to the hydrophilic bumps on the darkling beetle's back. After the fog harvesting behaviour of desert beetles had been known for 10 years, these authors claimed that their findings were the first proof of water collection directly from humid air which directly mimicked the mist in real-world conditions. Although a dewetting technique based on an annealing treatment is a more convenient solution for high-throughput preparation, we should notice that since dewetting is made to happen on a homogeneously hydrophobic surface, the resulting hydrophilic pattern should be random. However, the earlier research on fog harvesting by beetles showed that the ordered arrangement on their elytra helped the insects to collect water more efficiently. As a result, to improve the fog-collecting performance of the artificial material, ordered sticking points should be positioned on top of the hydrophobic layer prior to the annealing treatment, so that during the annealing process the dewetting hydrophilic polymer would preferentially retreat to these designated positions and form arrays.

Another example of enhancing water collection efficiency and increasing the heat transfer coefficient from hydrophilic/hydrophobic patterns is described in the work of Aritra Ghosh *et al.*[21] As shown in Figure 8.2d, laser cutting and metal passivation were used to create two types of patterns, stripes and interdigitated wedges. The striped pattern had a higher heat transfer coefficient because it used the hydrophilic stripes to transfer the mass, and at the same time used the hydrophobic stripes to convey the heat. These authors further optimized the pattern by using interdigitated wedge-shaped hydrophilic patches. In this way, the water mass transfer was more efficient, because while leaving a maximum area (the hydrophobic zone) on the surface for dropwise condensation, the water transferred from the hydrophobic areas was delivered to the reservoir by Laplace pressure.

8.2.3 Artificial Surfaces with Patterned Wettability Used for Liquid Collection *Via* Fog Deposition

An early attempt to prepare a hydrophilic pattern on a hydrophobic surface, mimicking the backs of desert beetles, can be found in the work of Lei Zhai *et al.*[22] In their research, the hydrophilic pattern was printed on a superhydrophobic polyelectrolyte surface *via* regional-selective electrostatic self-assembly of polyelectrolytes. The resulting pattern showed sharp wetting contrast (the water contact angle was ~20° on the hydrophilic spots and ~160° on the rest of the surface). Although the authors did not specifically investigate fog collection on such a surface, they applied a fine spray of water mist and found that the water droplets would bounce from the hydrophobic areas to the hydrophilic areas. As shown in Figure 8.3a, most of the water droplets coalesced and stuck to the hydrophilic sites, similar to the behaviour of fog water droplets on the beetle's back. The authors further pointed out that such a method could find applications in surfaces, controlled drug-release coatings, open-air microchannel devices, and lab-on-chip devices.

In 2007, R. P. Garrod *et al.* reported a plasma-chemical patterning technique to build up a fog harvester to collect fog at ambient temperature (Figure 8.3b).[23] In their study not only was the superhydrophilic/superhydrophobic pattern achieved, but a sophisticated fog collection system was set up. The authors

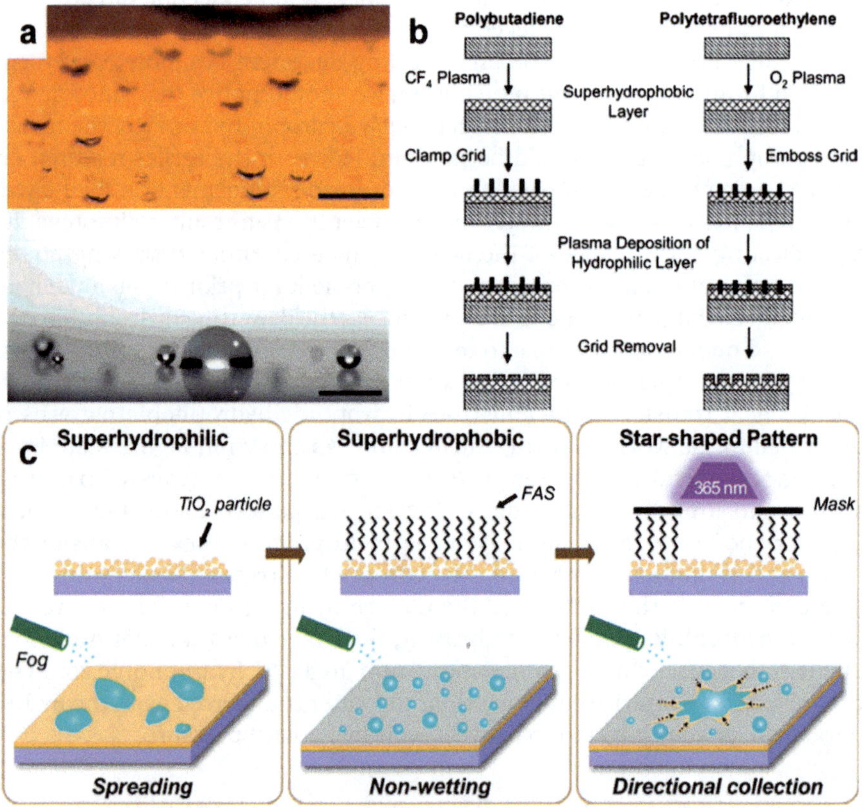

Figure 8.3 Different techniques for fabricating surfaces with patterned wettability for liquid collection. (a) Selective deposition of polyelectrolyte. (Adapted with permission from L. Zhai, *et al.*, *Nano Letters*, 2006, **6**, 1213–1217. Copyright (2006) American Chemical Society)[22] (b) Two-step plasma treatment. (Adapted with permission from R. P. Garrod, *et al.*, *Langmuir*, 2007, **23**, 689–693. Copyright (2007) American Chemical Society)[23] (c) Selective regional photo-degradation. (Adapted from ref. 24 with permission from John Wiley and Sons. Copyright © 2014 Wiley-VCH Verlag GmbH & Co. KGaA, Weinheim).

investigated the influence of wettability difference between the hydrophilic and hydrophobic areas, as well as the relative size ratio, on the fog collection efficiency. They found that the optimum fog collection efficiency occurred when the hydrophilic spot size (diameter) was 500 μm while the centre-to-centre distance between adjacent hydrophilic spots was around 1000 μm. The comparable size and distribution of the hydrophilic pattern to that on the back of *Physasterna cribripes* suggests that the original pattern distribution on the darkling beetle's back has been optimized through long-term natural selection. The authors also pointed out that in order to make the patterned surface suitable for fog collection, refreshment of the hydrophilic spots plays a key role.

Recently, Hao Bai *et al.* developed a method based on a photo mask combined with a photodegradation technique to fabricate surfaces with complex-shaped wettability contrast.[24] As shown in Figure 8.3c, to prepare superhydrophilic/superhydrophobic patterns, TiO_2 nanoparticles were deposited on a silicon surface, followed by surface fluorination using heptadecafluorodecyl-trimethoxysilane (FAS). A mask revealing the desired pattern was applied to the surface, which was then subjected to UV irradiation. The superhydrophilicity of the exposed area was achieved by the photodegradation of FAS catalysed by TiO_2. The fog collection behaviour of prepared surfaces with different patterns was investigated. This research answered two questions. First, why a patterned surface is superior in fog collection compared with non-patterned surfaces (superhydrophilic or superhydrophobic). Superhydrophilic and superhydrophobic surfaces each have their own advantages and disadvantages. Superhydrophilic surfaces adhere strongly to water, which helps to immobilize the incoming water, but at the same time the adhesion results in increasing evaporation of water due to delayed sliding off. Superhydrophobic surfaces have weak adhesion to water droplets, but the bead-shaped water droplets standing on the surface eventually become a hindrance to the further capture of tiny water droplets. Consequently, the patterned surface on one hand has the advantage of strong adhesion on the superhydrophilic spots and on the other hand uses the superhydrophobic regions to confine the spreading of water, making it easier to roll off. The second question is why some specific shaped patterns collect more water than others. In the study by Hao *et al.*,[24] a five-pointed star pattern was found to collect the most water. The authors explained that during fog collection, the hydrophilic region plays two paradoxical roles: it helps to capture the water but hinders transport of water. As a result, an optimal condition can be identified by tuning the wetting gradient and the shape gradient of the pattern.

8.3 Liquid Collection on Natural and Artificial Spider Silks

Spider silk has long been recognized for its superb mechanical properties. There is much research interest in spider silk-type tough materials from the perspective of protein folding patterns and the spinning mode of spiders.[25,26] But in addition to its mechanical properties, spider silk can also function as an excellent material for liquid collection.

8.3.1 Liquid Collection on Natural Spider Silks

Recently, Yongmei Zheng *et al.* found that the capture silk of the cribellate spider (*Uloborus walckenaerius*) would deform and rebuild into periodic spindle-knots and joints upon exposure to a humid atmosphere (Figure 8.4a).[27] Specifically, in dry conditions the spider silk is mainly comprised of two

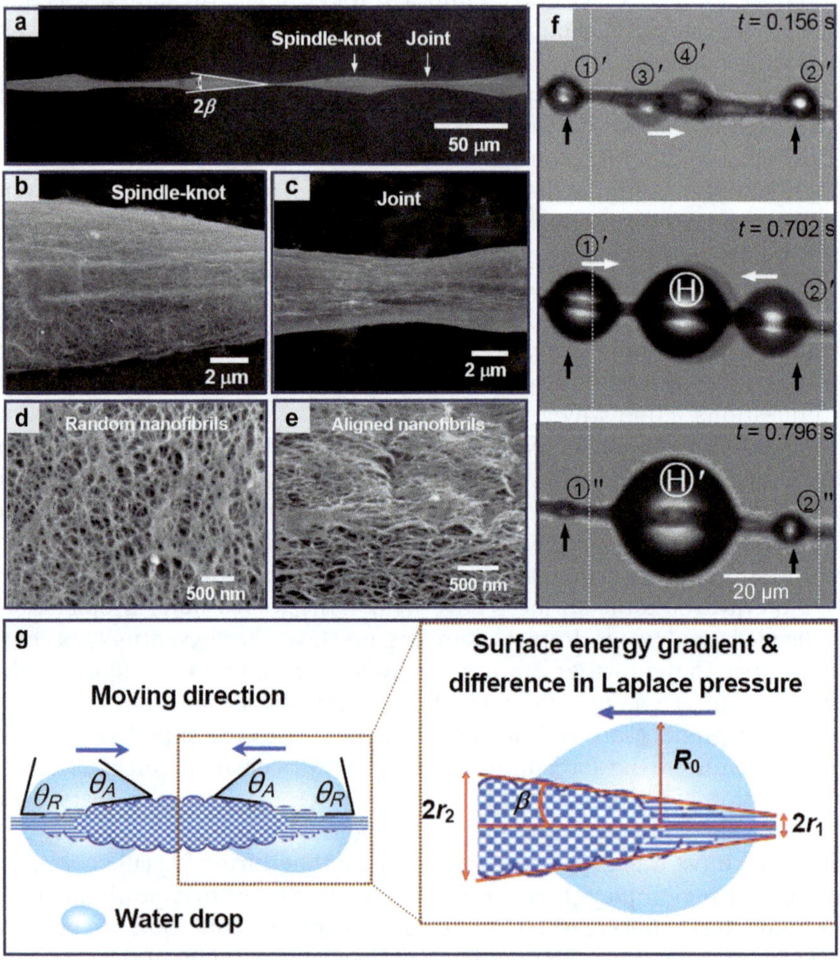

Figure 8.4 Microstructures and directional water collection ability of natural spider silk. (a–e) SEM images of the wet-rebuilt spider silk, showing periodic spindle-knots and joints. The spindle knots are mainly composed of random nanofibers while the joints are composed of aligned nanofibers. (f) Water droplets deposited on the spider silk perform directional movement and finally coalesce into a large drop. (g) The underlying mechanism is attributed to driving forces arising from the surface energy gradient and difference in Laplace pressure. (Adapted by permission from Macmillan Publishers Ltd: *Nature*.[27] Copyright (2010)).

main-axis fibres covered with periodic puffs, which are composed of random nanoscale fibrils.

The driving forces for this directional movement are believed to arise from two aspects, the surface free energy gradient and the Laplace pressure gradient, as shown in Figure 8.4g. The surface free energy gradient can be interpreted as follows: the different arrangements of nanofibrils on spindle-knots and joints produces a difference in roughness between these two regions, with greater roughness on the spindle-knots. According to Wenzel's equation,[28] $\cos\theta' = r\cos\theta$, where θ', r, and θ denote the apparent contact angle, roughness factor, and intrinsic contact angle, respectively. Therefore, the spindle-knots are more wettable than the joints. In other words, the surface free energy on the spindle-knots is lower than that on the joints, forming a surface free energy gradient from spindle-knots to joints. On the other hand, the radius of the spindle-knots is larger than that of the joints. Water droplets on these two surfaces are thus subject to different Laplace pressure (ΔP).[29] As Laplace pressure inside a droplet is inversely proportional to the local radius of the substrate (R), *i.e.* $\Delta P \sim 1/R$, the difference in radius of the spindle-knots and joints generates a difference of Laplace pressure on water droplets, with greater Laplace pressure in droplets on the joints. Since both the surface free energy gradient and the Laplace pressure gradient drive water droplets to move towards spindle-knots, water droplets on the spider silk thus move directionally under the integrated action of these two forces.

8.3.2 Liquid Collection on Artificial Spider Silks with Uniform Spindle-Knots

Inspired by the relationship between natural spider silk's unique structures and the resulting superwettability, various methods have been developed to fabricate artificial spider silks resembling both the microstructures and the directional water collection ability of natural spider silks.

First, a general dip-coating method was developed to prepare artificial spider silk.[27,30] As shown in Figure 8.5a, a nylon fibre was firstly immersed in a polymer solution, such as polymethylmethacrylate (PMMA), polystyrene (PS) in dimethylformamide (DMF) or polyvinylidene fluoride (PVDF) in dimethylacetamide (DMAC), and then drawn out horizontally at an appropriate velocity to form a cylindrical film of polymer solution on the fibre. Due to the Rayleigh instability effect, the polymer solution film then breaks up into periodic polymer droplets. After evaporation of the solvent under ambient conditions, artificial spider silk with uniform spindle-knots and joints can be obtained. In particular, these artificial spider silks show a stretched porous structure in the joint region and random porous structures on the spindle-knots, generating a similar surface roughness gradient to that on natural spider silks. Additionally, by controlling the polymer solution concentration and the drawing-out velocity, Hao Bai *et al.* succeeded in fabricating artificial spider silks with different sizes of spindle-knots, and found that the silks with larger spindle-knots are more efficient in collecting water.[31] Following

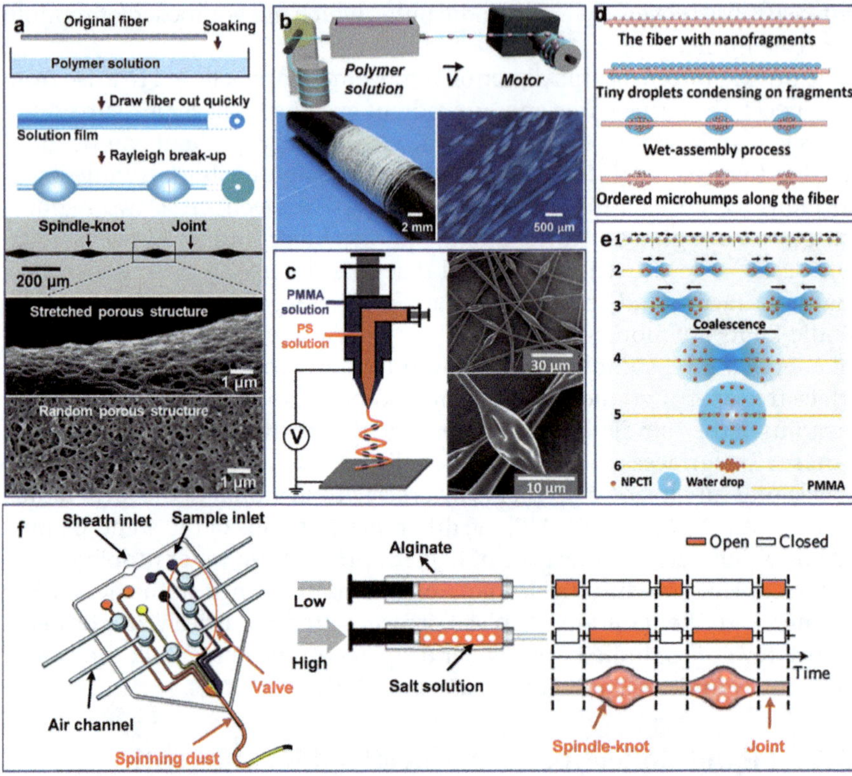

Figure 8.5 Different methods of fabricating artificial spider silks by means of dip-coating (a), fluid-coating (Adapted from ref. 30 with permission from John Wiley and Sons. Copyright © 2010 Wiley-VCH Verlag GMbH & Co. KGaA, Weinheim. Also adapted from ref. 42 with permission from John Wiley and Sons. Copyright © 2011 Wiley-VCH Verlag GmbH & Co. KGaA, Weinheim) (b), electrodynamic, (Adapted from ref. 33 with permission from John Wiley and Sons. Copyright © 2011 Wiley-VCH Verlag GmbH & Co. KGaA, Weinheim) electrospinning (c), (Adapted from ref. 34 with permission from John Wiley and Sons. Copyright © 2012 WILEY-VCH Verlag GmbH & Co. KGaA, Weinheim) combined with wet assembly (d and e) (Adapted from ref. 36 and 37 with permission from Royal Society of Chemistry) and microfluidics (f) (Adapted with permission from Macmillan Publishers Ltd: *Nature Materials*.[38] Copyright (2011)). Artificial spider silks resembling morphology (inset in b) and microstructures (insets in a and c) of natural spider silks can be prepared.

the same dip-coating concept in combination with the breath figure technique, Yongping Hou *et al.* controlled the atmospheric water content and the time range before polymer solidification and got a series of size-controllable porous artificial spider silks (smooth, less porous, homogenous porous, gradient porous, and dented microstructures).[32]

In addition to the dip-coating method, a fluidic-coating method suitable for large-scale fabrication of artificial spider silk was also developed.[33]

Specifically, a fibre with micrometre-scale radius was horizontally stretched out of a polymer reservoir guided by two capillary tubes fixed through holes on the side wall. A motor connecting one end of the fibre was used to drag the fibre at a specific velocity. By adjusting the velocity appropriately, artificial spider silks with morphology and fine structures similar to natural spider silks can be fabricated continuously on a large scale.

Electrodynamic techniques have also been harnessed to fabricate artificial spider silk. In 2012, Hua Dong *et al.* successfully used a coaxial electrospinning setup (Figure 8.5c).[34] In their experiment, diluted PMMA/DMF and concentrated PS/DMAC solutions are used as the outer and inner solution, respectively. After being pumped out from the injector and under a high voltage, the concentrated PS solution stretched out as a liquid thread whereas the outer layer of PMMA adhering to the PS thread broke up into isolated droplets and finally shrank to periodic spindle-knots. In another attempt to use an electrodynamic method, Xuelin Tian *et al.* combined the electrospinning and electrospray approach in one coaxial setup.[35] Using a spinnable PS solution with high viscosity as the inner solution and sprayable poly(ethylene glycol) (PEG) solution with low viscosity as the outer solution, a heterogeneous bead-on-string fibre can be made. Additionally, since PEG is a humidity-sensitive polymer, spindle-knots made of PEG can change volume at different humidity, so artificial spider silks prepared in this manner can be used as a humidity sensor.

Recently, Cheng Song *et al.* developed a method using wet-assembly technology to prepare artificial spider silks.[36] As shown in Figure 8.5d, a fibre with nanofragments obtained through electrospinning was placed in a foggy atmosphere. Tiny water droplets deposited on the fibre coalesced with increasing volume, so that the nanofragments inside the droplets clumped together. After water evaporation, periodic microhumps formed on the fibre, generating artificial spider silks. More important, through simply regulating the coalescence time of water droplets and the relative humidity during the assembly process, the size and separation of the spindle-knots can be controlled precisely. Apart from the wet assembly using nanofragments on nanofibres, Lin Zhao *et al.* utilized the assembly of $TiCl_4$ nanoparticles on PMMA nanofibre in foggy conditions and obtained periodic spindle-knots made of $TiCl_4$ (Figure 8.5e).[37] It is noteworthy that, due to the hydrophilic property of $TiCl_4$, the prepared artificial spider silk showed enhanced water collection ability.

In addition to the approaches described above which mimicked the silk-spinning process of spiders, a microfluidic system consisting of a digital and programmable flow controller was developed for continuous fabrication of artificial spider silk.[38] As can be seen from Figure 8.5f, by programming the flow through valves in two channels containing different solutions, the morphology and chemical composition of the resulting fibres can be strictly controlled. By introducing an alginate solution containing salt at relative high feeding rate and subsequently removing the salt, a typical kind of artificial spider silk with porous structured spindle-knots can be obtained.

For liquid droplets on natural or ordinary artificial spider silks, the movement direction is always from joints to spindle-knots. In some circumstance, the reverse movement, from spindle-knots to joints, is also necessary, such as in a smart catalysis system, in which the reaction product should move away from the catalysis fixed in the region of spindle-knots.[39] Researchers have developed various methods to control the movement direction of liquid droplets on artificial spider silks.

As can be seen from Figure 8.6a, by tuning the chemical composition and surface roughness of the spindle-knots, the movement direction of water droplets on artificial spider silks can be strictly controlled.[30] Water droplets on spindle-knots with the same surface roughness but different chemical composition, or on spindle-knots with the same chemical composition but different roughness, can move in a controlled direction. For example, water droplets deposited on artificial spider silk with smooth PMMA spindle-knots and smooth PS spindle-knots, or on artificial spider silk with rough and smooth PVDF spindle-knots, can move towards or away from the spindle-knots. This interesting phenomenon is believed to arise from the cooperation of three kinds of forces acting on the water droplets: (1) the chemical force induced by the chemical gradient deriving from the different compositions of the spindle-knots and joints; (2) the hysteresis force due to contact angle hysteresis which mainly stems from the surface roughness; (3) the Laplace force resulting from the local curvature or radius difference of the substrate. By changing parameters relating to these three kinds of forces, the movement direction of water droplets may be changed accordingly.

Besides the methods described above to tune droplet movement direction, Yongping Hou *et al.* achieved control of droplet movement direction *in situ* through coating the spindle-knots of an artificial spider silk with a stimulus-responsive polymer.[40,41] As shown in Figure 8.6b, coating the spindle-knots with a temperature-responsive polymer containing poly(*N*-isopropylacrylamide) (PNIPAAm) and then changing the experimental temperature below or above the lower critical solution temperature (LCST) of the polymer, the same water droplet on the spider silk can be controlled to move respectively towards or away from the spindle-knots at the same humidity. The change in wettability of the PNIPAAm-containing polymer below and above the LCST is responsible for this switch of movement direction. Moreover, water droplets on artificial spider silks with spindle-knots coated with a light-responsive polymer can also perform reversible motion; see Figure 8.6c. Here the spindle-knots are coated with azobenzene, a typical polymer that will transform between *trans* and *cis* configuration states under the stimulus of UV or visible light. If other experimental parameters are held constant, water droplets on this kind of spider silk will move away from the spindle-knots after visible light irradiation, whereas after UV irradiation the droplets will move towards the spindle-knots.

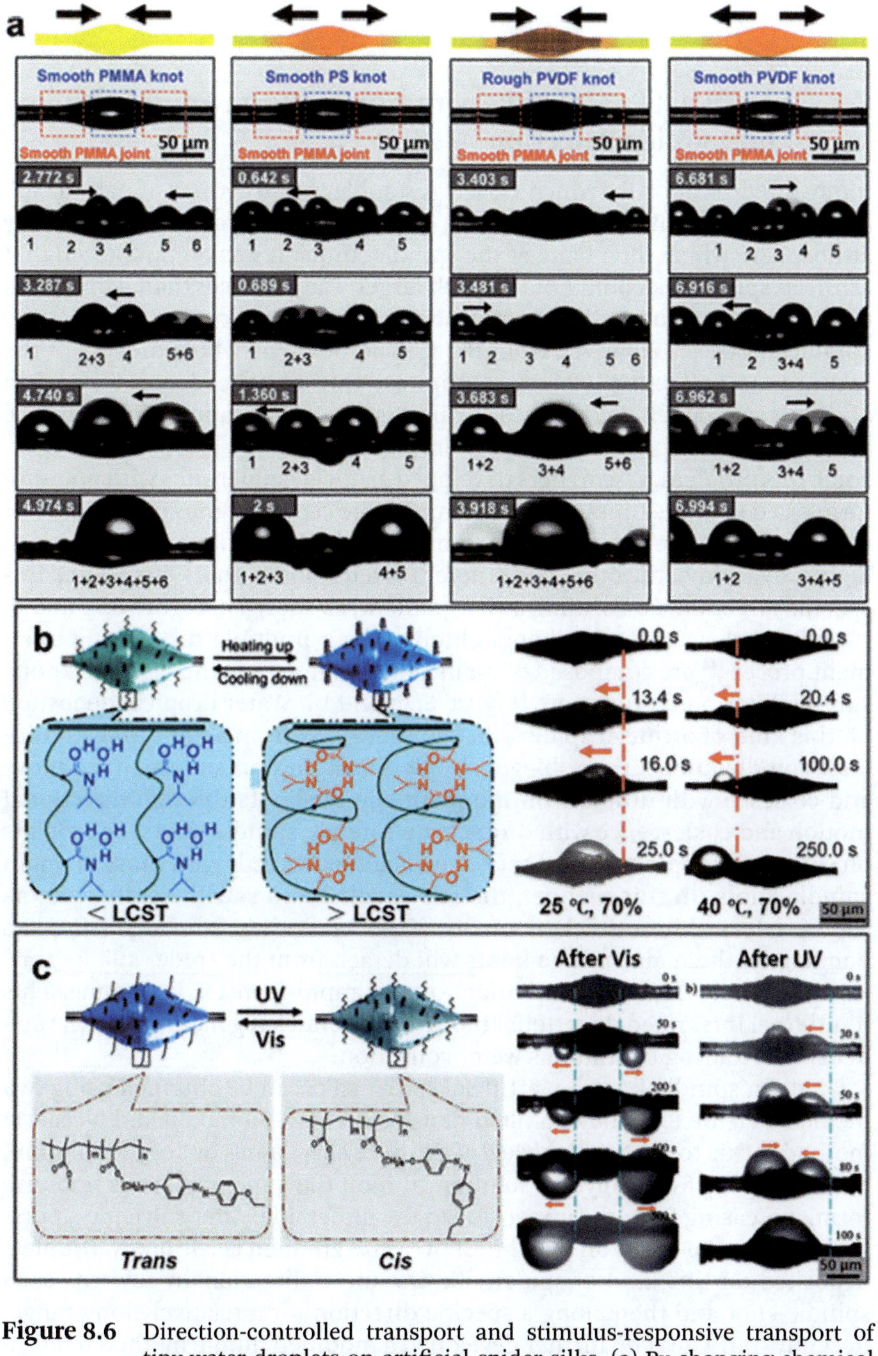

Figure 8.6 Direction-controlled transport and stimulus-responsive transport of tiny water droplets on artificial spider silks. (a) By changing chemical composition or surface roughness of the spindle-knots, droplets can move towards or away from the spindle-knots. (Adapted from ref. 30 with permission from John Wiley and Sons. Copyright (c) 2010 Wiley-VCH Verlag GMbH & Co. KGaA, Weinheim) (b and c) By modifying the spindle-knots with temperature-sensitive (b) and light-sensitive polymers (c), water droplets on the spider silk can move towards or away from the spindle-knots through changing the temperature and lighting condition, respectively. (Adapted from ref. 40 and 41 with permission from the Royal Society of Chemistry).

8.3.3 Artificial Spider Silks with Non-Uniform Spindle-Knots for Liquid Collection

Bioinspired spider silks which closely resemble the structures of natural spider silks have been shown to be able to collect water droplets from a foggy atmosphere which then hang at the spindle-knots. It was even reported that artificial spider silk could hold a much larger water droplet than a fibre with uniform radius, due to the chemical and curvature effects arising from the spindle-knots.[42,43] However, from the viewpoint of the whole process of the continuous collection of water droplets, this enhanced water hanging ability is not beneficial because it hinders the regeneration of the surface and delays the deposition of a new droplet in the same location.[39] To overcome this problem, researchers developed artificial spider silks with non-uniform-sized spindle-knots, which can mainly be classified into two kinds: one type has multi-level-sized spindle-knots alternately distributed, as shown in Figure 8.7a and b; the other has different-sized spindle-knots distributed in a specific direction, *i.e.* forming a size gradient, as shown in Figure 8.7c and d.

Multi-level-sized spindle-knots obtained by a multiple dip-coating treatment process[44] are composed of main spindle-knots, satellite spindle-knots, sub-satellites, and even more (Figure 8.7a and b). Water droplets deposited on this kind of artificial spider silk will move first from a joint to the multi-level spindle-knots. Then, bigger droplets hanging on sub-satellites move and coalesce with droplets on the main spindle-knots through directional motion and coalescence with droplets on satellite spindle-knots, or droplets on sub-satellite spindle-knots move and coalesce directly with those on main spindle-knots. In this manner, the sub-satellite and satellite spindle-knots can be released in a timely manner; at the same time, the larger droplets hanging on the main spindle-knots will detach from the spider silk by gravitational force more frequently due to their rapid increase in volume. This multi-level integrated directional motion accelerates regeneration of the spider silk, favouring continuous water collection.

Gradient spindle-knots on artificial spider silks can be obtained using two methods. Figure 8.7c shows a tilted drawing method—a modified dip-coating method.[45] Due to the tilted angle θ of the fibre as it comes out of the solution, the thickness of the polymer solution film on the support fibre is non-uniform, increasing from the upper side to the underside. After solvent evaporation, the spindle-knots on the fibre show a size gradient accordingly. Droplets deposited on this kind of spider silk first move directionally towards each spindle-knot and then along a specific direction for a relatively long range. Recently, Yan Xue *et al.* further developed a modified fluidic method to fabricate this type of artificial spider silk with gradient spindle-knots.[46] As shown in Figure 8.7d, by controlling the drawing velocity of the fibre with a motor, a thickness gradient of the liquid film can be generated. The subsequent development of the liquid film into periodic droplets and then to gradient spindle-knots is the same as the fluidic-coating method described previously, but the as-prepared spider silk exhibits directional liquid collection over a longer range.

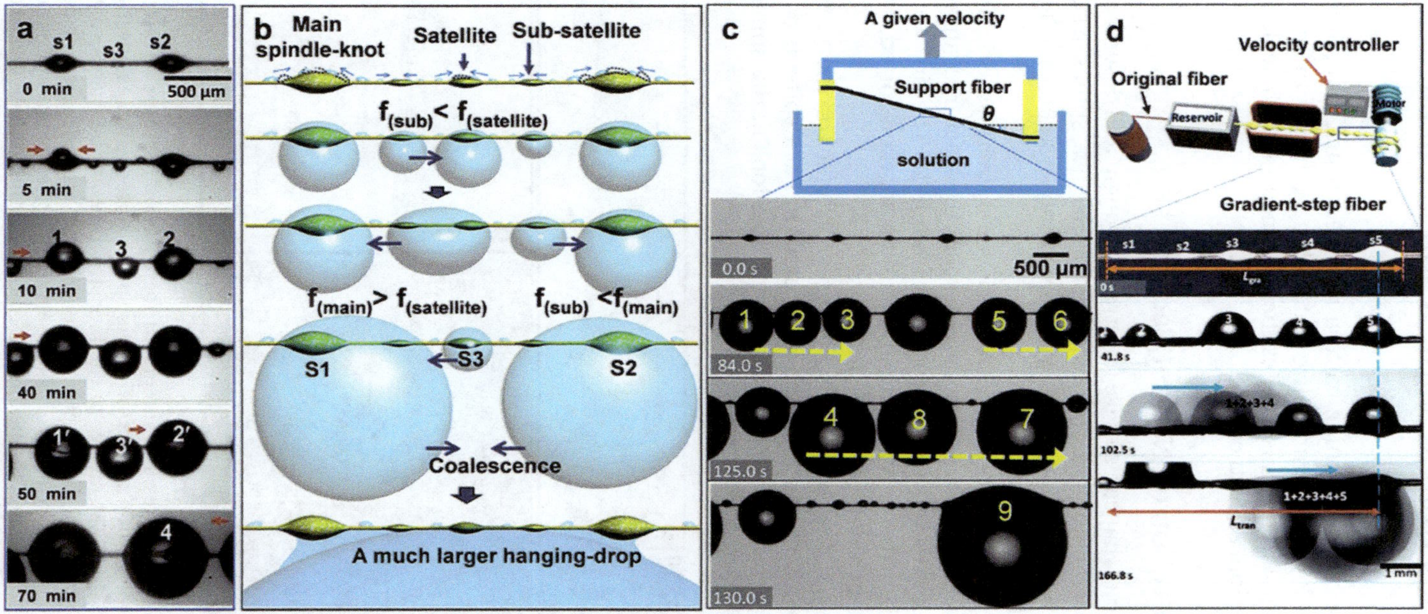

Figure 8.7 Preparation of artificial spider silks with non-uniform spindle-knots and liquid collection on them. Microscope image (a) and schematic illustration (b) of integrated motion of water droplets on artificial spider silk with non-uniform spindle-knots prepared *via* multiple dip-coating. (Adapted from ref. 44 with permission from the Royal Society of Chemistry) (c and d) Set-up for fabricating artificial spider silk with gradient spindle-knots using tilted dip-coating method (c) (Reprinted by permission from Macmillan Publishers Ltd: *Scientific Reports*,[45] Copyright (2013)) and velocity-changing fluidic-coating method (d) (Adapted from ref. 46 with permission from the Royal Society of Chemistry) as well as the directional movement of water droplets on them.

8.4 Liquid Collection on Natural and Artificial Cactus

The desert beetle uses its hydrophilic pattern to collect water, and spiders use silk with spindle-knots to immobilize water. But neither of them can accomplish long-range liquid transport (at the millimetre length scale) or continuous fog collection, as cacti can do.

8.4.1 Liquid Collection on Natural Cactus

Cacti are amazing plants that can survive in extreme drought and desert conditions. They have several biological features, such as needles, spines, a succulent stem covered by a cuticule, and a widespread root system, that help them to thrive in such extreme conditions.

Recently, Jie Ju *et al.* found that the cactus *Opuntia microdasys*, originating from the Chihuahua desert of Mexico, uses a specially structured surface of its above-ground part to actively harvest water from fog.[47] The tiny water droplets carried by the wind were captured, accumulated, transported, and absorbed by the plant in a well-organized manner.

As shown in Figure 8.8, the clusters of spines and trichomes on the cactus stem are evenly distributed in a hemispherical pattern. Each spine is conical, with a length of approximate 2 mm. Taking closer look at the spines,

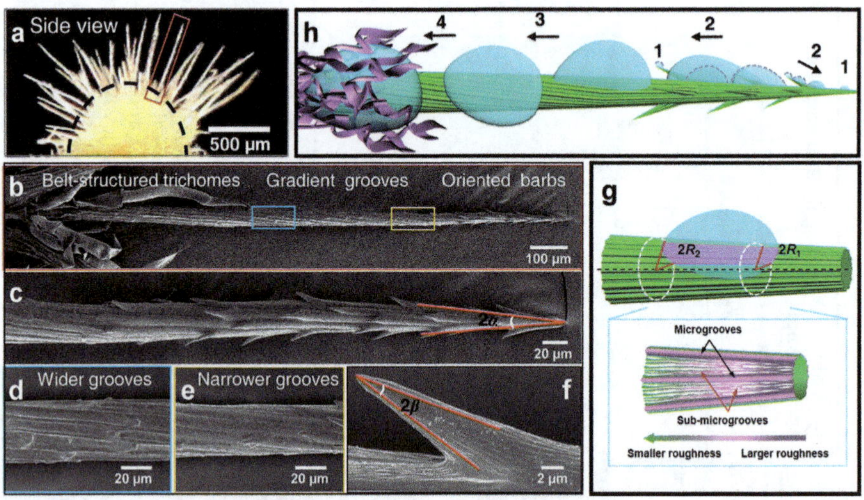

Figure 8.8 Multi-level structures and integrated fog collection system of cactus. (a) Morphology of a cluster of cactus spines with trichomes covering the base. (b–f) SEM images of the fine structures of the cactus spine. (g) Driving forces of the directional movement of water droplets on cactus spines. (h) Illustration of the integrated fog collection system on the cactus. (Reprinted by permission from Macmillan Publishers Ltd: *Nature Communications*.[47] Copyright 2012).

it can be seen clearly that they can be divided into three parts according to the difference in fine structure: the tip is covered with microscale oriented barbs; the middle part is sculptured with gradient grooves; and the base is covered with belt-structured trichomes (Figure 8.8b–f). These elaborate structures have different functions in the whole fog collection process. When a single spine with trichomes at its base is placed in a fog flow, tiny water droplets contained in the flow are first deposited on the very tip of the conical barbs in the tip region of the spine, due to the sharp effect of the tips. When the volume of these water droplets increases, they move along to the barbs to the base and coalesce with other water droplets on the surface of the spine. The resulting larger droplets then move towards the base of the spine through the grooves in middle part and are finally absorbed by the trichomes in the base. There are two driving forces in this water transportation process: the Laplace pressure gradient and the surface free energy gradient.

Typically, a Laplace pressure is generated inside a water droplet located on a conical surface; the whole droplet tends to move towards the side with larger radius of curvature, *i.e.* towards the base. During transport along the middle part of the spine, the wetting gradient derived from the gradient distribution of microgrooves, as shown in Figure 8.8g, further propels the droplets towards the base of the spine. Eventually, the droplets that reach the spine base are immediately absorbed by the trichomes and sucked into the cactus stem. From a more general point of view, we can see that large water droplets are driven to move towards the base not only by the Laplace pressure exerted by a single spine, but often by multiple adjacent spines in the same cluster as well.

A distinctive feature of fog collection on cactus spines is that each part of it is self-renewable for continuous collection cycles. This enables the whole spine cluster to operate with high efficiency as an integrated, long-range, continuous fog collection system.

Dew collection has also been studied on cactus species other than *Opuntia microdasys*. As recently reported by F. T. Malik *et al.*,[48] four cacti, *Copiapoa cinerea* var. *haseltoniana*, *Ferocactus wislizenii*, *Mammillaria columbiana* subsp. *yucatanensis*, and *Parodia mammulosa*, were tested in a dewing chamber. Through careful comparisons, a series of important conclusions have been drawn. First, there is a direct correlation between dew collection ability and the wettability of the spines. Generally, the more hydrophilic the spines, the more dew water they can collect. Second, the dew collection ability of a cactus is also related to the high emission coefficient of the radiative cooling of the spine clusters. The spines emit heat to the ambient so that the spine cluster area stays cool and becomes a condensation site on the cactus. Third, these authors also proved that the spines are used by the cacti as a tool to collect water from dew, because with the spines removed, the cacti collected a lot less water than they did with the spines in place. The reason for spines to be used as dew harvesters, as indicated by the authors, has something to do with the edge effect, which states that because edges and corners can make more contact with water vapour, these areas tends to collect more water.[49] These important scientific findings may become guiding reference for future studies.

Green bristlegrass *Setaria viridis* also has a fog collection function and generally shares the same driving force as the cactus *Opuntia microdasys* to transport captured water to storage. Yan Xue *et al.* reported that micro barbs were uniformly distributed on a single spine of the bristlegrass.[50] However, unlike the barbs on cactus spines, these micro barbs were aligned towards the tip of the spine. The authors found that the sole function of the barbs was to capture the tiny water droplets in fog and transport them to the main spine. The accumulated water on the main spine was transported towards the base by the Laplace pressure generated by the conical shape of the spine. The bristlegrass spines have nanogrooves, similar to those on the cactus spine, which show a size gradient along the axis of the spine. Moreover, the authors mentioned a second function of the grooves, which was to form a water film which then acted to reduce contact hysteresis the in later stages of water transport.

Natural examples like this can find applications in air filters and humidity sensors as well as fog collection.

8.4.2 Liquid Collection on Artificial Cactus

The way in which plants use physical principles such as the Laplace pressure gradient and wetting gradient to facilitate fog collection has inspired the invention of various fog harvesters as well as other useful tools for science and daily life.[51] For instance, Jie Ju *et al.* reported a cactus-inspired fog collector prepared by gradient electrochemical etching followed by gradient chemical modification (Figure 8.9a).[52] The prepared copper wire has a dual geometric and wettability gradient, with increasing wettability from the tip to the base. When this dual-gradient copper wire and a conical copper wire with pure hydrophobic or hydrophilic surfaces are placed in the same fog flow, they show different fog collection performance. As shown in Figure 8.9b, on the pure hydrophilic conical copper wire, water droplets grew slowly but were transported quickly to the base, while on the pure hydrophobic conical copper wire, water droplets grew quickly but were transported slowly. On the dual-gradient conical copper wire, however, water droplets had a relatively large growth rate and speed of motion towards the base. In other words, the dual-gradient copper wire combined the advantages of both the copper wires with pure wettabilities. Since droplet growth rate and motion speed are two crucial indicators of overall fog collection efficiency, the dual-gradient copper wire is thus the most efficient fog collector.

For higher fog collection efficiency, it is necessary to incorporate heterogeneous structures just as natural cacti do. A good example of this was given by Xin Heng *et al.*[53] They prepared a branched ZnO wire on a silicon substrate *via* a two-step vapour–solid method. As shown in Figure 8.9c, the main spine of the ZnO wire was on the ~1 mm scale, comparable to the single

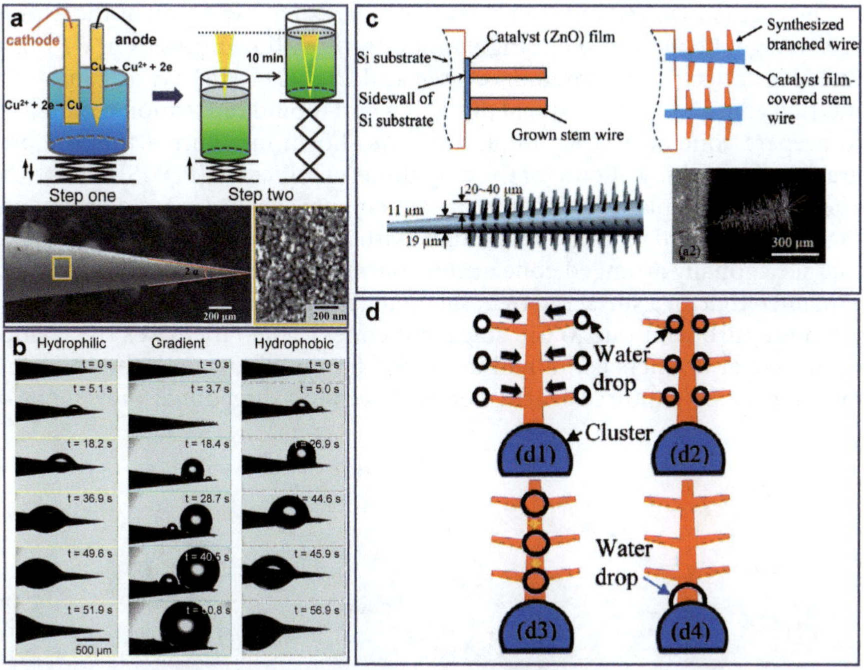

Figure 8.9 Cactus spine inspired single artificial spine and water collection behaviour on them. (a) A conical copper wire with increasing wettability from tip to base and with nanoscale roughness can be made using a typical gradient electrochemical corrosion followed by gradient chemical modification. (b) Compared to the conical copper wire with pure wettability, the dual-gradient copper wire show integrated advantages. (Adapted from ref. 52 with permission from John Wiley and Sons. Copyright © 2013 Wiley-VCH Verlag GmbH & Co. KGaA, Weinheim) (c) By exploiting the classical vapour–solid growth method, branched ZnO conical wires resembling the structure of cactus spines can be fabricated. (d) Directional collection of water droplets on the artificial cactus spine. (Adapted with permission from X. Heng, M. Xiang, Z. Lu and C. Luo, *ACS Applied Materials & Interfaces*, 2014, **6**, 8032–8041. Copyright (2014) American Chemical Society.)[53]

spine on *Opuntia microdasys*. Branches were grown on this main spine using a secondary seed-induced crystal growth *via* vapour deposition. The authors compared the fog collection efficiency of the artificial cactus and the natural cactus *Opuntia engelmannii* var. *lindheimeri*, which has similar microstructures to *Opuntia microdasys*. Their results showed that the artificial cactus can collect more water than its natural counterpart. In addition, the authors found that the direction of fog flow towards the sample affected the fog collection efficiency. An important modification was that the authors incorporated a hydrophilic tube in the artificial spine to, which acted to transport the collected water to a designated reservoir. This brings the fog collector one step closer to an integrated fog collection system.

The cactus-inspired fog collectors described so far were still at the exploratory stage, being tested as single units. In the following section, we introduce fog collectors prepared on a large scale. In 2014, Jie Ju *et al.* reported a method combining mechanical punching and mould replication technology to prepare cone arrays (Figure 8.10a).[54] As shown in Figure 8.10b, the programmed punching allows for the poly(dimethyl silicone) (PDMS) cones to be distributed on a plane in different arrangements. They also tested a smooth PDMS surface and natural cactus stem (with all spines removed). They found that hexagonally arranged cone arrays collected the most water. The authors explained that on a surface with hexagonally distributed cones, the flow fields are more turbulent due to the staggered cones, which increases the chance of tiny water droplets contained in the fog flow colliding with the solid surface and being captured by the cones. The rapid directional movement of

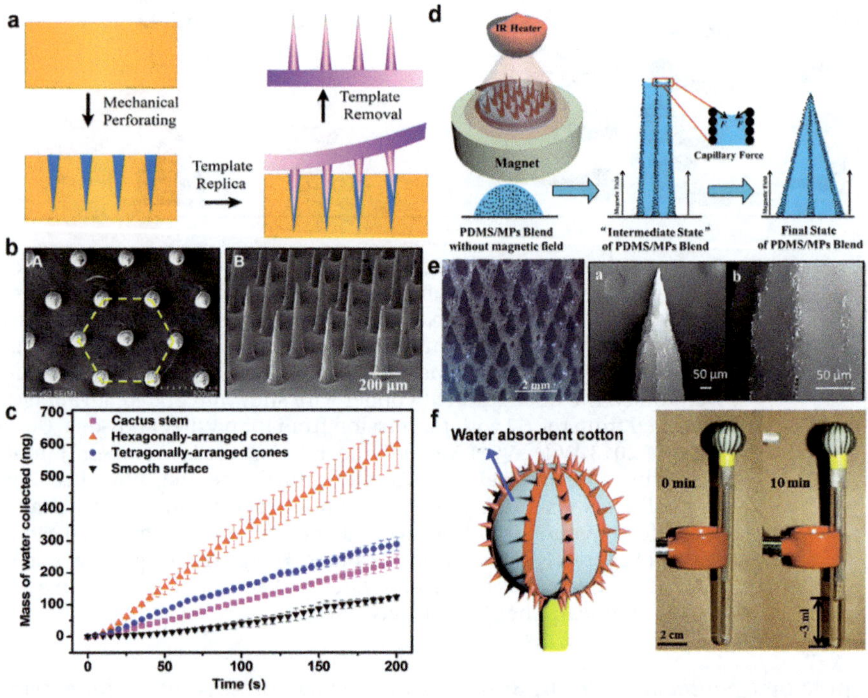

Figure 8.10 Cactus spine inspired artificial spine arrays and water collection ability on them. (a) Using a method combining mechanical lithography and mould replica technology, cone arrays with different arrangements can be fabricated (b,c) Water collection ability of surfaces with different fine structures. (Adapted from ref. 54 with permission from John Wiley and Sons. Copyright © 2014 Wiley-VCH Verlag GmbH & Co. KGaA, Weinheim) (d) By exploiting the self-assembly of magnetic particles in a magnetic field, cone arrays can be prepared on a large scale. (e,f), Cone arrays fixed on a hydrophilic substrate (water-absorbent cotton) can be used to collect water continuously. (Adapted from ref. 55 with permission from John Wiley and Sons. Copyright © 2014 Wiley-VCH Verlag GmbH & Co. KGaA, Weinheim).

water droplets along each cone also accounts for the efficient fog collection on cone-structured surfaces.

Moyuan Cao *et al.* proposed a more ingenious route to fabricate a fog collector that actually resembles the natural cactus in function and appearance.[55] As shown in the schematic drawing in Figure 8.10d, magnetic Co particles 2 μm in diameter were blended into the PDMS oligomer. The mixture was placed above a magnet, which induced the magnetic particles to assemble into cones along the direction of the magnetic field. On heating by infrared light the cones solidified. Large-scale cone arrays can be simply fabricated in this way (Figure 8.10e). On the SEM images, we can see that the cones have a very sharp apex and their surfaces are sculpted with microscale ridges and grooves parallel to the longitudinal direction of the cone. This asymmetric structure enables water droplets on the cone to be transferred from tip to base, directionally driven by a Laplace pressure gradient. When these rubber cones were assembled onto a sphere made of water-absorbent cotton fabric, the system functioned just like a real cactus to collect fog and absorb the water collected. As shown in Figure 8.10f, the cotton fabric can immediately absorb the water droplets that are transported to the base of the cones. The absorbed water converged at the bottom of the sphere and was guided into the water reservoir beneath. In this experiment, 3 mL of water collected after the artificial cactus was subjected to a fog flow with velocity of 45–50 cm s^{-1} for about 10 min.

8.4.3 Artificial Cactus for Oil/Water Separation

In the petroleum industry there is an increasing need to separate oil/water mixtures. Traditional materials which separate these mixtures based on flux through a membrane, such as oil-removing[56] and water-removing materials,[9] and bulk absorbing materials,[57] have dominated the market in coping with macro-sized oil/water mixtures. However, all of the current methods rely greatly on spontaneous phase separation and cannot deal with micro-sized oil droplets suspended in water, including a recently reported hygroresponsive membrane, which was able to separate oil/water mixtures in a single-unit operation with high separation efficiency.[58] As well as the common drawback of being unable to treat micro-sized oil droplets in the water phase, these methods are limited either by easy fouling or by difficulty in post-processing.

Inspired by fog collection behaviour, Li Kan *et al.* developed a cactus spine-like cone-structured material and fixed it to a custom-designed apparatus to collect the micro-sized oil droplets in an oil/water mixture.[59] By using the tip of a needle to mechanically puncture a plastic substrate, a plastic mould with negative conical void arrays was obtained. The cone arrays were then prepared by pouring PDMS oligomer into the mould, followed by degassing. In underwater condition, the PDMS surface shows relative affinity to oil due to its low surface tension.[60] As a result, when the substrate bearing the PDMS cone arrays was subject to a flow of water containing micro-sized oil droplets, the oil droplets tended to adhere to the surface of the cones after direct contact.

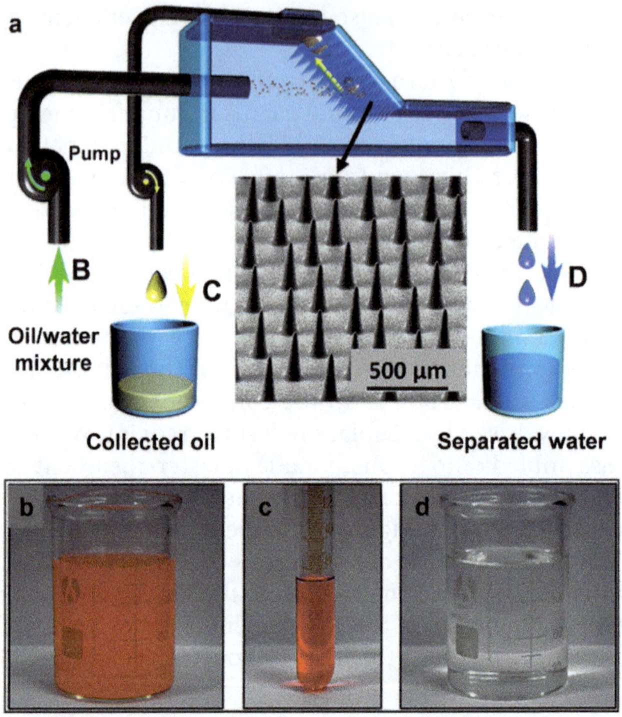

Figure 8.11 Oil/water separation by artificial cactus spines based on directional oil collection under water. (a) Sketch of the setup. The inset shows cone arrays used in the experiment. (b–d), Oil/water mixture before separation (b) and pure oil (c), pure water (d) after separation. (Adapted by permission from Macmillan Publishers Ltd: *Nature Communications*.[59] Copyright (2013)).

Importantly, the authors found that, as in water transport on cactus spines in air, oil droplets can also accumulate and be transferred to the base of those artificial spines under water. This provides the possibility of accumulating oil droplets into a continuous oil fluid that can be easily collected and stored, using the apparatus shown in Figure 8.11. Figure 8.11b–d show respectively the oil/water mixture before separation, pure oil collected and pure water remaining after separation using the apparatus as Figure 8.11a. The oil collection efficiency can be as high as 1.128 mL cm^{-2} s^{-1}, sufficient for practical use.

8.5 Other Kinds of Surfaces with Superwettability for Directional Liquid Collection

In addition to the desert beetles, spider silks, cacti, and the artificial surfaces inspired by them, there are still many other kinds of natural and artificial surfaces with superwettability exhibiting remarkable abilities to collect liquid efficiently. In this section, we focus on some typical examples and give a brief introduction to them.

8.5.1 Natural Surfaces with Superwettability for Liquid Collection

In nature, organisms have evolved unique abilities to adapt to their habitats during the long course of evolutionary history. As one of the most important elements for life, water is crucial to survival. Many organisms have therefore developed special structures to collect water in adverse conditions.

For example, the Australian lizard *Moloch horridus* has adapted to a very hostile environment where summer temperatures hover around 40 °C for weeks at a time and rainfall is always sporadic. *Moloch horridus* has evolved a suite of morphological features capable of making use of the scarcest water. Specifically, the lizard's outer skin is covered with scales in a honeycomb arrangement, each scale being sculpted with narrow grooves radiating from a central peak.[61,62] When water droplets, either from scarce rainfall or from dew condensation, are deposited on the scales, they are first guided to move down along the grooves and then are drawn to move towards the animal's mouth under the influence of a complex network of capillary forces stemming from the special arrangement of the scales. When the water finally reaches the animal's mouth, drinking is promoted by hygroscopic mucus secreted from ducts adjacent to the mouth.

In 2011, Australian researchers found that Australian green tree frogs *Litoria caerulea* manage to use condensation to help their hydration during the dry season.[63] The frogs hop around the chilly desert at night and then jump back to a relatively warmer and more humid tree hollow. Due to the "fogging up" effect, some water droplets will condense on their skin surface. The authors found that these frogs are able to collect 0.4 g of water, which is almost 1% of their total body weight, for one cycle of moisture capture while losing ~0.07 g of water. So, overall, they gain water by this behaviour. Like other amphibians, this frog can absorb water through its skin, and this accounts for nearly 60% of total water uptake by this species.

Some plants also show a remarkable ability to collect water efficiently, taking advantage of elaborate microstructures. The peanut *Arachis hypogaea* is a typical xerophyte, which can endure long-term absence of rainfall. During periods of drought, microstructures on the leaves help peanuts to collect vital water. Figure 8.12a shows some water droplets adhering to the leaves of peanut, demonstrating a typically superhydrophobic and high adhesive state.[64] When peanut leaves and lotus leaves (a kind of surface with superhydrophobicity and ultra-low adhesion) were placed in the same foggy atmosphere for a period of 60 min, the peanut leaves collected nearly as twice as much water as the lotus leaves, indicating the high moisture capture ability of the peanut leaves. To find the underlying reason for the superwettability of the peanut leaves, researchers investigated their microstructure. As can be seen clearly from Figure 8.12b and c, the leaf surface is densely covered

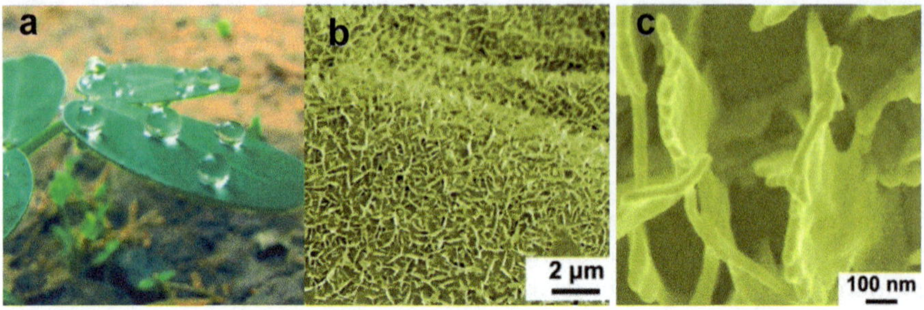

Figure 8.12 Peanut leaves with superwettability and liquid collection ability. (a) Spherical water droplets on a peanut leaf. (b and c) SEM images of the peanut leaves with different magnification. (Adapted from ref. 64 with permission from John Wiley and Sons. Copyright © 2013 Wiley-VCH Verlag GmbH & Co. KGaA, Weinheim).

with microscale slopes, with tops consisting of nanostructured papillae and smooth sidewalls forming quasi-square grids with separated ridges on a larger scale. When a water droplet is placed on the peanut leaf surface, a quasi-continuous and discontinuous triphase contact line (TCL) will be generated at the microscale and nanoscale, respectively, due to these peculiar structures. This special TCL pattern further results in superhydrophobicity with simultaneous high adhesion, which is favourable for liquid collection.

The Namib desert grass *Stipagrostis sabulicola*, living in the same dry conditions as the well-known desert beetle described earlier in the chapter, has also evolved a suite of special structures allowing it to collect water from the air and keep itself hydrated.[65,66] The culms of young *S. sabulicola* are tightly enveloped by involute leaves that end in acute tips. Droplets form on these involute leaves after exposure to fog. Microscale aligned ridges and grooves, parallel to the longitudinal direction of the leaves, result in the direct transportation of water droplets intercepted from the foggy wind towards the base of the plant, guaranteeing further efficient and rapid uptake by the shallow roots.

The California redwood *Sequoia sempervirens* is believed to be one of the oldest trees in the world. The mature trees have an average age of several hundred years. They usually stand 60–70 m tall with a diameter of 3–4.5 m, and some trees have been measured at more than 110 m. It is hard to imagine they can sustain such a height without another source of water in addition to uptake through the roots, especially because in their habitat, the north California coast, deep soil water often is unavailable due to lack of rainfall for several months in the summer. Fortunately there is an alternative water resource, the fog that frequently affects the north California coast during summer. The dense, needle-shaped leaves of the redwood efficiently intercept water droplets contained in the fog, and when these reach a critical size they drip down under gravitational force. According to the research of T. E. Dawson published in 1998, during the period of his study about 34% of the hydrologic input in the redwood forest came from water dripping off the redwood trees through intercepting water droplets from fog.[67]

8.5.2 Artificial Surfaces with Superwettability for Liquid Collection

In addition to surfaces with patterned wettability inspired by desert beetles, fibres with periodic spindle-knots and joints inspired by spider silks, and cones and cone arrays with different wettability inspired by cactus, there still many other surfaces with superwettability showing liquid collection ability, as shown in Figure 8.13.

Figure 8.13a is an SEM image of silicone nanofibres obtained *via* a simple hydrolysation and condensation process of methyltrimethoxysilane (MTMS).[68] Coating with these nanofibres renders the substrate beneath superhydrophobic. The inset image shows spherical water droplets condensed on this superhydrophobic surface under subcooling conditions. Compared to the glass slide without modification and with ordinary hydrophilicity, the prepared surface shows apparently higher water collection efficiency under the same experimental conditions and in the same time range (Figure 8.13b). This enhanced water collection ability mainly comes from the quicker regeneration of the condensation sites due to rapid release of water droplets from the low-adhesive superhydrophobic surface.

Apart from using a superhydrophobic surface to reduce water adhesion, Joanna Aizenberg *et al.* introduced a new kind of surface—the slippery lubricate-infused porous surface (SLIPS) that has remarkably low adhesion to water.[69] In this kind of surface, direct contact between the solid surface and the water droplets on it is prevented by the intermediate lubricated oil layer with ultra-low surface tension. The adhesion between surface and water droplets is thus very low, so water droplets on this kind of surface usually have high mobility and can slide off at very low tilt angles. Following this concept, Varanasi and Hashaikeh *et al.* first fabricated a nanomat made of PVDF-HFP using electrospinning technology (Figure 8.13d, upper left).[70] The electrospun nanomats have a porous structure consisting of different fibre diameters ranging from 100 to 500 nm. The nanomats were then infused with either total quartz oil or Krytox-1506, two kinds of lubricating oil with different physical parameters. The three types of surfaces have water contact angles of about $134 \pm 4°$, $94 \pm 4°$, and $116 \pm 4°$, respectively, as can be seen from the upper image of Figure 8.13d. These surfaces were then placed in a foggy atmosphere for a period of time. The amount of water collected on the three surfaces is shown in Figure 8.13d. The two surfaces with infused oil collected more water than the original surface. The smaller critical volume of water droplets starting to slide off the SLIPS (4 μl) compared to that on the plain nanomat (37 μl) is responsible for the higher water collection efficiency (Figure 8.13c). Confining a thin layer of lubricating oil on the top of the prepared surface thus confers enhanced water collection efficiency.

PNIPAAm is a polymer known for its temperature-sensitive characteristics. Combining the temperature-dependence of PNIPAAm around LCST and porous structures of cotton fibres, Esteves and Xin realized temperature-triggered collection and release of water from fogs (Figure 8.13e–h).[71]

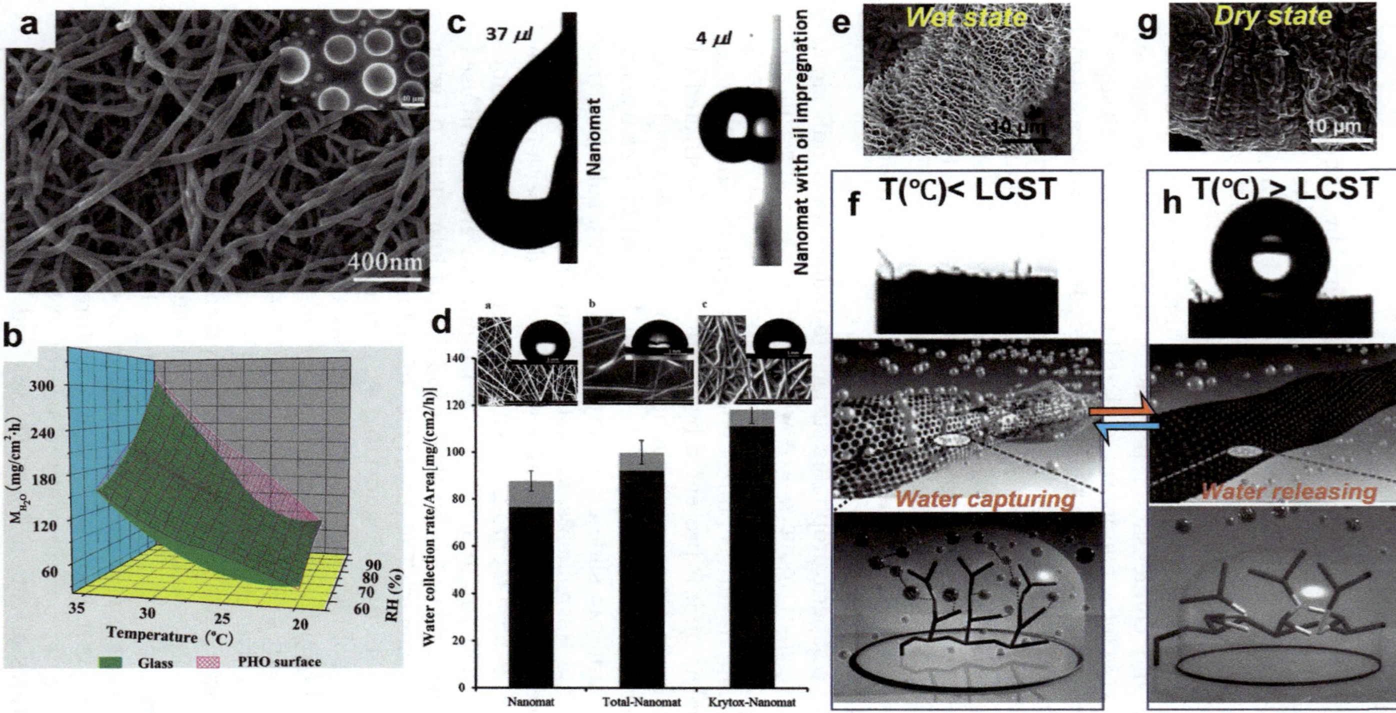

Figure 8.13 Other kinds of artificial surfaces with superwettability and capable of collecting liquid. (a and b), Surface modified with superhydrophobic silicon nanofibers (a) enhance water collection compared with bare glass surface (b). (Adapted with permission from R. Chen, *et al.*, *Journal of Physical Chemistry C*, 2009, **113**, 8350–8356. Copyright (2009) American Chemistry Society.)[68] (c and d) Nanomat infused with lubricated oil decreases critical volume of water droplets starting to slide down (c) and increases water collection efficiency. (Adapted with permission from B. S. Lalia, *et al.*, *Langmuir*, 2013, **29**, 13081–13088. Copyright (2013) American Chemistry Society.)[70] (e and h) Cotton fibre modified with temperature-responsive polymer can collect (f) and release (h) water upon temperature changes. (Adapted from ref. 71 with permission from John Wiley and Sons. Copyright © 2013 Wiley-VCH Verlag GmbH & Co. KGaA, Weinheim).

Figure 8.13e–g show the microstructures of the PNIPAAm-covered cotton fibres in the wet and dry state respectively. In dehydration conditions, PNIPA-Am-coated fibres were in a collapsed state; while in the wet state the fibres have a porous, sponge-like morphology, typical of a hydrogel network. This change of morphology according to temperature is a result of the differing wettability of PNIPAAm below and above the LCST. At a temperature below the LCST, the fabric has nearly zero water contact angle, whereas above the LCST, the water contact angle can be up to around 140°. This switch between surface morphology and wettability gives rise to a reversible conversion of the fabric between water-capturing and water-releasing states below and above the LCST. The concept of smart water capture and release in response to an external stimulus will surely guide a new direction in research relating to liquid collection.

8.6 Conclusion and Outlook

In this chapter, we have summarized some natural and artificial surfaces with superwettability capable of collecting liquid. All of these surfaces show different superwettability characteristics, for instance, desert beetles and surfaces inspired by them show patterned wettability, spider silks and fibres inspired by them show periodic wettability, cactus and cactus-inspired cones show gradient wettability, and so on. Despite the different forms of wettability, these surfaces share the same function—they collect liquid efficiently. As fresh water resources worldwide are inadequate, especially in some coastal, mountainous, and desert areas, many people lack access to adequate water. Water collection taking advantage of the special wettability of material surfaces is of great importance due to its simple, low cost, and easy handling properties. On the other hand, the increasing demand for oil gives rise to a growing number of accidental oil spills in the world's oceans, and efficient liquid collection is important here too.

More research attention should be directed to natural and artificial surfaces capable of collect liquid more efficiently. Also, the integration of superwettability into materials that can respond to external stimuli expands the range of potential applications even further.

References

1. L. Feng, S. H. Li, Y. S. Li, H. J. Li, L. J. Zhang, J. Zhai, Y. L. Song, B. Q. Liu, L. Jiang and D. B. Zhu, *Adv. Mater.*, 2002, **14**, 1857–1860.
2. T. Sun, L. Feng, X. Gao and L. Jiang, *Acc. Chem. Res.*, 2005, **38**, 644–652.
3. Y. Li, J. Zhang, S. Zhu, H. Dong, F. Jia, Z. Wang, Z. Sun, L. Zhang, Y. Li, H. Li, W. Xu and B. Yang, *Adv. Mater.*, 2009, **21**, 4731–4734.
4. H. Lee, M. L. Alcaraz, M. F. Rubner and R. E. Cohen, *ACS Nano*, 2013, **7**, 2172–2185.
5. R. Dou, J. Chen, Y. Zhang, X. Wang, D. Cui, Y. Song, L. Jiang and J. Wang, *ACS Appl. Mater. Interfaces*, 2014, **6**, 6998–7003.

6. S. W. Mingjie Liu, Z. Wei, Y. Song and L. Jiang, *Adv. Mater.*, 2009, **21**, 665–669.
7. S. Daniel, M. K. Chaudhury and J. C. Chen, *Science*, 2001, **291**, 633–636.
8. L. Feng, Z. Y. Zhang, Z. H. Mai, Y. M. Ma, B. Q. Liu, L. Jiang and D. B. Zhu, *Angew. Chem., Int. Ed.*, 2004, **43**, 2012–2014.
9. Z. Xue, S. Wang, L. Lin, L. Chen, M. Liu, L. Feng and L. Jiang, *Adv. Mater.*, 2011, **23**, 4270–4273.
10. http://www.worldwatercouncil.org/library/archives/water-crisis/.
11. A. Ritter and C. M. Regalado, *J. Hydrometeorol.*, 2008, **9**, 920–935.
12. S. Vogel and U. Müller-Doblies, *Flora*, 2011, **206**, 3–31.
13. A. Umur and P. Griffith, *J. Heat Transfer*, 1965, 275–282.
14. D. Beysens, *C. R. Phys.*, 2006, **7**, 1082–1100.
15. R. N. Leach, F. Stevens, S. C. Langford and J. T. Dickinson, *Langmuir*, 2006, **22**, 8864–8872.
16. A. R. Parker and C. R. Lawrence, *Nature*, 2001, **414**, 33–34.
17. T. Norgaard and M. Dacke, *Front. Zool.*, 2010, **7**, 23.
18. X. Chen, J. Wu, R. Ma, M. Hua, N. Koratkar, S. Yao and Z. Wang, *Adv. Funct. Mater.*, 2011, **21**, 4617–4623.
19. L. Zhang, J. Wu, M. N. Hedhili, X. Yang and P. Wang, *J. Mater. Chem. A*, 2015, **3**, 2844–2852.
20. S. C. Thickett, C. Neto and A. T. Harris, *Adv. Mater.*, 2011, **23**, 3718–3722.
21. A. Ghosh, S. Beaini, B. J. Zhang, R. Ganguly and C. M. Megaridis, *Langmuir*, 2014, **30**, 13103–13115.
22. L. Zhai, M. C. Berg, F. C. Cebeci, Y. Kim, J. M. Milwid, M. F. Rubner and R. E. Cohen, *Nano Lett.*, 2006, **6**, 1213–1217.
23. R. P. Garrod, L. G. Harris, W. C. E. Schofield, J. McGettrick, L. J. Ward, D. O. H. Teare and J. P. S. Badyal, *Langmuir*, 2007, **23**, 689–693.
24. H. Bai, L. Wang, J. Ju, R. Sun, Y. Zheng and L. Jiang, *Adv. Mater.*, 2014, **26**, 5025–5030.
25. F. G. Omenetto and D. L. Kaplan, *Science*, 2010, **329**, 528–531.
26. D. Porter and F. Vollrath, *Adv. Mater.*, 2009, **21**, 487–492.
27. Y. Zheng, H. Bai, Z. Huang, X. Tian, F.-Q. Nie, Y. Zhao, J. Zhai and L. Jiang, *Nature*, 2010, **463**, 640–643.
28. R. N. Wenzel, *Ind. Eng. Chem.*, 1936, **28**, 988–994.
29. É. Lorenceau and D. Quéré, *J. Fluid Mech.*, 2004, **510**, 29–45.
30. H. Bai, X. Tian, Y. Zheng, J. Ju, Y. Zhao and L. Jiang, *Adv. Mater.*, 2010, **22**, 5521–5525.
31. H. Bai, J. Ju, R. Sun, Y. Chen, Y. Zheng and L. Jiang, *Adv. Mater.*, 2011, **23**, 3708–3711.
32. S. Feng, Y. Hou, Y. Chen, Y. Xue, Y. Zheng and L. Jiang, *J. Mater. Chem. A*, 2013, **1**, 8363–8366.
33. H. Bai, R. Sun, J. Ju, X. Yao, Y. Zheng and L. Jiang, *Small*, 2011, **7**, 3429–3433.
34. H. Dong, N. Wang, L. Wang, H. Bai, J. Wu, Y. Zheng, Y. Zhao and L. Jiang, *Chemphyschem*, 2012, **13**, 1153–1156.
35. X. Tian, H. Bai, Y. Zheng and L. Jiang, *Adv. Funct. Mater.*, 2011, **21**, 1398–1402.

36. C. Song, L. Zhao, W. Zhou, M. Zhang and Y. Zheng, *J. Mater. Chem. A*, 2014, **2**, 9465–9468.
37. L. Zhao, C. Song, M. Zhang and Y. Zheng, *Chem. Commun.*, 2014, **50**, 10651–10654.
38. E. Kang, G. S. Jeong, Y. Y. Choi, K. H. Lee, A. Khademhosseini and S.-H. Lee, *Nat. Mater.*, 2011, **10**, 877–883.
39. H. Bai, J. Ju, Y. Zheng and L. Jiang, *Adv. Mater.*, 2012, **24**, 2786–2791.
40. Y. Hou, L. Gao, S. Feng, Y. Chen, Y. Xue, L. Jiang and Y. Zheng, *Chem. Commun.*, 2013, **49**, 5253–5255.
41. S. Feng, Y. Hou, Y. Xue, L. Gao, L. Jiang and Y. Zheng, *Soft Matter*, 2013, **9**, 9294–9297.
42. X. Tian, Y. Chen, Y. Zheng, H. Bai and L. Jiang, *Adv. Mater.*, 2011, **23**, 5486–5491.
43. Y. Hou, Y. Chen, Y. Xue, Y. Zheng and L. Jiang, *Langmuir*, 2012, **28**, 4737–4743.
44. Y. Chen, L. Wang, Y. Xue, Y. Zheng and L. Jiang, *Soft Matter*, 2012, **8**, 11450–11454.
45. Y. Chen, L. Wang, Y. Xue, L. Jiang and Y. Zheng, *Sci. Rep.*, 2013, **3**, 2927.
46. Y. Xue, Y. Chen, T. Wang, L. Jiang and Y. Zheng, *J. Mater. Chem. A*, 2014, **2**, 7156–7160.
47. J. Ju, H. Bai, Y. Zheng, T. Zhao, R. Fang and L. Jiang, *Nat. Commun.*, 2012, **3**, 1247.
48. F. T. Malik, R. M. Clement, D. T. Gethin, D. Beysens, R. E. Cohen, W. Krawszik and A. R. Parker, *Bioinspiration Biomimetics*, 2015, **10**, 036005.
49. M.-G. Medici, A. Mongruel, L. Royon and D. Beysens, *Phys. Rev. E*, 2014, **90**, 062403.
50. Y. Xue, T. Wang, W. Shi, L. Sun and Y. Zheng, *RSC Adv.*, 2014, **4**, 40837–40840.
51. J. Ju, Y. Zheng and L. Jiang, *Acc. Chem. Res.*, 2014, **47**, 2342–2352.
52. J. Ju, K. Xiao, X. Yao, H. Bai and L. Jiang, *Adv. Mater.*, 2013, **25**, 5937–5942.
53. X. Heng, M. Xiang, Z. Lu and C. Luo, *ACS Appl. Mater. Interfaces*, 2014, **6**, 8032–8041.
54. J. Ju, X. Yao, S. Yang, L. Wang, R. Z. Sun, Y. X. He and L. Jiang, *Adv. Funct. Mater.*, 2014, **24**, 6933–6938.
55. M. Cao, J. Ju, K. Li, S. Dou, K. Liu and L. Jiang, *Adv. Funct. Mater.*, 2014, **24**, 3235–3240.
56. L. Feng, Z. Zhang, Z. Mai, Y. Ma, B. Liu, L. Jiang and D. Zhu, *Angew. Chem.*, 2004, **43**, 2012–2014.
57. G. Hayase, K. Kanamori, M. Fukuchi, H. Kaji and K. Nakanishi, *Angew. Chem.*, 2013, **52**, 1986–1989.
58. A. K. Kota, G. Kwon, W. Choi, J. M. Mabry and A. Tuteja, *Nat. Commun.*, 2012, **3**, 1025.
59. K. Li, J. Ju, Z. Xue, J. Ma, L. Feng, S. Gao and L. Jiang, *Nat. Commun.*, 2013, **4**, 2276.
60. P. G. de Gennes, F. Brochard-Wyart and D. Quéré, *Capillarity and Wetting Phenomena*, Springer, 2004.

61. P. Comanns, C. Effertz, F. Hischen, K. Staudt, W. Böhme and W. Baumgartner, *Beilstein J. Nanotechnol.*, 2011, **2**, 204–214.
62. W. C. Sherbrooke, A. J. Scardino, R. d. Nys and L. Schwarzkopf, *Zoomorphology*, 2007, **126**, 89–102.
63. C. R. Tracy, N. Laurence and K. A. Christian1, *Am. Nat.*, 2011, **178**, 553–558.
64. S. Yang, J. Ju, Y. Qiu, K. Liu and L. Jiang, *Small*, 2014, **10**, 294–299.
65. A. Roth-Nebelsick, M. Ebner, T. Miranda, V. Gottschalk, D. Voigt, S. Gorb, T. Stegmaier, J. Sarsour, M. Linke and W. Konrad, *J. R. Soc., Interface*, 2012, 1965–1974.
66. M. Ebner, T. Miranda and A. Roth-Nebelsick, *J. Arid Environ.*, 2011, **75**, 524–531.
67. T. E. Dawson, *Oecologia*, 1998, **117**, 476–485.
68. R. Chen, X. Zhang, Z. Su, R. Gong, X. Ge, H. Zhang and C. Wang, *J. Phys. Chem. C*, 2009, **113**, 8350–8356.
69. T.-S. Wong, S. H. Kang, S. K. Y. Tang, E. J. Smythe, B. D. Hatton, A. Grinthal and J. Aizenberg, *Nature*, 2011, **477**, 443–447.
70. B. S. Lalia, S. Anand, K. K. Varanasi and R. Hashaikeh, *Langmuir*, 2013, **29**, 13081–13088.
71. H. Yang, H. Zhu, M. M. R. M. Hendrix, N. J. H. G. M. Lousberg, G. de With, A. C. C. Esteves and J. H. Xin, *Adv. Mater.*, 2013, **25**, 1150–1154.

CHAPTER 9

Wetting Properties of Surfaces and Drag Reduction

GLEN MCHALE[a]

[a]Smart Materials & Surfaces Laboratory, Faculty of Engineering & Environment, Northumbria University, Ellison Place, Newcastle upon Tyne, NE1 8ST, UK
*E-mail: glen.mchale@northumbria.ac.uk

9.1 Introduction

9.1.1 Superhydrophobicity, Leidenfrost Effect, and SLIPS/LIS Surfaces

In recent research, studies of the motion of simple Newtonian liquids past objects, or through objects, where the contact with their solid surfaces involves another fluid interface have become common. One of the most studied types of surface has been hydrophobic rough or textured surfaces with sufficiently strong aspect ratios for surface features to cause droplets of a liquid to almost completely ball-up. Droplets on these types of superhydrophobic surfaces can exist in a Cassie–Baxter "slippy" state, where the liquid bridges across gaps between surface features, or a Wenzel "sticky" state, where the liquid fully penetrates into the gaps, or a mixed state of partial penetration.[1] In the Cassie–Baxter state the surface is often considered as a composite of partial wetting areas characterized by a solid surface contact angle, θ_s, and complete non-wetting areas with $\theta_v = 180°$. The contact angle for the

RSC Soft Matter No. 5
Non-wettable Surfaces: Theory, Preparation, and Applications
Edited by Robin H. A. Ras and Abraham Marmur

Published by the Royal Society of Chemistry, www.rsc.org

surface is given by a Cassie solid surface area fraction, φ_s, weighted average of cosines, $\cos\theta_{CB} = \varphi_s\cos\theta_s + (1 - \varphi_s)\cos\theta_v$. This equation is often cited in the literature for simple surface textures involving features with flat tops, but can be used more widely by careful definition of the various symbols:[2] for example, Wenzel roughness can be taken into account by incorporating its effect into the solid surface contact angle rather than using the Young's law contact angle.[2] Such a viewpoint captures the idea of a surface with solid–liquid and liquid–vapour interfaces, and changes in wettability, but is an average viewpoint and provides little understanding of how a contact line may pin, unpin, and move. In the most extreme case, the solid surface fraction vanishes to give a complete non-wetting ("perfectly hydrophobic") surface and the liquid is separated from the solid at all points by a vapour layer. A physical example of this is the Leidenfrost effect, whereby as soon a droplet comes into contact with a hot surface a layer of the liquid is instantaneously vaporized to create a cushion of vapour upon which the droplet rests and becomes highly mobile.[3] Another possibility is that the texture of the surface causes hemi-wicking of an immiscible impregnating (infusing) liquid so that the surface seen by a droplet is a composite not of solid and vapour, but of solid and another impregnating liquid with quite different wetting properties. An impregnating liquid may not only hemi-wick into a surface, but can be chosen with suitable interfacial tensions such that it also coats the tops of the surface features and so forms a continuous film of liquid which is not displaced by other liquids. A droplet on such a lubricant impregnated/infused (porous) surface (often referred to as LIS or SLIPS) created using an immiscible and non-volatile lubricant liquid can be highly mobile.[4,5] All of these types of surfaces have become candidates for low friction/drag-reducing surfaces.

9.1.2 Importance of Vapour/Fluid Interfaces

It is a common expectation in fluid mechanics that simple Newtonian liquids flowing across a smooth solid surface will obey a no-slip boundary condition and that the flow will not be influenced by the contact angle, θ, the liquid makes with the surface. Classically, assuming a no-slip boundary condition, there is no change in drag as a liquid flows across a surface with mixed wetting regions of low contact angle and high contact angle (Figure 9.1a). In contrast, the motion of a droplet of the same liquid on the same smooth surface can be expected to depend on the boundary between these regions of different wettability because of the existence of a three-phase contact line. As a simple thought experiment, imagine tilting a surface on which a droplet rests (Figure 9.1b). Because of contact angle hysteresis, the droplet's contact angle on the lower side will increase and that on the upper side will decrease until the tilt angle is sufficiently large that the body forces due to gravity can overcome the capillary forces involved in contact line pinning, which are proportional to $\gamma_{LV}(\cos\theta_R - \cos\theta_A)$ where γ_{LV} is the liquid–vapour interfacial tension, and θ_A and θ_R are the advancing and receding contact angles (Figure 9.1c).[6,7] If we simplify to an idealized surface which has no contact

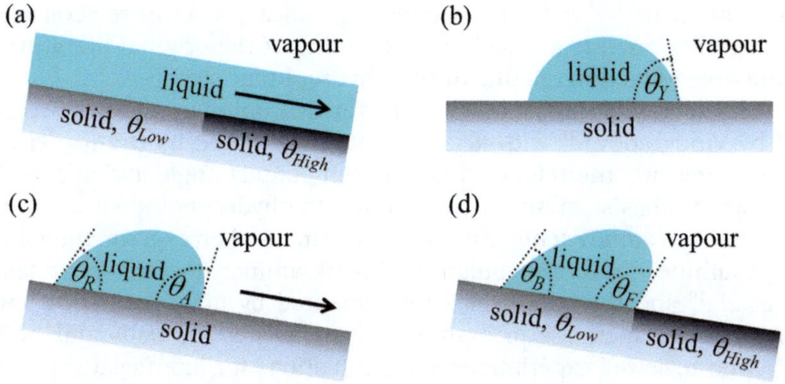

Figure 9.1 (a) Liquid flowing across a surface with mixed wettability, (b) droplet in a static equilibrium with a contact angle given by Young's law, (c) droplet sliding on a tilted surface of single wettability, (d) droplet pinning as it encounters a region of lower wettability.

angle hysteresis, and so is perfectly slippery, but now have the contact line at the droplet's lower side meeting a less-wetting boundary, we can expect there to be contact line pinning as the advancing droplet seeks to adopt the higher contact angle of the less-wetting region (Figure 9.1d). The motion of this droplet involves molecules of the liquid coming into contact with new solid surface on its lower side (and relinquishing contact on its upper side). In this thought experiment there is a distortion of the shape of the liquid–vapour surface.

Thus, there are concepts of adhesion and cohesion, vapour interfaces to the solid and the liquid, and the distortion of the shape of the liquid–vapour interface. The presence of vapour, or more generally a second fluid whether it is a gas or a liquid, can be expected to alter the ease of flow of a liquid across a surface. Moreover, even when the vapour does not initially exist, the possibility that it could be created, for example by cavitation or the Leidenfrost effect, alters our expectations of how flow across a surface might occur. Within our concepts of drag there are concepts relating to skin friction due to viscous shearing on the surface and form relating to the pressure distribution around the shape of an object. Both are relevant when considering the effects of the wetting properties of a surface. Linked to the idea of skin friction is the underlying question of whether the flow velocity vanishes at the surface or whether there is a finite (slip) velocity, v_s, at the surface.

9.1.3 Literature Reviews

The published literature on slip and drag reduction related to the wetting properties of surfaces has expanded rapidly over recent years as research on superhydrophobic and superoleophobic surfaces has provided new types of surfaces where the surface texture or roughness amplifies the effects of surface chemistry.[1,2,8–15] In this section we indicate a set of reviews which taken together provide a broad understanding of the original literature on slip and drag reduction relating to superhydrophobic surfaces.

An early review of slippage of water over hydrophobic surfaces was provided by Vinogradova,[16] with key concepts highlighted by Granick *et al.*,[17] and more recently the relationship between contact angle and slip, with a particular emphasis on simulations and superhydrophobic surfaces, was considered by Voronov *et al.*[18] An early review focused on experimental methods to examine slip in Newtonian liquids in laminar flow was provided by Neto *et al.*,[19] and this has been complemented by Bouzigues *et al.*[20] who described three methods to examine slip near smooth walls and by Lee *et al.*[21] who reviewed experiments and simulations for interfacial slip. A significant number of studies have also focused on microfluidics where flow is laminar and the reviews on boundary conditions and slip of Lauga *et al.*[22] and superhydrophobic textures of Vinogradova and Dubov[23] are particularly relevant to these studies. Motivated by the potential for microfluidics, Bocquet and Barrat[24] provided a review of the theoretical understanding of flow past solid interfaces at different length scales and, in particular, in terms of the relationship between slip length and friction. The review of Samaha *et al.*[25] discusses longevity of gas/air layers when immersed and Wang *et al.*[26] has sections on maintaining or generating gas/vapour layers. Samaha *et al.*'s most recent review[27] also includes discussion of the *Nepenthes* pitcher plant as the inspiration for SLIPS/LIS.

Rothstein's review of superhydrophobic surfaces is notable for including both laminar and turbulent flow and addressing skin friction (viscous shear) drag and form (pressure) drag.[6] Bhushan and coworkers have also provided a number of reviews of biomimetic and superhydrophobic surfaces and their applications for drag reduction including both laminar and turbulent flow.[28–32] The relationship between bubbles and drag reduction caused by superhydrophobic surfaces was discussed by McHale *et al.*[33] and this built upon a series of reviews addressing drag reduction related to droplets, bubbles, multiphase flows/compound objects and gas injection.[34–37] Core annular flows, which are important for understanding apparent slip, have been reviewed by Joseph *et al.*[38] The recent emerging areas of the Leidenfrost effect and SLIPS/LIS were included in the review of superhydrophobic surfaces for energy-related applications by Zhang *et al.*,[39] although the focus was not on the surfaces' drag-reducing properties. In this context, we refer to the original articles by Vakarelski *et al.*[40,41] using the Leidenfrost effect to delay the separation of the wake and so significantly reduce drag for flow past a sphere, and by Solomon *et al.*[42] for drag reduction by SLIPS/LIS.

9.1.4 Types of Experimental Methods

Measurements of slip and drag reduction have involved both direct methods of profiling the flow velocity and indirect methods.[22] Experiments have included ones focused on slip of simple Newtonian liquids at smooth hydrophobic walls, and ones focused on drag reduction in laminar and turbulent flow where the surfaces are textured and hydrophobic. Methods have varied dependent on whether flow is internal, *e.g.* through channels and pipes, or external, *e.g.* over cylinders, spheres, and plates. Moreover, the approaches have differed according to the scale of the surface, *e.g.* microchannels and hydrofoils.

Methods of directly measuring slip velocity near walls have included total internal reflection fluorescence recovery after photobleaching (TIRF/FRAP),[43] microparticle image velocimetry (μ-PIV),[44–47] PIV in turbulent channel flow,[48–50] PIV with rotational rheometry,[51] total internal reflection velocimetry (TIRV) with particle tracking velocimetry analysis (PTV),[52] double-focus spatial fluorescence cross correlation (DF-FCS),[53,54] and diffusion-based fluorescence correlation spectroscopy[55] (see also the comparative review by Bouzigues *et al.*[20]). Indirect methods focused on measuring slip across smooth surfaces have included dynamic surface force apparatus (SFA) using drainage between surfaces.[56–58] Common drag measurement methods have included flow rate *versus* pressure drop for channels/ducts[59–64] and pipes,[65–67] and rheometry of various types including, rotational coaxial cylinder-based,[68–70] cone-and-plate,[71–73] and rotational parallel plate.[74] These methods have spanned different flow regimes from laminar to turbulent. For drag reduction in the turbulent regime under external flow, methods have included force on a hydrofoil in a water tunnel,[75] PIV on a plate in a water tunnel,[76] force on cylinders in water tunnels and using PIV to measure vortex structures,[77,78] laser doppler anemometry (LDA) measurements on cylinders in recirculating water chambers,[79] plates in water tunnels using strain gauges and hot-film anemometers,[80–83] and terminal velocity experiments.[40,84]

9.1.5 Retention and Generation of Gas/Vapour Layers

In the context of superhydrophobic surfaces the ability to retain a layer of air at the solid surface when immersed (a "plastron")[85–87] is a critical part of the potential of the surface to reduce drag. In static wetting, surface tension is a force per unit length and so its force scales with linear dimension, whereas a gravitational force is product of density, volume, and acceleration due to gravity and so its force scales with cubic dimension. Surface tension forces dominate when gaps between surface features are significantly less than the capillary length, $\kappa^{-1} = (\gamma_{LV}/\rho g)^{1/2}$, which for water is ~2.73 mm. The Young's law contact angle at which it is energetically favourable for liquid to penetrate into the features of a surface, and how the penetration depends on re-entrant curvature, metastable states, and the possibility of a deformed meniscus contacting the surface at the bottom of any gap, together with gas dissolution and longevity of air layers have been widely discussed.[1,13,23,25,73,88] Simple energy balance considerations suggest robustness against penetration of liquid is favoured with high Young's law contact angle (intrinsic hydrophobicity), high surface tension, density of surface features to generate high perimeter liquid contact-to-area fractions, and re-entrant shaped features providing metastable states.

The extension to a flow situation introduces dynamic pressure considerations and more complex pressure distributions. The Weber number, We = $2R\rho u_\infty^2/\gamma_{bs}$, which is related to the ratio of dynamic pressure distorting the surface to surface tension stresses which resist distortion, then becomes important. The Weber number can also be written as We = Ca × Re, where Ca = $\eta u_\infty/\gamma_{bs}$ is a characteristic number for the balance between viscous and surface tension forces. Thus, higher speeds and sizes can be expected to cause distortions to fluid–fluid interfaces. In the context of a superhydrophobic surface, the liquid–air meniscus between surface features could distort sufficiently to cause penetration and a transition to a Wenzel state possessing complete contact between the liquid and the solid surface. In large surface area applications, any failure of a Cassie–Baxter state is expected to rapidly spread across the surface; breaker-ridge designs have been suggested to halt such failure.[77,89] Alternatives to maintain air/vapour layers include electrochemically generating gas,[90,91] actively bleeding/injecting air/gas into the surface region,[92] and collecting air/gas from the bulk flow;[26] blowing air/gas may also alter the boundary layer properties and hence drag.[93] In the context of liquid impregnated/infused textured surfaces, the incompressibility and immiscibility of the impregnating liquid is expected to provide higher stability to pressure under static wetting. However, it can be expected that flow could generate shear forces displacing infused liquid from either the tops of the texture or the gaps within the texture and possibly induce drainage. Recent work suggests control of surface chemistry to pattern the wettability may prevent drainage.[94]

9.2 Velocity Profiles Near Surfaces and Slip

9.2.1 Slip Velocity, Slip Length and Friction

The velocity profile, $u_x(y)$, near a surface at $y = 0$ for a liquid flowing in the x-direction can be written to first order using a Taylor series expansion,

$$u_x(y) = v_s + y\left(\frac{\partial u_x}{\partial y}\right)_{y=0} \tag{9.1}$$

where $v_s = u_x(y = 0)$ is its velocity at the surface and is referred to as the slip velocity. Eqn (9.1) provides a linear velocity gradient with distance from the surface (Figure 9.2a–c). Physically this reflects expectations arising from the viscosity, η_b, of the bulk liquid which determines how easily molecules in the planes of liquid above the surface can move across each other. Since the velocity is not assumed to vanish at the surface, a slip length, b, can be defined as the position, $y = -b$, below the surface at which the velocity profile would vanish, *i.e.* $u_x(y = -b) = 0$, and this gives the Navier slip length,

$$b = \frac{v_s}{\left(\frac{\partial u_x}{\partial y}\right)_{y=0}} \tag{9.2}$$

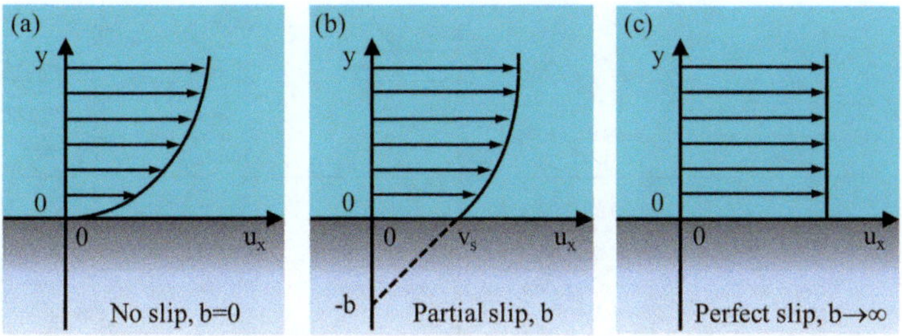

Figure 9.2 Three hydrodynamic boundary conditions for flow across a solid surface (solid–fluid interface at $y = 0$): (a) no slip, (b) partial slip, and (c) perfect slip. The slip length in (b) is given by extrapolating the slope of $u_x(y)$ as $y \rightarrow 0$ to find $u_x(-b) = 0$.

When the slip length vanishes, $b \rightarrow 0$, the fluid velocity approaches the surface with a linear profile and has a vanishing velocity at the surface. When the slip length tends to infinity, $b \rightarrow \infty$, the fluid velocity is constant independent of position above the surface representing plug flow, and is the same as slip velocity at the surface; this is also referred to as the shear-free condition. The slip velocity can also be viewed as a response to the shear stress exerted by the fluid on the surface, $\tau_{xy}(y = 0)$, where the strength of the response is characterized by a solid–liquid coefficient of friction κ defined *via* $\tau_{xy}(y = 0) = \kappa v_s$. Thus, the slip length is given by the ratio of the viscosity, η_b, to the coefficient of friction,[24]

$$b = \frac{\eta_b}{\kappa} \tag{9.3}$$

9.2.2 Apparent Slip and Lubricating Surface Flows

In reality, a non-zero slip velocity may be an artefact of an unrecognised layer of fluid of lower viscosity, η_s, close to the surface. If a surface has such a surface layer of fluid of thickness h, on top of which there is the bulk fluid, the shear stresses, $\tau_{xy} = \eta(\partial u_x/\partial y)$, representing the force per unit area exerted by each fluid, should match at the interface between the fluids at $y = h$ (Figure 9.3a), *i.e.*

$$\eta_s\left(\frac{u_x^{\,s}(h)}{h}\right) = \eta_b\left(\frac{u_x^{\,b}(h)}{h + b}\right) \tag{9.4}$$

Since the velocity profiles in the surface layer fluid, $u_x^{\,s}(y)$, and the bulk fluid $u_x^{\,b}(y)$ are equal at $y = h$, rearranging eqn (9.4) gives the apparent slip length in terms of the ratio of bulk to surface viscosities and the surface layer fluid thickness,

$$b = \left(\frac{\eta_b}{\eta_s} - 1\right)h \tag{9.5}$$

Eqn (9.5) is Vinogradova's formula for apparent slip across a superhydrophobic surface with a vanishing solid surface fraction $\varphi_s \rightarrow 0$.[16,95] The ratio of

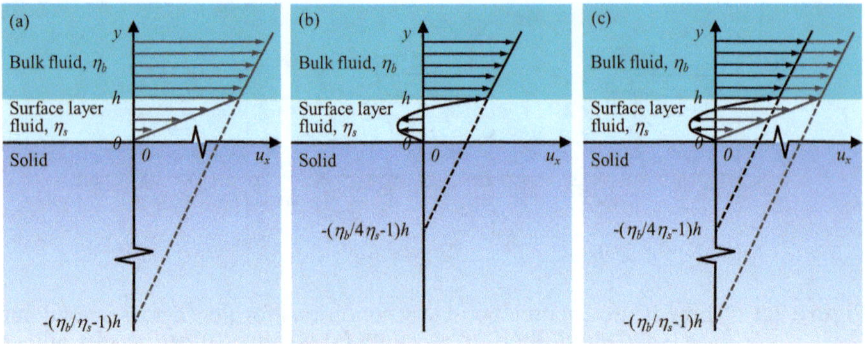

Figure 9.3 Fluid velocity profiles for flow with a low-viscosity fluid close to the surface. (a) Flow when the pressure gradient in the bulk fluid and the surface layer fluid are the same gives a linear velocity profile in the low-viscosity surface layer, (b) a net zero mass flow rate (ZMF) condition results in a backflow, and (c) combined representation showing the reduction in apparent slip length from $b = -(\eta_b/\eta_s - 1)h$ to $b = -(\eta_b/4\eta_s - 1)h$ when the linear velocity profile is replaced by the ZMF condition.

viscosity of air to water $(\eta_{water}/\eta_{air}) \sim 55$ and so the slip length for water flowing over an air-lubricated surface is substantially larger than the thickness of the layer of air and lubricates the flow of the water over the solid surface. A similar effect can occur in flow of oil where a low-viscosity fraction can separate out to the surface and then lubricate the flow of a heavier fraction to give a core annular flow.[38,96]

The key assumption leading to the slip length in eqn (9.5) is that the surface layer fluid has a linear velocity profile, which vanishes at the surface $y = 0$ and matches the bulk fluid velocity at $y = h$, *i.e.*

$$u_s(y) = v_s\left(1 + \frac{h\tau_b(h)}{\eta_b v_s}\right)\frac{y}{h} \tag{9.6}$$

where $\tau_b(h)$ is the shear stress from the bulk flow at $y = h$, and the bulk fluid velocity is,

$$u_b(y) = v_s + \left(\frac{\tau_b(h)}{\eta_b}\right)y \tag{9.7}$$

It is, however, known in fluid flow that there can be a backflow close to a surface and this would be inconsistent with the assumption of a linear velocity profile for $u_s(y)$ (eqn (9.6)). One alternative is to use an assumption that there is net zero mass flow rate (ZMF) within the surface layer flow together with a vanishing surface layer fluid velocity at $y = 0$,[97] *i.e.*

$$u_s(y) = Ay\left(\frac{3y}{2h} - 1\right) \tag{9.8}$$

where $A = \tau_b(h)/2\eta_s$ is a constant determined by matching the shear stresses at the fluid–fluid interface at $y = h$. This surface flow velocity profile allows for a backflow and has a flow reversal occurring at a distance $y = 2h/3$ from the surface (Figure 9.3b). Matching the fluid velocities at $y = h$ and using $u_b(y = -b) = 0$ to convert from slip velocity to slip length gives,

$$b = \left(\frac{\eta_b}{4\eta_s} - 1\right)h \tag{9.9}$$

In this case, assuming the bulk viscosity is much larger than the surface viscosity (*i.e.* $\eta_b/\eta_s \gg 1$), the slip length is around one quarter of that given by Vinogradova's formula (eqn (9.5)) (Figure 9.3c).

9.2.3 Molecular Slip and Equilibrium/Dynamic Contact Angles

An alternative to apparent slip is that the shear stress exerted by the fluid at the surface leads to a surface mobility, μ_s, of the molecules of the liquid that is higher than the mobility, μ_b, of the molecules in the bulk liquid (Figure 9.4).[19,98,99] In the Tolstoi–Blake argument the mobility is a thermally activated process governed by the energy, W, required to create a "hole" into which a molecule can move and this is different for a molecule in the bulk compared to one at the surface. The ratio of molecular mobilities is approximated by

$$\frac{\mu_s}{\mu_b} = \exp\left(\frac{W_b - W_s}{k_B T}\right) \tag{9.10}$$

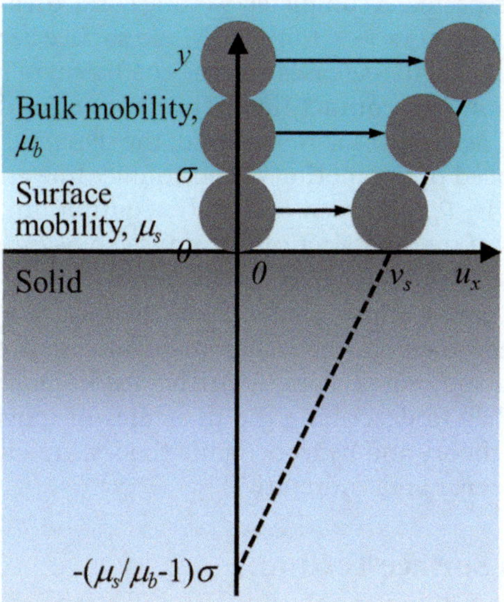

Figure 9.4 Motion of molecules in a fluid with molecular slip occurring in a layer close to the liquid–solid interface.

where k_BT is the thermal energy. In this view, the energy to create a hole per unit molecular surface area in the bulk liquid is γ_{LV}, and to create a hole on a surface is $\alpha(\gamma_{SV} - \gamma_{SL}) + (1 - \alpha)\gamma_{LV}$, where γ_{SV} and γ_{SL} are the solid–vapour and solid–liquid interfacial tensions, and α is the surface area fraction of the hole in contact with the surface. The energy difference in eqn (9.10) is therefore proportional to the difference between the work of cohesion and the work of adhesion, *i.e.* $W_b - W_s \propto \gamma_{SL} + \gamma_{LV} - \gamma_{SV} = \gamma_{LV}(1 - \cos\theta_Y)$, where the replacement by the equilibrium (Young's law) contact angle, θ_Y, is valid when $W_b \geq W_s$. To find the relationship between the contact angle and slip length, the mobilities are assumed proportional to the respective velocity gradients. This gives the ratio of surface to bulk mobility as $\mu_s/\mu_b = (1 + b/\sigma)$, where σ is the centre-to-centre separation between molecules. The slip length is then

$$b = \left(\frac{\mu_s}{\mu_b} - 1\right)\sigma \tag{9.11}$$

or in terms of the Young's law contact angle,

$$b = \sigma\left(\exp\left(\frac{\alpha A\gamma_{LV}\left(1 - \cos\theta_Y\right)}{k_BT}\right) - 1\right) \tag{9.12}$$

where αA is the surface area of the hole in contact with the surface and is $\sim\alpha\sigma^2$.

The molecular mobilities are equal when the surface is completely wetting ($\theta_Y \rightarrow 0°$) and the slip length vanishes. When the surface is partially wetting ($180° > \theta_Y > 0°$), the surface mobility is larger than the bulk mobility and a finite slip length can be expected. A completely wetting surface with $b \rightarrow 0$ is also one with a coefficient of friction $\kappa \rightarrow \infty$. Conversely, as a surface becomes increasingly non-wetting, the slip length increases and the coefficient of friction decreases.

For forced motion of a contact line between two fluids (*e.g.* between a liquid and vapour) as it moves across a solid, the dynamic contact angle, θ_D, depends on the speed of the motion of the contact line and is not the value of the contact angle, θ_e, observed in a static equilibrium situation. In this situation the ideas of thermally activated jumps can be extended to include a driving force due to the unbalanced capillary force $\propto\gamma_{LV}(\cos\theta_e - \cos\theta_D)$. Moreover, the dynamic contact angle depends on the direction of the motion of contact line and can therefore be either an advancing (wetting) or a receding (dewetting) dynamic contact angle. Further information on the two most common approaches to describing the dynamics of moving contact lines, molecular-kinetic theory and hydrodynamic theory, are given in the reviews of Blake[100] and Snoeijer and Andreotti.[101]

9.2.4 Slip and Surface Texture

One of the simplest models used to understand wetting on superhydrophobic surfaces uses a schematic with a side profile cross-section showing a periodic array of rectangular cross-section posts characterized by a Cassie solid surface area fraction, φ_s. The simplicity of this schematic does not indicate the complexity that may occur for a three-dimensional surface. Such a surface could, for example, consist of ridges and grooves aligned with or across the flow, or a set of separated posts with square, circular, or other cross-section where the gaps are connected (Figure 9.5a) or disconnected (Figure 9.5b). In the context of gaps between posts filled by a fluid, *e.g.* a gas or an impregnating immiscible liquid, one can expect the detail of connectivity to alter the slip of a second immiscible fluid flowing across the surface.

At length scales similar to the texture one might model a textured surface using periodic regions of no-slip (or low slip) and infinite slip (or high slip) (Figure 9.5c), which when viewed at some appropriate length scale as providing an average effect on the core flow that can be characterized by an effective slip length, b_{eff}.[23,24,102] Thus, the topography of the surface is converted into a model that is analogous to a smooth (non-topographically structured) surface, but with a texture described by regions of differing slip with the Cassie solid surface area fraction, φ_s, reinterpreted as characterizing the fractional area over which slip occurs. The effective slip can therefore represent quite different physical mechanisms. These may or may not involve molecular slip, flow within gaps in topography that are aligned at all depths with the direction of the core flow, or a recirculating flow pattern within gaps in the topography (Figure 9.5d). It could also be the case that the recirculating fluid within gaps has a concave or convex meniscus to the bulk flow, thereby reducing or increasing the drag dependent on the detailed shape of the fluid–fluid interface (*e.g.* bubble mattress type interfaces[103–105]) or that a thin layer of the lower viscosity fluid from the gaps flows across both the tops and gaps of the topographic features.[106–108] Although not involving solid surface texture, the occurrence of an overall flow of a lubricating fluid together with recirculating flows in dynamically created texture at the oil-water interface is known in studies on flying core flow (Figure 9.5e).[38]

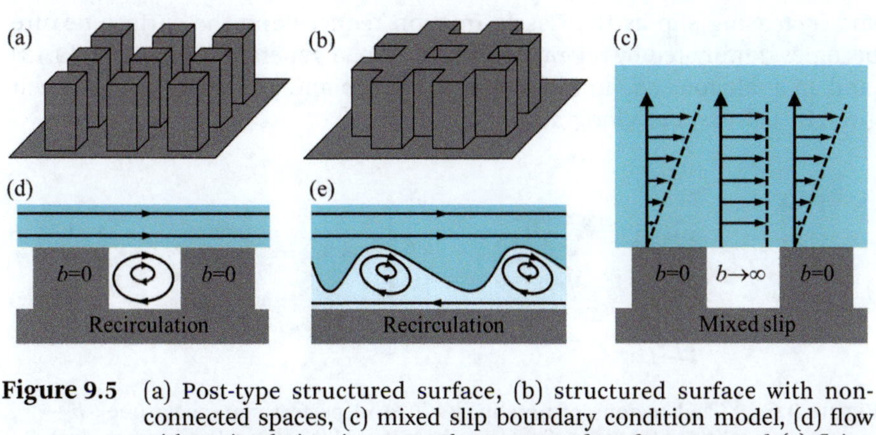

Figure 9.5 (a) Post-type structured surface, (b) structured surface with non-connected spaces, (c) mixed slip boundary condition model, (d) flow with recirculation in spaces between surface features, and (e) flying core of oil-on-water dynamically inducing a wavy fluid–fluid interface.

9.2.5 Effective Slip and Mixed Boundary Conditions

Experiments applying a pressure gradient and measuring a flow rate, or *vice versa*, result in estimates of a large-scale effective slip length, b_{eff}, rather than providing details of microscopic mechanisms. Two common models for pipe flow assume a heterogeneous surface with periodic regions of no slip ($b = 0$) and perfect slip ($b \to \infty$) aligned either across the flow direction (Figure 9.6a) or along the flow direction (Figure 9.6b).[102] Perfect slip ($b \to \infty$) can also be referred to as a shear-free model, or from eqn (9.3), vanishing coefficient of friction ($\kappa \to 0$).

For flow parallel to stripes of period L, the exact analytical solution for low Re number is[109,110]

$$b_{\text{eff}}^{//} = \frac{L}{\pi}\log_e\left(\sec\left[\frac{\pi}{2}(1-\phi_s)\right]\right) \approx \frac{L}{\pi}\left[\log_e\left(\frac{2}{\pi\phi_s}\right) + \left(\frac{\pi\phi_s}{24}\right)^2 + O(\phi_s^{\,4})\right] \tag{9.13}$$

where $\varphi_s = 1 - l/L$ is the Cassie surface area fraction for which a no-slip boundary condition applies. For superhydrophobic surfaces this corresponds to the Cassie solid surface fraction although the models in Figure 9.6 map topographic changes onto changes in slip/no-slip boundary conditions. The first term in the expansion in eqn (9.13) approximates the exact expression to within 10% up to $\varphi_s = 37.5\%$. The effective slip length becomes comparable to the periodicity, L, when $\varphi_s = 2/(\pi\exp(\pi)) = 2.75\%$. A generalisation of the effective slip length (eqn (9.13)) to surfaces with regions of no slip and partial slip has also been derived.[23,111] Simulations,[24,112] supported by analytical calculations,[113] suggest the small Cassie fraction limit for a square array of square (or circular) posts is

$$b_{\text{eff}} \approx \frac{L}{\pi\sqrt{\phi_s}} - 0.44 \tag{9.14}$$

For the case of pipe flow across stripes (Figure 9.6a) with small periodicity, L, the effective slip length is $b_{\text{eff}}^{\perp} = b_{\text{eff}}^{//}/2$.[22,102] These models predict a linear dependence on the periodicity, L, of the pattern for a given Cassie fraction and increasing slip as the Cassie fraction reduces and the surface texture becomes dominated by regions of perfect slip. A generalisation of eqn (9.13) (and an analogous equation for $b_{\text{eff}}^{\perp}$) to no slip and partial slip, b, has been given by Belyaev and Vinogradova.[111]

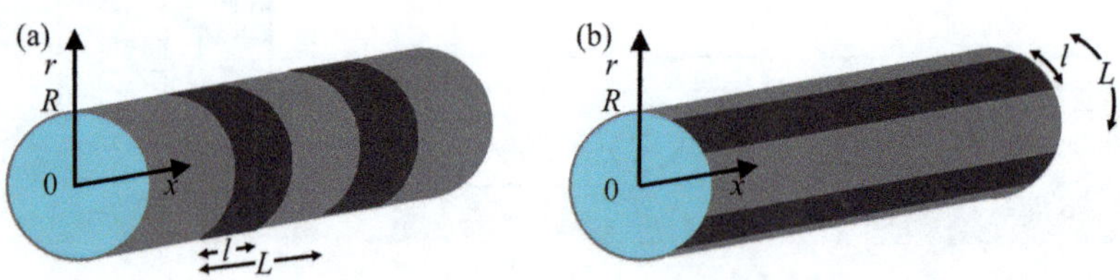

Figure 9.6 Mixed regions of no slip, $b = 0$, and perfect slip (shear-free), $b \to \infty$, aligned (a) across flow (transverse), and (b) along flow (streamwise).

Recent work has numerically computed the effective slip length for Re ≤ 1000 for water–air systems with two-dimensional walls patterned with ribs aligned perpendicular to the flow direction.[83,114] The model assumed the meniscus between the main flow of water and the air within cavities was flat. One conclusion of this study was that provided the cavities were sufficiently deep for a recirculating flow of air to be developed, it was possible to fit the results using Lauga's analytical solution as Re → 0. Moreover, over the Re number range studied the $b_{eff}^{\perp}$ modified by an overall semi-empirical factor dependent on the Re number was able to fit the data for the full range of Cassie fractions.

9.3 Internal Flow Through Pipes

9.3.1 Navier–Stokes Equations and Reynolds Number

Incompressible fluid flow is governed by the Navier–Stokes equations

$$\rho\left[\frac{\partial \mathbf{u}}{\partial t} + (\mathbf{u}.\nabla)\mathbf{u}\right] = -\nabla P + \eta\nabla^2\mathbf{u} \tag{9.15}$$

where $\mathbf{u}$ is the fluid velocity, P is the pressure, ρ is the density, and η is the viscosity of the fluid. For incompressible flow the conservation of mass requires,

$$\nabla.\mathbf{u} = 0 \tag{9.16}$$

Essentially eqn (9.15) is a version of Newton's law, which incorporates mass per unit volume times acceleration balanced by the forces per unit volume. The viscosity has units $[\eta] = [\mathrm{Pa\ s}] = [\mathrm{N\ s\ m^{-2}}] = [\mathrm{kg\ m^{-1}\ s^{-1}}]$ and is the physical property of the fluid which relates the force per unit area or shear stress, τ, transmitted across the fluid to the velocity gradient across the flow, *i.e.* the fluid friction. A linear relationship between shear stress and velocity gradient implies a Newtonian fluid.

The ratio per unit volume of inertial to viscous forces using characteristic scales for velocity, U, length, L, and time, T, gives the Reynolds number, Re,

$$\mathrm{Re} = \frac{\rho\left[U/T + U^2/L\right]}{\eta U/L^2} = \frac{\rho UL}{\eta} \tag{9.17}$$

where $T \sim L/U$ has been assumed. The ratio of viscosity and density, $\nu = \eta/\rho$, is the kinematic viscosity with units $[\mathrm{m^2\ s^{-1}}]$ and has a particular importance when comparing inertial to viscous forces. In low Re flow (also called Stokes flow or creeping flow) inertia can be ignored and eqn (9.15) reduces to Stokes equation,

$$\nabla P = \eta\nabla^2\mathbf{u} \tag{9.18}$$

The Navier–Stokes equations can be solved for laminar (streamline) flow when a fluid flows in parallel layers, with no disruption between the layers. However, as Re increases and inertia becomes important, instabilities can grow, mixing across layers due to transverse flows can occur and flow can become turbulent. In the turbulent case, the velocity and pressure can be decomposed into a time-averaged and a fluctuating component to provide Reynolds-averaged Navier–Stokes (RANS) equations which can be used in computational fluid dynamics (CFD) calculations.[83,115,116]

9.3.2 Poiseuille Flow and Friction Factor

For low Re flow in a circular cross-section pipe of radius R (Figure 9.7a), using cylindrical coordinates and with $G = -\mathrm{d}P/\mathrm{d}x$ as a fixed pressure gradient along the axis of the pipe, the Stokes equation (eqn (9.18)) becomes

$$-G = \frac{\eta}{r}\frac{\partial}{\partial r}\left(r\frac{\partial u_x}{\partial r}\right) \tag{9.19}$$

where the velocity along the pipe is $u_x(r)$. Assuming no-slip boundary conditions and axial symmetry, the Hagen–Poiseuille solution is found:

$$u_x(r) = u_{\max}\left(1 - \frac{r^2}{R^2}\right) \tag{9.20}$$

where the maximum flow rate, $u_{\max}$, is along the centre of the pipe and is given by $u_{\max} = GR^2/4\eta$ (Figure 9.7b). The flow velocity averaged across the cross-sectional area of the pipe is $u_{\text{ave}} = u_{\max}/2$ and the volumetric flow rate is,

$$Q = 2\pi\int_0^R u_x(r)r\mathrm{d}r = \pi R^2 u_{\text{ave}} \tag{9.21}$$

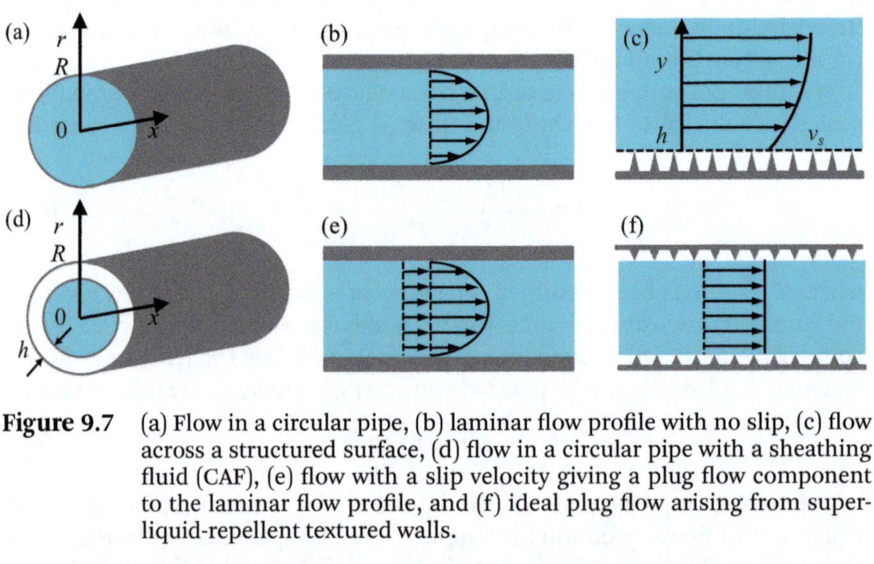

Figure 9.7 (a) Flow in a circular pipe, (b) laminar flow profile with no slip, (c) flow across a structured surface, (d) flow in a circular pipe with a sheathing fluid (CAF), (e) flow with a slip velocity giving a plug flow component to the laminar flow profile, and (f) ideal plug flow arising from super-liquid-repellent textured walls.

To characterize friction, the ratio of shear stress at the wall of the pipe to dynamic pressure can be compared to give a dimensionless pipe Fanning (skin) friction factor, C_F,

$$C_F = \frac{\tau_w (r = R)}{\rho u_{ave}^2 / 2} = \frac{16}{Re} \tag{9.22}$$

where $Re = 2R\rho u_{ave}/\eta$. Experimentally, for a given pipe radius and fluid density, the friction factor can be deduced from the volumetric flow rate, Q, at a fixed pressure drop, Δp, for a pressure gradient $G = -\Delta p/L$ across a length of pipe, L. Specifically, in the definition of the friction factor eqn (9.22), $\tau_w = R\Delta p/2L$ and u_{ave} in the dynamic pressure is given from the measured Q using eqn (9.21). However, this assumes the flow is fully developed and any entrance effects have subsided, which for laminar pipe flow requires a distance from the entrance ~$0.12R\text{Re}$.[116] For non-circular pipes and channels an equivalent pipe diameter (hydraulic diameter) $D_H = 4 \times$ cross-sectional area/wetted perimeter can be defined and C_F then includes geometric factors. A hydraulic resistance, R_{hyd}, can also defined by $\Delta p = R_{hyd}Q$, and this has been tabulated for various cross-sectional shapes.[117]

Eqn (9.22) generally applies for Reynolds numbers up to Re ~ 2300, but above this critical value the flow enters a transition zone and then becomes turbulent. In this regime the flow profile is no longer parabolic (eqn (9.20)), but has a viscous sublayer close to the surface and a turbulent core, joined by an overlap layer, leading to a much more uniform profile across the pipe due to the transverse mixing of momentum.[116] In considering turbulent flows the velocity is often scaled by a characteristic number with units of velocity called the wall friction (or wall shear) velocity, $u_\tau = (\langle \tau_w \rangle / \rho)^{1/2}$, where $\langle \tau_w \rangle$ is the time-averaged wall shear stress. Similarly, distances from the wall are scaled by a characteristic length $\eta/\rho u_\tau$; details of flow profiles in the turbulent regime are given in various ref. 93 and 116. The experimental data for the friction factor in turbulent flow in a pipe with rough surfaces has been described by the Colebrook–White formula,[118]

$$\frac{1}{\sqrt{C_F}} = -4\log_{10}\left(\frac{k_s/2R}{3.7} + \frac{1.26}{\text{Re}\sqrt{C_F}}\right) \tag{9.23}$$

where k_s is the scale of surface roughness ("equivalent sand size") and typically applies for $4000 < \text{Re} < 10^7$. In turbulent pipe flow the entrance effect is shorter than in laminar flow and is ~$8.8R\text{Re}^{1/6}$.[116] At very high Re the roughness dominates and there is little dependence on Re. Often eqn (9.23) is plotted on a Moody diagram (Figure 9.8) using the Darcy friction factor $C_D = 4C_F$.[119] The theory and experiments leading to eqn (9.22) and eqn (9.23) did not examine the effects of surface chemistry and wetting of the surface roughness on the friction factor.

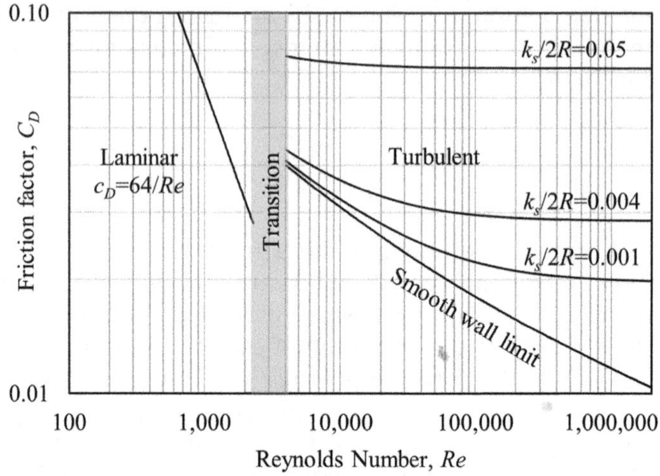

Figure 9.8 Darcy friction factor, C_D, for flow through a pipe. Both laminar and turbulent regimes are shown. The ratio of effective roughness scale to pipe diameter, $k_s/2R$, determines the extent to which roughness dominates in the turbulent regime.

9.3.3 Apparent Slip, Core Annular Flow, and Net ZMF Condition

The relaxation of the no-slip boundary condition using a slip length b modifies the flow profile in a pipe to

$$u_x(r) \approx u_{\text{max}}\left(1 - \frac{r^2}{R^2} + \frac{2b}{R}\right) \tag{9.24}$$

where the maximum flow rate, u_{max}, is along the centre of the pipe and is given by $u_{\text{max}} = G(R+b)^2/4\eta_b \approx GR^2(1+2b/R)/4\eta_b$. The effect of slip is to introduce an additional plug flow component with slip velocity of $v_s = GRb/2\eta_b$ to the velocity profile. The flow velocity averaged across the cross-sectional area of the pipe becomes $u_{\text{ave}} \approx u_{\text{max}}(1+2b/R)/2$ when $b/R \ll 1$. The volumetric flow rate is $Q = \pi R^2 u_{\text{ave}}$, which is a factor of $(1+4b/R)$ greater than for the same pipe subject to the same pressure gradient, but without slip. This implies that for small slip, $b/R \ll 1$, the experimentally measured friction factor is reduced by a factor of $(1-4b/R)$.

A possible cause of apparent slip could be the presence of super-liquid repellent texture (Figure 9.7c) or a layer of fluid of low viscosity, η_s, along the surface of the pipe lubricating the motion of a core bulk fluid of higher viscosity, η_b (Figure 9.7d). The effect of this can be modelled by assuming the layer has a constant thickness, h. In this model the Stokes equation (eqn (9.18)) is solved for the flow velocity of fluid in both the core and the surface layer, and the velocities and shear stresses are matched at $r = R - h$ to ensure continuity of the solutions. The difference in assumptions is from either requiring the same pressure gradients in the core and for the flow along the surface (core annular flow, CAF) or by requiring the fluid along the surface to have a net ZMF.[97] For superhydrophobic surfaces drag reduction due to the retention of a layer of air at the solid surface has been called "plastron" drag reduction.[33,67,79,85,97,114,120,121] In the ZMF case, the flow along the surface has a reversal in direction at a distance $2h/3R$ from the surface indicating a back-flow as part of a recirculating flow pattern.

In the pure CAF case, the flow velocity profile of the core bulk fluid is

$$u_x(r) = \frac{GR^2}{4\eta_b}\left(1 - \left(\frac{r}{R}\right)^2 + \left(\frac{\eta_b}{\eta_s} - 1\right)\zeta(2-\zeta)\right) \equiv \frac{GR^2}{4\eta_b}\left(1 - \left(\frac{r}{R}\right)^2 + 2\lambda_{CAF}\right) \quad (9.25)$$

where $\zeta = h/R$ and the equation is used to define a function λ_{CAF}.[97] Comparing with eqn (9.24) for small h/b and h/R, gives Vinogradova's slip length $b = (\eta_b/\eta_s - 1)h$. In the latter, ZMF, case the flow velocity profile is,[97]

$$u_x(r) = \frac{GR^2}{4\eta_b}\left(1 - \left(\frac{r}{R}\right)^2 + \left(\frac{\eta_b f(\zeta)}{\eta_s} - 1\right)\zeta(2-\zeta)\right) \equiv \frac{GR^2}{4\eta_b}\left(1 - \left(\frac{r}{R}\right)^2 + 2\lambda_{ZMF}\right) \quad (9.26)$$

where the equation is used to define a function λ_{ZMF} and the function $f(\zeta)$ has been defined as,

$$f(\zeta) = \frac{2(1-\zeta)^2\left[(2-\zeta)\zeta + (2 - 2\zeta + \zeta^2)\log_e(1-\zeta)\right]}{\zeta(4 - 14\zeta + 12\zeta^2 - 3\zeta^3) + 4(1-\zeta)^4 \log_e(1-\zeta)} \approx \frac{1}{4} \quad (9.27)$$

Comparing with eqn (9.24) for small h/b and h/R, gives a slip length $b = (\eta_b/4\eta_s - 1)$, which is identical to the result in eqn (9.9). The solutions in eqn (9.25) and (9.26) are not restricted to small h/R. The corrections to the flow velocity profiles given by $\lambda = \lambda_{CAF}$ or $\lambda = \lambda_{ZMF}$ depend on η_b/η_s and h/R, and give slip velocities of $v_s = GR^2\lambda/2\eta_b$ representing a plug flow correction to the no-slip case (Figure 9.7e). In the extreme case, for example a perfectly super-liquid repellent texture, fully developed plug flow may occur (Figure 9.7f).

The volumetric flow rate is calculated as

$$Q = 2\pi\int_0^{R-h}\left(\frac{GR^2}{4\eta_b}\right)\left(1 - \left(\frac{r}{R}\right)^2 + 2\lambda\right)r\mathrm{d}r = \frac{\pi GR^2}{8\eta_b}(1-\zeta)^2(1+4\lambda) \quad (9.28)$$

and as h increases this contains two competing effects due to the presence of a lubricating layer of fluid adjacent to the surface. The first is an increasingly effective lubrication leading to higher flow velocities and improved mass transport of the core fluid. The second is an increasing constriction of the cross-sectional area of the core as its radius, $R - h = R(1 - \zeta)$, reduces. Figure 9.9 shows the ratio of the volumetric flow rate for water with an air-lubricated boundary compared to flow without an air-lubricated boundary for a range of h/R. For perfect CAF, the core mass flux is maximized when the core to pipe radii are in the ratio $(R - h)/R \equiv (1 - \zeta) = (2 - \eta_s/\eta_b)^{-1/2}$.[38,97] When the lubricating fluid has a significantly lower viscosity than the viscosity of the core, this ratio is $1/\sqrt{2} = 0.707$. For the net ZMF case, there is also an optimum ratio which maximizes the transport.[97]

Busse *et al.* give full details of flow velocity profiles in both the bulk fluid and surface fluid layer, and changes in slip length and drag properties for laminar flow in pipes and channels for both CAF and ZMF.[97] The flow geometries they considered included Couette flow, symmetric pressure-driven channel flow, one-sided pressure-driven channel flow, and pipe flow. Their work was motivated by modelling flow bounded by perfectly superhydrophobic surfaces, *i.e.* ones where the solid surface texture may be ignored other than for its effect in providing a fixed-thickness air layer. Since their results are general for a fluid sheathed from a solid surface by a second immiscible fluid, they also apply to idealized LIS surfaces (see also Schönecker and Hardt[122]) and surfaces possessing a Leidenfrost induced vapour layer. A number of the literature reviews in Section 9.1.3 address skin friction (viscous shear) drag reduction for turbulent flow in channels with superhydrophobic surfaces (*e.g.* Rothstein[6] and references therein; also see Jeffs *et al.*[123] and Martell *et al.*[124]).

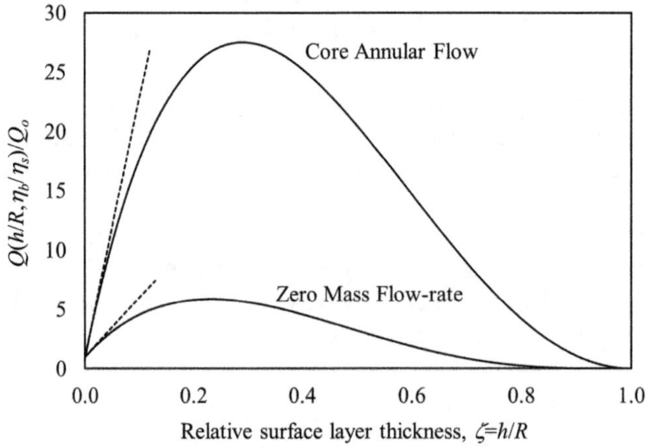

Figure 9.9 The relative increase in volumetric flow rate for the core annular flow (CAF) and net zero mass flow rate (ZMF) models of boundary lubrication calculated for $\eta_b/\eta_s = 55$ (water to air). The dashed lines are the $h/R \ll 1$ limits corresponding to a small slip length.

9.4 External Flow Past Cylinders and Spheres

9.4.1 Pressure and Form Drag

In internal flow through a pipe of constant cross-section the only type of drag is the skin friction drag due to the viscous shearing of the fluid. The skin friction drag due to the wall stress, τ_w, acting on the wetted surface area gives the drag force. The pressure arising from the conversion of kinetic energy of the flow is the dynamic pressure and so the coefficient of frictional drag is defined as the drag force/(dynamic pressure × wetted area) and is $\tau_w/(\rho u^2_{ave}/2)$, which is the same as the pipe friction factor. When an external flow occurs around a bluff object, such as a cylinder or sphere, the frictional boundary layer where viscosity is important adjacent to the surface may eventually break away. This is because across the front part of the surface there is an accelerated flow and a pressure drop, but across the back part of the surface there is a decelerated flow with a pressure increase.[93] The difference in pressures at the front and at the rear of the object leads to form (pressure) drag. In flow occurring distant from the surface, pressure is transformed to kinetic energy and then back into pressure. Flow near the surface is acted upon by the pressure distribution, but loses energy due to friction. This difference can lead to a negative pressure and the flow near the surface reversing direction (Figure 9.10a) thereby causing vortices and a turbulent wake behind the object (Figure 9.10b). The separation point occurs when the wall shear stress vanishes, $\tau_w = 0$, and when this is towards the front of the object form drag dominates, whereas when it is towards the rear there is also a contribution from skin friction drag. The drag force due to pressure is the drag force/(dynamic pressure × projected area), where the projected area is the area of the object perpendicular to the flow. The total drag force includes contributions from both the skin friction (viscous shear) drag and the form (pressure) drag and the overall coefficient of drag is defined as

$$C_D = \frac{\text{Resistance force}}{\text{Dynamic pressure} \times \text{projected area}} = \frac{2F_R}{\rho u_\infty^2 A_p} \tag{9.29}$$

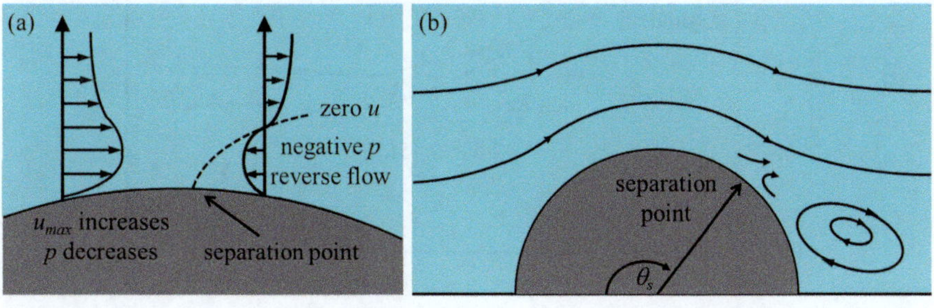

Figure 9.10 Origin of form drag. (a) Development of negative pressure and reverse flow, and (b) schematic of separation of flow at an angle θ_s and formation of a vortex.

where F_R is the resistance force, u_∞ is the free stream velocity, and A_p is the projected area.

The boundary layer concept allows flow at high Re to be modelled as a flow close to the wall which is influenced by the viscosity, and hence is a frictional or boundary layer, and an outer inviscid flow away from the walls where viscosity can be neglected. Flow in a boundary layer can be laminar or turbulent. A simple estimate of the boundary layer thickness in laminar flow is given by equating the inertia over a characteristic length x along the flow to the wall shear stress, *i.e.* $\rho(\mathbf{u}.\nabla)\mathbf{u} \sim \rho U_\infty^2/x$ to $\eta\partial^2 u_x/\partial y^2 \sim \eta U_\infty/\delta_x^2$, which gives $\delta_x/x \sim A/\mathrm{Re}_x^{1/2}$ where $\mathrm{Re}_x = \rho x U_\infty/\eta$; more precise calculations for flow across a flat plate suggest $A = 5$. The boundary layer therefore becomes thinner as Re increases or as the viscosity decreases. In the turbulent boundary layer case when $5 \times 10^5 < \mathrm{Re} < 10^7$, $\delta_x/x \sim 0.16/\mathrm{Re}_x^{1/7}$ is often quoted for a flat plate although $\delta_x/x \sim 0.382/\mathrm{Re}_x^{1/5}$ has also been suggested. Generally, the distance from a wall required for the velocity of the fluid to achieve a value 99% of the free stream value is called the boundary layer thickness, δ. Boundary layer theory can also be applied to how the layer grows in thickness with the distance downstream from the leading edge of a bluff body (or a pipe entrance).[115,116]

9.4.2 Coefficient of Drag and Types of Flow Patterns

For a smooth solid cylinder the projected area is $A_p = 2RL$ where R is the radius and L is the length. The drag undergoes a complex set of changes as Re increases (Figure 9.11) with vortices forming in the wake and the boundary layer eventually transitioning from laminar to turbulent.[125] At low Re (Re → 0), the flow is steady and never separates from the cylinder (Figure 9.12a), skin

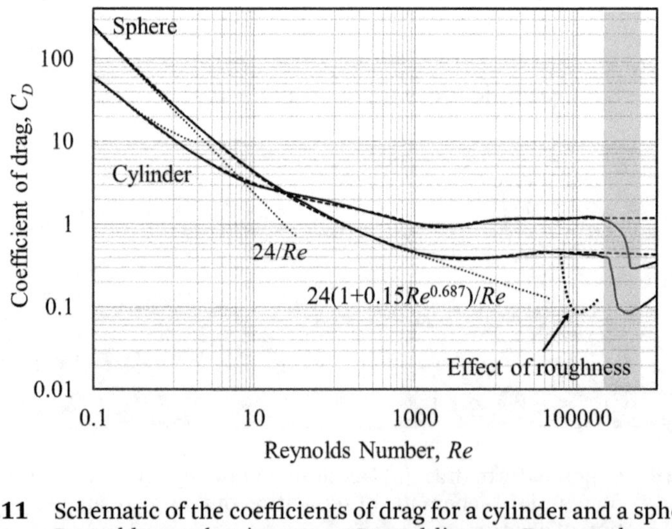

Figure 9.11 Schematic of the coefficients of drag for a cylinder and a sphere as the Reynolds number increases. Dotted lines as Re → 0 show analytical results. Dashed lines are experimental interpolations which describe the trends up to the drag crises (shaded range).

friction drag dominates and drag is at its highest. At higher Re ($3 < \text{Re} < 40$) the flow is steady, but the boundary layer breaks away at a separation angle between 180° and 130° and symmetric vortices form leading to $C_D \sim 4.5–1.5$. The vortices then elongate, the separation angle moves to ~115° and the wake becomes unstable indicating the onset of the von Karman vortex street ($40 < \text{Re} < 90$) leading to $C_D \sim 1.2$. Subsequently ($90 < \text{Re} < 300$), a pure Karman vortex street forms with vortices breaking away alternately from the top and bottom of the cylinder with a characteristic frequency, characterized by the Strouhal number $\text{Sr} = 2Rf/U_\infty$ ($0.14 < Sr < 0.21$). Further increases in Re ($300 < \text{Re} < 1.3 \times 10^5$) causes the flow to enter a subcritical regime with a separation angle $\theta_s \sim 80°$, vortex street instabilities (Sr = 0.21) and $C_D \sim 1.2$. In the critical regime ($1.3 \times 10^5 < \text{Re} < 3.5 \times 10^6$) there is turbulent separation and reattachment, and a turbulent wake with θ_s between 80° and 140°, and C_D falling from 1.2 to 0.2. The delay in the separation of flow from the cylinder surface is because of a transition from laminar to turbulent flow in the boundary layer. This causes the wide turbulent wake to switch to a narrower turbulent wake with an associated rapid fall in drag coefficient to a minimum and is referred to as a "drag crisis". For the highest values of Re ($\text{Re} > 3.5 \times 10^6$) the

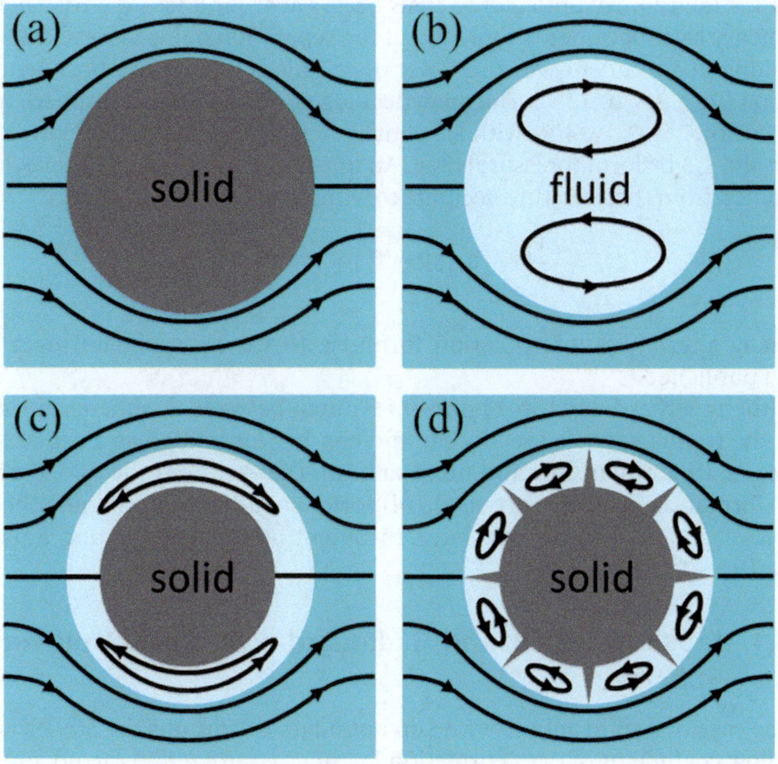

Figure 9.12 Schematic of Stokes equation solutions for creeping flow past spheres of different types. (a) solid sphere, (b) fluid sphere, (c) fluid/gas sheathed solid sphere (*e.g.* solid sphere with a plastron), and (d) superhydrophobic sphere.

flow enters a supercritical regime with turbulent separation characterized by $0.25 < \mathrm{Sr} < 0.30$, $\theta_s \sim 115°$, and a $C_D \sim 0.6$. The analytical result for the drag of a cylinder of infinite length in Stokes flow is[93]

$$C_D = \frac{8\pi}{\mathrm{Re}\left(0.5 - \gamma - \log_e(\mathrm{Re}/8)\right)} \tag{9.30}$$

or equivalently $8\pi/(\mathrm{Re} \times \log_e(7.406/\mathrm{Re}))$, where $\gamma = 0.5772$ is Euler's constant. An empirical interpolation modified from the literature,[126] which is accurate to within 10% over the range $0.2 < \mathrm{Re} < 2 \times 10^5$, is

$$C_D = 1.17 + \frac{5.47}{\mathrm{Re}} + \frac{3.65}{\mathrm{Re}^{0.42}} - \frac{0.0005\mathrm{Re}}{1 + 3.64 \times 10^{-7}\,\mathrm{Re}^2} \tag{9.31}$$

For a smooth solid sphere the projected area is $A_p = \pi R^2$, where R is the spherical radius. The coefficient of drag has a similar sequence of changes with Stokes flow ($\mathrm{Re} < 0.2$), followed by boundary layer separation and the formation of a ring (toroidal) vortex with the separation point moving to a stable $\theta_s \sim 80°$ when $\mathrm{Re} \sim 1000$. The coefficient of drag is then approximately constant $C_D \sim 0.4$–0.45 for $10^3 < \mathrm{Re} < 2 \times 10^5$ until a drag crisis occurs at around $\mathrm{Re} \sim 4 \times 10^5$ when the boundary layer becomes turbulent and the separation point moves backwards resulting in a smaller wake and lower drag. This is followed by turbulent separation ($\mathrm{Re} > 3 \times 10^6$).[93,127] The analytical result for the drag of a sphere in the Stokes flow[93] is $C_D = 24/\mathrm{Re}$ with 2/3 contributed by skin friction (viscous shear) drag and 1/3 by form (pressure) drag. An improved experimental interpolation modified from the literature accurate to within 10% for $\mathrm{Re} < 2 \times 10^5$ is[36,128]

$$C_D = \frac{24}{\mathrm{Re}}\left(1 + 0.15\mathrm{Re}^{0.687}\right) + \frac{0.38}{1 + 42500/\mathrm{Re}^{1.16}} \tag{9.32}$$

Many alternative interpolation formulae to the experimental data have been published.[129–131]

Although the details in Figure 9.11 differ between a sphere and cylinder, the trends with Re are similar and can be interpreted as a sequence of changes: (i) laminar flow and no separation ($\mathrm{Re} \to 0$), (ii) steady separation with formation of stable vortice(s), (iii) vortex shedding, (iv) a laminar boundary layer and wide wake, and (vi) a turbulent boundary layer and a narrow wake ($\mathrm{Re} \sim 10^5$–10^6).

9.4.3 Stokes with Slip and Hadamard–Rybczinski Drag for Spheres

In the case of a solid sphere with slip boundary conditions the Stokes creeping flow coefficient of drag correction factor, ξ, compared to the no-slip case is defined by $C_D = 24\xi/\mathrm{Re}$, and is[132]

$$\xi = \frac{1 + 2b/R}{1 + 3b/R} \approx 1 - \frac{b}{R} \tag{9.33}$$

where the expansion is valid for small slip lengths compared to the radius of the sphere, *i.e.* $b \ll R$.

In the case of external fluid flow past a fluid sphere, the no-slip boundary condition is replaced by a continuity of shear stress boundary condition at the fluid–fluid interface. Unless the interface is rigidified by contaminants or surfactants, this induces motion of the fluid within the sphere (Figure 9.12b) which itself must be conserved in a similar manner to the net ZMF condition in Section 9.2.2.[33,120] The creeping flow solution of Stoke's equation (eqn (9.18)), assuming no distortion of the spherical shape by the external flow, gives the Hadamard–Rybczinski drag correction,[133–135]

$$\xi_{HR} = \frac{\eta_{sb} + 2/3}{\eta_{sb} + 1} \tag{9.34}$$

where $\eta_{sb} = \eta_s/\eta_b$ is a viscosity ratio with η_s the viscosity of the fluid within the sphere and η_b the viscosity of the external bulk fluid. In the limit $\eta_{sb} \ll 1$, such as for a bubble of air in water, $\xi \to 2/3$ and the drag is reduced due to the lubricating effect of the internal fluid circulation. In the limit $\eta_{sb} \gg 1$, such as for a solid sphere in water, $\xi \to 1$ and the Stokes coefficient of drag is recovered. Corrections to eqn (9.34) for higher Re for viscous spheres are summarized by Feng *et al.*[136] In contrast to a solid sphere, a spherical bubble in a liquid at $\eta_{sb} \to 0$ with an uncontaminated interface will have an internal recirculation, which eliminates any wake separation of the external liquid, at all Reynolds numbers.[36] Distortions of the shape of a fluid sphere become important at higher We.

9.4.4 Plastron Drag Reduction for Spheres

In the case of a superhydrophobic surface, the surface texture maintains a layer of air when the surface is immersed in water. The idealized case of a perfectly hydrophobic sphere assumes the Cassie solid surface area fraction, $\varphi_s \to 0$, and there is a constant thickness layer of air surrounding the solid sphere (a plastron).[120] The general problem of Stokes flow past a sphere of one fluid encapsulated by a concentric sphere of a second fluid has a known solution.[137,138] This solution can be applied to create a plastron model of drag on a perfectly hydrophobic sphere provided it is recognised that the coefficient of drag is then for the fluid sheathed solid sphere complete with its encapsulating fluid layer (Figure 9.12c). The drag coefficient reduction factor for this plastron model is

$$\xi_p = \frac{2}{3\varepsilon}\left[\frac{1 + 3\eta_{pb}F(\varepsilon)}{1 + 2\eta_{pb}F(\varepsilon)}\right] \tag{9.35}$$

where $\varepsilon = 1/(1 + h/R)$ and $\eta_{pb} = \eta_p/\eta_b$ is a viscosity ratio with η_p the viscosity of the surface-retained sheathing fluid (the plastron) and η_b is the viscosity of the external bulk fluid. The function $F(\varepsilon)$ is defined to be

$$F(\varepsilon) = \frac{(1 + \varepsilon)(2\varepsilon^2 + \varepsilon + 2)}{(1 - \varepsilon)(4\varepsilon^2 + 7\varepsilon + 4)} \tag{9.36}$$

The drag reduction factor, ξ, has a minimum dependent on the ratio of plastron thickness to solid radius, h/R (Figure 9.13), and with a depth that increases with decreasing plastron to bulk fluid viscosity ratio, η_{pb}. For a plastron of air retained on a solid sphere in water the maximum reduction in drag is ~19% at $h = 0.1R$.[120] Physically, the effect of the sheathing layer of fluid is twofold: (i) lubrication of the solid–external fluid interface by the sheathing fluid (which is air for a superhydrophobic surface), and (ii) increasing the effective cross-section of the sphere (which becomes the sphere diameter plus twice the thickness of the plastron). If the plastron is very thin it is difficult for the external flow to induce an internal recirculating flow within the plastron (Figure 9.13, inset), but if the plastron becomes very thick, it increasingly obstructs the external bulk flow; the limiting case of $h \gg R$ gives the drag for a large bubble of radius $(R + h)$ (slope → 2/3 in Figure 9.13). This is analogous to the situation with ZMF in a pipe where a sufficiently thick plastron is required for a recirculating flow in the plastron to provide effective lubrication, but if the thickness continues to increase then it eventually constricts the channel and obstructs the internal bulk flow. The limiting case for eqn (9.35) assuming a finite η_{pb}, but small h/R is

$$\xi_p \approx 1 - \left(\frac{\eta_{bp}}{4} - 1\right)\frac{h}{R} + \dots \tag{9.37}$$

and is shown by the dotted line in Figure 9.13. From a comparison to eqn (9.33) we deduce that this drag reduction factor is equivalent to that from a solid sphere without a plastron, but modelled as having a slip length $b = (\eta_b/4\eta_p - 1)h$. This slip length is the same as from the ZMF condition for flow

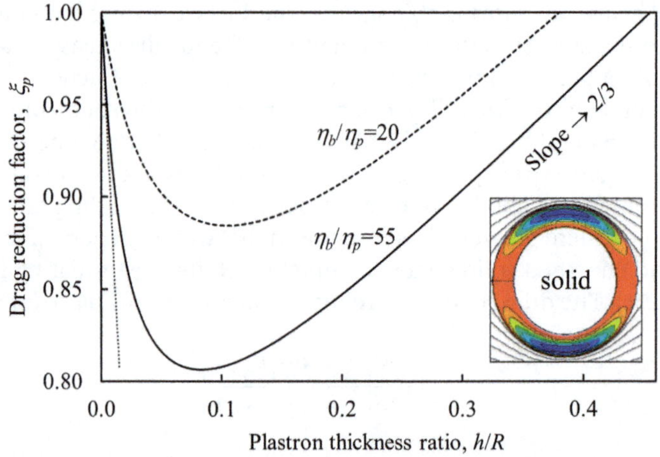

Figure 9.13 Drag reduction factor, ξ_p, in Stokes flow for fluid encapsulated spheres of viscosity ratio $\eta_{pb} = 1/55$ (air to water) and $\eta_{pb} = 1/20$; the analytical result for small h/R with a finite η_{pb} is shown by the dotted line. The inset shows streamlines for the internal recirculation of the sheathing fluid induced by the shear stress from the external flow.

across a surface (eqn (9.9)) and is also the same as the slip length for the ZMF condition for laminar flow in a pipe. On a more realistic superhydrophobic surface the plastron would not be a continuous sheathing layer of air. However, if we break the plastron into cells, we might then expect individual recirculation of the air in each cell depending on the details of the surface texture (Figure 9.12d). The plastron model may also potentially describe a surface hot enough to create a single continuous encapsulating (sheathing) Leidenfrost vapour layer or an object with a surface-retained fluid other than a gas, such as occurs with LIS surfaces.

9.4.5 Plastrons and Vortex Suppression

For higher Re, up to Re = 100, the problem of drag for flow of water past an air encapsulated sphere, as a model of a perfectly hydrophobic surface, has been considered using CFD.[83,114] This showed that large decreases in drag occurred in the range Re > 10 with a reduction in drag of ~50% for $h = 0.1R$ being noted at Re = 100 (Figure 9.14). A suppression of separation of flow and of the attached vortex regime was found in the range 30 < Re < 100, resulting in a narrower wake (Figure 9.14a). A drag reduction of ~10% was achieved for the thinnest plastron tested, $h = 0.01R$, and the vortex was still suppressed. The possible effect of an internal recirculation eliminating any wake separation of the external fluid has previously been noted for bubbles of air with non-contaminated surfaces (see Section 9.4.3).[36] Significant reduction in drag, correlated to a delayed flow separation and a narrower

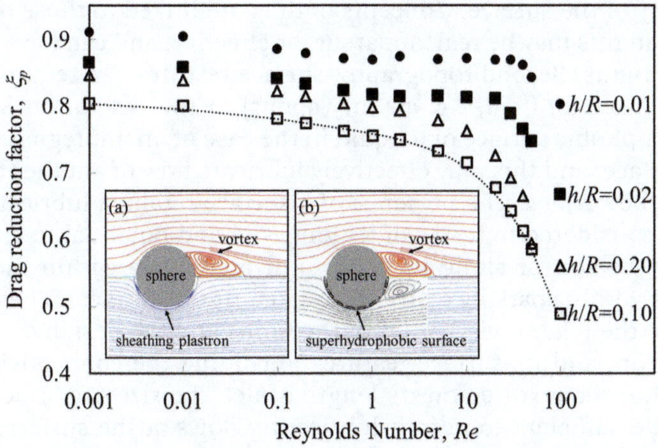

Figure 9.14 Drag reduction factor, ξ_p, calculated numerically for Re ≤ 100 for h/R = 0.01, 0.02, 0.2, and 0.1 (data from Gruncell;[83,114] dotted line is a guide for the trend). Insets show flow of water at Re = 100 past a solid sphere with no slip (upper halves of insets) and a solid sphere with (a) with a sheathing layer of air (plastron), and (b) an overlay of inset (a) with an axisymmetric superhydrophobic ridge model (showing internal recirculation within individual cells defined by the ridge structure).[114]

more streamlined wake, has been observed for falling spheres in a perfluorinated liquid and water using the Leidenfrost effect.[40,41] A reduction in drag and a delay in the onset of vortex shedding in a two-dimensional wake behind a cylinder with slip was previously shown by direct numerical simulations with Re ≤ 800.[139]

Gruncell *et al.* also considered a model of a sphere with axisymmetric ridges aligned perpendicular to the flow as an improved model of a superhydrophobic surface with a finite Cassie solid surface area fraction, φ_s.[83,114] In the limit of infinitely thin ridges the internal recirculation of air within the plastron becomes a series of internal recirculation of air within each cell defined by the ridges and the vortex regime remains suppressed. However, for finite-thickness ridges, the vortex regime is no longer completely suppressed (Figure 9.14b). Three related effects contributing to the overall drag were noted: (i) the slip at the air–water interface reduced the shear stress and hence the skin friction (viscous shear) drag, (ii) the combination of surface areas of no slip and slip changed the flow separation location and, hence, the form (pressure) drag, and (iii) increased φ_s resulted in an increase in the relative blockage of the sphere approaching the drag of a solid sphere of radius $(R + h)$. Cassie fractions above 10% were observed to increase, rather than reduce, drag.

9.5 Summary

The wetting properties of surfaces are determined not simply by the surface chemistry and the interfacial tensions of the fluids, but also by the topography of the surface. Concepts of drag are linked to those of interfacial slip and this may be real, apparent, or effective, and can have different physical origins. Beyond topography, surface textures can retain a second (often immiscible) fluid, *e.g.* air (or vapour) in the case of an immersed superhydrophobic surface or a liquid in the case of an impregnated (LIS or SLIPS) surface, and this may effectively lubricate flow of another fluid (liquid) over the surface. The presence of a surface-retained lubricating fluid has to be considered in terms of its longevity and potential to either flow along the surface (or along its texture) or recirculate within the surface texture (or within spatially restricted regions at the surface). These factors depend on the relative viscosities of the fluids and the design or structure of the texture and its surface features, including the connectivity of the space within the texture and its length scale. The size of surface texture needs to be sufficient to permit lubricating flows at the surface, but not so large as to create a dominating obstruction to the bulk flow or a collapse of topography induced wetting effects. The effect on drag reduction depends on whether the bulk flow is internal or external and the relative dominance of skin friction (viscous shear) drag and form (pressure) drag. Ideas of slip and drag reduction presented within this chapter for pipes, cylinders, and spheres can be extended to internal flows through channels and external flows past more general shaped objects. The recent focus on development of superhydrophobic surfaces to control wetting and drag reduction is likely to be complemented by increasing interest in new methods that combine surface texture with active gas/vapour generation, such as the Leidenfrost effect, or their infusion/impregnation with lubricating liquids.

Acknowledgements

The author acknowledges research grant funding from the UK Engineering and Physical Sciences Research Council (EPSRC) which supported the development of many of the plastron concepts in collaboration with co-investigators, including Dr A. Busse, Dr M. R. Flynn, Dr B. R. K. Gruncell, Dr M. I. Newton, and Professor N. D. Sandham.

References

1. D. Quéré, *Annu. Rev. Mater. Res.*, 2008, **38**, 71–99.
2. N. J. Shirtcliffe, G. McHale, S. Atherton and M. I. Newton, *Adv. Colloid Interface Sci.*, 2010, **161**, 124–138.
3. D. Quéré, *Annu. Rev. Fluid Mech.*, 2013, **45**, 197–215.
4. T.-S. Wong, S. H. Kang, S. K. Y. Tang, E. J. Smythe, B. D. Hatton, A. Grinthal and J. Aizenberg, *Nature*, 2011, **477**, 443–447.
5. J. D. Smith, R. Dhiman, S. Anand, E. Reza-Garduno, R. E. Cohen, G. H. McKinley and K. K. Varanasi, *Soft Matter*, 2013, **9**, 1772–1780.
6. J. P. Rothstein, *Annu. Rev. Fluid Mech.*, 2010, **42**, 89–109.
7. L. Gao and T. J. McCarthy, *Langmuir*, 2006, **22**, 6234–6237.
8. T. Sun, L. Feng, X. Gao and L. Jiang, *Acc. Chem. Res.*, 2005, **38**, 644–652.
9. P. Roach, N. J. Shirtcliffe and M. I. Newton, *Soft Matter*, 2008, **4**, 224–240.
10. M. Nosonovsky and B. Bhushan, *Curr. Opin. Colloid Interface Sci.*, 2009, **14**, 270–280.
11. N. J. Shirtcliffe, G. McHale and M. I. Newton, *J. Polym. Sci., Part B: Polym. Phys.*, 2011, **49**, 1203–1217.
12. A. Marmur, *Soft Matter*, 2013, **9**, 7900–7904.
13. A. K. Kota, G. Kwon and A. Tuteja, *NPG Asia Mater.*, 2014, **6**, e109.
14. T. Darmanin and F. Guittard, *Mater. Today*, 2015, **18**, 273–285.
15. T. Darmanin and F. Guittard, *J. Mater. Chem. A*, 2014, **2**, 16319–16359.
16. O. I. Vinogradova, *Int. J. Miner. Process.*, 1999, **56**, 31–60.
17. S. Granick, Y. Zhu and H. Lee, *Nat. Mater.*, 2003, **2**, 221–227.
18. R. S. Voronov, D. V. Papavassiliou and L. L. Lee, *Ind. Eng. Chem. Res.*, 2008, **47**, 2455–2477.
19. C. Neto, D. R. Evans, E. Bonaccurso, H.-J. Butt and V. S. J. Craig, *Rep. Prog. Phys.*, 2005, **68**, 2859–2897.
20. C. I. Bouzigues, L. Bocquet, E. Charlaix, C. Cottin-Bizonne, B. Cross, L. Joly, A. Steinberger, C. Ybert and P. Tabeling, *Philos. Trans. R. Soc., A*, 2008, **366**, 1455–1468.

21. T. Lee, E. Charrault and C. Neto, *Adv. Colloid Interface Sci.*, 2014, **210**, 21–38.
22. E. Lauga, M. P. Brenner and H. A. Stone, in *Handbook of Experimental Fluid Dynamics*, ed. C. Tropea, A. L. Yarin and J. F. Foss, Springer, New York, 2007, pp. 1219–1240.
23. O. I. Vinogradova and A. L. Dubov, *Mendeleev Commun.*, 2012, **22**, 229–236.
24. L. Bocquet and J.-L. Barrat, *Soft Matter*, 2007, **3**, 685–693.
25. M. A. Samaha, H. V. Tafreshi and M. Gad-el-Hak, *C. R. Mec.*, 2012, **340**, 18–34.
26. J. Wang, B. Wang and D. Chen, *Friction*, 2014, **2**, 295–309.
27. M. Samaha and M. Gad-el-Hak, *Polymers (Basel, Switz.)*, 2014, **6**, 1266–1311.
28. B. Bhushan and Y. C. Jung, *Prog. Mater. Sci.*, 2011, **56**, 1–108.
29. G. D. Bixler and B. Bhushan, *Crit. Rev. Solid State Mater. Sci.*, 2014, **40**, 1–37.
30. Y. C. Jung and B. Bhushan, *J. Phys.: Condens. Matter*, 2010, **22**, 035104.
31. G. D. Bixler and B. Bhushan, *Nanoscale*, 2013, **5**, 7685–7710.
32. G. D. Bixler and B. Bhushan, *Adv. Funct. Mater.*, 2013, **23**, 4507–4528.
33. G. McHale, M. I. Newton and N. J. Shirtcliffe, *Soft Matter*, 2010, **6**, 714–719.
34. R. E. Johnson and S. S. Sadhal, *Annu. Rev. Fluid Mech.*, 1985, **17**, 289–320.
35. P. M. Gresho, *Fusion Technol.*, 1999, **35**, 157–188.
36. E. Loth, *Int. J. Multiphase Flow*, 2008, **34**, 523–546.
37. S. L. Ceccio, *Annu. Rev. Fluid Mech.*, 2010, **42**, 183–203.
38. D. D. Joseph, R. Bai, K. P. Chen and Y. Y. Renardy, *Annu. Rev. Fluid Mech.*, 1997, **29**, 65–90.
39. P. Zhang and F. Y. Lv, *Energy*, 2015, **82**, 1068–1087.
40. I. U. Vakarelski, J. O. Marston, D. Y. C. Chan and S. T. Thoroddsen, *Phys. Rev. Lett.*, 2011, **106**, 214501.
41. I. U. Vakarelski, D. Y. C. Chan and S. T. Thoroddsen, *Soft Matter*, 2014, **10**, 5662–5668.
42. B. R. Solomon, K. S. Khalil and K. K. Varanasi, *Langmuir*, 2014, **30**, 10970–10976.
43. R. Pit, H. Hervet and L. Léger, *Tribol. Lett.*, 1999, **7**, 147–152.
44. D. C. Tretheway and C. D. Meinhart, *Phys. Fluids*, 2002, **14**, 9–12.
45. P. Joseph and P. Tabeling, *Phys. Rev. E*, 2005, **71**, 035303.
46. J. Ou and J. P. Rothstein, *Phys. Fluids*, 2005, **17**, 103606.
47. P. Joseph, C. Cottin-Bizonne, J.-M. Benoît, C. Ybert, C. Journet, P. Tabeling and L. Bocquet, *Phys. Rev. Lett.*, 2006, **97**, 156104.
48. R. J. Daniello, N. E. Waterhouse and J. P. Rothstein, *Phys. Fluids*, 2009, **21**, 085103.
49. K. V. Sharp and R. J. Adrian, *Exp. Fluids*, 2004, **36**, 741–747.
50. B. Woolford, J. Prince, D. Maynes and B. W. Webb, *Phys. Fluids*, 2009, **21**, 085106.

51. R. Truesdell, A. Mammoli, P. Vorobieff, F. van Swol and C. J. Brinker, *Phys. Rev. Lett.*, 2006, **97**, 1–4.
52. S. Jin, P. Huang, J. Park, J. Y. Yoo and K. S. Breuer, *Exp. Fluids*, 2004, **37**, 825–833.
53. D. Lumma, A. Best, A. Gansen, F. Feuillebois, J. O. Rädler and O. I. Vinogradova, *Phys. Rev. E*, 2003, **67**, 056313.
54. O. I. Vinogradova, K. Koynov, A. Best and F. Feuillebois, *Phys. Rev. Lett.*, 2009, **102**, 1–4.
55. L. Joly, C. Ybert and L. Bocquet, *Phys. Rev. Lett.*, 2006, **96**, 046101.
56. J. Baudry and E. Charlaix, *Langmuir*, 2001, 5232–5236.
57. Y. Zhu and S. Granick, *Phys. Rev. Lett.*, 2001, **87**, 096105.
58. C. Cottin-Bizonne, B. Cross, A. Steinberger and E. Charlaix, *Phys. Rev. Lett.*, 2005, **94**, 056102.
59. K. Watanabe, Yanuar and H. Mizunuma, *JSME Int. J., Ser. B*, 1998, **41**, 525–529.
60. C.-H. Choi, K. J. A. Westin and K. S. Breuer, *Phys. Fluids*, 2003, **15**, 2897–2902.
61. K. Watanabe, S. Ogata, A. Hirose and A. Kimura, *Trans. Jpn. Soc. Mech. Eng. Ser. B*, 2005, **71**, 2843–2848.
62. J. Ou, J. B. Perot, J. P. Rothstein, B. Perot and J. P. Rothstein, *Phys. Fluids*, 2004, **16**, 4635–4643.
63. C.-H. Choi, U. Ulmanella, J. Kim, C. M. Ho and C.-J. 'CJ' Kim, *Phys. Fluids*, 2006, **18**, 1–8.
64. G. D. Bixler and B. Bhushan, *J. Colloid Interface Sci.*, 2013, **393**, 384–396.
65. K. Watanabe, Y. Udagawa and H. Udagawa, *J. Fluid Mech.*, 1999, **381**, 225–238.
66. K. Watanabe, *Am. J. Sci. Technol.*, 2015, **2**, 106–110.
67. N. J. Shirtcliffe, G. McHale, M. I. Newton and Y. Zhang, *ACS Appl. Mater. Interfaces*, 2009, **1**, 1316–1323.
68. K. Watanabe and T. Akino, *J. Fluids Eng.*, 1999, **121**, 541–547.
69. K. Watanabe, T. Takayama, S. Ogata and S. Isozaki, *AIChE J.*, 2003, **49**, 1956–1963.
70. S. Srinivasan, J. A. Kleingartner, J. B. Gilbert, R. E. Cohen, A. J. B. Milne and G. H. McKinley, *Phys. Rev. Lett.*, 2015, **114**, 014501.
71. C. Lee, C. Choi and C.-J. 'CJ' Kim, *Phys. Rev. Lett.*, 2008, **101**, 1–4.
72. C.-H. Choi and C.-J. 'CJ' Kim, *Phys. Rev. Lett.*, 2006, **96**, 066001.
73. C. Lee and C.-J. 'CJ' Kim, *Langmuir*, 2009, **25**, 12812–12818.
74. S. Srinivasan, W. Choi, K.-C. Park, S. S. Chhatre, R. E. Cohen and G. H. McKinley, *Soft Matter*, 2013, **9**, 5691–5702.
75. S. Gogte, P. Vorobieff, R. Truesdell, A. Mammoli, F. van Swol, P. Shah and C. J. Brinker, *Phys. Fluids*, 2005, **17**, 051701.
76. A. K. Balasubramanian, A. C. Miller and O. K. Rediniotis, *AIAA J.*, 2004, **42**, 411–414.
77. P. Muralidhar, N. Ferrer, R. J. Daniello and J. P. Rothstein, *J. Fluid Mech.*, 2011, **680**, 459–476.

78. R. J. Daniello, P. Muralidhar, N. Carron, M. Greene and J. P. Rothstein, *J. Fluids Struct.*, 2013, **42**, 358–368.
79. J. C. Brennan, N. R. Geraldi, R. H. Morris, D. J. Fairhurst, G. McHale and M. I. Newton, *Sci. Reports*, 2015, **5**, 10267.
80. R. A. Bidkar, L. Leblanc, A. J. Kulkarni, V. Bahadur, S. L. Ceccio and M. Perlin, *Phys. Fluids*, 2014, **26**, 085108.
81. E. Aljallis, M. A. Sarshar, R. Datla, V. Sikka, A. Jones and C.-H. Choi, *Phys. Fluids*, 2013, **25**, 025103.
82. H. Haibao, D. Peng, Z. Feng, S. Dong and W. Yang, *Exp. Therm. Fluid Sci.*, 2015, **60**, 148–156.
83. B. R. K. Gruncell, Ph.D Thesis, Univ. Southampton, 2014, 1–271.
84. G. McHale, N. J. Shirtcliffe, C. R. Evans and M. I. Newton, *Appl. Phys. Lett.*, 2009, **94**, 064104.
85. N. J. Shirtcliffe, G. McHale, M. I. Newton, C. C. Perry and F. B. Pyatt, *Appl. Phys. Lett.*, 2006, **89**, 104106.
86. A. Marmur, *Langmuir*, 2006, **22**, 1400–1402.
87. M. R. Flynn and J. W. M. Bush, *J. Fluid Mech.*, 2008, **608**, 275–296.
88. A. Balmert, H. Florian Bohn, P. Ditsche-Kuru and W. Barthlott, *J. Morphol.*, 2011, **272**, 442–451.
89. J. P. Rothstein and R. J. Daniello, *U.S. Pat. Appl.* 20100330340, 2010.
90. M. E. McCormick and R. Bhattacharyya, *Nav. Eng. J.*, 1973, **85**, 11–16.
91. C. Lee and C.-J. 'CJ' Kim, *Phys. Rev. Lett.*, 2011, **106**, 014502.
92. B. R. Elbing, E. S. Winkel, K. a. Lay, S. L. Ceccio, D. R. Dowling and M. Perlin, *J. Fluid Mech.*, 2008, **612**, 201–236.
93. J. Schlichting and K. Gersten, *Boundary Layer Theory*, Springer-Verlag, 8th edn, 2001.
94. J. S. Wexler, A. Grosskopf, M. Chow, Y. Fan, I. Jacobi and H. A. Stone, *Soft Matter*, 2015, **11**, 5023–5029.
95. O. I. Vinogradova, *Langmuir*, 1995, **11**, 2213–2220.
96. D. D. Joseph, R. Bai, C. Mata, K. Sury and C. Grant, *J. Fluid Mech.*, 1999, **386**, 127–148.
97. A. Busse, N. D. Sandham, G. McHale and M. I. Newton, *J. Fluid Mech.*, 2013, **727**, 488–508.
98. T. D. Blake, *Colloids Surf.*, 1990, **47**, 135–145.
99. J. S. Ellis, G. McHale, G. L. Hayward and M. Thompson, *J. Appl. Phys.*, 2003, **94**, 6201–6207.
100. T. D. Blake, *J. Colloid Interface Sci.*, 2006, **299**, 1–13.
101. J. H. Snoeijer and B. Andreotti, *Annu. Rev. Fluid Mech.*, 2013, **45**, 269–292.
102. E. Lauga and H. A. Stone, *J. Fluid Mech.*, 2003, **489**, 55–77.
103. A. Steinberger, C. Cottin-Bizonne, P. Kleimann and E. Charlaix, *Nat. Mater.*, 2007, **6**, 665–668.
104. E. Karatay, A. S. Haase, C. W. Visser, C. Sun, D. Lohse, P. A. Tsai and R. G. H. Lammertink, *Proc. Natl. Acad. Sci. U. S. A.*, 2013, **110**, 8422–8426.

105. A. M. J. Davis and E. Lauga, *Phys. Fluids*, 2009, **21**, 011701.
106. P. Gao and J. J. Feng, *Phys. Fluids*, 2009, **21**, 102102.
107. C. Schönecker, T. Baier and S. Hardt, *J. Fluid Mech.*, 2014, **740**, 168–195.
108. C. Schönecker and S. Hardt, *J. Fluid Mech.*, 2013, **717**, 376–394.
109. J. R. Philip, *Z. Angew. Math. Phys.*, 1972, **23**, 353–372.
110. J. R. Philip, *Z. Angew. Math. Phys.*, 1972, **23**, 960–968.
111. A. V. Belyaev and O. I. Vinogradova, *Soft Matter*, 2010, **6**, 4563–4570.
112. C. Ybert, C. Barentin, C. Cottin-Bizonne, P. Joseph and L. Bocquet, *Phys. Fluids*, 2007, **19**, 123601.
113. A. M. J. Davis and E. Lauga, *J. Fluid Mech.*, 2010, **661**, 402–411.
114. B. R. K. Gruncell, N. D. Sandham and G. McHale, *Phys. Fluids*, 2013, **25**, 043601.
115. J. D. Anderson, *Computational Fluid Dynamics*, McGraw-Hill, Internatio., 1995.
116. F. M. White, *Fluid Mechanics*, McGraw-Hill, New York, 6th edn, 2008.
117. H. Bruus, *Theoretical Microfluidics*, Oxford University Press Inc., New York, 2008.
118. C. F. Colebrook and C. M. White, *Proc. R. Soc. London, Ser. A*, 1937, **161**, 367–381.
119. L. F. Moody, *Trans. ASME*, 1944, **66**, 671–684.
120. G. McHale, M. R. Flynn and M. I. Newton, *Soft Matter*, 2011, **7**, 10100–10107.
121. J. C. Brennan, D. J. Fairhurst, R. H. Morris, G. McHale and M. I. Newton, *J. Phys. D: Appl. Phys.*, 2014, **47**, 205302.
122. C. Schönecker and S. Hardt, *Microfluid. Nanofluid.*, 2015, **19**, 199–207.
123. K. Jeffs, D. Maynes and B. W. Webb, *Int. J. Heat Mass Transfer*, 2010, **53**, 786–796.
124. M. B. Martell, J. P. Rothstein and J. B. Perot, *Phys. Fluids*, 2010, **22**, 065102.
125. C. H. K. Williamson, *Annu. Rev. Fluid Mech.*, 1996, **28**, 477–539.
126. D. Sucker and H. Brauer, *Wäerme- Stoffüebertrag.*, 1975, **8**, 149–158.
127. C. E. Brennen, *Fundamentals of Multiphase Flow*, Cambridge University Press, 2005.
128. R. Clift and W. H. Gauvin, *Proceedings of the CHEMECA'70*, Butterworth, Melbourne, 1970, vol. 1, pp. 14–28.
129. F. A. Morrison, *An Introduction to Fluid Mechanics*, Cambridge University Press, New York, 2013.
130. N. S. Cheng, *Powder Technol.*, 2009, **189**, 395–398.
131. W. J. Whiten and C. E. Özer, *Miner. Process. Extr. Metall. Rev.*, 2015, **36**, 92–102.
132. A. B. Basset, *A Treatise on Hydrodynamics*, Dover, New York, 1961, vol. 2.
133. J. S. Hadamard, *C. R. Hebd. Seances Acad. Sci.*, 1911, **152**, 1735–1738.
134. W. Rybczynski, *Bull. Int. Acad. Sci. Cracovie*, 1911, 40–46.
135. J. C. Boussinesq, *C. R. Hebd. Seances Acad. Sci.*, 1913, **156**, 983–989.

136. Z.-G. Feng, E. E. Michaelides and S. Mao, *Fluid Dyn. Res.*, 2012, **44**, 025502.
137. E. Rushton and G. G. A. Davies, *Int. J. Multiphase Flow*, 1983, **9**, 337–342.
138. S. S. Sadhal and H. N. Oguz, *J. Fluid Mech.*, 1985, **160**, 511–529.
139. D. Legendre, E. Lauga and J. Magnaudet, *J. Fluid Mech.*, 2009, **633**, 437–447.

CHAPTER 10

Lubricant-Impregnated Surfaces

BRIAN R. SOLOMON[a], SRINIVAS BENGALURU SUBRAMANYAM[a], TAYLOR A. FARNHAM[a], KARIM S. KHALIL[a], SUSHANT ANAND[a] AND KRIPA K. VARANASI*[a]

[a]Department of Mechanical Engineering, Massachusetts Institute of Technology, Cambridge, MA 02139, USA
*E-mail: varanasi@mit.edu

10.1 Introduction

Lubricant-impregnated surfaces (sometimes abbreviated to LIS) are composed of a liquid lubricant that is stabilized in a porous or textured solid by capillary forces.[1–5] Drops can exhibit high mobility and remarkably low contact angle hysteresis (<1°) on stable lubricant-impregnated surfaces. Many research groups have taken an interest in this technology and have expanded its use to many applications.

The following section outlines how to achieve a stable lubricant-impregnated surface and discusses its basic features including the wetting ridge and lubricant cloak. Next, applications of lubricant-impregnated surfaces are detailed including condensation, anti-icing, anti-fouling, fluid mobility, optics, and active surfaces. In each application, the design of a lubricant-impregnated surface has particularly relevant criteria. For example, cloaking of condensed water drops can adversely affect condensation heat transfer, and ice adheres more weakly to surfaces with more densely packed textures.

RSC Soft Matter No. 5
Non-wettable Surfaces: Theory, Preparation, and Applications
Edited by Robin H. A. Ras and Abraham Marmur

Published by the Royal Society of Chemistry, www.rsc.org

10.2 Fundamentals

A lubricant-impregnated surface is composed of a liquid lubricant that is stabilized in a porous or textured solid by capillary forces. The interface between a lubricant-impregnated surface and an immiscible liquid coming into contact with it (the working fluid) gives rise to many novel properties. If the system is properly designed (discussed below), the working fluid will not displace the lubricant. This section discusses the thermodynamics and the morphology of drops on lubricant-impregnated surfaces. It details how to achieve stability and important features such as the lubricant cloak and lubricant ridge that can form on working fluid drops.

Whereas a drop on a solid surface forms a single three-phase contact line between the liquid, air, and solid, the boundary between a working fluid drop and a lubricant-impregnated surface is more complex. Smith *et al.* describe the thermodynamics of drops on lubricant-impregnated surfaces and show that a drop on a lubricant-impregnated surface can exist in one of 12 different thermodynamic states depending on the properties of the working fluid droplet, impregnating lubricant, solid texture, and surrounding environment (see Figure 10.1).[5] The thermodynamic phase diagram is constructed

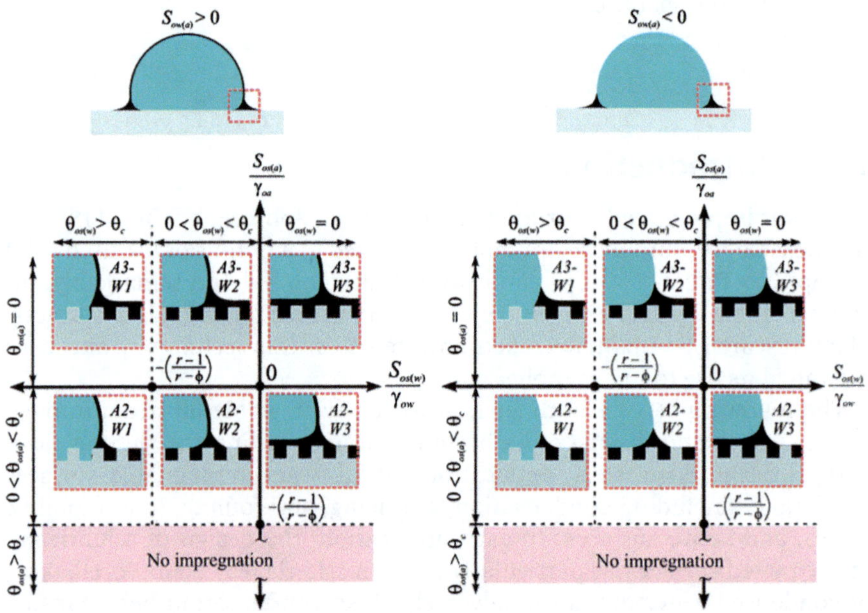

Figure 10.1 Possible thermodynamic states of a water drop placed on a lubricant-impregnated surface. The top two schematics show cases where the drop does and does not get cloaked by the lubricant. For each case, there are six possible states depending on how the lubricant wets the texture in the presence of air (the vertical axis) and the working fluid (horizontal axis). Reproduced from ref. 5 with permission from the Royal Society of Chemistry.

by considering the various possible configurations that are described below. Based on the properties of the materials involved (working fluid, lubricant, solid), three distinctly different morphologies of both the interfaces outside and underneath the drop are possible, as summarized in Table 10.1.[5] Considering the interface underneath the drop, in the first case, the working fluid displaces the lubricant, and the working fluid makes direct contact with the solid everywhere under the drop. This Wenzel-like state is referred to as the *impaled* state. In the second case, called the *impregnated*, *emerged* state, the lubricant remains contained in the solid and the working fluid contacts only the exposed features. In the third case, the working fluid makes no contact with the solid and sits on a thin equilibrium film (in contrast to an excess film[3]) of lubricant with its thickness set by the balance of intermolecular forces. This *encapsulated* state exhibits the most slipperiness (as quantified by the roll-off angle) whereas the impaled state, similar to the Wenzel state on a superhydrophobic surface, has poor slipperiness.[5]

These three distinct morphologies underneath a working fluid drop on a liquid-impregnated surface are quantified in terms of total interfacial energy (Table 10.1). The state with the lowest interfacial energy will be the one observed in steady state for a chosen working fluid, lubricant, and solid combination. The solid surface is characterized by the roughness r that denotes the total surface area of the solid per projected area, and the solid fraction ϕ denotes the fraction of the solid that makes contact with the working fluid in the impregnated, emerged state. Accounting for each interface beneath the drop gives the total interfacial energy per area of $E_{w1} = r\gamma_{sw}$ for the impaled state, $E_{w2} = (r - \phi)\gamma_{os} + \phi\gamma_{ws} + (1 - \phi)\gamma_{ow}$ for the impregnated, emerged state, and $E_{w3} = \gamma_{ow} + r\gamma_{os}$ for the encapsulated state. γ_{ij} denotes the interfacial tension between components i and j (w for working fluid, o for lubricant, and s for solid). The morphology with the lowest total interfacial energy per area will exist in steady state.

Equivalently, the encapsulated state is observed over the emerged state ($E_{w3} < E_{w2}$) when $S_{os(w)} = \gamma_{ws} - \gamma_{ow} - \gamma_{os} > 0$. Here $S_{os(w)}$ is the spreading coefficient of the lubricant (o) on the solid (s) in the presence of the working fluid (w). Spreading coefficients can be experimentally deduced by observing if a drop of lubricant spreads on a flat solid when in an environment of the working fluid. The drop size R should be small relative to the capillary length $\ell_c = (\gamma/\rho g)^{1/2}$ to ensure the spreading is not driven by gravity. Here, γ is the surface tension of the spreading liquid, ρ its density, and g is the gravitational acceleration.[6] Observing spreading of a lubricant indicates that $S_{ow(s)} > 0$ and the encapsulated state will exist for a textured solid of the same chemistry. These conclusions are summarized in Table 10.1.

As described by the above equations, lubricants and solids with low surface energies tend to form the most stable lubricant-impregnated surfaces. Oftentimes fluorocarbons (*e.g.* Krytox, FC-70), fatty alcohols (*e.g.* decanol), hydrocarbons, and silicone oils are used. Because evaporation of lubricant causes a lubricant-impregnated surface to lose its properties,[7] ionic liquids (*e.g.* BMIIm) have been proposed due to their extremely low vapour pressure

Table 10.1 Schematics of wetting configurations and interface energies. The total interface energies per unit area (column 3) are calculated for configurations outside and underneath a drop (column 2) by summing the individual interfacial energy contributions. Equivalent requirements for stability of each configuration are provided in column 4. In this table the lubricant is called "oil" and the working fluid is called "water".[a]

Interface	Configuration	Total interface energy per unit area	Equivalent criteria		
Oil-Solid-Air 	*A1* Dry 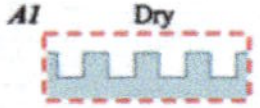	$E_{A1} = r\gamma_{sa}$	$E_{A1} < E_{A2}, E_{A3}$	$S_{os(a)} < -\gamma_{oa}\left(\frac{r-1}{r-\phi}\right)$	$\theta_{os(a)} > \theta_c$
	A2 Impregnated, emerged 	$E_{A2} = (r-\phi)\gamma_{os} + \phi\gamma_{sa} + (1-\phi)\gamma_{oa}$	$E_{A2} < E_{A1}, E_{A3}$	$-\gamma_{oa}\left(\frac{r-1}{r-\phi}\right) < S_{os(a)} < 0$	$0 < \theta_{os(a)} < \theta_c$
	A3 Encapsulated 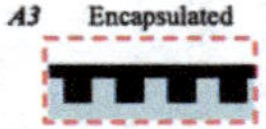	$E_{A3} = \gamma_{oa} + r\gamma_{os}$	$E_{A3} < E_{A2}, E_{A1}$	$S_{os(a)} \geq 0$	$\theta_{os(a)} = 0$
Oil-Solid-Water 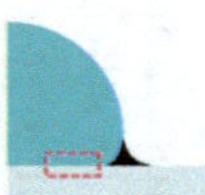	*W1* Impaled 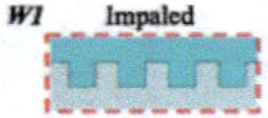	$E_{w1} = r\gamma_{sw}$	$E_{w1} < E_{w2}, E_{w3}$	$S_{os(w)} < -\gamma_{ow}\left(\frac{r-1}{r-\phi}\right)$	$\theta_{os(w)} > \theta_c$
	W2 Impregnated, emerged 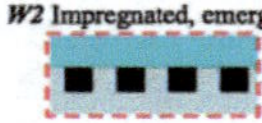	$E_{w2} = (r-\phi)\gamma_{os} + \phi\gamma_{sw} + (1-\phi)\gamma_{ow}$	$E_{w2} < E_{w1}, E_{w3}$	$-\gamma_{ow}\left(\frac{r-1}{r-\phi}\right) < S_{os(w)} < 0$	$0 < \theta_{os(w)} < \theta_c$
	W3 Encapsulated 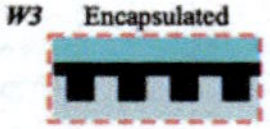	$E_{w3} = \gamma_{ow} + r\gamma_{os}$	$E_{w3} < E_{w1}, E_{w2}$	$S_{os(w)} \geq 0$	$\theta_{os(w)} = 0$

[a]Reproduced from ref. 5 with permission from the Royal Society of Chemistry.

but tend to exist in the impregnated, emerged state rather than the encapsulated state.[5]

Porous or textured solids can either be inherently low-energy materials (*e.g.* PTFE membranes) or other materials (*e.g.* silicon, SU-8, aluminium) treated to be hydrophobic. Fabricating a texture for creating a lubricant-impregnated surface is identical to fabricating a texture to make a superhydrophobic surface. See Li *et al.*[8] and Roach *et al.*[9] for comprehensive reviews of superhydrophobic surfaces that include techniques to create textured surfaces. Researchers have successfully demonstrated lubricant-impregnated surfaces created by photolithography,[5,10] wet etching,[11–13] sol–gel synthesis,[14–18] layer-by-layer assembly,[19–22] and other techniques.[13,23–27]

A convenient way of filling a texture is to withdraw the textured or porous solid from a bath of lubricant. A lubricant will spontaneously wick into a texture provided its contact angle $\theta_{os(a)}$ on a smooth substrate of the same chemistry is below a critical angle. The critical angle θ_c is defined by $\cos\theta_c = (1-\phi)/(r-\phi)$ where ϕ is the solid fraction and r the total area divided by the projected area of a texture (Table 10.1).[1,28]

In withdrawing a substrate from a bath of lubricant, Seiwert *et al.* showed that a solid with well-defined micropillars entrains no excess lubricant provided the capillary number $\mathrm{Ca} = \mu_o U/\gamma < 10^{-4}$ where μ_o is the viscosity of the lubricant, U is the withdrawal speed, and γ is the surface tension of the lubricant.[29] Such a dip-coating method is a popular technique to ensure the lubricant-impregnated surface has no excess film.

10.2.1 The Cloak

The lubricant can spread over the top of a working fluid drop and form a cloak (Figure 10.1). A cloak will form provided the spreading parameter of the lubricant on the working fluid in the presence of air (a) is greater than zero. Consideration of the cloak extends the possible morphologies of a drop on lubricant-impregnated surfaces to 12 possible states as shown in Figure 10.1.

The thickness of the lubricant cloak is set by a balance between the repulsive disjoining pressure and the Laplace pressure due to curvature. The Laplace pressure can be written as $2\gamma/R$ where γ is the surface tension of the lubricant and R the drop radius. The disjoining pressure $A_H/(6\pi h^3)$ is a function of the film thickness h and the Hamaker constant A_H which quantifies the interaction between air and working fluid molecules across the lubricant film. By balancing the Laplace pressure and disjoining pressure, Schellenberger *et al.* estimate that a cloak of the fluorocarbon FC-70 on a 1 mm drop of water is on the order of 20 nm thick.[10] Rykaczewski *et al.* and Anand *et al.* have confirmed cloaks over water drops by SEM,[30,31] and Schellenberger *et al.* have additionally confirmed the cloak by confocal microscopy (Figure 10.2).[32] Cheng *et al.* have also demonstrated high resolution X-ray tomography to visualize the water–lubricant interface.[33]

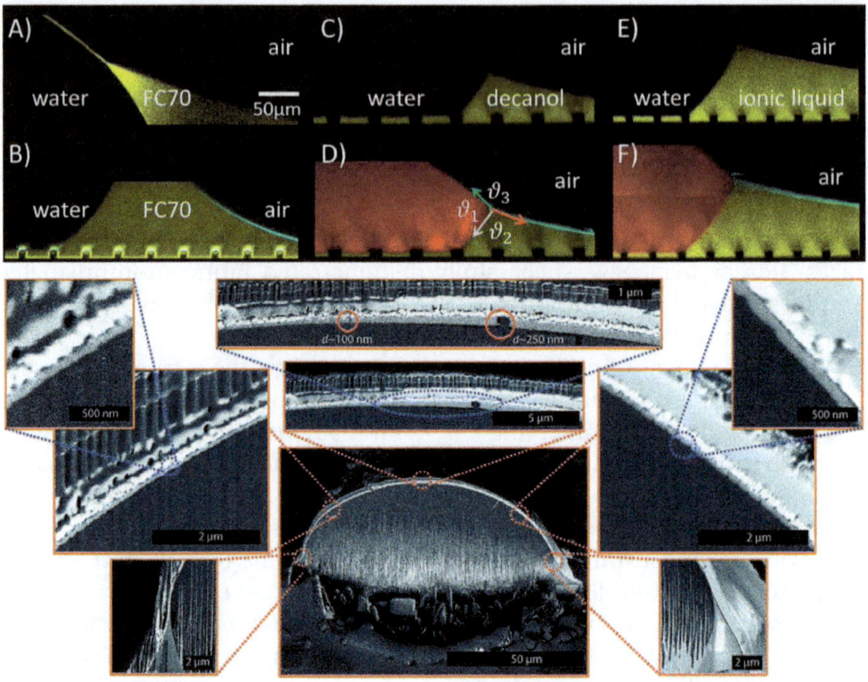

Figure 10.2 Cloaking of drops and the liquid–three-phase contact line. (top) Confocal images of vertical sections through a water drop placed on lubricant-impregnated micropillar array. FC70, decanol, and ionic liquid are used as lubricants. (bottom) Silicone oil cloak around a condensed drop suspended on a lubricant-impregnated surface obtained by a cryo-FIB-SEM process. The film is light grey sandwiched between the dark grey (water) and white (platinum). Reproduced from ref. 10 and 31 with permission from the Royal Society of Chemistry.

Anand *et al.* suggest that the formation of the cloak occurs in two steps.[31] First, a monolayer front spreads and its position follows $R = (4S_{ow(a)}/3(\mu_o\rho_o)^{1/2})^{1/2}t^{3/4}$ which is deduced from a balance between surface tension gradients and shear stress of the lubricant monolayer spreading on the liquid drop.[31,34] Next, a thicker film spreads. The relative scale of surface tension, viscous, and inertial forces in the spreading film is captured by the Ohnesorge number $\mathrm{Oh} = \mu_o(\rho_o R\gamma_{oa})^{-1/2}$ where μ_o is the lubricant viscosity, ρ_o the lubricant density, R the working fluid drop radius, and γ_{oa} the lubricant surface tension. Carlson *et al.* observe that the time it takes a drop to detach from a needle when brought into contact with an oil film is $\tau_\rho = (\rho_o R^3/\gamma_{oa})$ for Oh < 1 and $\tau_\mu = \mu_o R/\gamma_{oa}$ for Oh > 1. These timescales give an estimate for the time it takes for the thicker film to completely cover the drop presuming detaching from the needle is a result of the complete spreading. Further work is needed to understand the dynamics of the spreading of liquids on liquids.

10.2.2 Wetting Ridge

When a working drop contacts a lubricant-impregnated surface, a ridge of lubricant forms around the drop[5] that is similar to ridges that can develop on soft solids.[35,36] Schellenberger *et al.* analysed the wetting ridge of several lubricant-impregnated surfaces (Figure 10.2).[32] They find that the height z of the wetting ridge follows from a balance between Laplace pressure and hydrostatic pressure, and its solution is a modified Bessel function of the second kind approximated by $z = \exp(-r/\ell_c)$ where r is the radial position and ℓ_c the capillary length. The wetting ridge is important because most of the viscous dissipation in a mobile drop occurs in the wetting ridge, as discussed later in this chapter.[5] Neeson *et al.* present a useful analysis on the morphology of drops with immiscible fluids that is relevant to drops on lubricant-impregnated surfaces.[37]

10.2.3 Excess Films and Steady State

At equilibrium, the configurations in which the tops of the solid texture underneath the drop is covered by a thin lubricant film (*e.g.* states *A3–W3* and *A2–W3* in Table 10.1) are only possible when the spreading coefficient $S_{os(w)}$ is positive. In all other cases the solid textures come into contact with the working fluid. The texture can be also overfilled,[3] however, as depicted in Figure 10.3. In addition, Schellenberger *et al.* demonstrated that overfilling

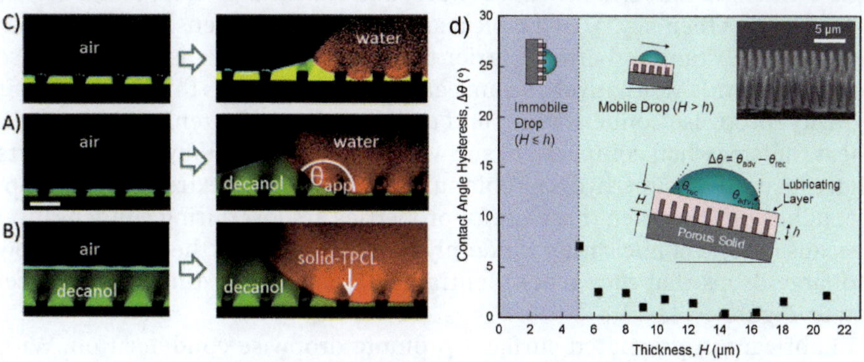

Figure 10.3 Influence of the filling height. (a–c) Image of a micropillar array infiltrated with decanol before and after a water drop is deposited. The height of the lubricant film is adjusted (a) to be underfilled, (b) to match the height of the posts, and (c) to be overfilled. Colour code: red, water; yellow/green, decanol; black, air or solid. Black shadows extend from the pillars because the sample was imaged from underneath. (d) Contact angle hysteresis of a 5 μL water drop on a lubricant-impregnated surface as a function of the lubricant height. (a)–(c) Reproduced from ref. 10 with permission from the Royal Society of Chemistry. (d) Reprinted by permission from Macmillan Publishers Ltd: *Nature*,[3] Copyright 2011.

a texture leads to different wetting ridge morphology,[32] and researchers have shown that the contact angle hysteresis of a water drop on a lubricant-impregnated surface as well as its sliding speed is significantly improved if an excess film is present.[12,38] While excess oil films may be beneficial (and also better in reducing ice adhesion[39]), the excess lubricant is not stabilized by capillary forces and can readily drain by gravity and other forces, thereby compromising slippery properties.

Carlson *et al.* demonstrate that a water drop can sit atop a thin oil film on a substrate for a finite time before the film drains in a system when the film drainage is favoured.[40] Similarly, drops on lubricant-impregnated surfaces may be mobile in transient states but behave differently once the lubricant-impregnated surface reaches its equilibrium state. The dip-coating method studied by Seiwert *et al.* and mentioned earlier in this chapter[29] is a convenient way to achieve a lubricant-impregnated surface with no excess lubricant.

10.3 Applications

10.3.1 Condensation

Approximately 40% of water usage worldwide is related to energy generation,[41] a demand primarily from power plants, most of which use steam cycles in their operation. A steam cycle comprises many components, but the condenser consumes the most water and contributes most to the overall steam cycle efficiency.[42] On a typical surface, steam condenses as a film that acts as a substantial thermal barrier to subsequent condensation (filmwise condensation). Alternatively, steam can condense as drops that roll off under gravity (dropwise condensation) and can provide up to a tenfold increase in heat transfer when compared to surfaces that condense filmwise.[42,43] Efforts have focused on superhydrophobic surfaces that exhibit extremely low droplet adhesion. However, these useful properties are lost during condensation because droplets nucleating randomly within textures of the solid can grow to large drops that may remain entrained within the textures in a Wenzel state (Figure 10.4a–c).

Lubricant-impregnated surfaces promote dropwise condensation. With a lubricant-impregnated surface, the lubricant contained in the surface prevents water from condensing within the texture (Figure 10.4e–f). In addition, the presence of the lubricant imparts exceptional mobility to condensed drops. On a conventional superhydrophobic surface, the critical size for drops to shed from the surface is on the order of a few millimetres (Figure 10.4g). On a lubricant-impregnated surface, Anand *et al.* observed that drops as small as 20 μm were mobile (speeds ~1 mm s^{-1}) on the surface (Figure 10.4g).[44] In addition, researchers have observed that under identical conditions, the onset of water condensation on lubricant-impregnated surfaces is faster than on superhydrophobic surfaces, suggesting

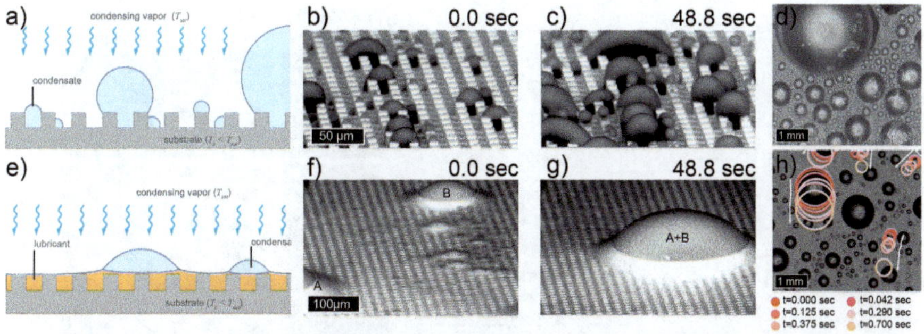

Figure 10.4 Comparison of condensation of water vapour on superhydrophobic (top row) and lubricant-impregnated surfaces (bottom row) with identical textures. (a) Schematic of condensation on superhydrophobic surfaces showing that water can nucleate within the texture and (b–c) timelapse ESEM images. (d) Drops grow and coalesce and eventually form large Wenzel drops that are pinned on the surface. (e) Schematic of condensation on lubricant impregnated surfaces showing condensation on top of the lubricant and (f–g) timelapse ESEM images. (h) Drops that condense and grow on a vertical lubricant impregnated surface are highly mobile. Reproduced with permission from Anand, S., Paxson, A. T., Dhiman, R., Smith, J. D. & Varanasi, K. K. Enhanced Condensation on Lubricant-Impregnated Nanotextured Surfaces. *ACS Nano* **6**, 10122–10129 (2012).[44] Copyright (2012) American Chemical Society.

that the lubricant-impregnated surfaces have a lower energy barrier for nucleation.[44,45]

The enhanced shedding and nucleation of drops indicates the potential of lubricant-impregnated surfaces for high condensation heat transfer. Xiao *et al.* quantify the condensation heat transfer of lubricant-impregnated surfaces as twice that of conventional hydrophobic and superhydrophobic surfaces in conditions comparable to those of industrial condenser operation (Figure 10.5a).[45] A lubricant-impregnated surface maintained dropwise condensation of steam up to the highest supersaturation tested whereas a superhydrophobic surface transitioned to filmwise condensation at high supersaturation.

Initially, Anand *et al.* had postulated that the nucleation of steam into water on lubricant-impregnated surfaces occurs at the lubricant–air interface,[44] whereas Xiao *et al.* had suggested that nucleation occurs at the solid surface beneath the lubricant.[45] In a later work, Anand *et al.* rationalize that water vapour has limited absorption into the lubricant and cannot achieve supersaturation to allow for nucleation within the lubricant, which suggest nucleation should occur only at the lubricant–air interface.[31] Further, they show that depending on the lubricant's surface tension and interfacial tension with water, nucleation on a lubricant can have a significantly lower energy barrier compared to that on solids. Based on such an analysis, they

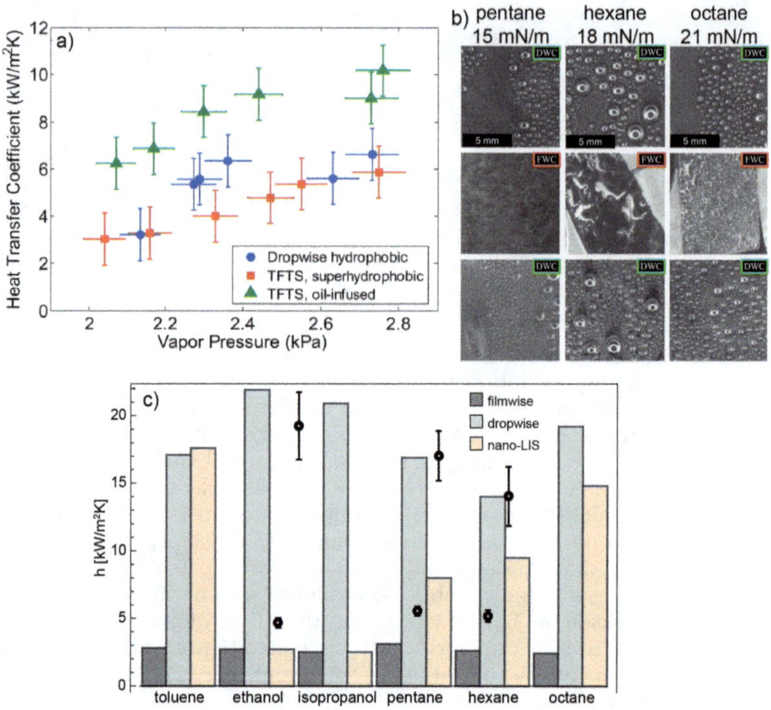

Figure 10.5 (a) Measured heat transfer coefficients for a flat hydrophobic surface, superhydrophobic surface, and Krytox-impregnated surface with varying vapour pressures. The Krytox-impregnated surface shows roughly twice the heat transfer of the other surfaces. (b) Images of the condensate of three low surface tension fluids on three different surfaces. Both dropwise condensation (DWC) and filmwise condensation (FWC) is observed. (c) Heat transfer coefficients for the condensation of low surface tension liquids on a flat silicon surface, a flat hydrophobic surface, and a Krytox-impregnated surface. Bars are generated from modelling whereas points indicate measured values. (a) Reprinted by permission from Macmillan Publishers Ltd: *Scientific Reports* (ref. 45), Copyright 2013. (b)–(c) Reprinted by permission from Macmillan Publishers Ltd: *Scientific Reports* (ref. 49), Copyright 2014.

constructed a regime to aid selection of lubricants that may lead to enhanced nucleation.[31]

Because condensers operate at low pressures, lubricants used for condensation applications must have low vapour pressure so that they are not rapidly lost. For condensation the existence of a lubricant cloak plays a special role. As discussed in Section 10.2, a lubricant can cloak over the top of a working fluid drop deposited on a lubricant-impregnated surface (Figure 10.2). The lubricant will cloak provided the spreading coefficient of the lubricant on the working fluid in the presence of air is positive ($S_{ow(a)} > 0$). Anand *et al.* show that the rate of condensed water drop growth on lubricant surfaces is significantly reduced on lubricants that form a cloak over the condensed water drops as compared

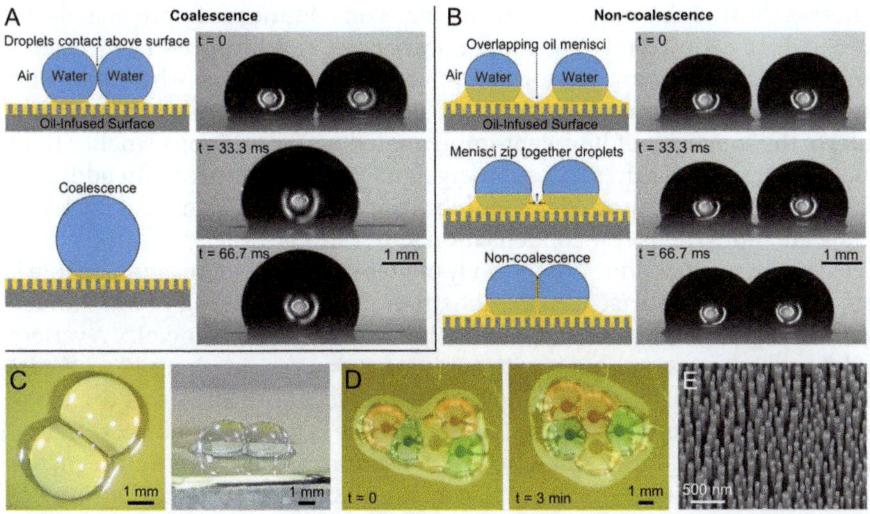

Figure 10.6 Interactive behaviour of water droplets on a lubricant impregnated surface. (a) Droplets colliding at the working fluid–air interfaces exhibited coalescence. (b) When the lubricant menisci of two drops overlap, a lubricant film formed between the droplets prevents coalescence. (c) Photographs of non-coalescing drops. (d) Multiple drops could be connected into a network. These networks spontaneously rearrange over time to minimize their surface energy. (e) SEM of nanopillared substrate. The oils used were (a,b) Krytox 100 and (c,d) Fomblin 25/6. Food colouring was used in (d). Reproduced from ref. 46 with permission of The National Academy of Sciences.

to those that do not. In a later work, Anand *et al.* show that a lubricant cloak forms almost immediately after water drops nucleate at the lubricant–air interface.[31] As a consequence, the capillary forces of the lubricant tend to submerge the drop after its formation. This prediction is confirmed by observing submerged microscopic droplets after condensation on a thin film of lubricant using a cryogenic focused ion beam (FIB)–SEM technique.

Figure 10.6 shows the non-coalescence of drops on a lubricant-impregnated surface. Boreyko *et al.* find that the non-coalescence is due to the lubricant wetting ridge.[46] Surprisingly, the time for two water drops to coalesce on a lubricant-impregnated surface is greater by an order of magnitude than for two water drops in a bath of the same lubricant, and increases with the lubricant viscsoity.[46] For example, the time for two 5 μL water drops to coalesce is roughly 1 day when the lubricant is 500 cSt silicone oil but only 1 s when the lubricant is 10 cSt silicone oil. Furthermore, they show that mixing phospholipids into the water drops creates lipid bilayers that prevent coalescence indefinitely. Barman *et al.* demonstrate that the coalescence process can be rapidly accelerated by applying a voltage between the two drops.[47]

Drops that condense and grow on a lubricant-impregnated surface with 1000 cSt silicone oil as the lubricant tend to grow in a narrow size distribution whereas those that condense and condense on a lubricant-impregnated surface with 10 cSt oil tend to be more polydisperse.[31] While preventing coalescence can create controlled formation of condensed drops (which may find use in breath figure templating[48]), cloaking of water droplets by lubricant affects the longevity of lubricant-impregnated surfaces. Drops smaller than the solid texture features can submerge and displace lubricant. In addition, shedding of cloaked drops depletes lubricant. As a result, non-cloaking lubricants are more robust for condensation applications.

Rykaczewski *et al.* added to the body of literature on condensation on lubricant-impregnated surfaces to demonstrate the condensation of low surface tension liquids such as pentane and toluene (Figure 10.5b and c).[49] A variety of low surface tension liquids with surface tension ranging from 12 to 28 mN m^{-1} are tested on a lubricant-impregnated surface with Krytox as the lubricant and compared with flat and re-entrant textured oleophobic surfaces. Some liquids exhibited filmwise condensation on lubricant-impregnated surfaces as a consequence of displacement of Krytox by the condensing liquid, but most liquids exhibited sustained dropwise condensation. Up to an eight times increase in heat transfer resulted from promoting dropwise condensation of the low surface tension liquids.[49]

The condensation heat transfer discussed up until this point involves a vapour transitioning to a liquid when it cools. A related problem is the generation of water from fog comprised of liquid water drops (rather than water vapour) where phase change does not occur. The efficiency of fog collection is greatly dependent on the mobility of collected water drops on the collecting surface which can be improved by using lubricant-impregnated surfaces.[50] However, experiments on fog collection by Park *et al.* using Krytox-impregnated surfaces show gradual loss of lubricant with time.[51] Later, Boor *et al.* studied fog collection using superhydrophobic electrospun surfaces and compared fog collection with and without impregnated liquids.[52] Their results showed that a lubricant-impregnated surface with Krytox increased the water collection rate from fog as compared to a superhydrophobic surface by about 130%. Furthermore, they also investigated the leaching of oil from the surfaces and showed that the under their experimental conditions 3–5 μL of lubricant was detected per litre of collected water.

10.3.2 Anti-Icing

Lubricant-impregnated surfaces have been demonstrated as a passive means of reducing ice adhesion, which is a significant issue spanning multiple industries including transportation, agriculture, energy, and construction. Whereas superhydrophobic surfaces have been explored for anti-icing,[53,54] Varanasi *et al.* showed that frost that forms on superhydrophobic surfaces leads to strong ice adhesion.[55]

Kim *et al.* demonstrate reduced ice and frost adhesion and accumulation on a Krytox-impregnated surface with excess lubricant.[56] As shown in Figure 10.7, during frost tests the impregnated aluminium sample tilted at 75° show

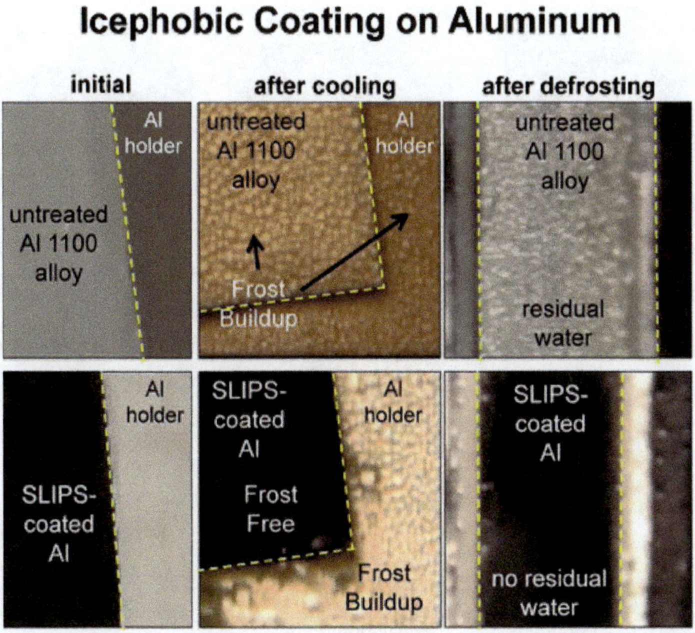

Figure 10.7 Comparison between an untreated aluminium 1100 sample (top) and lubricant-impregnated surface (bottom) comprised of textured alumina and Krytox. The samples were cooled to −10 °C at 60% relative humidity and defrosted at room temperature. Both samples are tilted at 75° for the duration of the experiment. Reproduced with permission from Kim, P. *et al.* Liquid-Infused Nanostructured Surfaces with Extreme Anti-Ice and Anti-Frost Performance. *ACS Nano* **6**, 6569–6577 (2012).[56] Copyright (2012) American Chemical Society.

no frost formation and no residual water after defrosting when compared to untreated samples. The lubricant-impregnated surface has less ice accumulation because condensed water can roll off easily at the 75° tilt angle before it freezes, and any accumulated ice easily rolls off the surface when defrosted. Additionally, measurements of ice adhesion strength show a reduction in adhesion strength of almost two orders of magnitude on surfaces with an excess lubricant film.

Subramanyam *et al.* extended these results by comparing a lubricant-impregnated surface with an excess film to one with an equilibrium film obtained by a controlled dip-coating process.[39] Compared to a lubricant-impregnated surface with excess lubricant, a lubricant-impregnated surface with the same texture and no excess lubricant shows higher ice adhesion (Figure 10.8). Optimizing the texture density of the lubricant-impregnated surface with no excess film results in lower ice adhesion than one of the lowest surface energy materials (80/20 PEMA-FluoroPOSS). Surprisingly, ice adhesion is lower on lubricant-impregnated surfaces that have higher texture densities. The researchers suggest that ice fractures more easily from

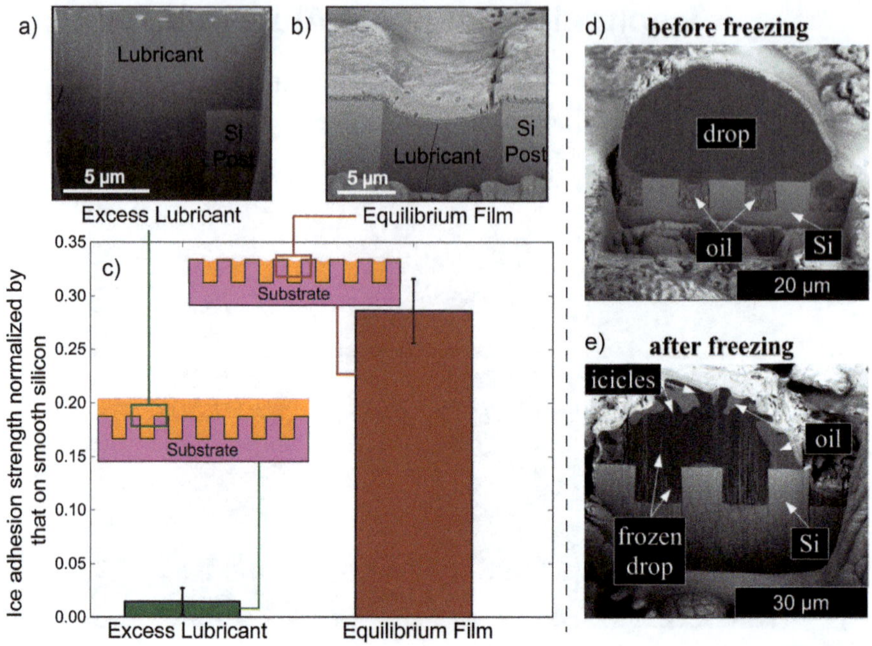

Figure 10.8 Cryo-SEM images of a cross-sectioned LIS with (a) excess lubricant and (b) no excess lubricant. (c) Comparison of the ice adhesion strength on LIS comparing excess and equilibrium lubricant films. The ice adhesion strength is normalized to that on a smooth, uncoated silicon surface. The textured surface of the LIS consists of lithographically textured silicon with 10 μm square posts with an edge-to-edge spacing of 50 μm. (d,e) Cross-section SEM images drops before and after freezing. The textured surface of Krytox-impregnated surface consists of lithographically textured silicon with 10 μm square posts with an edge-to-edge spacing of 10 μm. In the frozen drop the lubricant has migrated out of the texture and covers icicles. (a)–(c) Reproduced with permission from Subramanyam, S. B., Rykaczewski, K. & Varanasi, K. K. Ice Adhesion on Lubricant-Impregnated Textured Surfaces. *Langmuir* **29**, 13414–13418 (2013).[39] Copyright (2013) American Chemical Society. (d,e) Reproduced with permission from Rykaczewski, K., Anand, S., Subramanyam, S. B. & Varanasi, K. K. Mechanism of Frost Formation on Lubricant-Impregnated Surfaces. *Langmuir* **29**, 5230–5238 (2013).[122] Copyright (2013) American Chemical Society.

surfaces with higher texture densities because there is higher density of stress concentration sites.

Even with a lubricant in thermodynamic equilibrium, the lubricant can be depleted due to cloaking, solubility, evaporation, and other effects. In particular during ice and frost formation, Rykaczewski *et al.* show using cryogenic FIB-SEM that the lubricant can wick into icicles that accumulate on frozen water on lubricant-impregnated surfaces (Figure 10.8d and e).[57] When the ice is removed from the surface it takes lubricant with it. Depending on the selection of lubricant and texture, significant depletion can occur in just a single frost–defrost cycle. Once the lubricant is depleted the ice adhesion will closely mirror the performance of the underlying superhydrophobic surface, which can be worse than that of an untreated surface.[55]

Accordingly, active research focuses on developing lubricant-impregnated surfaces for anti-icing with enhanced durability.[11,19,58,59] In one approach, Yin *et al.* introduce nanoparticles into the lubricant layer to selectively heat the lubricant-impregnated surface under near-infrared irradiation.[60] This allows for thermal deicing while still maintaining many of the benefits inherent in a lubricant-impregnated surface.

10.3.3 Anti-Fouling

10.3.3.1 Self-Cleaning

A surface from which contaminants such as dust can easily be removed by a liquid is referred to as a self-cleaning surface. Self-cleaning has been achieved using superhydrophilic surfaces that rely on film flow or using superhydrophobic surfaces with low contact angle hysteresis on which drops can easily roll off, taking contaminants with them.[61] Unfortunately, the durability of such surfaces to a wide range of contaminants and fluids limits their widespread use.

Lubricant-impregnated surfaces are well-suited for self-cleaning applications because of their extremely low contact angle hysteresis and ability to repel a wide variety of liquids.[50,59,62,63] Additionally, the angle at which a drop of a given size rolls off a lubricant-impregnated surface is much smaller compared to that on smooth, low surface energy solid materials. Furthermore, identically sized drops will have a larger area of contact on a lubricant-impregnated surface compared to a superhydrophobic surface, making lubricant-impregnated surfaces a useful tool for self-cleaning (Figure 10.9a).

The slippery nature of lubricant-impregnated surfaces also alters the drying pattern of droplets containing particles. Figure 10.9b shows a typical deposition pattern observed on a surface that has been called the "coffee ring effect." When an evaporating drop becomes pinned, particles migrate to the contact line and deposit forming a ring pattern. This behaviour is compared with the drying pattern of the same drop on a lubricant-impregnated surface. The lack of pinning during evaporation[64] on liquid-impregnated surfaces allows for a more uniform and localized deposition of particles.[50,62] Yang *et al.* have taken advantage of this effect to concentrate analytes in an evaporating liquid drop to improve Raman characterization.[65]

10.3.3.2 Biofilm Formation

Biofouling is widely prevalent in numerous industrial and medical applications. The formation of biofilms hampers the operation of marine vessels and desalination plants and can be catastrophic in catheter tubes and implants. Bacteria can also evolve to resist antimicrobial agents and hence

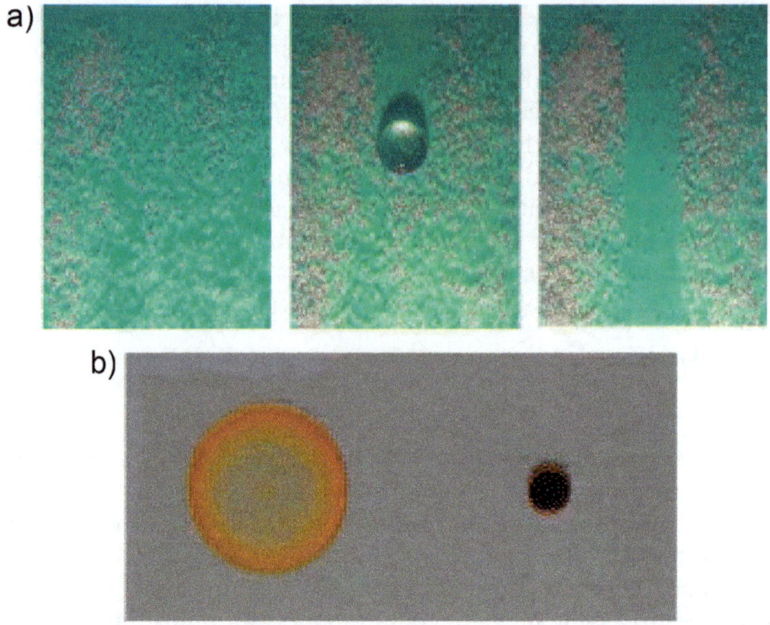

Figure 10.9 (a) A drop of water deposited on a silicone oil-impregnated surface cleans away silica dust particles. The surface is tilted at 20°. (b) Evaporated coffee drop on a plastic surface (left) in contrast to a coffee drop that evaporated on a silicone oil-impregnated surface. Reproduced with permission from *EPL*, 2011, 56001.[4]

other solutions are needed. The design of non-fouling surfaces has primarily focused on preventing protein adsorption and bacterial adhesion using functional groups including poly(ethylene glycol), zwitterions, and other hydrophilic groups that remain free of biofilm formation for only a limited time. Initial experiments showing lubricant-impregnated surfaces can repel blood and prevent bacterial attachment prompted further investigation into biofilm prevention.[63,66]

Epstein *et al.* showed that the slippery nature of lubricant-impregnated surfaces reduced bacterial accumulation and the overall adhesion of films under mild flow conditions.[67] Figure 10.10a compares the accumulation of *Staphylococcus aureus* on a PTFE substrate with a lubricant-impregnated surface. The lubricant-impregnated surface has 97.2% less bacteria after 48 h under flow. Similar reduction has been observed for *Escherichia coli*,[68] *Pseudomonas aeruginosa*,[68,69] and *Chlorella vulgaris*[70] as well as bacteria of the genus *Desulfovibrio*.[71]

Selective cell-repellency, which is important in biosensing and microfluidics, has also benefitted from liquid-impregnated surfaces. Ueda and Levkin have repelled cells into well-defined regions by patterning liquid-impregnated regions of a substrate.[72] Figure 10.10b shows fluorescent human cervical tumour cells separated by lubricant-impregnated regions in which cells cannot attach.

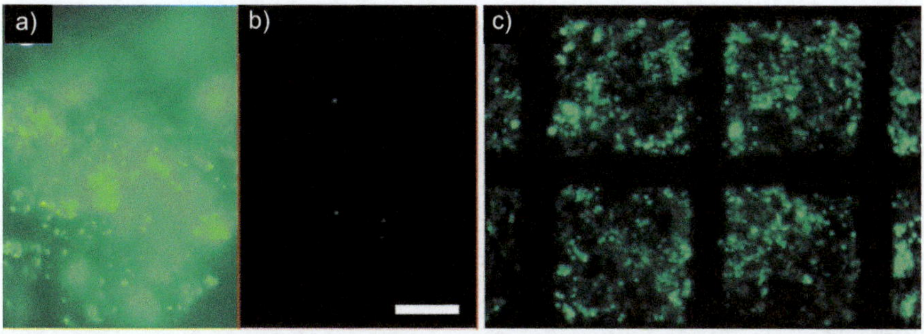

Figure 10.10 *Staphylococcus aureus* bacteria attachment (a) to a PTFE substrate and (b) to a lubricant-impregnated surface. Scale bar is 30 μm. (c) Selective repellency of human cervical tumour cells from the hydrophobic liquid barriers and preferentially attachment to the square hydrophilic areas. The width of each square well is 500 μm. Reproduced with permission from Epstein, A. K., Wong, T.-S., Belisle, R. A., Boggs, E. M. & Aizenberg, J. Liquid-infused structured surfaces with exceptional anti-biofouling performance. *Proc. Natl. Acad. Sci.* (2012). doi:10.1073/pnas.1201973109.[67] (c) Reproduced from ref. 72 with permission from John Wiley and Sons. Copyright © 2013 Wiley-VCH Verlag GmbH & Co. KGaA, Weinheim.

In marine vessels fouling can also occur from plants and animals that adhere to the hull. Xiao *et al.* investigate the attachment and adhesion of motile spores of the seaweed *Ulva linza*.[69] After 2 h, the number of *U. linza* spores that had attached to a Krytox-impregnated surface is significantly lower than for a control glass surface. The spores that attach have comparable adhesion strength to those on glass, however, as indicated by their resilience in staying attached under a shear water flow. In separate assays, these authors also demonstrated that the coverage of the larvae of the barnacle *Balanus amphitrite* was up to an order of magnitude less on Krytox-impregnated surfaces than on glass or polystyrene.

10.3.3.3 Scale Fouling

Fouling and corrosion of heat exchangers, oil and gas pipelines, and turbine systems lead to increased maintenance and losses in production. For heat exchanger scaling alone, the costs associated with operational losses, energy requirements, and maintenance is on the order of 0.25% of the GDP of industrialized countries.[73] Mechanical and chemical cleaning methods are economically or environmentally expensive. Low surface energy coatings have been shown to provide a passive route for limiting the fouling of surfaces but lack robustness in harsh conditions.

Liquid-impregnated surfaces are promising as a robust alternative to other low surface energy coatings. Reducing the nucleation rate of scale requires a lower density of nucleation sites and a high activation barrier. A low surface tension liquid entrained within the solid texture offers a molecularly smooth surface with a large activation barrier that can be used to lower the nucleation rate. Figure 10.11a and b shows a comparison between an untreated surface and a silicone oil-impregnated surface immersed in a gypsum (calcium sulfate) solution for >3 days.[74] Subramanyam *et al.* showed that an optimal design with respect to the liquid surface tension and spreading coefficient can result in up to ten times lower scale formation on the impregnated surface compared to an untreated surface.[74] They also show that lubricant-impregnated surfaces can be fabricated on steel, which is used in industrial applications. Charpentier *et al.* extended this work to show a ten times decrease in scale deposits of calcium carbonate on liquid-impregnated surfaces.[75]

The corrosion of a material can also be slowed down with a liquid-impregnated surface. The use of superhydrophobic surfaces in marine environments to lower corrosion rates has limited success because, over time, the trapped air is lost. Figure 10.11c compares the corrosion of bare steel, hydrophobic steel, and a liquid-impregnated steel with varying amounts of infused liquid after 3 days of immersion in 3.5% NaCl solution.[76] On

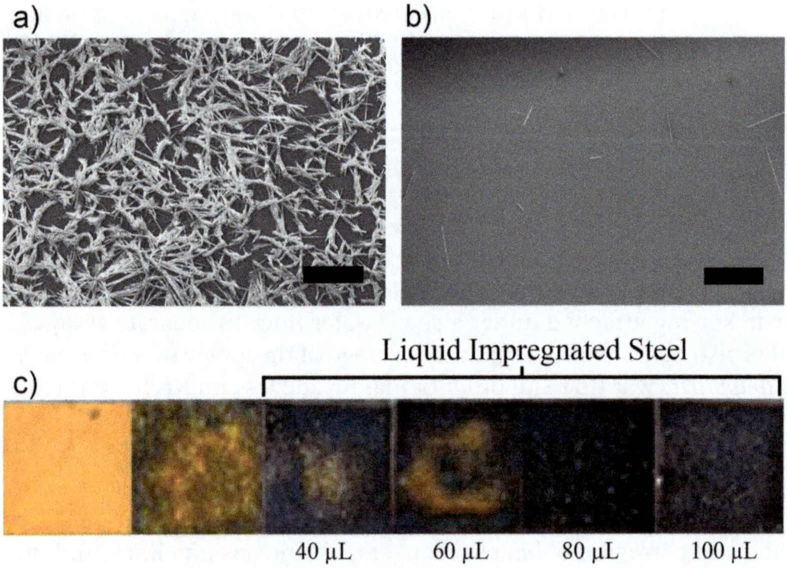

Figure 10.11 Gypsum scale formation after 80 h in a salt solution on (a) an untreated smooth silicon surface (b) a silicone oil-impregnated surface. Scale bar is 1 mm. (c) The corrosion of low alloy steel after 3 days of immersion in 3.5 wt% NaCl solution. From left to right: bare steel, hydrophobic steel, lubricant-impregnated steel where the volume of lubricant is varied. Reproduced from ref. 74 with permission from John Wiley and Sons. Copyright © Wiley-VCH Verlag GmbH & Co. KGaA, Weinheim, and reprinted from *Applied Surface Science*, **328**, S Yang *et al.*, Slippery liquid-infused porous surface based on perfluorinated lubricant/iron tetradecanoate: Preparation and corrosion protection application, 491–500, Copyright 2015 with permission from Elsevier.[76]

the liquid-impregnated surface, the perfluoropolyether lubricant acts as a protective layer and lowers the corrosion rate. Song *et al.* further quantify this corrosion resistance using electrochemical impedance spectroscopy to show that lubricant-impregnated surfaces do not corrode even after 76 days immersed in 3.5 wt% NaCl.[77,78]

10.3.4 Fluid Mobility

A drop placed on a lubricant-impregnated surface is mobile at very low angles. Smith *et al.* concluded that drops roll on lubricant-impregnated surfaces rather than slide, by balancing shear forces in the lubricant film beneath a drop and those within a drop.[5] Looking at the lubricant–working fluid interface beneath a drop on a lubricant-impregnated surface, the shear forces on the lubricant side scale as $\mu_o V_i/t$ where μ_o is the lubricant viscosity, V_i the interface velocity, and t the lubricant film thickness beneath the drop. The shear stress on the working fluid side scales as $\mu_w(V - V_i)/h_{cm}$ where μ_w is the working fluid viscosity, V the velocity of the centre of mass of the drop, and h_{cm} the height of the centre of mass. At the interface the shear stresses must balance, giving $V_i \sim V(1 + \mu_o h_{cm}/\mu_w t)^{-1}$. For the experiments of Smith *et al.* $V_i \ll V$, which indicates that the drop rolls. They confirm the rolling motion of the drop using tracer particles (Figure 10.12c).

Next, the speed of a drop in steady state on an inclined lubricant-impregnated surface is determined by balancing gravitational, pinning, and viscous forces. There are three possible regions of viscous dissipation resisting a drop's motion: in the rolling drop, in the lubricant beneath the drop, and

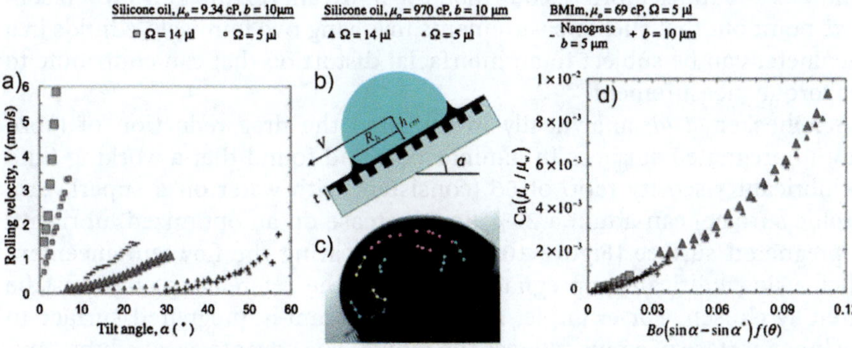

Figure 10.12 (a) Measured velocities of water droplets as a function of substrate tilt angle for various lubricant viscosities, textures, and drop sizes. (b) Schematic of a water droplet moving on a lubricant-impregnated surface showing the various parameters of consideration. (c) Trajectories of tracer particles measured relative to the water droplet reveal that the drop rolls rather than slides on LIS. (d) Non-dimensional plot collapsing the datasets shown in (a) onto a single curve. Reproduced from ref. 5 with permission from the Royal Society of Chemistry.

in the wetting ridge. For a water drop on a silicone oil-impregnated surface with no excess film, Smith *et al.* show that dissipation in the wetting ridge is the most dominant term and explains observed velocities for a wide range of lubricants, tilt angles, and drop sizes (Figure 10.12d).[5]

Abstracting away from drop level experiments, introducing a slippery surface has many applications in established industries. Slip in the walls of a pipe, for example, will increase the flow rate or decrease the power required to pump a fluid.

At a fluid–solid interface the most universally accepted boundary condition is no-slip, meaning the velocity of the fluid and solid must be matched. The no-slip condition has been experimentally validated under most normal flow conditions.[79] The interface between a lubricant-impregnated surface and working fluid requires special consideration, however, because the working fluid interfaces with either only the lubricant or a combination of lubricant and solid. In such a situation an apparent slip can arise although no-slip can still hold microscopically.

Such drag reduction has been studied for superhydrophobic surfaces by modelling the air–working fluid interface as shear free. Several studies have extended such work to incorporate the viscosity of air into models.[80–82]

While lubricant-impregnated surfaces are more stable than superhydrophobic surfaces and can repel a wider variety of liquids, lubricant-impregnated surfaces should provide less drag reduction than superhydrophobic surfaces since the viscosity of the lubricant is greater than that of air. Solomon *et al.* use a rheometer to measure the drag reduction on lubricant-impregnated surfaces in laminar flow with varying working fluid viscosities and attained a drag reduction of 16% in a 1 mm geometry when the working fluid was 260 times more viscous than the lubricant (Figure 10.13a).[83] Jacobi *et al.* point out that such measurements involving two immiscible fluids in a rheometer can be subject to an interfacial distortion that can contribute to the torque measurement.[84]

Schönecker *et al.* analytically investigated the drag reduction of lubricant-impregnated surfaces in laminar flow and found that a working fluid to lubricant viscosity ratio of 56 (consistent with water on a superhydrophobic surface) can attain a 20% flow increase on an optimized lubricant-impregnated surface (Figure 10.14). In comparing the flow enhancement when using lubricant-impregnated surfaces, the reference plane must be carefully chosen. For example, adding a lubricant-impregnated surface to the inner surface of a pipe reduces the radius. The advantages of a lubricant-impregnated surface would have to be greater than the added resistance of a decreased pipe radius.

Rosenberg *et al.* extended findings on drag reduction to higher Reynolds numbers.[85] Using a Taylor–Couette geometry they measure 10% drag reduction on superhydrophobic surfaces and 14% drag reduction on a heptane-impregnated surface which was constant over the Reynolds number range 7000–9000 (Figure 10.14b). Other work suggests that drag reduction can also be achieved in turbulent flows on lubricant-impregnated surfaces,[86] as has

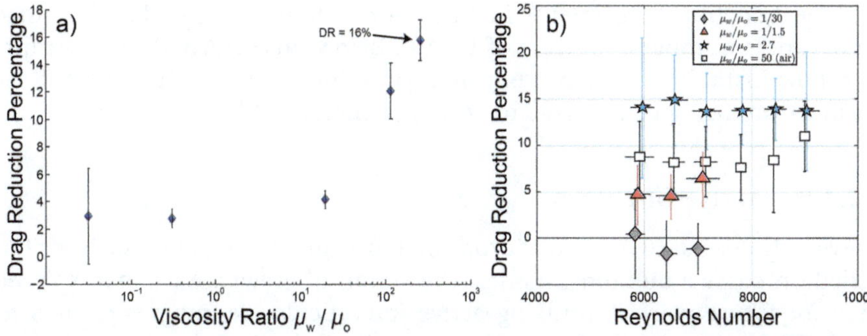

Figure 10.13 (a) Plot of the drag reduction percentage *vs.* the ratio of the working fluid to lubricant viscosity for a laminar flow. Mixtures of water and glycerol are used as the working fluid and a laser-textured sample impregnated with silicone oil is used as the lubricant-impregnated surface. Experiments are conducted in a parallel plate geometry. (b) Plot of drag reduction percentage *vs.* Reynolds number for a variety of working fluid to lubricant viscosity ratios at higher Reynolds numbers. The least viscous lubricant heptane gives a drag reduction percentage of 14%. Experiments are conducted in a Taylor–Couette geometry. Reproduced with permission from Solomon, B. R., Khalil, K. S. & Varanasi, K. K. Drag Reduction using Lubricant-Impregnated Surfaces in Viscous Laminar Flow. *Langmuir* **30**, 10970–10976 (2014).[83] Copyright (2014) American Chemical Society. Reproduced from ref. 85 with permission from AIP Publishing.

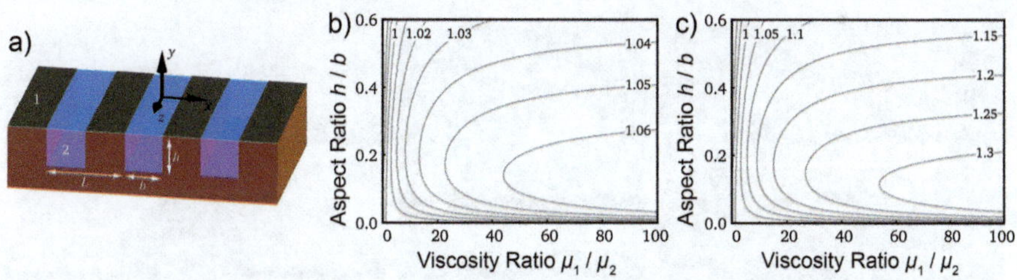

Figure 10.14 (a) Schematic of the analysed lubricant-impregnated surface. Influence of the viscosity ratio on the enhancement factor (given by contours) for longitudinal flow over open grooves with $b/L = 0.98$ where the ratio between the period of the grooves and radius of the channel is (b) 0.02 and (c) 0.1. The enhancement factor is the increased flow rate provided by adding a lubricant-impregnated surface compared to a no-slip condition. C. Schönecker *et al.*, Influence of the enclosed fluid on the flow over a microstructured surface in the Cassie state, *The Journal of Fluid Mechanics*, **740**, 168–195, reproduced with permission.[82]

been measured and rationalized on superhydrophobic surfaces.[87] Jackson points out using simulations that confining lubricants on small scales may lead to additional drag reduction benefits.[88] Wang *et al.* set up an alternate way of measuring drag by spraying surfaces with a controlled flow to measure up to a 7% decrease in drag force on a lubricant-impregnated surfaces.[59]

Drag reduction experiments in Taylor–Couette flows should be extended to Reynolds numbers in excess of 1×10^4. Below this critical Reynolds number, flow instabilities are present in Taylor–Couette flows that prevent the results from being generalized to other geometries.[87,89]

10.3.5 Active Surfaces

Previously discussed motion of drops on lubricant-impregnated surfaces has relied on passive methods, namely the action of gravity. An underexplored area for future work is utilizing active force fields (*i.e.* magnetic, electric, thermal) to manipulate drops. Chen *et al.* have demonstrated that the low hysteresis of lubricant-impregnated surfaces enables drops with magnetic particles to be moved with a magnetic field.[90] More recently, Khalil *et al.* (Figure 10.15a) have created a lubricant-impregnated surface where the lubricant is a superparamagnetic ferrofluid.[91] The lubricant is designed to cloak over drops deposited on the surface, allowing the drops to be manipulated without introducing magnetic particles directly into them. This technique also allows for a wide variety of fluids to be manipulated as long as the lubricant chosen can cloak the working fluid.

Electrowetting, which modifies wetting properties by applying a voltage bias, has emerged as a versatile tool to manipulate droplets of various sizes in a controlled fashion and has been applied to electronic displays, energy

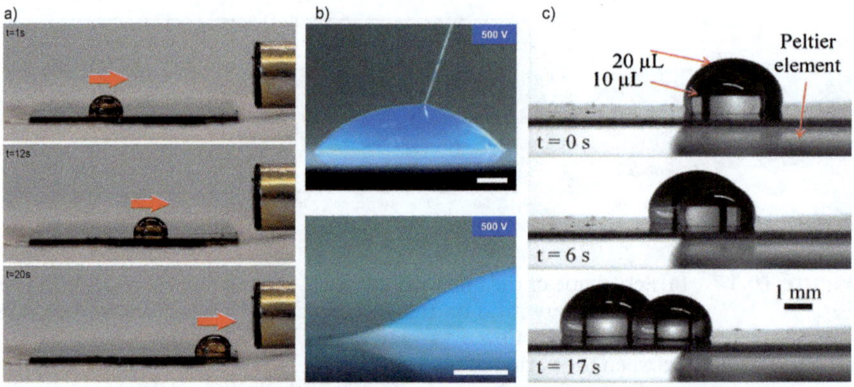

Figure 10.15 (a) A water droplet on a surface impregnated with a ferrofluid moves in response to a magnet. (b) Images of a water drop electrowetted on a lubricant-impregnated surface with an applied voltage of 500 V displaying an apparent contact angle (CA) of ~53°. The bottom image shows the wetting ridge that forms. The scale bar is 400 μm on the top image and 200 μm on the bottom. (c) Simultaneous thermocapillary motion of 10 and 20 μL droplets on a surface impregnated with silicone oil. Reproduced from ref. 91 with permission of AIP Publishing. Reproduced from ref. 95 under the CC BY 4.0 licence. Reproduced from ref. 98 with permission from John Wiley and Sons. Copyright © 2014 Wiley-VCH Verlag GmbH & Co. KGaA, Weinehim.

generation, and microfluidic systems. In a typical system, a voltage between a conducting drop and substrate covered with a dielectric film is applied. Upon application of the voltage, the contact angle of a water drop decreases.[92] When the voltage is removed, contact angle hysteresis can prevent the drop from recovering. Also, severe and uncontrollable droplet oscillations are often encountered. Lubricant-impregnated surfaces have been shown to reduce these undesirable effects as well as reduce any contact angle hysteresis to minimize the reversibility issues, as shown in Figure 10.15b.[47,93–97] Barman *et al.* have also demonstrated that electrowetting two adjacent water drops on lubricant-impregnated surfaces decreases the time it takes them to coalesce by an order of magnitude.[47]

Because liquid surfaces provide extremely low contact angle hysteresis, thermocapillarity can drive the drop along surfaces as shown in Figure 10.15c.[98] Eifert *et al.* remark that the motion can have contributions from thermocapillary forces in the working fluid drop as well as bulk motion of the lubricant but deduce that the motion must be due to thermocapillary forces within the working fluid drop by varying the drop size. Drops on lubricant-impregnated surfaces have also been controlled by changing the local wettability of lubricant-impregnated surfaces.[21,99] The techniques mentioned promise useful for the fabrication of microfluidic designs where a pre-existing microchannel design is not needed.

10.3.6 Optics

Coupling transparency and slipperiness is important for applications ranging from solar modules to commercial eyewear. Superhydrophobic surfaces comprising subwavelength features in transparent materials have shown high transparency while being non-wetting.[100] Lubricant-impregnated surfaces also enable highly transparent and slippery surfaces by reducing the refractive index contrast at the lubricant–air interface in comparison to the original solid–air interface. Several researchers have reported enhancement in broadband optical transmission compared to textured surfaces.[23,26,38,59,90,101,102] Vogel *et al.* demonstrate that the optical transmission through lubricant-impregnated surfaces can surpass that through a glass slide.[23]

Manabe *et al.* point out that even for surfaces with nanoscale features where total transmittance is similar with and without impregnation, lubricant-impregnated surfaces significantly reduce the amount of light scattered.[58] Yao *et al.* impregnated an elastic matrix with a lubricant and showed that the optical transmission can be tuned by deforming the lubricant-impregnated surface.[103]

10.3.7 Infused Gels

A related technology is achieved when for example silicone oil is absorbed by cross-linked solid poly(dimethylsiloxane) (PDMS). The resulting organogel shows slippery properties similar to liquid-impregnated surfaces owing to the absorbed silicone oil, but unlike a lubricant-impregnated surface the solid texture does not stabilize the silicone oil. Such an infused gel has confusingly been called a lubricant-impregnated surface, but its principle of construction and operation is different from lubricant-impregnated surfaces as described earlier in this chapter.

Surfaces where an organic liquid (*e.g.* silicone oil or Krytox) is infused into a polymer are referred to as organogels.[104–112] Analogously, water can be absorbed by hygroscopic polymers to form a hydrogel.[113–115] It is also possible to create ionic liquid-infused polymer gels.[116,117]

Liquid-infused gels exhibit many similar properties to lubricant-impregnated surfaces including low roll-off of drops,[106,109,110,112,117] anti-icing,[105,113–115,118] enhanced condensation,[119] and anti-biofouling.[107] For example MacCallum *et al.* prevent biofilm accumulation by infusing silicone oil into silicone tubing. They demonstrate by flowing a cultured bacteria solution that the infused silicone tube is devoid of *P. aeruginosa* cells while the control has a large amount of accumulation.[107] Leslie *et al.* also report an infused organogel that prevents thrombosis.[108] Chen *et al.* have also created a hydrogel gel surface and show ice adheres more weakly than superhydrophilic, superhydrophobic, and flat surfaces (Figure 10.16).[113]

10.3.8 Durability

Lubricant-impregnated surfaces show promise in a wide range of applications. For these surfaces to become practically relevant, their durability and robustness should be carefully tested under conditions required by the applications. For example while drop impact has been investigated on

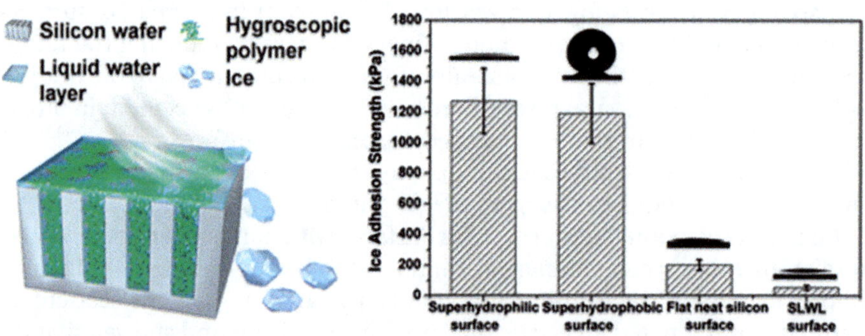

Figure 10.16 Depiction of a microporous silicon filled with a hydrogel. The self-lubricating liquid water layer (SLWL) of the hydrogel surface adheres ice more weakly than superhydrophilic, superhydrophobic, and flat surfaces. Reproduced with permission from Chen, J. *et al.* Robust Prototypical Anti-icing Coatings with a Self-lubricating Liquid Water Layer between Ice and Substrate. *ACS Appl. Mater. Interfaces* 5, 4026–4030 (2013).[113] Copyright (2013) American Chemical Society.

lubricant-impregnated surfaces,[120,121] under high enough impact conditions an impinging water drop will displace the lubricant and cause the surface to fail. Also, lubricant can be lost from frosting[122] and other phase transitions, but a lubricant source that can replenish the surface over time could overcome this challenge.

In addition, in shear flow Howell *et al.* report that Krytox-impregnated surfaces remain stable under shear flow.[123] However, the work of Wexler, Jacobi, and Stone revealed two mechanisms by which lubricant-impregnated surfaces can fail. In the first, shear forces by an imposed flow can overcome the capillary forces holding the lubricant in place, but cleverly placing non-wetting regions prevent this phenomena from occurring (Figure 10.16a).[124,125] Also, a flow can cause the lubricant to overflow and eventually leave the surface (Figure 10.16b).[126]

As with superhydrophobic surfaces, lubricant-impregnated surfaces are also vulnerable to mechanical failure depending on the strength of the underlying texture. Shillingford *et al.* observe mechanical damage to lubricant-impregnated surfaces made from silica or alumina particles when mechanically abraded that compromises their repellency,[16] hence underlying structures that are robust can overcome this challenge.

LiquiGlide has demonstrated robust lubricant-impregnated surfaces to overcome these durability challenges and has commercialized this technology (see Figure 10.17).[127]

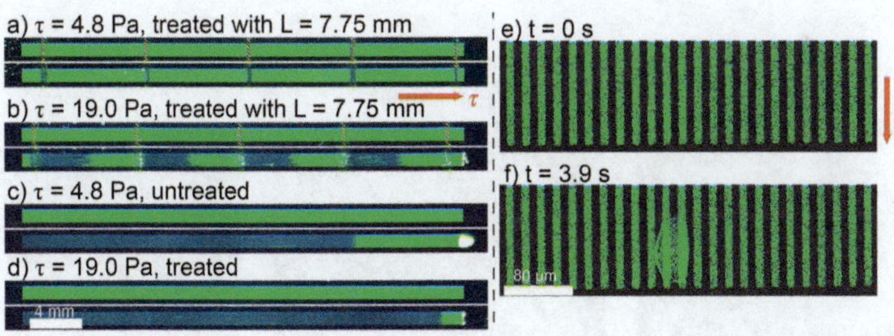

Figure 10.17 Experiments in which a water–glycerol mixture flows over grooves filled with a fluorescent green lubricant. The red arrow indicates the flow direction. The grooves are 9 μm wide, 10 μm deep, and 35 mm long. (a–d) Regions with periodicity L that are non-wetting to the lubricant interrupt grooves aligned with the flow and in (a) prevent drainage of the lubricant. For each experiment, the top image shows the initial state and the bottom the steady state. Drained portions of the groove reflect blue light while excess lubricant appears white. The strength of the imposed flow is characterized by a shear stress τ. (e,f) In a separate set of experiments, at a sufficient stress the flow over lubricant filled groups induces overflow of the lubricant at a shear stress of $\tau = 2.58$ Pa. Reproduced from ref. 125 with permission from The Royal Society of Chemistry. Reprinted from ref. 126 with the permission of AIP Publishing.

10.4 Conclusion and Outlook

Researchers have also begun creating systems iterating on lubricant-impregnated surface technologies to extend their use in new ways. Hou *et al.* have created a liquid-impregnated surface that functions as a membrane to selectively pass gases and liquids,[128] while Sun *et al.* have created a surface that secretes antifreeze upon contact with ice.[129] Dai *et al.* report on a slippery surface comprised of a thin layer of oil trapped in nanoscale textures on a surface of microscale textures. A water drop placed on the surface impales into the texture in a Wenzel-like state but is surprisingly still highly mobile,[130] similar to liquid marbles. McHale and Newton hypothesised that it should be possible to fabricate a liquid-impregnated surface around a liquid drop that would serve microfluidic applications.[131]

Lubricant-impregnated surfaces are a versatile platform that show promise in condensation, anti-icing, anti-fouling, fluid mobility, optics, and drop control. In each application, the design of a lubricant-impregnated surface has particularly relevant criteria of which some are not intuitive. Lubricant-impregnated surfaces can reach commercial applications where superhydrophobic surfaces have fallen short, and may also benefit a multitude of yet undiscovered applications. For example, using the thermodynamic framework,[5] LiquiGlide has designed robust lubricant-impregnated surface coatings and has recently commercialized the technology for manufacturing applications (Figure 10.18).[127]

Figure 10.18 Comparison of paint dispensing from a 100 gallon (~400 L) paint mixing tank without (top row) and with (bottom row) the LiquiGlide LIS coating. LiquiGlide coatings demonstrate complete dispensing of the product, saving significant yield loss and reduction in wash water required to clean such tanks. Figure courtesy of LiquiGlide Inc.[127]

References

1. D. Quérék, Non-sticking drops, *Rep. Prog. Phys.*, 2005, **68**, 2495.
2. J. D. Smith, R. Dhiman and K. Varanasi, Liquid-encapsulating surfaces: overcoming the limitations of superhydrophobic surfaces for robust non-wetting and anti-icing surfaces, *64th Annu. Meet. APS Div. Fluid Dyn.*, 2011, **56**, S4.00001.
3. T.-S. Wong, *et al.* Bioinspired self-repairing slippery surfaces with pressure-stable omniphobicity, *Nature*, 2011, **477**, 443–447.
4. A. Lafuma and D. Quéré, Slippery pre-suffused surfaces, *EPL Europhys. Lett.*, 2011, **96**, 56001.
5. J. D. Smith, *et al.* Droplet mobility on lubricant-impregnated surfaces, *Soft Matter*, 2013, **9**, 1772–1780.
6. D. Bonn, J. Eggers, J. Indekeu, J. Meunier and E. Rolley, Wetting and spreading, *Rev. Mod. Phys.*, 2009, **81**, 739–805.
7. J. Zhang, *et al.* Evaporation-induced transition from nepenthes pitcher-inspired slippery surfaces to lotus leaf-inspired superoleophobic surfaces, *Langmuir*, 2014, **30**, 14292–14299.
8. X.-M. Li, D. Reinhoudt and M. Crego-Calama, What do we need for a superhydrophobic surface? A review on the recent progress in the preparation of superhydrophobic surfaces, *Chem. Soc. Rev.*, 2007, **36**, 1350–1368.
9. P. Roach, N. J. Shirtcliffe and M. I. Newton, Progess in superhydrophobic surface development, *Soft Matter*, 2008, **4**, 224–240.
10. F. Schellenberger, *et al.* Direct observation of drops on slippery lubricant-infused surfaces, *Soft Matter*, 2015, **11**, 7617–7626.
11. P. W. Wilson, *et al.* Inhibition of ice nucleation by slippery liquid-infused porous surfaces (SLIPS), *Phys. Chem. Chem. Phys.*, 2012, **15**, 581–585.
12. P. Kim, M. J. Kreder, J. Alvarenga and J. Aizenberg, Hierarchical or Not? Effect of the Length Scale and Hierarchy of the Surface Roughness on Omniphobicity of Lubricant-Infused Substrates, *Nano Lett.*, 2013, **13**, 1793–1799.
13. J. Yang, H. Song, H. Ji and B. Chen, Slippery lubricant-infused textured aluminum surfaces, *J. Adhes. Sci. Technol.*, 2014, **28**, 1949–1957.
14. Y. Tsuruki, M. Sakai, T. Isobe, S. Matsushita and A. Nakajima, Static and dynamic hydrophobicity of alumina-based porous ceramics impregnated with fluorinated oil, *J. Mater. Res.*, 2014, **29**, 1546–1555.
15. Q. Wei, *et al.* Supramolecular Polymers as Surface Coatings: Rapid Fabrication of Healable Superhydrophobic and Slippery Surfaces, *Adv. Mater.*, 2014, 7358–7364.
16. C. Shillingford, N. MacCallum, T.-S. Wong, P. Kim and J. Aizenberg, Fabrics coated with lubricated nanostructures display robust omniphobicity, *Nanotechnology*, 2014, **25**, 014019.

17. M. Raimondo, M. Blosi, A. Caldarelli, G. Guarini and F. Veronesi, Wetting behavior and remarkable durability of amphiphobic aluminum alloys surfaces in a wide range of environmental conditions, *Chem. Eng. J.*, 2014, **258**, 101–109.
18. Y. Takada, M. Sakai, T. Isobe, S. Matsushita and A. Nakajima, Preparation and hydrophobicity of solid–liquid bulk composite using porous glass and fluorinated oil, *J. Mater. Sci.*, 2015, **50**, 7760–7769.
19. Q. Liu *et al.*, Durability of a lubricant-infused Electrospray Silicon Rubber surface as an anti-icing coating. *Appl. Surf. Sci.*, 2015, **346**, 68–76.
20. X. Huang, J. D. Chrisman and N. S. Zacharia, Omniphobic Slippery Coatings Based on Lubricant-Infused Porous Polyelectrolyte Multilayers, *ACS Macro Lett.*, 2013, **2**, 826–829.
21. U. Manna and D. M. Lynn, Fabrication of Liquid-Infused Surfaces Using Reactive Polymer Multilayers: Principles for Manipulating the Behaviors and Mobilities of Aqueous Fluids on Slippery Liquid Interfaces, *Adv. Mater.*, 2015, **27**, 3007–3012.
22. S. Sunny, N. Vogel, C. Howell, T. L. Vu and J. Aizenberg, Lubricant-Infused Nanoparticulate Coatings Assembled by Layer-by-Layer Deposition, *Adv. Funct. Mater.*, 2014, 6658–6667.
23. N. Vogel, R. A. Belisle, B. Hatton, T.-S. Wong and J. Aizenberg, Transparency and damage tolerance of patternable omniphobic lubricated surfaces based on inverse colloidal monolayers, *Nat. Commun.*, 2013, **4**.
24. E. Jenner and B. D'Urso, Wetting states on structured immiscible liquid coated surfaces, *Appl. Phys. Lett.*, 2013, **103**.
25. M. Röhrig, *et al.* Nanofur for Biomimetic Applications, *Adv. Mater. Interfaces*, 2014, **1**, 1300083.
26. P. Zhang, H. Chen, L. Zhang, T. Ran and D. Zhang, Transparent self-cleaning lubricant-infused surfaces made with large-area breath figure patterns, *Appl. Surf. Sci.*, 2015, **355**, 1083–1090.
27. R. Qiu *et al.*, Fabrication of slippery liquid-infused porous surface based on carbon fiber with enhanced corrosion inhibition property, *Colloids Surf. A*, 2014, **453**, 132–141.
28. J. Bico, U. Thiele and D. Quéré, Wetting of textured surfaces, *Colloids Surf., A*, 2002, **206**, 41–46.
29. J. Seiwert, C. Clanet and D. Quéré, Coating of a textured solid, *J. Fluid Mech.*, 2011, **669**, 55–63.
30. K. Rykaczewski, T. Landin, M. L. Walker, J. H. J. Scott and K. K. Varanasi, Direct Imaging of Complex Nano- to Microscale Interfaces Involving Solid, Liquid, and Gas Phases, *ACS Nano*, 2012, **6**, 9326–9334.
31. S. Anand, K. Rykaczewski, S. B. Subramanyam, D. Beysens and K. K. Varanasi, How droplets nucleate and grow on liquids and liquid impregnated surfaces, *Soft Matter*, 2015, **11**, 69–80.
32. F. Schellenberger, Direct observation of drops on slippery lubricant-infused surfaces, *Soft Matter*, 2015, **11**, 7617–7626.

33. Y. Cheng *et al.*, Direct three-dimensional imaging of polymer–water interfaces by nanoscale hard X-ray phase tomography, *Soft Matter*, 2014, **10**, 2982–2990.
34. V. Bergeron and D. Langevin, Monolayer Spreading of Polydimethylsiloxane Oil on Surfactant Solutions, *Phys. Rev. Lett.*, 1996, **76**, 3152–3155.
35. M. E. R. Shanahan and A. Carre, Viscoelastic Dissipation in Wetting and Adhesion Phenomena, *Langmuir*, 1995, **11**, 1396–1402.
36. A. Carre, J.-C. Gastel and M. E. R. Shanahan, Viscoelastic effects in the spreading of liquids, *Nature*, 1996, **379**, 432–434.
37. M. J. Neeson, R. F. Tabor, F. Grieser, R. R. Dagastine and D. Y. C. Chan, Compound sessile drops, *Soft Matter*, 2012, **8**, 11042–11050.
38. J. Zhang, A. Wang and S. Seeger, Nepenthes Pitcher Inspired Anti-Wetting Silicone Nanofilaments Coatings: Preparation, Unique Anti-Wetting and Self-Cleaning Behaviors, *Adv. Funct. Mater.*, 2014, **24**, 1074–1080.
39. S. B. Subramanyam, K. Rykaczewski and K. K. Varanasi, Ice Adhesion on Lubricant-Impregnated Textured Surfaces, *Langmuir*, 2013, **29**, 13414–13418.
40. A. Carlson, P. Kim, G. Amberg and H. A. Stone, Short and long time drop dynamics on lubricated substrates, *EPL Europhys. Lett.*, 2013, **104**, 34008.
41. W. H. Wiser, *Energy Resources: Occurrence, Production, Conversion, Use*, Springer Science & Business Media, 2012.
42. J. W. Rose, Dropwise condensation theory and experiment: A review, *Proc. Inst. Mech. Eng., Part A*, 2002, **216**, 115–128.
43. W. M. Rohsenow, J. P. Hartnett and Y. I. Cho, *Handbook of heat transfer*, McGraw-Hill, New York, 1998, vol. 3.
44. S. Anand, A. T. Paxson, R. Dhiman, J. D. Smith and K. K. Varanasi, Enhanced Condensation on Lubricant-Impregnated Nanotextured Surfaces, *ACS Nano*, 2012, **6**, 10122–10129.
45. R. Xiao, N. Miljkovic, R. Enright and E. N. Wang, Immersion Condensation on Oil-Infused Heterogeneous Surfaces for Enhanced Heat Transfer, *Sci. Rep.*, 2013, **3**.
46. J. B. Boreyko, G. Polizos, P. G. Datskos, S. A. Sarles and C. P. Collier, Air-stable droplet interface bilayers on oil-infused surfaces, *Proc. Natl. Acad. Sci.*, 2014, **111**, 7588–7593.
47. J. Barman, R. Pant, A. K. Nagarajan and K. Khare, Electrowetting on dielectrics on lubricating fluid-infused smooth/rough surfaces with negligible hysteresis, *J. Adhes. Sci. Technol.*, 2016, 1–12.
48. A. Zhang, H. Bai and L. Li, Breath Figure: A Nature-Inspired Preparation Method for Ordered Porous Films, *Chem. Rev.*, 2015, **115**, 9801–9868.
49. K. Rykaczewski, *et al.* Dropwise Condensation of Low Surface Tension Fluids on Omniphobic Surfaces, *Sci. Rep.*, 2014, **4**.
50. M. Cao *et al.*, Water-Repellent Properties of Superhydrophobic and Lubricant-Infused 'Slippery' Surfaces: A Brief Study on the Functions and Applications, *ACS Appl. Mater. Interfaces*, 2016, **8**, 3615–3623.

51. K.-C. Park, S. S. Chhatre, S. Srinivasan, R. E. Cohen and G. H. McKinley, Optimal Design of Permeable Fiber Network Structures for Fog Harvesting, *Langmuir*, 2013, **29**, 13269–13277.
52. B. S. Lalia, S. Anand, K. K. Varanasi and R. Hashaikeh, Fog-Harvesting Potential of Lubricant-Impregnated Electrospun Nanomats, *Langmuir*, 2013, **29**, 13081–13088.
53. L. Cao, A. K. Jones, V. K. Sikka, J. Wu and D. Gao, Anti-Icing Superhydrophobic Coatings, *Langmuir*, 2009, **25**, 12444–12448.
54. S. A. Kulinich and M. Farzaneh, Ice adhesion on super-hydrophobic surfaces, *Appl. Surf. Sci.*, 2009, **255**, 8153–8157.
55. K. K. Varanasi, T. Deng, J. D. Smith, M. Hsu and N. Bhate, Frost formation and ice adhesion on superhydrophobic surfaces, *Appl. Phys. Lett.*, 2010, **97**, 234102.
56. P. Kim, *et al.* Liquid-Infused Nanostructured Surfaces with Extreme Anti-Ice and Anti-Frost Performance, *ACS Nano*, 2012, **6**, 6569–6577.
57. K. Rykaczewski, S. Anand, S. B. Subramanyam and K. K. Varanasi, Mechanism of Frost Formation on Lubricant-Impregnated Surfaces, *Langmuir*, 2013, **29**, 5230–5238.
58. K. Manabe, S. Nishizawa, K.-H. Kyung and S. Shiratori, Optical Phenomena and Antifrosting Property on Biomimetics Slippery Fluid-Infused Antireflective Films *via* Layer-by-Layer Comparison with Superhydrophobic and Antireflective Films, *ACS Appl. Mater. Interfaces*, 2014, **6**, 13985–13993.
59. Y. Wang, H. Zhang, X. Liu and Z. Zhou, Slippery liquid-infused substrates: a versatile preparation, unique anti-wetting and drag-reduction effect on water, *J. Mater. Chem. A*, 2016, **4**, 2524–2529.
60. X. Yin, *et al.* Integration of Self-Lubrication and Near-Infrared Photothermogenesis for Excellent Anti-Icing/Deicing Performance, *Adv. Funct. Mater.*, 2015, **25**, 4237–4245.
61. R. Blossey, Self-cleaning surfaces—virtual realities, *Nat. Mater.*, 2003, **2**, 301–306.
62. A. Lafuma and D. Quéré, Slippery pre-suffused surfaces, *Europhys. Lett.*, 2011, **96**, 56001.
63. T.-S. Wong, *et al.* Bioinspired self-repairing slippery surfaces with pressure-stable omniphobicity, *Nature*, 2011, **477**, 443–447.
64. J. H. Guan *et al.,* Evaporation of Sessile Droplets on Slippery Liquid-Infused Porous Surfaces (SLIPS), *Langmuir*, 2015, **31**, 11781–11789.
65. S. Yang, X. Dai, B. B. Stogin and T. S. Wong, Ultrasensitive surface-enhanced Raman scattering detection in common fluids, *Proc. Natl. Acad. Sci.*, 2016, **113**, 268–273.
66. A. B. Tesler, *et al.* Extremely durable biofouling-resistant metallic surfaces based on electrodeposited nanoporous tungstite films on steel, *Nat. Commun.*, 2015, **6**, 8649.
67. A. K. Epstein, T.-S. Wong, R. A. Belisle, E. M. Boggs and J. Aizenberg, Liquid-infused structured surfaces with exceptional anti-biofouling performance, *Proc. Natl. Acad. Sci.*, 2012, **109**, 13182–13187.

68. A. K. Epstein, T.-S. Wong, R. A. Belisle, E. M. Boggs and J. Aizenberg, Liquid-infused structured surfaces with exceptional anti-biofouling performance, *Proc. Natl. Acad. Sci.*, 2012, **109**, 13182–13187.
69. L. Xiao, *et al.* Slippery Liquid-Infused Porous Surfaces Showing Marine Antibiofouling Properties, *ACS Appl. Mater. Interfaces*, 2013, **5**, 10074–10080.
70. P. Wang, D. Zhang and Z. Lu, Slippery liquid-infused porous surface bio-inspired by pitcher plant for marine anti-biofouling application, *Colloids Surf., B*, 2015, **136**, 240–247.
71. P. Wang, D. Zhang, Z. Lu and S. Sun, Fabrication of Slippery Lubricant-Infused Porous Surface for Inhibition of Microbially Influenced Corrosion, *ACS Appl. Mater. Interfaces*, 2016, **8**, 1120–1127.
72. E. Ueda and P. A. Levkin, Micropatterning Hydrophobic Liquid on a Porous Polymer Surface for Long-Term Selective Cell-Repellency, *Adv. Healthcare Mater.*, 2013, **2**, 1425–1429.
73. R. Steinhagen, H. Müller-Steinhagen and K. Maani, Problems and costs due to heat exchanger fouling in New Zealand industries, *Heat Transfer Eng.*, 1993, **14**, 19–30.
74. S. B. Subramanyam, G. Azimi and K. K. Varanasi, Designing Lubricant-Impregnated Textured Surfaces to Resist Scale Formation, *Adv. Mater. Interfaces*, 2014, **1**, 1300068.
75. T. V. J. Charpentier, *et al.* Liquid infused porous surfaces for mineral fouling mitigation, *J. Colloid Interface Sci.*, 2015, **444**, 81–86.
76. S. Yang, *et al.* Slippery liquid-infused porous surface based on perfluorinated lubricant/iron tetradecanoate: Preparation and corrosion protection application, *Appl. Surf. Sci.*, 2015, **328**, 491–500.
77. T. Song, *et al.* Multiple sheet-layered super slippery surfaces based on anodic aluminium oxide and its anticorrosion property, *RSC Adv.*, 2015, **5**, 70080–70085.
78. T. Song, *et al.* Fabrication of super slippery sheet-layered and porous anodic aluminium oxide surfaces and its anticorrosion property, *Appl. Surf. Sci.*, 2015, **355**, 495–501.
79. J. P. Rothstein, Slip on Superhydrophobic Surfaces, *Annu. Rev. Fluid Mech.*, 2010, **42**, 89–109.
80. C. Ybert, C. Barentin, C. Cottin-Bizonne, P. Joseph and L. Bocquet, Achieving large slip with superhydrophobic surfaces: Scaling laws for generic geometries, *Phys. Fluids*, 2007, **19**, 123601.
81. A. Busse, N. D. Sandham, G. McHale and M. I. Newton, Change in drag, apparent slip and optimum air layer thickness for laminar flow over an idealised superhydrophobic surface, *J. Fluid Mech.*, 2013, **727**, 488–508.
82. C. Schönecker, T. Baier and S. Hardt, Influence of the enclosed fluid on the flow over a microstructured surface in the Cassie state, *J. Fluid Mech.*, 2014, **740**, 168–195.
83. B. R. Solomon, K. S. Khalil and K. K. Varanasi, Drag Reduction using Lubricant-Impregnated Surfaces in Viscous Laminar Flow, *Langmuir*, 2014, **30**, 10970–10976.

84. I. Jacobi, *et al.* Stratified thin-film flow in a rheometer, *Phys. Fluids*, 2015, **27**, 052102.
85. B. J. Rosenberg, T. V. Buren, M. K. Fu and A. J. Smits, Turbulent drag reduction over air- and liquid- impregnated surfaces, *Phys. Fluids*, 2016, **28**, 015103.
86. M. K. Fu and M. Hultmark, *A Model for Turbulent Drag Reduction Over Liquid Infused Micro-textured Surfaces*, European Drag Reduction and Flow Control Meeting – EDRFCM 2015, Cambridge, UK., March 23–26, 2015.
87. S. Srinivasan, *et al.* Sustainable Drag Reduction in Turbulent Taylor-Couette Flows by Depositing Sprayable Superhydrophobic Surfaces, *Phys. Rev. Lett.*, 2015, **114**, 014501.
88. R. L. A. Jackson, Scale Dependent Simulation of Liquid Lubricated Textured Surfaces, *J. Tribol.*, 2010, **132**, 022001–022001.
89. D. P. Lathrop, J. Fineberg and H. L. Swinney, Transition to shear-driven turbulence in Couette-Taylor flow, *Phys. Rev. A*, 1992, **46**, 6390–6405.
90. L. Chen, A. Geissler, E. Bonaccurso and K. Zhang, Transparent Slippery Surfaces Made with Sustainable Porous Cellulose Lauroyl Ester Films, *ACS Appl. Mater. Interfaces*, 2014, **6**, 6969–6976.
91. K. S. Khalil, S. R. Mahmoudi, N. Abu-dheir and K. K. Varanasi, Active surfaces: Ferrofluid-impregnated surfaces for active manipulation of droplets, *Appl. Phys. Lett.*, 2014, **105**, 041604.
92. F. Mugele and J.-C. Baret, Electrowetting: from basics to applications, *J. Phys.: Condens. Matter*, 2005, **17**, R705–R774.
93. H. J. J. Verheijen and M. W. J. Prins, Reversible Electrowetting and Trapping of Charge: Model and Experiments, *Langmuir*, 1999, **15**, 6616–6620.
94. T. Krupenkin, S. Yang and P. Mach, Tunable liquid microlens, *Appl. Phys. Lett.*, 2003, **82**, 316–318.
95. C. Hao, *et al.* Electrowetting on liquid-infused film (EWOLF): Complete reversibility and controlled droplet oscillation suppression for fast optical imaging, *Sci. Rep.*, 2014, **4**, 6846.
96. E. Bormashenko, R. Pogreb, Y. Bormashenko, R. Grynyov and O. Gendelman, Low voltage reversible electrowetting exploiting lubricated polymer honeycomb substrates, *Appl. Phys. Lett.*, 2014, **104**, 171601.
97. M. G. Pollack, A. D. Shenderov and R. B. Fair, Electrowetting-based actuation of droplets for integrated microfluidics, *Lab Chip*, 2002, **2**, 96–101.
98. A. Eifert, D. Paulssen, S. N. Varanakkottu, T. Baier and S. Hardt, Simple Fabrication of Robust Water-Repellent Surfaces with Low Contact-Angle Hysteresis Based on Impregnation, *Adv. Mater. Interfaces*, 2014, **1**, 1300138.
99. I. You, T. G. Lee, Y. S. Nam and H. Lee, Fabrication of a Micro-omnifluidic Device by Omniphilic/Omniphobic Patterning on Nanostructured Surfaces, *ACS Nano*, 2014, **8**, 9016–9024.
100. J.-G. Kim, *et al.* Multifunctional Inverted Nanocone Arrays for Non-Wetting, Self-Cleaning Transparent Surface with High Mechanical Robustness, *Small*, 2014, **10**, 2487–2494.

101. W. Ma, Y. Higaki, H. Otsuka and A. Takahara, Perfluoropolyether-infused nano-texture: a versatile approach to omniphobic coatings with low hysteresis and high transparency, *Chem. Commun.*, 2013, **49**, 597–599.
102. I. Okada and S. Shiratori, High-Transparency, Self-Standable Gel-SLIPS Fabricated by a Facile Nanoscale Phase Separation, *ACS Appl. Mater. Interfaces*, 2014, **6**, 1502–1508.
103. X. Yao, *et al.* Adaptive fluid-infused porous films with tunable transparency and wettability, *Nat. Mater.*, 2013, **12**, 529–534.
104. H. Liu, P. Zhang, M. Liu, S. Wang and L. Jiang, Organogel-based Thin Films for Self-Cleaning on Various Surfaces, *Adv. Mater.*, 2013, **25**, 4477–4481.
105. L. Zhu, *et al.* Ice-phobic coatings based on silicon-oil-infused polydimethylsiloxane, *ACS Appl. Mater. Interfaces*, 2013, **5**, 4053–4062.
106. A. Eifert, D. Paulssen, S. N. Varanakkottu, T. Baier and S. Hardt, Simple Fabrication of Robust Water-Repellent Surfaces with Low Contact-Angle Hysteresis Based on Impregnation, *Adv. Mater. Interfaces*, 2014, 1300138.
107. N. MacCallum, *et al.* Liquid-Infused Silicone As a Biofouling-Free Medical Material, *ACS Biomater. Sci. Eng.*, 2014, **1**, 43–51.
108. D. C. Leslie, *et al.* A bioinspired omniphobic surface coating on medical devices prevents thrombosis and biofouling, *Nat. Biotechnol.*, 2014, **32**, 1134–1140.
109. J. Cui, D. Daniel, A. Grinthal, K. Lin and J. Aizenberg, Dynamic polymer systems with self-regulated secretion for the control of surface properties and material healing, *Nat. Mater.*, 2015, **14**, 790–795.
110. C. Urata, G. J. Dunderdale, M. W. England and A. Hozumi, Self-lubricating organogels (SLUGs) with exceptional syneresis-induced anti-sticking properties against viscous emulsions and ices, *J. Mater. Chem. A*, 2015, **3**, 12626–12630.
111. V. G. Damle, *et al.* 'Insensitive' to Touch: Fabric-Supported Lubricant-Swollen Polymeric Films for Omniphobic Personal Protective Gear, *ACS Appl. Mater. Interfaces*, 2015, **7**, 4224–4232.
112. L. Wang and T. J. McCarthy, Covalently Attached Liquids: Instant Omniphobic Surfaces with Unprecedented Repellency, *Angew. Chem., Int. Ed.*, 2016, **55**, 244–248.
113. J. Chen, *et al.* Robust Prototypical Anti-icing Coatings with a Self-lubricating Liquid Water Layer between Ice and Substrate, *ACS Appl. Mater. Interfaces*, 2013, **5**, 4026–4030.
114. J. Chen, Z. Luo, Q. Fan, J. Lv and J. Wang, Anti-Ice coating inspired by ice skating, *Small*, 2014, **10**, 4693–4699.
115. R. Dou, *et al.* Anti-icing Coating with an Aqueous Lubricating Layer, *ACS Appl. Mater. Interfaces*, 2014, **6**, 6998–7003.
116. Y. Ding, *et al.* Ionic-Liquid-Gel Surfaces Showing Easy-Sliding and Ultra-durable Features, *Adv. Mater. Interfaces*, 2015, **2**.

117. D. F. Miranda, *et al.* Physically and chemically stable ionic liquid-infused textured surfaces showing excellent dynamic omniphobicity, *APL Mater.*, 2014, **2**, 056108.
118. Y. Wang, *et al.* Organogel as durable anti-icing coatings, *Sci. China Mater.*, 2015, **58**, 559–565.
119. V. G. Damle, X. Sun and K. Rykaczewski, Can Metal Matrix-Hydrophobic Nanoparticle Composites Enhance Water Condensation by Promoting the Dropwise Mode?, *Adv. Mater. Interfaces*, 2015, **2**.
120. C. Lee, H. Kim and Y. Nam, Drop Impact Dynamics on Oil-Infused Nanostructured Surfaces, *Langmuir*, 2014, **30**, 8400–8407.
121. C. Hao, *et al.* Superhydrophobic-like tunable droplet bouncing on slippery liquid interfaces, *Nat. Commun.*, 2015, **6**.
122. K. Rykaczewski, S. Anand, S. B. Subramanyam and K. K. Varanasi, Mechanism of Frost Formation on Lubricant-Impregnated Surfaces, *Langmuir*, 2013, **29**, 5230–5238.
123. C. Howell, *et al.* Stability of Surface-Immobilized Lubricant Interfaces under Flow, *Chem. Mater.*, 2015, **27**, 1792–1800.
124. J. S. Wexler, I. Jacobi and H. A. Stone, Shear-Driven Failure of Liquid-Infused Surfaces, *Phys. Rev. Lett.*, 2015, **114**, 168301.
125. J. S. Wexler, *et al.* Robust liquid-infused surfaces through patterned wettability, *Soft Matter*, 2015, **11**, 5023–5029.
126. I. Jacobi, J. S. Wexler and H. A. Stone, Overflow cascades in liquid-infused substrates, *Phys. Fluids*, 2015, **27**, 082101.
127. *LiquiGlide Inc*, available at: http://liquiglide.com/.
128. X. Hou, Y. Hu, A. Grinthal, M. Khan and J. Aizenberg, Liquid-based gating mechanism with tunable multiphase selectivity and antifouling behaviour, *Nature*, 2015, **519**, 70–73.
129. X. Sun, V. G. Damle, S. Liu and K. Rykaczewski, Bioinspired Stimuli-Responsive and Antifreeze-Secreting Anti-Icing Coatings, *Adv. Mater. Interfaces*, 2015, **2**, 1400479.
130. X. Dai, B. B. Stogin, S. Yang and T.-S. Wong, Slippery Wenzel State, *ACS Nano*, 2015, **9**, 9260–9267.
131. G. McHale and M. I. Newton, Liquid marbles: topical context within soft matter and recent progress, *Soft Matter*, 2015, **11**, 2530–2546.

CHAPTER 11

Fundamentals of Anti-Icing Surfaces

ALIDAD AMIRFAZLI*[a] AND CARLO ANTONINI*[b]

[a]Department of Mechanical Engineering, York University, Toronto, ON, M3J13P, Canada; [b]Applied Wood Materials – Functional Cellulose Materials, EMPA, Swiss Federal Laboratories for Materials Science and Technology, 8600 Dübendorf, Switzerland
*E-mail: alidad2@yorku.ca; carlo.antonini@empa.ch

11.1 Introduction

Icing has long been recognized as a serious hazard for safety and for functioning of systems in diverse areas such as transportation (*e.g.* icing of wings or instrumentation for aircraft, icing of ship decks and navigation systems, as well as weapons systems in naval ships), power systems (*e.g.* icing of high-tension power lines or wind turbines), communication systems (*e.g.* antennas and dishes), various infrastructures (*e.g.* offshore platforms, locks on waterways, railroad switches, or ice formed on the curtain wall of tall buildings), and even domestic or commercial appliances (*e.g.* refrigerators or ice-making machines). One of the most tragic examples of a safety risk induced by icing is the loss of Air France flight 447 from Rio de Janeiro to Paris in June 2009, in which 228 people were killed: the final report from the French aeronautical authorities[1] (BEA – Bureau d'Enquêtes et d'Analyses pour la sécurité de l'aviation civile) highlighted that a partial obstruction of the total pressure probes in icing conditions was one of the main reason

RSC Soft Matter No. 5
Non-wettable Surfaces: Theory, Preparation, and Applications
Edited by Robin H. A. Ras and Abraham Marmur

Published by the Royal Society of Chemistry, www.rsc.org

for inconsistent velocity measurements, which consequently led to aircraft aerodynamic stall, height loss, and eventually to the crash into the Atlantic Ocean.

Traditionally, mechanical means (*e.g.* striking the iced part to remove ice), or thermal means (*i.e.* heating the iced components using electrothermal elements or hot fluids) are used to combat ice accumulation. Also, chemicals that suppress the freezing temperature of water (antifreezes) are employed to either avoid icing or remove it once it is formed. A combination of such approaches is also an option (*e.g.* during on-ground de-icing of aircraft a heated antifreeze liquid is applied to the aircraft skin to remove ice).

In the past 10 years, however, attention has focused increasingly on the use of non-wetting coatings as a way of either avoiding icing of surfaces altogether, or reducing the accumulation of ice, or reducing ice adhesion to a surface. This focus on coating systems to avoid icing is largely the result of intense studies of a class of non-wetting surfaces known (primarily) as superhydrophobic surfaces in the past 20 years. Superhydrophobic surfaces repel water through a combination of suitable surface texture and chemistry. The high contact angle on such surfaces (usually >150°) is known to reduce drop adhesion to a surface when contact angle hysteresis is low (<10°). Low adhesion of water drops onto a surface means that water may be easily removed from the surface before it can freeze and form ice. A detailed discussion of the thermodynamic aspects of wetting of such surfaces can be found in earlier chapters.

Considering that here the discussion is on the use of non-wetting or superhydrophobic surfaces to avoid or reduce icing, there is a need to clarify what is meant by icing of a surface. Ice is, of course, water in solid form/phase. There are two paths to formation of solid-phase water: either through freezing of water into ice, or through deposition of water vapour into ice. In the icing of surfaces, the latter is generally known as frost, and the former is simply called ice (although the final state in both cases is ice); hence one can find that these terms are used interchangeably in the literature, especially when initial frosting of the surface accelerates or facilitates ice formation on a surface through liquid phase change. The emphasis of this chapter is on icing as defined above, *i.e.* when a liquid is turned into ice. This is an important distinction to make, as will be discussed later; if a non-wetting coating is used properly, it can avoid water accumulation, and hence avoid icing of a surface/structure. Furthermore, as in many applications mentioned above, icing is a result of drops accumulating on a surface, the focus will be on drop interactions with surfaces under icing conditions.

Another point that needs clarification is the issue of ice adhesion, defined as the force (or energy) required to dislodge a piece of ice from a surface/coating. In thermodynamic terms, ice adhesion strength is a manifestation of surface energies at the ice–surface interface, *i.e.* the difference between the energy of ice–surface interface, and the summation of surface energies of ice–air and surface–air. As there is no liquid involved in such a case, the matter is beyond the scope of this chapter. We will nonetheless discuss aspects of ice adhesion, where researchers have tried to find practical correlations between wetting of a surface as an indication of surface energy and ice adhesion strength or force.

Considering all of the above, this chapter aims to discuss the fundamentals of science behind the application of non-wetting surfaces to mitigate icing. To achieve this, it is important to look at the strategies that can be deployed to take advantage of a non-wetting (*e.g.* superhydrophobic) coating to reduce or eliminate icing. Equally important is to understand the fundamental concepts of ice nucleation to allow insights into why and how a superhydrophobic surface (*i.e.* a textured surface) can function. Discussions using nucleation theory can also show how drop (or environmental) conditions, *e.g.* the degree of supercooling or lack thereof, may influence the effectiveness of non-wetting surfaces in icing conditions. However, discussion of the fundamentals of wetting for superhydrophobic surfaces is avoided since it is discussed in earlier chapters. This chapter instead focuses on the effects of surface wetting and topography on ice nucleation, on the wetting behaviour of superhydrophobic surfaces in icing conditions, and on ice adhesion, highlighting the role of environmental conditions in icing conditions where relevant. A brief discussion of practical issues, challenges, and innovations in using non-wetting coatings is presented in the final section. In addition, we refer the interested reader to reviews of the most recent literature regarding anti-icing surfaces.[2–4]

11.2 How Surfaces Can Be Used to Help with Icing—Icephobicity *Versus* Superhydrophobicity

Functional anti-icing surfaces would be extremely desirable for their potential to decrease, delay, or inhibit ice accretion on a solid surface. In particular, the integration of suitable coatings on aerodynamic and structural surfaces can either enhance the effectiveness of standard anti/de-icing system requirements, or lead to substantial reduction of the energy consumption of present systems, with the additional benefit that surface modification does not require any modification of bulk material properties. To provide an example, traditional ice protection system can be classified into three different categories, which are characterized by different energy requirements: (i) de-icing systems, which allow partial ice accretion and operate cyclically to limit power consumption; (ii) running wet anti-icing systems, by which water is maintained liquid, typically by heating areas where drops impact, and (iii) evaporative anti-icing systems, which totally prevent ice and liquid water accumulation in the drop collection area and hence also avoid runback ice (*i.e.* ice formed by liquid water flowing onto unprotected areas). Although such thermal systems are generally effective, they can have high energy demands,[5] with heat fluxes up to 40 kW m^{-2}. In the ideal case, a novel approach based on anti-icing surfaces would be implemented as a standalone passive strategy, eliminating the need for a thermal system.

Developing an anti-icing surface is, however, a complex interdisciplinary scientific and technological challenge, at the interface between interfacial thermodynamics, surface chemistry, and micro- and nanoengineering. This is not an easy task, since ice formation is a complex multiphase process, which can occur under at least three different paths (vapour to solid, vapour to liquid to solid, and liquid to solid) and is very sensitive to environmental and process conditions, such as temperature, humidity, pressure, and cooling rate. Also, water has the unusual property of being able to remain supercooled in a relatively wide temperature range, from 0 °C to −40 °C (at atmospheric pressure), which complicates our understanding of the nucleation process and ice growth: homogenous nucleation is a difficult phenomenon to capture, since water typically freezes by heterogeneous nucleation due to presence of contaminants, or at the interface with a solid.

In the literature so far, studies have focused either on the fundamental science of water and ice interaction at the interface with a solid surface,[6–12] or on scaled-up engineering testing in realistic icing conditions, such as those simulated in icing wind tunnel tests.[13–15] A wide variety of test methods and environmental conditions can be found to evaluate anti-icing potential of surfaces. For ice adhesion tests, ice can be formed on the surface either by accreting ice by accumulation of impacting water drops, or forming ice blocks by freezing liquid water in a mould placed on a substrate. To test the effectiveness of surfaces for reduction of icing, tests have been done by evaluating the behaviour of sessile liquid drops,[9,16,17] impacting water drops,[8,10,18] or sessile drops exposed to a shear flow.[19,20] Given the complexity of the physical phenomena involved in icing, and the variety of approaches used, it is thus not surprising that sometimes the results from various studies on the potential and effectiveness of anti-icing surfaces may appear contradictory, at least at a first glance.

In the literature, two main approaches and strategies can be identified (see Figure 11.1). In the first approach, studies have focused on achieving icephobicity with the goal of: (i) lowering ice adhesion, (ii) reducing heterogeneous nucleation temperature, and/or (iii) increasing the freezing delay, *i.e.* the time needed before a supercooled drop freezes on the surface. In the second approach, studies have focused on superhydrophobic surfaces, taking advantage of the ability of superhydrophobic surfaces to repel liquid water. This repellent property of superhydrophobic surfaces facilitates drop shedding by means of drop rebound and drop roll-off, before water can freeze on the solid surface.

It is thus important to stress that in this chapter we use the general term "anti-icing surfaces" to identify all surfaces that have a potential to combat icing, which include the attempts to develop both icephobic surfaces, characterized by low ice adhesion, low heterogeneous nucleation and/or high freezing delay, and superhydrophobic surfaces, capable of promoting drop rebound and shedding at temperatures below freezing, as schematically represented in Figure 11.1.

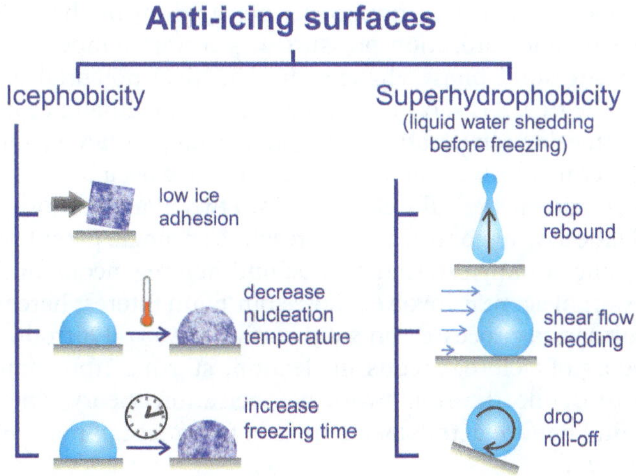

Figure 11.1 Existing approaches and strategies for anti-icing surfaces are based on icephobicity and/or superhydrophobicity. Adapted from ref. 21.

11.3 Fundamental Concepts of Ice Nucleation

The physics of ice and nucleation is complex, due to a plethora of phenomena at different time and space scales, ranging from fundamental interactions between atoms at the nano- and microscale, to "macroscopic" effects (in the order of 1 mm), a scale at which capillary interactions come into play. In addition, nucleation has an inherently stochastic nature, influenced by a variety of factors, and is not initiated in a continuous, deterministic manner. This complicates the performance and interpretation of results from experiments, which have to be conducted under careful control of the thermodynamic variables (temperature and humidity), processes, such as cooling rate, and other influencing factors, such as water purity.

One of the main challenges in understanding nucleation is associated with the existence of metastable states, such as supercooled water, which can persist for a long time below the expected equilibrium freezing (melting) temperature: it is well known[22] that pure water drops can remain liquid at temperatures as low as ~−40 °C at ambient pressure, well below the expected freezing temperature of 0 °C. Metastability is associated with the process of phase change, in which a new phase with highly ordered molecules at lower entropy (solid ice) has to be formed from a less ordered phase (liquid water or water vapour) at higher entropy. For the new (ice) phase to occur, entropy must decrease: this is reflected in the existence of an energy barrier between the metastable state and the equilibrium state. Overcoming the energy barrier requires a thermodynamic driver, which is the chemical potential difference between the phases involved in the phase-change process, $\Delta\mu = \mu_f - \mu_i$ (where f and i indicate the final and initial phases, respectively). In the case of vapour-to-solid phase change, the chemical potential difference is directly related to vapour supersaturation, $S = p_v/p_s(T)$, given by the ratio between vapour pressure and saturation pressure at a certain temperature T; in the case of liquid-to-solid phase change, the chemical potential is the liquid supercooling degree, $\Delta T_s = T_f(p) - T$, expressed as the difference between the equilibrium freezing temperature, $T_f(p)$, at a given pressure p, and the effective freezing temperature, T, at which phase change occurs.

Phase nucleation is typically classified into two main categories: homogeneous nucleation, when only water is present as a single parent phase, without the presence of any external agents; and heterogeneous nucleation, in which an external agent is involved. Since our main interest here is to understand ice formation processes on solid surfaces, we will introduce here the main concepts of heterogeneous nucleation, starting from homogeneous nucleation to define the framework of nucleation theory. The interested reader can find a more detailed and specific treatment of nucleation theory in ref. 23 and 24.

Ice can be in principle formed by liquid water or water vapour. In particular, vapour may either directly deposit as ice (vapour to solid), or may first condense into a liquid form and subsequently freeze (vapour to liquid to solid). In equilibrium conditions, the occurrence of either of the pathways is clearly defined: direct vapour deposition occurs if water vapour pressure is lower than triple point pressure, $p_v < p_{TP} = 611.7$ Pa, whereas the two-step condensation–freezing process will occur for higher vapour pressure, $p_v > p_{TP}$. However, in reality, and depending on environmental conditions, condensation–freezing may occur instead of direct vapour deposition even for $p_v < p_{TP}$, since it is not entropically favourable for molecules in the disordered, high-entropy vapour phase to directly transform into a highly ordered, low-entropy ice phase. This tendency, which makes the two-step phase-change process more favourable for $T > -100$ °C in homogeneous nucleation conditions, is also known in the field of atmospheric as the Ostwald's "rule of stages". Since liquid-to-solid freezing is a preferential condition for ice nucleation, even when vapour is the starting phase, it is appropriate to focus on it, without loss of generality. In particular, the focus here is on the case of ice germ formation at an interface, including the interface between liquid water and a solid surface.

The analysis is based on the Gibbs free energy, as for example in ref. 23, which is appropriate for systems maintained at a constant temperature and pressure. Alternative formulations using Helmholtz free energy (constant temperature and volume) are also possible,[24] and more appropriate when Laplace pressure difference across a vapour–liquid interface, due to capillary effects, needs to be taken into consideration.

11.3.1 Homogeneous Freezing

To explain freezing, two main concepts need to be introduced: (i) a thermodynamic quantity, the Gibbs free energy barrier of freezing, ΔG_f, and (ii) the ice nucleation rate, J, associated with the kinetics of ice nuclei formation.

The Gibbs free energy barrier is calculated considering the change in free energy associated with the formation of an ice embryo, *i.e.* a cluster of water molecules in solid phase, inside a system consisting of supercooled liquid water. ΔG_f is the sum of two terms: (i) a negative volume term, driving towards phase change, $-V_i\Delta\mu$, where V_i is the ice embryo volume, and $\Delta\mu$ is the chemical potential, proportional to latent heat of freezing, h_f, and liquid supercooling degree as $\Delta\mu = h_f\Delta T_s/T_f$; and (ii) an opposing positive surface term, associated to the energy required to create a new liquid water–ice interface, $-A_i\sigma_{i,l}$, where A_i is the ice embryo surface and $\sigma_{i,l}$ is the ice–liquid water surface tension. The surface term is predominant up to a critical ice embryo radius $r_i^* = 2\sigma_{i,l}/h_f\left(\Delta T_s/T_f\right)$, at which the free energy barrier reaches its maximum (see Figure 11.2), meaning that an ice embryo is unstable below this critical value, and needs to overcome this critical size in order to grow further; for water, the critical ice embryo size is of the order of 10 nm. To grow, it has to overcome the critical Gibbs free energy barrier:

$$\Delta G_f^* = \frac{16\pi\sigma_{i,l}^3}{3h_f^2\left(\Delta T_s/T_f\right)^2} \tag{11.1}$$

Note that the energy barrier strongly depends on the degree of supercooling, since it is inversely proportional to ΔT_s^2: the larger the supercooling, the easier it is for an ice embryo to overcome the barrier and grow. This also confirms why and how ΔT_s is a thermodynamic driver for ice nucleation.

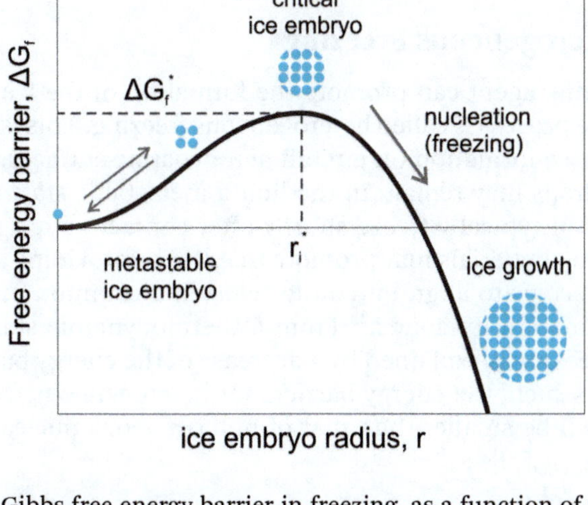

Figure 11.2 Gibbs free energy barrier in freezing, as a function of the ice embryo radius. The ice embryo is metastable below a critical size, above which ice growth (*i.e.* freezing) occurs.

The ice nucleation rate, J [$m^{-3}s^{-1}$], is the rate at which molecules can add up to an ice embryo and is thus a measure of the probability of freezing to occur. Its estimation is given by the kinetic theory, and can be expressed as:

$$J \propto \frac{k_B T}{h} \exp\left(-\frac{\Delta G_f^*}{k_B T} - \frac{\Delta g_{act}}{k_B T}\right) \tag{11.2}$$

where k_B is the Boltzmann constant, h is Planck's constant, and Δg_{act} is the activation energy barrier, associated to the transfer of water molecules across the ice–water interface. From eqn (11.2), two important comments can be made. First, the ice nucleation rate is extremely dependent on temperature: an increase of 1 °C in supercooling increases the ice nucleation rate, J, by 2–3 orders of magnitude. Second, the two terms in the exponent identify the existence of two energy barriers: the first term is related to the Gibbs free energy barrier, ΔG_f^*, and the second term is associated to Δg_{act}, an energy barrier associated with self-diffusion in water: the latter term accounts for the reduced mobility of molecules at the interface, due to viscous effects, slowing down the transport of molecules leaving the liquid phase and joining the ice phase. For low supercooling ($\Delta T_s < 30$ °C),[24] ΔG_f^* is the dominating factor, whereas for high supercooling the second term needs to be included as well. The values of J allows deriving the probability of nucleation (see complete derivation in ref. 23 and 24), identifying the threshold for ice homogeneous nucleation of supercooled drops from 1 μm to 1 mm in the range 30 °C $< \Delta T_s <$ 40 °C. Note that ice nucleation is dominated by the most active nucleation site:[25] once an ice cluster above the critical radius has been formed, ice formation in the liquid continues spontaneously. As such, the probability of nucleation at a given ice nucleation rate, J, is lower for smaller drops, due to their smaller volume.

11.3.2 Heterogeneous Freezing

An ice nucleating agent can promote the formation of the ice phase: when this occurs, the process is called heterogeneous freezing. This is, for example, the reason for ice formation on aircraft aerodynamic surfaces: atmospheric supercooled drops may remain in the liquid metastable state for long time in the cloud, but typically freeze shortly after contact with a solid surface. Some crystalline materials may promote freezing by providing a template for the water molecules to align into an ice cluster, a common mechanism for ice crystal formation in clouds.[23] From a thermodynamic standpoint, heterogeneous freezing is explained by a decrease of the energy barriers needed for freezing. As such, the energy barriers for heterogeneous freezing, $\Delta G_{f,het}^*$ and $\Delta g_{act,het}$, will be smaller than that of homogeneous nucleation barriers, $\Delta G_{f,hom}^*$ and $\Delta g_{act,hom}$, *i.e.*:

$$\Delta G_{f,het}^* < \Delta G_{f,hom}^*, \Delta g_{act,het} < \Delta g_{act,hom} \tag{11.3}$$

To account for surface effects, it is common[9,23,24] to express $\Delta G^*_{\mathrm{f,het}} = f\left(\cos\theta_{\mathrm{iw}}, r/r^*_{\mathrm{i}}\right)\Delta G^*_{\mathrm{f,hom}}$, where the function *f*, defined analytically as:

$$f\left(\cos\theta_{\mathrm{iw}}, r/r_{\mathrm{i}}^*\right) = 1 + \left(\frac{1-\cos\theta_{\mathrm{iw}}\, r/r_{\mathrm{i}}^*}{g}\right)^3 + \left(r/r_{\mathrm{i}}^*\right)^3\left[2 - 3\left(\frac{r/r_{\mathrm{i}}^* - \cos\theta_{\mathrm{iw}}}{g}\right) + \left(\frac{r/r_{\mathrm{i}}^* - \cos\theta_{\mathrm{iw}}}{g}\right)^3\right] + 3\cos\theta_{\mathrm{iw}}\left(r/r_{\mathrm{i}}^*\right)^2\left(\frac{r/r_{\mathrm{i}}^* - \cos\theta_{\mathrm{iw}}}{g} - 1\right) \tag{11.4}$$

accounts for surface wetting and texture effects, through the value of the ice–water contact angle, θ_{iw}, and local values of surface radius of curvature, r, non-dimensionalized through the critical ice embryo radius, r_{i}^*, and the function $g = \left(1 + \left(r/r_{\mathrm{i}}^*\right)^2 - 2\cos\theta\, r/r_{\mathrm{i}}^*\right)^{0.5}$.

11.4 The Role of Surface Properties and of the Environment in Icing

Considering the brief discussion of the nucleation theory in the last section, it is important now to understand how to relate surface characteristics, such as topography for a textured surface or/and chemical nature of a surface, to the θ_{iw} and r_{i}^*. This is a necessary step to interpret ice nucleation tests involving drops (millimetre and submillimetre sized) on a surface, to assess the potential anti-icing properties of the surface correctly. This is not an easy task, since in addition to the intrinsic stochastic nature of freezing, there are difficulties related to controlling impurities in water, and potential heterogeneities of the surfaces (such as local texture or chemical defects), which can significantly affect data reproducibility. Also, results from different tests may not be directly compared due to differences in environmental conditions (such as cooling rate or humidity)[20]. Nonetheless, a few attempts have been made in recent years in this direction,[7,9,17,20,25,26] and will be used in the following discussion to elucidate the anti-icing potential of surfaces and identify possible strategies to mitigate freezing on surfaces. We focus in particular on clarifying the role of surface wetting, surface topography, and environmental conditions.

11.4.1 Surface Wetting

The first surface parameter affecting ice nucleation is surface wetting. Li *et al.*[26] investigated the freezing behaviour of sessile water drops on two smooth surfaces, one hydrophilic (unfunctionalized silicon wafer) and one hydrophobic (grafting of a fluoroalkylslane, FAS-17, on silicon), by observing freezing of sessile microdrops cooled at a rate of ~0.08 °C s^{-1}. Reported values of contact angles and root-mean-square roughness were $\theta = 55°$ and $R_{rms} = 0.51 \pm 0.21$ nm for the hydrophilic surface, $\theta = 114°$ and $R_{rms} = 1.21 \pm 0.07$ nm for the hydrophobic surface.[25] The most relevant results are illustrated in Figure 11.3: microdrops remained liquid down to −38 °C, but started to nucleate below this temperature; also, data show that the ice nucleation temperature is slightly lower on the hydrophilic unmodified surface than on the hydrophobic surface. This result looks surprising at first glance, since one would expect that hydrophobicity should make freezing unfavourable, as is the case for vapour-to-liquid condensation, where hydrophobicity has been shown to decrease the energy barrier.[23,24] However, this result could be explained on the basis of a previous work by Goertz *et al.*,[27] who demonstrated that the viscosity of the interfacial liquid layer is higher than that of bulk water, the effect being prominent on hydrophilic surfaces. The increased viscosity on hydrophilic surfaces decreases the diffusion of water molecules at the interface, causing an increase of the activation energy barrier Δg_{act} (which becomes relevant for supercooling higher than 30 °C, as explained earlier) and thus a decrease in the nucleation rate, J (see eqn (11.2)), with respect to the hydrophobic surfaces. As such, hydrophilicity delays nucleation rate in this temperature range (high supercooling, $\Delta T_s > 30$ °C). We will see in the next section that this is not necessarily the case for lower supercooling values ($\Delta T_s < 30$ °C).

The results from ref. 26 also help by introducing an important concept in freezing: the quasi-liquid layer at the liquid–solid interface. Although

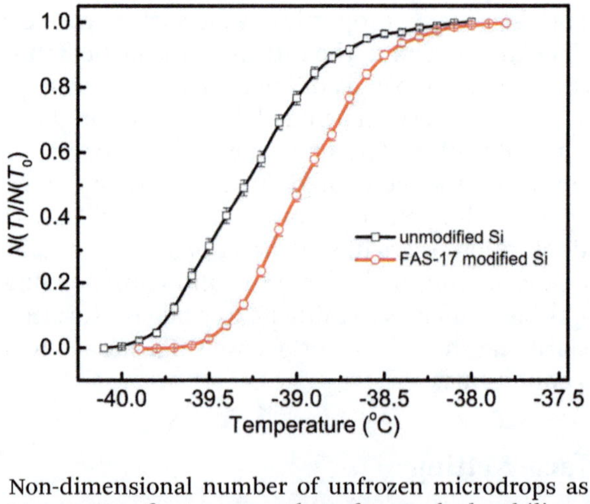

Figure 11.3 Non-dimensional number of unfrozen microdrops as a function of temperature for two smooth surfaces: a hydrophilic surface (unfunctionalized silicon wafer) and a hydrophobic surface (grafting of a fluoroalkylslane, FAS-17, on silicon wafer). Experiments were performed at a cooling rate of ~0.08 °C s^{-1}. Reproduced with permission from K. Li, *et al.*, *Appl. Phys. Lett.*, 2014, 104, 101605. Copyright [2014], AIP publishing LLC.[26]

generally an interface between different phases is considered as a mathematical discontinuity, with a sharp transition of properties, in reality an interface is a transition zone, with properties changing through a strong, but finite, gradient; this is true for a liquid–vapour interface, and also for a liquid–solid interface. As a result, water properties, such as viscosity, may change significantly at the interface with a solid surface and/or with an ice embryo, affecting the freezing. The introduction of the concept of the quasi-liquid layer is not new (as reconstructed by Jellinek)[28]: the idea can be traced back to Faraday in 1859, and gained renewed attention in the 1950s, when an experimental study of the contact interaction between ice spheres suggested the existence of a liquid-like layer at the ice–ice interface for temperatures down to −25 °C. Based on additional ice adhesion tests performed on different surfaces, through both tensile and shear stresses, Jellinek[28] in 1960 provided an estimate of the thickness of the quasi-liquid layer, in the order of 10 nm (recall that the critical ice embryo size is also of the same order), and of its viscosity, ranging from 10^2 to 10^4 mPa s, thus several orders of magnitude higher than bulk water. The existence of a quasi-liquid layer in the framework of anti-icing surfaces has been recently discussed while investigating the effect of surface topography on nucleation temperature and freezing delay by Heydari *et al.*[9] and Eberle *et al.*,[17] as discussed in the next session.

11.4.2 Textured or Rough Surfaces

It is well known that liquids confined in pores experience a depression in freezing point, through the Gibbs–Thomson effect: the variation of the chemical potential between two phases across a curved interface increases the energy required to form small particles with high curvature. This leads to freezing point depression, ΔT_f, which in an infinite cylindrical pore can be estimated as:[29]

$$\Delta T_f\left(r_{\text{pore}}\right) = T_f - T_f\left(r_{\text{pore}}\right) = -T_f \frac{2\sigma_{\text{sl}}\cos\theta}{h_f \rho_s} \frac{1}{r_{\text{pore}}} \tag{11.5}$$

where T_f is the bulk freezing temperature, $T_f(r_{\text{pore}})$ is the freezing temperature in the pore with radius r_{pore}, σ_{sl} is the solid–liquid interfacial energy, and ρ_s is the density of the solid. For water, freezing point depression of the order of 10 °C can be obtained for porosity of the order of 10 nm. The Gibbs–Thomson effect is at the base of thermoporometry techniques, such as differential scanning calorimetry (DSC), where freezing point shift is used as an indirect measurement to recover information of pore-size distribution in porous materials. It is thus clear that topography can affect freezing and can be used to hinder it. The question then becomes how to promote the same effect on surfaces, and design optimal anti-icing surfaces with extreme freezing point depression.

If one focuses on the "low" supercooling regime ($\Delta T_s < 30$ °C), where the contribution of Δg_{act} to ice nucleation rate is negligible (see eqn (11.2)) and freezing is mainly dominated by the Gibbs free energy barrier, $\Delta G^{*}_{f,het}$, it is useful, as proposed in ref. 9 and 17, to study the role of the ice–water contact angle and surface local radius of curvature surface through the factor $f\left(\cos\theta_{iw}, r/r_i^{*}\right)$. Two main trends can be observed: first, in this "low" supercooling regime, $f\left(\cos\theta_{iw}, r/r_i^{*}\right)$ increases with increasing contact angles of the ice nucleus, and second (see Figure 11.4) for hydrophobic surfaces with roughness in the nanoscale regime, *i.e.* $r/r_i^{*} < 10$, one can expect that freezing is likely to occur in (concave) nanopits, rather than on (convex) nanobumps. How is this reflected in the freezing of a millimetric drop?

Heydari *et al.*[9] and Eberle *et al.*[17] independently investigated sessile drop freezing on surfaces with different morphology, to understand the effect of nanoscale roughness on the nucleation temperature and the freezing time delay. In both studies it was observed that, in fact, topography does not significantly affect nucleation temperature. That is why experiments on nanotextured surfaces, spanning a wide range of root-mean-square roughness (from ~0.1 nm to 100 nm), showed that all surfaces have a constant nucleation temperature of ~ −24 °C.[17] This is because any real surface has a distribution of curvatures, and is inevitably constituted of both pits and bumps, so that freezing will be initiated there, where the energy barrier is the lowest.

The result may appear discouraging, since it would suggest that there is no chance to fabricate superhydrophobic surfaces with antifreezing capability. However, this is not the case. First, hierarchical structures with micro- and nanotexturing exhibited a lower freezing temperature than the corresponding substrates having only the nanostructure,[17] as a result of a reduction in the effective drop surface contact area: for a given ice nucleation rate per unit area, ice nucleation can be effectively reduced by minimizing the liquid–solid

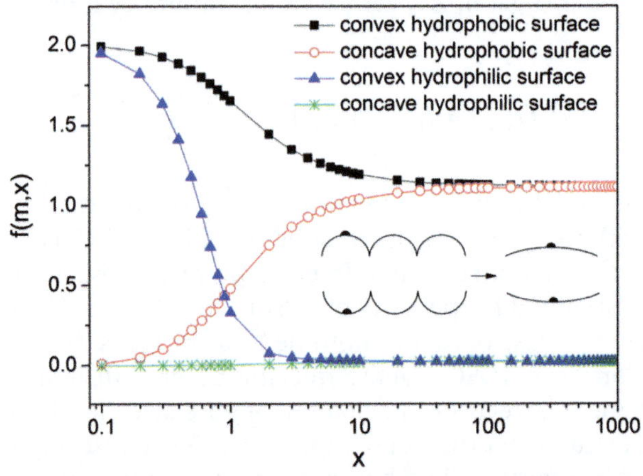

Figure 11.4 The function $f\left(\cos\theta_{iw}, r/r_i^{*}\right)$ plotted as function of the roughness parameter $x = r/r_i^{*}$, for a hydrophilic and a hydrophobic surface, with contact angles of 30° and 94°, respectively. Values of the function f are given for both (concave) pits and (convex) bumps.

contact area. Indeed, Eberle *et al.*[17] reported a reduction of nucleation temperature from nanostructured to hierarchical (micropillars and nanostructures) surfaces of ~2 °C, corresponding to a nucleation temperature of ~ −26 °C. Second, superhydrophobic surfaces significantly promote water shedding and rebound of impinging drops, so that a drop has the time to leave the surface before freezing,[8,10] as discussed in detail in the following section. Finally, the freezing delay time can be significantly increased at temperatures slightly higher than the nucleation temperature.[4,30] The freezing delay time, t_d, is inversely proportional to the nucleation rate, $t_d \propto J^{-1}$, with J depending exponentially on supercooling degree (see eqn (11.2)) according to nucleation theory. This has two important consequences: (i) even a relative small reduction of ~2–3 °C for nucleation temperature may extend the freeze delay time[7] at a prescribed temperature; (ii) superhydrophobic surfaces operating at temperatures slightly above the ice nucleation temperatures can lead to extreme freezing delay. Experiments[17] have confirmed that a supercooled drops may remain liquid on a surface operating at −21 °C, only 3 °C above the surface nucleation temperature, for as long as 21 h. A point that needs attention when comparing the literature results for rough surfaces from various sources is the presence of chemical heterogeneity. As shown in a study by Fang and Amirfazli,[31] the presence of chemical heterogeneities can significantly mask the effect of surface roughness (this issue is of practical importance when dealing with surfaces/coatings to be deployed in the field).

As an additional remark regarding the quasi-liquid layer, it has already been highlighted[4,17] that its existence may play a role in defining the value of the nucleation temperature. The reported value of nucleation for nanostructured surfaces in stationary conditions (~−24 °C) is very close to the lower limit identified for the existence of a liquid-like layer at the ice–ice interface for temperatures down to −25 °C in static conditions (note that the liquid layer may exist at even lower temperature in transient conditions, *e.g.* as demonstrated by experiments from ref. 26, discussed earlier). The concept of the quasi-liquid layer will be further discussed below, in the context of ice adhesion.

11.4.3 Environmental Conditions

Environmental conditions clearly play an important role in the freezing of liquids on solid surfaces. Most freezing experiments designed to study freezing rate, freezing probability, and freezing delay are performed in saturated conditions to avoid drop evaporation effects,[7,9,17,20] such as drop volume change and evaporative cooling. In a few cases, saturation (or supersaturation) has also been used to generate a distribution of micrometric and submillimetric drops through condensation before the actual freezing test.[25,26] Also, tests are performed with homogeneous temperature distribution, where environment (air), drops and the surface are kept at the same temperature.

However, it is important to underline that environmental conditions can play a clear role on nucleation for realistic applications. It is well known that condensation or frosting of the surface may compromise superhydrophobic surface properties:[31,32] condensate and frost form indiscriminately on any surface location, thus not only on the top of surface asperities, but also within the pits, occupied by water drops. In the cases where superhydrophobic surfaces are used as a mean to reduce icing, frosting in the indiscriminate form mentioned above, can trigger a transition from Cassie–Baxter to Wenzel state. As such, a significant increase of drop adhesion can be seen when supercooled drops contact a surface.[21]

More interestingly, Jung *et al.*[20] investigated the effect of environmental conditions, highlighting the primary role played by humidity in the shedding and freezing mechanism of supercooled drops on superhydrophobic and hydrophobic surfaces, when exposed to air flow at temperatures below freezing. In saturated conditions (relative humidity φ = 100%), freezing of supercooled drops took place according to the expected and commonly observed heterogeneous nucleation at the substrate–water interface, whereas at low-humidity conditions (φ = 30% in the study),[20] freezing was engendered by homogeneous nucleation at the drop-free surface. This is because the exposure of supercooled drops to a low-humidity environment introduces evaporative cooling, which decreases drop local temperature at the gas–liquid interface and allows homogeneous nucleation as primary mechanism for drop freezing. A fast imaging technique confirmed that homogeneous nucleation was promoted at the upstream region of the free interface, where the evaporation rate is expected to be higher.

11.5 Water and Ice Interaction with Surfaces in Icing Conditions

In this section, the physics of water–surface interaction during drop impact, drop shedding and self-propulsion, are discussed first, highlighting the peculiarities of such phenomena in icing conditions, compared to room temperature. Then, the ice–surface interaction will be examined to understand conditions under which anti-icing surface can be used to minimize ice adhesion.

11.5.1 Dynamic Water–Surface Interaction in Icing Conditions

Superhydrophobicity is associated with the Cassie–Baxter wetting state, in which gas pockets are present at the liquid–solid interface; if the Cassie–Baxter state is not stable over time, the water penetrates into the crevices of a textured surface, leading to a transition to the Wenzel state. The Wenzel state represents complete wetting of the solid substrate and liquid/drop mobility can be significantly reduced. As such, maintaining stable air pockets, *e.g.* under increased pressure during drop impact, is thus a general requirement for superhydrophobic surfaces operating in any condition. In the following sections, first the mechanism and parameters controlling the repellency and mobility of water drops on solid surfaces in general are introduced; this will cover the fundamentals of drop shedding, self-propulsion, and drop rebound in general (note that most studies are performed at room temperature). Second, the issue of how operating in icing conditions may affect drop–surface interaction is discussed, highlighting wherever possible those strategies that have been identified as effectively using superhydrophobic surfaces in icing conditions.

11.5.1.1 Drop Shedding and Self-Propulsion

To determine whether the drop will move under external forces, such as gravity or aerodynamic loads, it is necessary to predict the drop capillary adhesion force, F_{adh}. The adhesion force is the manifestation of surface tension acting at the contact line, and depends thus on the contact line shape and contact line distribution.[33] In general, for a drop of a given volume, drop mobility is enhanced on surfaces where contact angles are high, due to minimization of the contact line length, and contact angle hysteresis is low.[34] Care must be used in the measurement of contact angles: studies on drop shedding on tilted surfaces and by airflow[19] have pointed out that the values of maximum contact angle, θ_{max}, observed downhill/downstream, and minimum contact angle, θ_{min}, observed uphill/upstream, measured at the moment of incipient motion, may differ from the values of advancing, θ_A, and receding, θ_R, contact angles measured quasi-statically on a horizontal surface. Given the above understanding of the drop adhesion, then it is straightforward to understand why superhydrophobic surfaces represent a good choice to minimize the capillary adhesion force under normal laboratory conditions. But the question is whether the drop shedding capability for superhydrophobic surfaces is maintained in icing conditions.

Recent studies[35,36] have shown that, in icing conditions, superhydrophobic surfaces can still be effective, similar to the room temperature tests, but the degree of their effectiveness may be diminished as the critical air velocity at which a drop starts to shed, generally increases under icing conditions. A closer look at the results in these studies show that the contact angle of drops in icing conditions are different from those in room temperature tests. For a given drop volume, lower contact angles (especially θ_{min}) seen under icing conditions also mean a larger contact line; the combination of these two factors will then lead to an increased adhesion force. The difference in contact angles has to do with environmental conditions, such as humidity and possibility of early formation of micro frost crystals on a rough surface—note that icing conditions usually represent a more saturated atmosphere than room temperature tests. In addition, increased shearing air velocity is related to the reduced drag coefficient on the drop under icing conditions, because of its modified profile (lower contact angle, larger contact line), compared to the room temperature conditions.[35,36]

The above discussion indicates that environmental conditions such as atmospheric saturation (humidity) can play an important role in the performance of superhydrophobic surfaces. If the supercooled drop nucleation theory is considered naively, *i.e.* only on the basis of contact angle value being high for superhydrophobic surfaces, one may conclude that superhydrophobic surfaces may perform better at high humidity conditions, *i.e.* close to saturation. This would be a benefit in applications such as aircraft icing, since icing typically occurs when the aircraft flies through a cloud, where a saturation condition applies. However, as stated above, humidity can negatively affect the adhesion/shedding of drops on superhydrophobic surfaces under shear flow. In a detailed study of this topic Jung *et al.*[20] showed that for a superhydrophobic surface exposed to a supersaturated environment, which causes condensation on the surface, drop adhesion increased with respect to dry surfaces, as observed by the increase of critical shearing gas velocity required to initiate drop shedding (critical velocity increased by a factor ~2–3 for a 5 μL drop in unsaturated and saturated conditions)—see schematic in Figure 11.5. In contrast, the effect of humidity on drop adhesion is almost negligible on a smooth hydrophobic (*e.g.* Teflon) surface.[35] This occurs because on most superhydrophobic surfaces the condensate forms on the entire surface and not only on the top of surface asperities, causing the drop wetting transition from Cassie–Baxter state to partial or complete Wenzel state. As such, air pockets are replaced by condensed drops and capillary adhesion is higher.

The above studies for single drops on a surface have also shown that if the surface (or environment) temperature is too low (say below −10 °C) and/or drop resides on the surface for prolonged periods of time, it will eventually freeze. Studies done in an icing wind tunnel have also shown that when a cloud of drops impact a cold superhydrophobic surface,[14,15] ice starts to accumulate readily. So it seems that superhydrophobic surfaces, or any other type of surfaces (observations by authors) cannot be used as a passive anti-icing strategy. The question is, however, whether surface treatments in

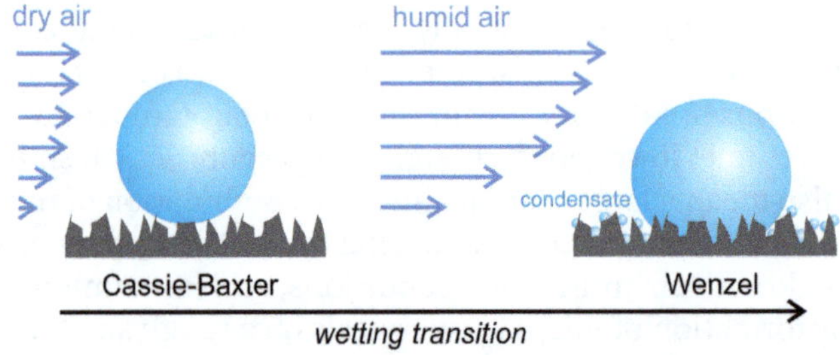

Figure 11.5 Effect of environmental conditions on superhydrophobicity in humid conditions, relevant to icing applications. Drop adhesion is higher in a humid supersaturated environment than in dry conditions, due to the presence of condensate on the surface, promoting wetting transition from Cassie–Baxter to partial or complete Wenzel state. As a result, the critical shearing gas velocity needed to shed the drop is higher in humid conditions.

combination with traditional methods of combating icing, *e.g.* heating of surfaces, can show any benefits. This idea is discussed below.

The above idea of combining a superhydrophobic surface and a thermal system was examined in a simulated icing condition study in an icing wind tunnel.[13,15] Efficient and safe operation of aerodynamic surfaces, such as wings, typically require that anti-icing systems have to operate as fully evaporative systems: impinging liquid water drops in the collection zone, typically corresponding to the leading edge area on a wing, need to be completely evaporated by a heating system, to avoid formation of runback ice by freezing of water flowing downstream to unheated areas. Providing latent heat for impinging water evaporation is a strategy that demands high energy, which may in addition pose problems due to thermal stresses and fatigue, particularly critical for composite materials. If superhydrophobic surfaces are used in the drop collection area,[13] much less energy is required, since heat is only needed to keep the drop collection zone slightly above the freezing point, to avoid frosting and liquid freezing, while drop removal is promoted after impact by superhydrophobic surfaces. Figure 11.6 shows

Figure 11.6 Icing mitigation strategy, based on combination of coating and leading edge heating system: standard hydrophilic surface (left) *vs.* anti-icing superhydrophobic surface (right). Top row images show a schematic of icing processes on the two wings: on the hydrophilic surface, no ice forms in the leading edge area, but impacting drops can slide and freeze downstream, to form runback ice; on the superhydrophobic surface, drops rebound after impact and are shed from the surface without sliding downstream. Bottom row pictures give a top view of wings during icing wing tunnel tests, with runback ice on the hydrophilic wing, and the ice-free superhydrophobic wing in the same environmental conditions.

schematically how water accumulation is reduced in the collection zone by drops either being shed or rebounding (see below) from the superhydrophobic coated zone. Figure 11.6 also shows images of bare aluminium and superhydrophobic coated test articles where runback ice (ice usually accumulating in the unheated areas of a wing) was also absent when superhydrophobic coating was used since drops were re-entrained into the air stream. A further additional benefit has been observed that on superhydrophobic surfaces isolated ice structures form, rather than compact ice, as on hydrophilic surfaces.[15] As a result, detachment of the ice on superhydrophobic surfaces is facilitated, due to three possible factors: (i) lower ice surface contact area, (ii) increased drag force exerted on the ice islands on superhydrophobic surfaces, because of ice shape with respect to the ice formation on the hydrophilic surface, and (iii) weaker adhesion force on surfaces with low wettability.[37]

Aside from shedding of drops from superhydrophobic surfaces by way of a shear flow or gravity, two other developing strategies are discussed below.

On a two-tier superhydrophobic surface, with carbon nanotubes deposited on silicon micropillars, Boreyko *et al.*[38] demonstrated that coalescence of neighbouring condensate drops can lead to jumping and autonomous removal of drops from surfaces,[39] without external forces. This mechanism allows quick removal of micrometric drops, in the order of 10 μm, and reduces surface coverage up to ~40%.[40] In icing conditions, the mechanism of self-propelled jumping drops can be beneficial directly,[12] to promote rapid removal of drops before freezing, and also indirectly, since the drop jumping effect was shown to maximize the separation between drops, thus minimizing ice bridging between drops and decreasing the intra-drop frost propagation speed by a factor of 3, compared to smooth hydrophobic surfaces.[40]

In a subsequent study, Boreyko *et al.*[11] also investigated the cyclic frosting and defrosting of nanotextured superhydrophobic surfaces, using a nanopillar array with characteristic pitch of the order of 100 nm. After accreting frost on the superhydrophobic surface, the nanotextured surface was defrosted, by heating the surface slightly above 0 °C (see Figure 11.7): during this process, frost melts and spontaneously dewets the surface, with the slush (mixture of ice and water) showing a high mobility, as typically observed for water drops in the Cassie–Baxter state. This demonstrates that nanoscale texturing of surfaces may limit the problems stated above related to condensate formation; however, more investigation is needed to understand whether this effect is due to the preferential formation of frost at the top of surface asperities on nanostructures, or is enabling a wetting transition from partial or complete Wenzel state to Cassie–Baxter state during the defrost process.

The dynamic interaction of a drop with a surface is also of importance for icing applications (*e.g.* see Figure 11.6). As such, in the next section, drop impact onto superhydrophobic surfaces under icing conditions is discussed.

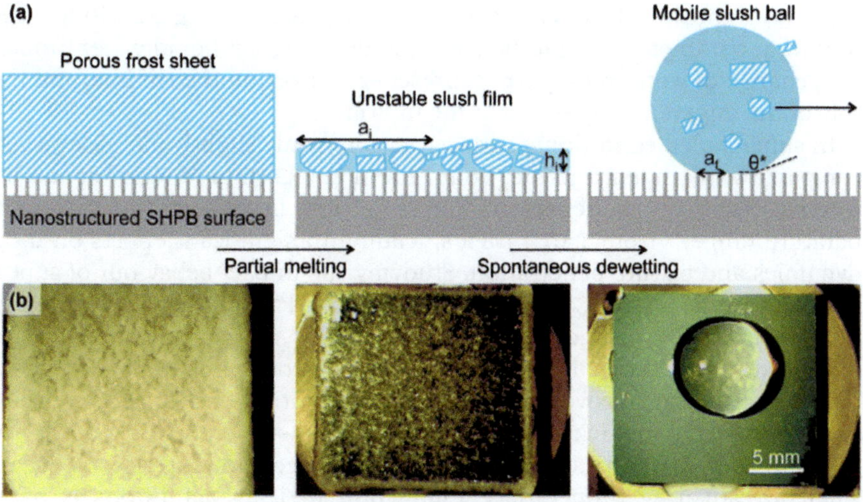

Figure 11.7 (a) Side-view schematic of the dynamic defrosting process on a nanostructured superhydrophobic surface. (b) Top-down imaging of dynamic defrosting. A porous frost sheet of thickness 2.0 mm melts into a film of slush, exhibiting an initial height of 0.6 mm and contact radius of 9.2 mm that spontaneously dewets to a mobile drop with radius of 3.4 mm. Reproduced from J. B. Boreyko, B. R. Srijanto, T. D. Nguyen, C. Vega, M. Fuentes-Cabrera and C. P. Collier, *Langmuir*, 2013, 29, 9516–24. Copyright (2013) American Chemical Society.[11]

11.5.1.2 Drop Impact

The impact of drops onto superhydrophobic surfaces has been a relevant topic in the past decade,[41] with the goal of elucidating the important parameters related to drop impact dynamics, such as the maximum drop spreading and drop contact time with the surface between impact and rebound[42] (also defined as rebound time). Superhydrophobic surfaces are beneficial, since they minimize both the water nucleation rate, due to low solid fraction and minimization of contact area upon spreading, and the contact time, allowing rapid dewetting of the surface, before water can freeze and stick to the substrate.

One additional essential parameter is the critical velocity of impalement,[43,44] V_c, above which the drop remains impaled on the surface, unable to rebound, as a result of liquid meniscus penetration into the texture upon impact. High critical velocity for impalement can be achieved by increasing surface resistive capillary pressure, p_c, which is proportional to water surface tension, σ, and to the advancing contact angle of the corresponding smooth surface, θ_A^s, and inversely proportional to the surface characteristic pore size, r_{pore}, *i.e.* $p_c \propto \sigma \cos\theta_A^s / r_{pore}$. On a micropillar based surface, V_c can be relatively low,[43,44] *i.e.* in the order of few m s^{-1}. This is the reason why nanotextured surfaces, or hierarical surfaces with nanoscale features, are necessary to promote stable superhydrophobicity under drop impact conditions at high speed.[30]

Looking specifically at supercooled drop impact in icing conditions, the following parameters should be considered: change in liquid water properties, change of wetting properties, presence of frost on the surface, and the role of the intervening gas layer at the interface.

In supercooled conditions, surface tension[45] only slightly increases when compared to room temperature, from 72 N m^{-1} at 23 °C to 78 N m^{-1} at −20 °C, whereas viscosity experiences an almost fourfold increase in the same temperature range,[46] from 1 to 4 mPa s, leading to significant effects on drop dynamics and texture penetration. Studying the impact behaviour of supercooled drops down to −17 °C, it has been observed[10] that the full-penetration velocity threshold was increased markedly, *i.e.* by ~25% for the tested micropillar-textured surfaces. Nonetheless, higher viscosity in supercooling conditions can hinder recovery from partial penetration into surface crevices, causing the drop to stick on the surface, and consequently being unable to rebound completely. As such, the importance of considering viscous effects when investigating drop/surface interaction in icing conditions is paramount: although viscous effects are often disregarded in drop impact studies on superhydrophobic surfaces at room temperature and for the case of no-penetration condition, since *e.g.* they do not influence the contact time, which scales as $\tau \propto \left(\rho D_0^3/\sigma\right)^{0.5}$, they cannot be neglected when considering the supercooled drops.

The effect of liquid supercooling on wetting properties, measured through the contact angles, depends strongly on environmental conditions: Maitra *et al.*[10] reported that contact angles were not affected by the temperature in dry conditions, at relative humidity RH ≈ 0%. However, contact angles may change due to condensation effects at higher humidity conditions, as observed directly by measurements of contact angles,[9,35,47] as well as indirectly,[20] through an increase of drop capillary adhesion, which can be explained by an increase of contact angle hysteresis. In addition to the advancing and receding contact angles, measured by a quasi-static process, the values of dynamic contact angles can change in supercooled conditions: the dynamic contact angle,[48] $\theta_D = f(\theta_A, \theta_R, \mathrm{Ca})$, is indeed a function of both the advancing and receding contact angles, and the capillary number, Ca $= V\mu/\sigma$, which depends linearly on liquid viscosity. In particular, during the recoil phase of drop impact phenomenon, higher viscosity, *i.e.* higher capillary numbers, lead to a decrease of the dynamic contact angle, slowing down the recoil process. In addition, Bahadur *et al.*[49] developed a model for drop dynamics, based on the assumption that ice nucleation causes a reduction of the receding contact angle and thus of the retraction force, responsible for dewetting, consequently delaying or preventing drop recoil and rebound.

Frost formation on the surface can be particularly critical for superhydrophobic surfaces to prevent drop rebound, as also mentioned above in the context of drop shedding: Varanasi *et al.*[32] demonstrated that on a frosted micropillar-textured superhydrophobic surface, on which ice nucleates indiscriminately over the entire surface, drops lose their ability to rebound, since frost changes surface wettability and causes drop freezing during drop impact. The influence of nucleation sites on the dynamics of the contact line in the recoiling stage was further highlighted by studying the ideal case of contactless impact on sublimating and evaporating surfaces.[50] Even under extreme freezing conditions, down to cryogenic temperatures, a water drop can impact and rebound without freezing, due to the presence of an intervening vapour layer induced by sublimation or evaporation at the interface. If a vapour layer can be continuously sustained at the interface between the water drop and the substrate, as made possible in the presence of phase-change effects, a water drop will not freeze on surfaces at temperatures as low as −196 °C. Although such systems may not be practical in the context of anti-icing systems and are only of scientific interest, it has been shown[51] that in a low-pressure environment the increased vaporization rates experienced by the drops can boost the levitation process, leading to spontaneous removal of liquid drops, even at the moment of freezing.

Overall, the discussion above highlights that using a purely passive anti-icing strategy based on surface superhydrophobicity, at zero energy input, may delay the freezing of sessile drops, but still cannot hinder the inevitable freezing. This is why taking advantage of superhydrophobic properties is not an alternative to current systems, but should be used *in combination* with classical anti-icing strategies (*e.g.* thermal systems), for example to reduce energy consumption from an application perspective.

11.5.2 Ice Adhesion on Anti-Icing Surfaces

One of the possible strategies for anti-icing surface is to develop surfaces with minimal ice–surface interaction and adhesion strength. On such surfaces, ice may eventually accrete as a result of liquid water freezing or frosting, but could be shed if low adhesion forces are overcome by external forces, such as gravitational, aerodynamic or centrifugal forces.

Meuler *et al.*[37] conducted a phenomenological study to find a relationship between ice adhesion and surface wettability for smooth surfaces with Wenzel roughness $r_W < 1.01$, where r_W is the ratio between the actual surface area and the projected area. It was found that ice adhesion strength correlates strongly with the so-called "practical work of adhesion", defined as $\gamma(1 + \cos\theta_R)$, required to remove a liquid water drop from each test surface. The data of ice adhesion strength measured at −10 °C were well captured by the experimentally derived fitting curve $\tau_{ice} = 340 \pm 40$ kPa$(1 + \cos\theta_R)$, as can be seen in Figure 11.8. This correlation also seemed to fit the data from two other studies; so although the casual effect of relation of ice adhesion to receding contact angle may be debatable, this correlation can present a practical tool. For example, by comparing a surfaces with $\theta_R = 0°$ $(1 + \cos\theta_R = 2)$, with the most smooth hydrophobic material, whose receding contact angle does not exceed $\theta_R = 120°$ $(1 + \cos\theta_R = 0.5)$, ice adhesion can be reduced by a factor of 4 by tuning wettability. Using smooth materials, one can thus conclude that hydrophobicity well correlates with icephobicity, when measured in terms of ice adhesion. Nonetheless, ice adhesion remains on all materials in the same order of magnitude.

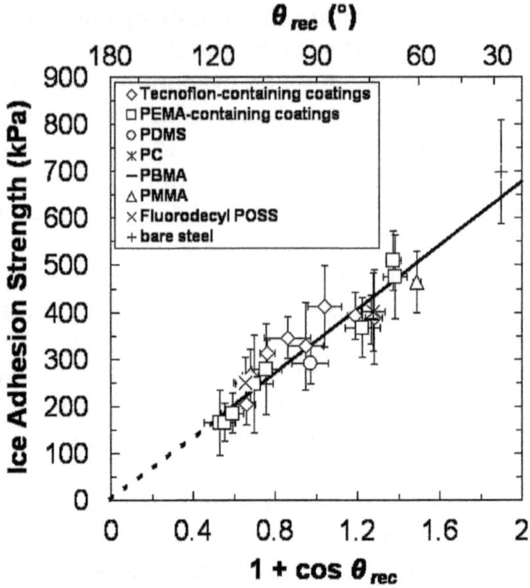

Figure 11.8 Relationship between ice adhesion and wetting properties: average ice adhesion strength, τ_{ice}, as function of work of adhesion for liquid water, $1 + \cos\theta_R$. Dashed portion of the best fitting curve, $\tau_{ice} = 340 \pm 40$ kPa$(1 + \cos\theta_R)$, represents the extrapolation to the origin. Reproduced with permission from A. J. Meuler, J. D. Smith, K. K. Varanasi, J. M. Mabry, G. H. McKinley and R. E. Cohen, *ACS Appl. Mater. Interfaces*, 2010, 2, 3100–10. Copyright (2010) American Chemical Society.[37]

If the correlation identified for smooth materials holds also for textured surfaces, then fabricating a surface with $\theta_R \rightarrow 180°$ would virtually lead to an adhesion strength $\tau_{ice} \rightarrow \sim 0$. However, for textured surfaces the application of such a correlation will not be straightforward, since other factors such as surface roughness and environmental conditions, as well as test procedures, can dramatically affect the outcome of the ice adhesion test. As such, whether or not superhydrophobicity implies icephobicity and *vice versa* remains a debatable topic. It is generally a misunderstanding to believe that a surface with high contact angles should consequently lead to an icephobic surface.[52]

Varanasi *et al.*[32] studied ice adhesion strength on micropillar-textured surfaces and highlighted that ice adhesion strength was even larger than that of a smooth surface of the same material. Scaling of ice adhesion values from textured surfaces with those from smooth surfaces suggested that ice was contacting all available area of the textured surface (note that the surface area of a rough surface is much larger than a smooth one). Kulinich and coworkers[53,54] specifically addressed ice adhesion on textured superhydrophobic surfaces and identified that ice adhesion on the tested surfaces, having random texturing but similar chemistry, correlated well with the value of contact angle hysteresis. This is not necessarily in conflict with the correlation stated above, as surfaces with high hysteresis usually have low receding contact angles (a direct comparison between ice adhesion and receding contact angle was not given in ref. 53). In a subsequent work, Kulinich *et al.*[54] reported that low ice adhesion properties of the tested surface deteriorated even after a few icing/de-icing cycles, due to gradual damaging of the surface, especially on top of the asperities: this highlights a very critical issue, *i.e.* the durability of superhydrophobic surfaces (see below). In addition, it was also shown that ice adhesion was increasing in a humid atmosphere, as a result of condensate formation, in agreement with other studies, discussed above.

Maitra *et al.*[52] analysed the reduction of ice adhesion for different superhydrophobic surfaces, highlighting that best superhydrophobic properties do not necessarily correlate with ice adhesion reduction. The superhydrophobic properties were evaluated by means of resistive capillary pressure during drop impact to prevent transition from Cassie–Baxter to Wenzel state. The authors showed that ice adhesion strongly depends on the applied stress orientation: in particular, the simultaneous effect of shear and tensile stresses needs to be accounted for, to properly evaluate the ice adhesion to complex textured surfaces.

In many practical situations it may be difficult to have a surface that is frost free (without any external energy supply, *e.g.* as discussed above having a thermal energy supplied in combination with the superhydrophobic coating). Then the question is: how should a surface be designed to have low ice adhesion properties? A possible solution can come from surfaces with an infused liquid layer, which can be either an immiscible oil[55–57] or an aqueous layer.[58–60]

The concept of slippery liquid-infused porous surfaces[55] (SLIPS), liquid impregnated surfaces[56] (LIS), and slippery pre-suffused surfaces[57] has been recently proposed as a way to promote a non-wetting state. Taking inspiration from the *Nepenthes* pitcher plants,[55] chemically functionalized nano/microtextured substrates are used to lock an infused lubricating fluid in place. The SLIPS/LIS based strategy, in addition to decreasing the risk of liquid impalement into the texture, since the infused lubricating fluid layer is ideally more difficult to displace than a gas layer, has been shown to bring some advantages also in reducing ice adhesion.[61] However, the main technological limit in view of practical applications is related to the loss of the infused lubricating fluid over time,[62] which at present makes surfaces effective for only a few hours. The other issue may be the contamination of the infused liquid by pollutants present in air (*e.g.* particles) that can stick to the oil layer and create either nucleating points for ice formation or pinning points to hinder drop mobility.

As discussed in Section 11.3, the presence of a water quasi-liquid layer can be helpful not only to retard nucleation, but also to reduce the adhesion of ice to the solid substrate.[58–60] Chen *et al.*[58] demonstrated that a hygroscopic polymer on a solid interface can deliquesce and swell due to water absorption or condensation, allowing formation of a self-lubricating liquid water layer at the interface, even in freezing conditions. The presence of such a layer, as confirmed by ice adhesion measurements, enables the reduction of the ice adhesion strength by an order of magnitude. This approach was proved effective for temperatures down to ~ −25 °C, which is consistent with the minimum temperature at which a quasi-liquid layer exist in stationary conditions. Below this temperature, the self-lubricating liquid water layer disappears due to the phase transition of the liquid water to the ice, and ice adhesion can increase significantly.

11.6 Alternative Routes: Soft Surfaces and Biomimicry of the Antifreeze Protein

Aside from well-established and classical surface properties discussed above, in this section we allude to other alternative routes to develop anti-icing surfaces that involve the use of surface compliance, or of biomolecules.

Surfaces with viscoelastic properties have been attracting increasing attention for their hybrid solid–liquid behaviour. With respect to icing, it has been reported[63] in an experimental study on frost growth mechanism that soft deformable surfaces may delay frost formation, compared to other solid hard substrates. This opens up a new possible route for research on icephobicity.

Another fundamentally different approach inspired by nature is based on the use of antifreeze proteins (AFP). It has been known[64] since the 1970s that AFP are able to lower the temperature at which ice growth occurs, without affecting the temperature at which ice melts, since AFP are capable to adsorbing to the ice surface and preventing it from growing. The interested reader is referred to ref. 65 for a detailed description of the antifreezing mechanism of AFP, based on molecular dynamic simulations. In recent years a few attempts have been made to attach AFP to polymer chains[66] and use the polymer–protein conjugate to coat a glass substrate, and to stabilize proteins to prevent denaturation, allowing formation of a stable coating (~12 days) on an aluminium substrate.[67] Designing and developing anti-icing surfaces using AFP is a challenging route, requiring a strong collaboration of biology, nucleation thermodynamics, and surface chemistry and engineering that should continue to garner attention in the future.

11.7 Surface Durability Considerations

In view of practical application, one essential aspect of the development of anti-icing surfaces is surface durability. Although this chapter does not address specifically how anti-icing surface durability for practical cases can be improved, it should be emphasized that ultimately the durability of such coatings is what will determine their wide usage in industry. Development of long-lasting surfaces, over a time scale of days or years, depending on the specific application, currently represents one of the major bottlenecks for applications of non-wetting surfaces in general, and specifically in the context of anti-icing applications. Surfaces should be able to maintain their anti-icing properties by resistance to mechanical abrasion, chemical attack, atmospheric drop impact, extreme temperature cycles, and icing/de-icing cycles, as well as stresses induced by ice shedding or associated to phase-change processes, as in boiling applications.

At present, no universal standard exists for assessing durability of non-wetting and anti-icing surfaces; one reason being that durability requirements also depend on the specific application. However, a few attempts have been made recently to define a standard protocol for non-wetting surfaces,[68,69] and anti-icing durability has been evaluated in the contest of ice shedding[54,70] and de-icing cycles. In particular, mechanical damaging of the tips of surface asperities during consecutive cycles of icing and ice-shedding phases has been highlighted.[54] However, a more recent study[71] based on a multi-tier hierarchical surface showed that surfaces were less sensitive to ice-shedding damage, and no significant deterioration in case of consecutive icing and ice-melting cycles was seen. This is promising for the use of anti-icing surfaces in combination with an intermittent de-icing heating system.[11] Hierarchically structured surfaces, combining topography at different scales as well as multi-tier functionalization layers,[72] are beneficial not only for non-wetting, but also for improved mechanical robustness.[73]

11.8 Conclusions

The current state of the understanding and application of non-wetting (or superhydrophobic) surfaces, as anti-icing surfaces, is very promising. There are multiple mechanisms by which a non-wetting surface can help with mitigating icing depending on the environmental conditions and application (*e.g.* if a shearing air flow exists). There are also indications that adhesion strength of ice to non-wetting surfaces can be lower than for traditional surface coatings. However, it is also clear that a coating is likely not a standalone solution but a pathway to reduce the use of thermal energy, or chemicals, or mechanical force needed to clear a surface of ice. Furthermore, this chapter has shown that depending on the environmental conditions (*e.g.* humidity, frost, or presence of supercooled drops rather than just cold drops), or dynamical conditions of drop–surface interactions (*e.g.* impacting drops or sessile drops) for the system of interest, there will not be a one-size-fits-all solution as some of the literature may suggest.

References

1. BEA, Final Report *On the accident on 1st June 2009 to the Airbus A330–203 registered F-GZCP operated by Air France flight AF 447*, Rio de Janeiro – Paris, 2012.
2. L. B. Boinovich and A. M. Emelyanenko, *Mendeleev Commun.*, 2013, **23**, 3–10.
3. J. Lv, Y. Song, L. Jiang and J. Wang, *ACS Nano*, 2014, **8**, 3152–3169.
4. T. M. Schutzius, S. Jung, T. Maitra, P. Eberle, C. Antonini, C. Stamatopoulos and D. Poulikakos, *Langmuir*, 2015, **31**, 4807–4821.

5. A. M. Palacios, J. L. Palacios and L. Sánchez, *Expert Syst. Appl.*, 2012, **39**, 10212–10225.
6. L. Cao, A. K. Jones, V. K. Sikka, J. Wu and D. Gao, *Langmuir*, 2009, **25**, 12444–12448.
7. S. Jung, M. Dorrestijn, D. Raps, A. Das, C. M. Megaridis and D. Poulikakos, *Langmuir*, 2011, **27**, 3059–3066.
8. T. Maitra, M. K. Tiwari, C. Antonini, P. Schoch, S. Jung, P. Eberle and D. Poulikakos, *Nano Lett.*, 2014, **14**, 172–182.
9. G. Heydari, E. Thormann, M. Järn, E. Tyrode and P. M. Claesson, *J. Phys. Chem. C*, 2013, **117**, 21752–21762.
10. T. Maitra, C. Antonini, M. K. Tiwari, A. Mularczyk, Z. Imeri, P. Schoch and D. Poulikakos, *Langmuir*, 2014, **30**, 10855–10861.
11. J. B. Boreyko, B. R. Srijanto, T. D. Nguyen, C. Vega, M. Fuentes-Cabrera and C. P. Collier, *Langmuir*, 2013, **29**, 9516–9524.
12. Q. Zhang, M. He, J. Chen, J. Wang, Y. Song and L. Jiang, *Chem. Commun.*, 2013, **49**, 4516–4518.
13. C. Antonini, M. Innocenti, T. Horn, M. Marengo and A. Amirfazli, *Cold Reg. Sci. Technol.*, 2011, **67**, 58–67.
14. S. Tarquini, C. Antonini, A. Amirfazli, M. Marengo and J. Palacios, *Cold Reg. Sci. Technol.*, 2014, **100**, 50–58.
15. D. Mangini, C. Antonini, M. Marengo and A. Amirfazli, *Cold Reg. Sci. Technol.*, 2015, **109**, 53–60.
16. P. Tourkine, M. Le Merrer and D. Quéré, *Langmuir*, 2009, **25**, 7214–7216.
17. P. Eberle, M. K. Tiwari, T. Maitra and D. Poulikakos, *Nanoscale*, 2014, **6**, 4874–4881.
18. A. Alizadeh, M. Yamada, R. Li, W. Shang, S. Otta, S. Zhong, L. Ge, A. Dhinojwala, K. R. Conway, V. Bahadur, A. J. Vinciquerra, B. Stephens and M. L. Blohm, *Langmuir*, 2012, **28**, 3180–3186.
19. A. J. B. Milne and A. Amirfazli, *Langmuir*, 2009, **25**, 14155–14164.
20. S. Jung, M. K. Tiwari, N. V. Doan and D. Poulikakos, *Nat. Commun.*, 2012, **3**, 615.
21. C. Antonini, A. Amirfazli and M. Marengo, in *15th International Heat Transfer Conference, IHTC-15*, 2014.
22. E. B. Moore and V. Molinero, *Nature*, 2011, **479**, 506–508.
23. D. Lamb and J. Verlinde, *Physics and Chemistry of Clouds*, 2011.
24. H. R. Pruppacher and J. D. Klett, *Microphysics of Clouds and Precipitation*, 1997.
25. K. Li, S. Xu, W. Shi, M. He, H. Li, S. Li, X. Zhou, J. Wang and Y. Song, *Langmuir*, 2012, **28**, 10749–10754.
26. K. Li, S. Xu, J. Chen, Q. Zhang, Y. Zhang, D. Cui, X. Zhou, J. Wang and Y. Song, *Appl. Phys. Lett.*, 2014, **104**, 101605.
27. M. P. Goertz, J. E. Houston and X.-Y. Zhu, *Langmuir*, 2007, **23**, 5491–5497.
28. H. H. G. Jellinek, *J. Appl. Phys.*, 1961, **32**, 1793.
29. S. Jähnert, F. Vaca Chávez, G. E. Schaumann, a. Schreiber, M. Schönhoff and G. H. Findenegg, *Phys. Chem. Chem. Phys.*, 2008, **10**, 6039–6051.
30. T. Maitra, M. K. Tiwari, C. Antonini, P. Schoch, S. Jung, P. Eberle and D. Poulikakos, *Nano Lett.*, 2014, **14**, 172–182.

31. G. Fang and A. Amirfazli, *Surf. Innovations*, 2014, **2**, 94–102.
32. K. K. Varanasi, T. Deng, J. D. Smith, M. Hsu and N. Bhate, *Appl. Phys. Lett.*, 2010, **97**, 234102.
33. C. Antonini, F. J. Carmona, E. Pierce, M. Marengo and A. Amirfazli, *Langmuir*, 2009, **25**, 6143–6154.
34. D. Quéré, *Rep. Prog. Phys.*, 2005, **68**, 2495–2532.
35. D. K. Mandal, A. Criscione, C. Tropea and A. Amirfazli, *Langmuir*, 2015, **31**, 9340–9347.
36. I. V. Roisman, A. Criscione, C. Tropea, D. K. Mandal and A. Amirfazli, *Phys. Rev. E*, 2015, **92**, 023007.
37. A. J. Meuler, J. D. Smith, K. K. Varanasi, J. M. Mabry, G. H. McKinley and R. E. Cohen, *ACS Appl. Mater. Interfaces*, 2010, **2**, 3100–3110.
38. J. Boreyko and C.-H. Chen, *Phys. Rev. Lett.*, 2009, **103**, 184501.
39. M. He, X. Zhou, X. Zeng, D. Cui, Q. Zhang, J. Chen, H. Li, J. Wang, Z. Cao, Y. Song and L. Jiang, *Soft Matter*, 2012, **8**, 6680.
40. J. B. Boreyko and C. P. Collier, *ACS Nano*, 2013, **7**, 1618–1627.
41. M. Marengo, C. Antonini, I. V. Roisman and C. Tropea, *Curr. Opin. Colloid Interface Sci.*, 2011, **16**, 292–302.
42. D. Richard, C. Clanet and D. Quéré, *Nature*, 2002, **417**, 811.
43. D. Bartolo, F. Bouamrirene, E. Verneuil, A. Buguin, P. Silberzan and S. Moulinet, *Europhys. Lett.*, 2006, **74**, 299–305.
44. M. Reyssat, A. Pépin, F. Marty, Y. Chen and D. Quéré, *Europhys. Lett.*, 2006, **74**, 306–312.
45. M. A. Floriano and C. A. Angell, *J. Phys. Chem.*, 1990, **94**, 4199–4202.
46. J. Hallett, *Proc. Phys. Soc.*, 1963, **82**, 1046–1050.
47. I. V. Roisman, A. Criscione, C. Tropea, D. K. Mandal and A. Amirfazli, *Phys. Rev. E*, 2015, **92**, 023007.
48. S. Šikalo, H.-D. Wilhelm, I. V. Roisman, S. Jakirlic' and C. Tropea, *Phys. Fluids*, 2005, **17**, 062103.
49. V. Bahadur, L. Mishchenko, B. Hatton, J. A. Taylor, J. Aizenberg and T. Krupenkin, *Langmuir*, 2011, **27**, 14143–14150.
50. C. Antonini, I. Bernagozzi, S. Jung, D. Poulikakos and M. Marengo, *Phys. Rev. Lett.*, 2013, **111**, 014501.
51. T. M. Schutzius, S. Jung, T. Maitra, G. Graeber, M. Köhme and D. Poulikakos, *Nature*, 2015, **527**, 82–85.
52. T. Maitra, S. Jung, M. E. Giger, V. Kandrical, T. Ruesch and D. Poulikakos, *Adv. Mater. Interfaces*, 2015, **2**, 1500330.
53. S. A. Kulinich and M. Farzaneh, *Langmuir*, 2009, **25**, 8854–8856.
54. S. A. Kulinich, S. Farhadi, K. Nose and X. W. Du, *Langmuir*, 2011, **27**, 25–29.
55. T.-S. Wong, S. H. Kang, S. K. Y. Tang, E. J. Smythe, B. D. Hatton, A. Grinthal and J. Aizenberg, *Nature*, 2011, **477**, 443–447.
56. K. Rykaczewski, S. Anand, S. B. Subramanyam and K. K. Varanasi, *Langmuir*, 2013, **29**, 5230–5238.
57. A. Lafuma and D. Quéré, *Nat. Mater.*, 2003, **2**, 457–460.
58. J. Chen, R. Dou, D. Cui, Q. Zhang, Y. Zhang, F. Xu, X. Zhou, J. Wang, Y. Song and L. Jiang, *ACS Appl. Mater. Interfaces*, 2013, **5**, 4026–4030.

59. J. Chen, Z. Luo, Q. Fan, J. Lv and J. Wang, *Small*, 2014, **10**, 4413.
60. R. Dou, J. Chen, Y. Zhang, X. Wang, D. Cui, Y. Song, L. Jiang and J. Wang, *ACS Appl. Mater. Interfaces*, 2014, **6**, 6998–7003.
61. P. Kim, T.-S. Wong, J. Alvarenga, M. J. Kreder, W. E. Adorno-Martinez and J. Aizenberg, *ACS Nano*, 2012, **6**, 6569–6577.
62. J. D. Smith, R. Dhiman, S. Anand, E. Reza-Garduno, R. E. Cohen, G. H. McKinley and K. K. Varanasi, *Soft Matter*, 2013, **9**, 1772.
63. J. Petit and E. Bonaccurso, *Langmuir*, 2014, **30**, 1160–1168.
64. J. A. Raymond and A. L. Devriesf, *Proc. Natl. Acad. Sci. U. S. A.*, 1977, **74**, 2589–2593.
65. D. R. Nutt and J. C. Smith, *J. Am. Chem. Soc.*, 2008, **130**, 13066–13073.
66. A. P. Esser-Kahn, V. Trang and M. B. Francis, *J. Am. Chem. Soc.*, 2010, **132**, 13264–13269.
67. Y. Gwak, J.-I. Park, M. Kim, H. S. Kim, M. J. Kwon, S. J. Oh, Y.-P. Kim and E. Jin, *Sci. Rep.*, 2015, **5**, 12019.
68. I. Malavasi, I. Bernagozzi, C. Antonini and M. Marengo, *Surf. Innovations*, 2014, **3**, 49–60.
69. T. Maitra, C. Antonini, M. Auf der Mauer, C. Stamatopoulos, M. K. Tiwari and D. Poulikakos, *Nanoscale*, 2014, **6**, 8710–8719.
70. Y. Wang, J. Xue, Q. Wang, Q. Chen and J. Ding, *ACS Appl. Mater. Interfaces*, 2013, **5**, 3370–3381.
71. L. Zhu, J. Xue, Y. Wang, Q. Chen, J. Ding and Q. Wang, *ACS Appl. Mater. Interfaces*, 2013, **5**, 4053–4062.
72. T. Maitra, C. Antonini, M. Auf der Mauer, C. Stamatopoulos, M. K. Tiwari and D. Poulikakos, *Nanoscale*, 2014, **6**, 8710–8719.
73. W. Li and A. Amirfazli, *Soft Matter*, 2008, **4**, 462.

CHAPTER 12

Oil–Water Separation with Selective Wettability Membranes

ETHAN POST[†a,b], GIBUM KWON[†b,c] AND ANISH TUTEJA*[a,b,c,d]

[a]Macromolecular Science and Engineering, University of Michigan – Ann Arbor, MI 48109, USA; [b]Biointerfaces Institute, University of Michigan – Ann Arbor, MI 48109, USA; [c]Department of Materials Science and Engineering, University of Michigan – Ann Arbor, MI 48109, USA; [d]Department of Chemical Engineering, University of Michigan – Ann Arbor, MI 48109, USA
*E-mail: atuteja@umich.edu

12.1 Introduction

Incredibly large volumes of oil–water mixtures are produced worldwide in a wide variety of industries. These mixtures range from free oil and water to surfactant-stabilized oil–water emulsions, which are particularly difficult and expensive to separate. Oil–water separation is a widely used unit operation in many industries including textile and leather processing, rendering, metal fabrication and machining, wastewater treatment, petroleum drilling and refining, and fracking.[1] Estimates show that for every barrel of oil extracted, 3–10 barrels of water are produced as a byproduct.[2] The limitations

[†]Equal contributing authors.

RSC Soft Matter No. 5
Non-wettable Surfaces: Theory, Preparation, and Applications
Edited by Robin H. A. Ras and Abraham Marmur

Published by the Royal Society of Chemistry, www.rsc.org

on oil and grease content set by the United States Environmental Protection Agency have become increasingly stringent over the years. The best available technology (BAT) limit on oil and grease discharge in produced water is now 42 mg L^{-1} for any one day, with a 30 consecutive-day average of 29 mg L^{-1}.[3] Depending on the industry, the oil and grease concentration in the untreated effluent can typically range from a few hundred to 200 000 mg L^{-1}.[4] The large volumes of contaminated mixtures, including effluents from accidents such as the Deepwater Horizon spill, necessitate the development of durable, cost-effective means of selectively and quickly separating oil–water mixtures.

Numerous methods, including gravity separation, flotation, oil-absorbing materials, electrocoagulation, and flocculation, have traditionally been used to separate oil–water mixtures.[1,5–10] Several issues are encountered while using these methods including: unsuitability in separating emulsions,[6] lack of selectivity and low separation efficiency,[11,12] high energy consumption,[9] or secondary pollution.[13] Membrane-based methods for oil–water separation are now gaining increased attention due to their energy efficiency, versatility in treating a variety of industrial waste streams, and consistent performance.[1,6] However one of the biggest, current challenges with membranes, limiting their widespread usage, is fouling due to surfactant adsorption or pore plugging by oil droplets, resulting in significantly diminished permeate flux.[14,15]

Research on surfaces with selective wettabilities promises to improve the efficiency of and imbue anti-fouling properties to membranes for oil–water separation.[1,16] If a membrane demonstrates a differing wettability between water and oil, it may be useful for the extremely efficient separation of oil–water mixtures.[17] This idea has led to a large number of membranes with selective wettability being developed for separating a range of different oil–water mixtures. In this chapter, we briefly discuss the design strategies for membranes with selective wettability. This includes the parameterization of two important physical characteristics: the surface porosity and the breakthrough pressure. We also discuss how they are related for membranes with a periodic geometry. On the basis of this understanding, we explore principles that allow for the systematic design of membranes with selective wettability.

12.2 Fundamentals of Wettability

A surface's wettability is commonly characterized by a contact angle.[18] On a non-textured (or smooth) surface, a liquid's equilibrium contact angle θ is given by Young's relation:[19]

$$\cos\theta = \frac{\gamma_{SV} - \gamma_{SL}}{\gamma_{LV}} \tag{12.1}$$

here, γ is the interfacial tension between two phases and S, L, and V refer to the solid, liquid, and vapour phases, respectively. Thus, γ_{LV} is the interfacial tension between the liquid and vapour phases, and it is commonly called the liquid surface tension. γ_{SV} is typically referred to as the solid surface energy. Based on previous literature,[20–22] the wettability of the solid surface can be classified into four regimes using contact angles for water: superhydrophilic ($\theta_{water} \sim 0°$), hydrophilic (HL) ($\theta_{water} < 90°$), hydrophobic (HP) ($\theta_{water} > 90°$), and superhydrophobic ($\theta_{water} > 150°$). Similarly, for low surface tension liquids such as oils or alcohols, surfaces are considered superoleophilic ($\theta_{oil} \sim 0°$), oleophilic (OL) ($\theta_{oil} < 90°$), oleophobic (OP) ($\theta_{oil} > 90°$), and superoleophobic ($\theta_{oil} > 150°$). Superhydrophobic or superoleophobic surfaces are commonly referred to as super-repellent surfaces.

On a textured (or rough) surface, a liquid's apparent contact angle (θ^*) can be significantly different from the Young's contact angle θ. A liquid droplet on a textured surface may realize either the "fully-wetted" Wenzel[23] state or the Cassie–Baxter[24] state, forming a composite solid–liquid–air interface (Figure 12.1a). In the Wenzel state, the overall free energy reaches its minimum when the apparent contact angle becomes θ^*, given by the Wenzel relation as:[23]

$$\cos\theta^* = r\cos\theta \tag{12.2}$$

The roughness, r, is the ratio of the actual surface area (Figure 12.2a) to the projected surface area and is greater than 1 by definition. Consequently, roughness yields a lower apparent contact angle for a liquid with $\theta < 90°$ and a higher apparent contact angle if $\theta > 90°$.

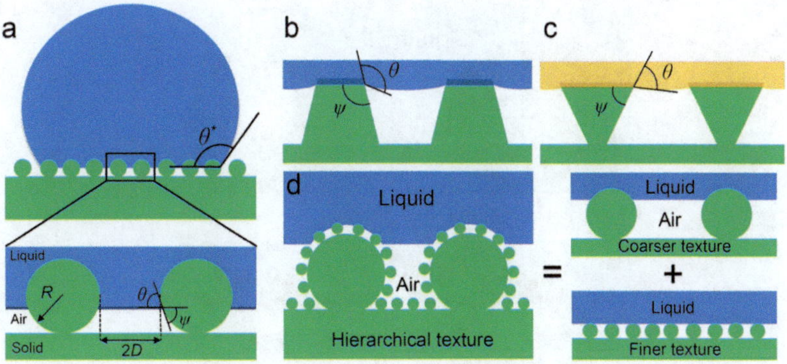

Figure 12.1 Liquid droplets on textured surfaces. (a) The Cassie–Baxter state is shown. The Wenzel state is similar, but with the liquid completely filling in the pores and wetting the solid. In the diagram, R is the feature radius, $2D$ is the inter-feature spacing, θ is the equilibrium contact angle, θ^* is the apparent contact angle, and ψ is the texture angle. (b) A Cassie–Baxter state on a concave texture with $\psi > 90°$ and $\theta > 90°$. (c) A similar state exists with a lower surface tension liquid ($\theta < 90°$) on convex, re-entrant texture ($\psi < 90°$). (d) A hierarchical texture combines coarser and finer textures to maximize the solid–air interface. Adapted from Kota *et al.*[22] © 2014 with permission from Nature Publishing Group.

The Cassie–Baxter relation for a composite solid–liquid–air interface is given as:

$$\cos\theta^* = f_{SL}\cos\theta + f_{LV}\cos\pi = f_{SL}\cos\theta - f_{LV} \tag{12.3}$$

This relation shows the impact of the local areal fractions of the solid–liquid (f_{SL}) and liquid–air (f_{LV}) interfaces, in the vicinity of the triple-phase (solid–liquid–air) contact line, on the apparent contact angle.[25] The fraction f_{SL} is defined as the actual solid–liquid area divided by a projected unit area (normal to the surface) and f_{LV} is the actual liquid–vapour area per projected unit area (Figure 12.2b). For most surfaces, the local and global areal fractions are equivalent due to homogeneity.

Eqn (12.2) and (12.3) show that high apparent contact angles ($\theta^* > 150°$) can be observed in either the Wenzel state, if $\theta > 90°$ and $r \gg 1$, or in the Cassie–Baxter state, if $f_{SL} \ll 1$. However, contact angle hysteresis (*i.e.* the difference between the advancing, maximum contact angle on a given surface and receding, minimum contact angle on a given surface) is minimal for super-repellent surfaces. Hysteresis arises due to the presence of multiple, metastable energy states on real, heterogeneous surfaces.[26] Typically, the contact angle hysteresis is larger in the Wenzel state due to solid–liquid interfacial pinning on the fully wetted, textured surface.[27] By contrast, a composite solid–liquid–air interface in the Cassie–Baxter state leads to lower contact angle hysteresis and higher apparent contact angles due to decreased contact area between

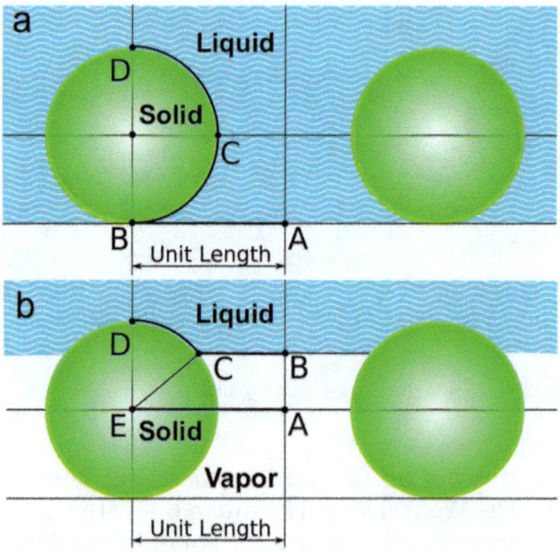

Figure 12.2 Fractional notation for the Wenzel and Cassie–Baxter relations using example cases. (a) For the Wenzel state, the roughness, r, is the arc length BCD + AB divided by the unit length AB. (b) For the Cassie–Baxter state, the areal fraction of solid–liquid (f_{SL}) is arc length DC divided by the unit length EA and the areal fraction of liquid–vapour (f_{LV}) is CB divided by EA. By definition, $f_{SL} + f_{LV} \geq 1$. Adapted from ref. 24 with permission of the Royal Society of Chemistry.

the solid and the liquid.[27,28] Consequently, composite interfaces are essential for fabricating super-repellent surfaces.

12.3 Design Strategies for Composite Membranes with Selective Wettability

Although composite interfaces are necessary for engineering super-repellent surfaces, the details of the surface texture can significantly affect the stability or robustness of a composite interface. Previous literature[29–32] revealed that a stable Cassie–Baxter state may be formed if the Young's contact angle θ is greater than or equal to the local texture angle ψ. In Figure 12.1b, the texture angle $\psi > 90°$ can lead to the formation of a composite interface when $\theta \geq \psi$. If $\theta < 90°$, which is common for oils and other low surface tension liquids, a stable composite interface cannot be maintained, regardless of its surface energy or composition. However, for the same low surface tension liquid with $\theta < 90°$, it is possible to support a composite interface as long as $\theta \geq \psi$. A surface geometry $\psi < 90°$ is said to possess re-entrant texture (Figure 12.1c). Surfaces with re-entrant texture enable the formation of composite interfaces with low surface tension liquids, and thereby allowing for the possibility of OP (or superoleophobic) properties.

Systematic design of membranes for oil–water separation requires the parametrization of surface porosity and breakthrough pressure.[33] Surface porosity affects the rate of liquid permeation through the membrane, which must be matched with the flow rate/flux requirements of a particular application at a given pressure. The Hagen–Poiseuille relation[34] shows that the volumetric flow rate $Q \propto r^4$ (where r is the pore radius), with all other parameters held constant. Smaller pore sizes increase the viscous resistance to fluid flow and decrease Q, but are necessary for separating smaller-sized emulsions. Although the Hagen–Poiseuille relation provides a correlation between the flow rate and the pore size, it does not incorporate the effect of pore spacing on flux. Previous work[20,31,35,36] discussed the spacing ratio, D^*, a dimensionless measure of surface porosity that considers both the pore size and spacing. For surfaces with cylindrical texture, such as interwoven meshes or fabrics, $D^*_{\text{cylinder}} = (R+D)/R$, where R is the cylinder radius and $2D$ is the inter-cylinder spacing. As D increases for a constant R, the membrane pore size and surface porosity increase. Therefore, membranes with a higher D^* will show greater permeation rates for a given contacting liquid.

The other critical parameter is the breakthrough pressure ($P_{\text{breakthrough}}$), which is the maximum transmembrane pressure withstood before a given liquid permeates a membrane. Previous work[31,35–37] parametrized $P_{\text{breakthrough}}$ with the robustness factor, A^*. This dimensionless value is obtained by scaling $P_{\text{breakthrough}}$ with respect to a reference pressure $P_{\text{ref}} = 2\gamma_{\text{LV}}/l_{\text{cap}}$. Here, $l_{\text{cap}} = \sqrt{\gamma_{\text{LV}}/\rho g}$ is the capillary length of a liquid, ρ is the liquid density, and g is the acceleration due to gravity. P_{ref} is close to the minimum possible pressure differential across a millimetre-sized liquid droplet or a puddle.[31]

Consequently, large values of A^* ($A^* \gg 1$) correlate to robust composite interfaces with high $P_{breakthrough}$ for a given contacting liquid. For $A^* \leq 1$, a composite interface cannot be supported, as the contacting liquid penetrates into the pores and is fully imbibed. The robustness factor, for a surface possessing predominantly cylindrical texture, is given as:[35,37]

$$A^*_{cylinder} = \frac{P_{breakthrough}}{P_{ref}} = \frac{l_{cap}}{R(D^*_{cylinder} - 1)} \frac{(1 - \cos\theta)}{(D^*_{cylinder} - 1 + 2\sin\theta)} \quad (12.4)$$

For the effective separation of oil and water, membranes must be designed for a high permeation rate of one phase (*e.g.* water) and simultaneously, a high breakthrough pressure for the other phase (*e.g.* oil). This can be achieved by maximizing A^* for the repelled phase, and maintaining a large D^* to achieve high permeation rate/flux for the second phase.

D^* and A^* are strongly coupled for membranes with a periodic, cylindrical geometry,[31,35,37] as is evident from eqn (12.4). The value of D^* can be increased by either reducing R or increasing D, both of which lead to a decrease in A^*. As discussed above, it is crucial to increase A^* without affecting D^*. Such an enhancement can be achieved by adding low surface energy materials to the membrane, which increases the values of Young's contact angle θ. This increases A^* and the breakthrough pressure without changing the membrane geometry. However, significant lowering of the solid surface energy may result in omniphobic surfaces, which repel both water and oil[31,38,39] and are incapable of separations in most circumstances (an exception is the HP/OP membrane described later). As described by Kota *et al.*,[22] there are other design methods for increasing A^* without affecting D^* and *vice versa.* By reducing the length scale of the texture in such a way that both R and D are decreased, while keeping D^* constant, A^* increases according to eqn (12.4). Conversely, D^* can be increased, while A^* remains constant, by adding hierarchical scales of texture (Figure 12.1d). Composite interfaces are the least stable on the largest scale of texture, so $A^*_{hierarchical} \approx A^*_{micro}$ (if both micro and nanostructures are present), while $D^*_{hierarchical} \gg D^*_{micro}$ due to the extra air space within the multiple texture scales. Both A^* and D^* can be maximized in this way for designing robust, repellent membranes with a high permeation rate of a desired liquid.

Membranes possessing a high A^* value for one phase ($A^*_{liquid1} >> 1$), as well as a small A^* value for the second phase ($A^*_{liquid2} \leq 1$), allow for selective permeation of one liquid over the other. Such membrane surfaces display significant differences between θ^*_{water} and θ^*_{oil}. In general, membranes can be categorized into four groups based on their contact angles with oil and water (Figure 12.3): HP/OL, HL/OL, HL/OP, and HP/OP. Figure 12.4 shows how these membranes can allow selective separation of immiscible oil–water mixtures based on the differing component contact angles. In the following sections, we discuss recent developments and the merits of each membrane type used for the separation of oil and water, based on where they fall on the wettability landscape shown in Figure 12.3. As will be evident, the fabrication methods of various membranes, with different wettabilities, share many common aspects such as controlling the surface texture and surface energy.

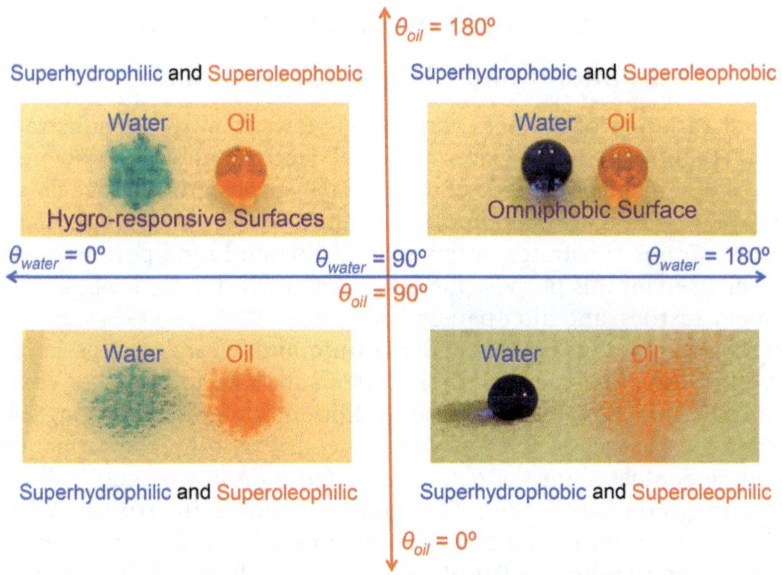

Figure 12.3 Membrane wettability classifications with oil and water. A membrane is HL/OP when $\theta^*_{water} < 90°$ and $\theta^*_{oil} > 90°$, HP/OP when $\theta^*_{water} > 90°$ and $\theta^*_{oil} > 90°$, HL/OL when $\theta^*_{water} < 90°$ and $\theta^*_{oil} < 90°$, and HP/OL when $\theta^*_{water} > 90°$ and $\theta^*_{oil} < 90°$.[16] Reprinted with permission from G. Kwon, E. Post and A. Tuteja: Membranes with selective wettability for the separation of oil–water mixtures. *MRS Communications* **5**, 475 (2015). Copyright 2015 Cambridge University Press.

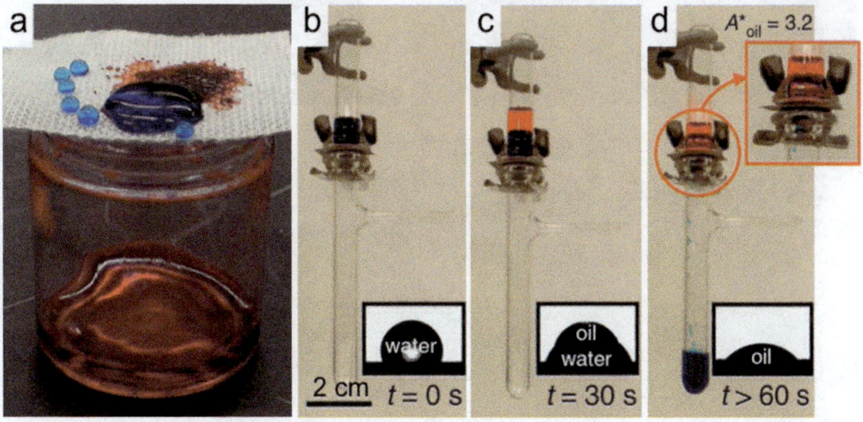

Figure 12.4 Principle of separation for selective wettability membranes. (a) A hydrophobic and oleophilic membrane separates oil (red) and water (blue) by allowing oil to permeate through, while repelling water. From A. Tuteja, W. Choi, M. L. Ma, J. M. Mabry, S. A. Mazzella, G. C. Rutledge, G. H. McKinley and R. E. Cohen: Designing superoleophobic surfaces. *Science* **318**, 1618 (2007). Reprinted with permission from AAAS. (b–d) A hydrophilic and oleophobic membrane, with similar liquids, allows water to permeate, while repelling oil. Reprinted from Kota *et al.*[33] © 2012 with permission from Nature Publishing Group.

12.4 Membranes with Selective Wettability

12.4.1 Hydrophobic and Oleophilic Membranes

The first variation of surfaces with selective wettability is a substrate that is both HP and OL. Lotus leaves and duck feathers are well known natural surfaces that display this type of wettability. Various groups have developed HP/OL membranes for oil–water separation by coating a hydrophobic material onto porous substrates. A range of flexible and rigid porous substrates has been used for this purpose including: stainless steel and copper meshes, polymers, textiles, and filter papers.

Meshes inherently have a regular texture and several groups have used steel and stainless steel wire meshes as a substrate to be either coated or chemically modified. In most cases, additional, hierarchical roughness is added to the existing mesh while simultaneously tailoring the surface energy to achieve contact angles of $\theta^*_{water} > 150°$ and $\theta^*_{oil} \approx 0°$ (to enable oil permeation through the membrane). These HP/OL surfaces, in turn, allow for high efficiency separations of oil and water mixtures. The different coating methods employed to fabricate HP/OL membranes include: spray coating with a polytetrafluoroethylene (PTFE) emulsion (Figure 12.5a–d),[40] growing rough ZnO crystals on the mesh and coating with PTFE,[41] and electrospinning fluorodecyl polyhedral oligomeric silsesquioxane-poly(methyl methacrylate)

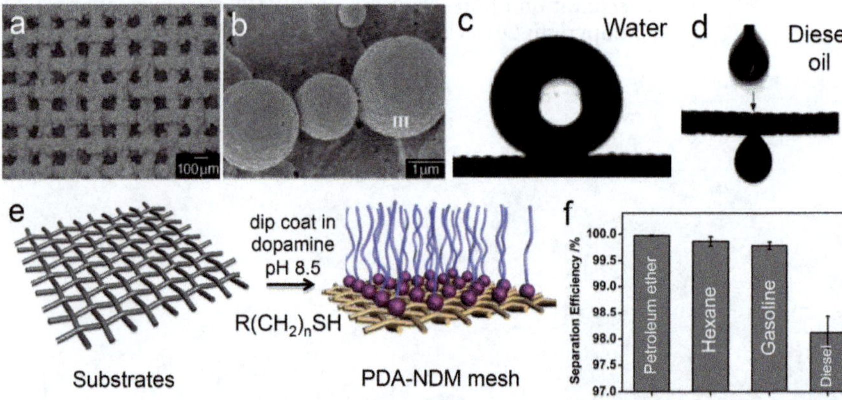

Figure 12.5 Coated stainless steel meshes with HP/OL selective wettability. (a–b) Textured Teflon coating on a stainless steel wire mesh. (c–d) The superhydrophobic and superoleophilic nature of the mesh. Reproduced from Feng *et al.*[40] © 2004 Wiley-VCH Verlag GmbH & Co. KGaA, Weinheim with permission from John Wiley & Sons, Inc. (e) The fabrication of an *n*-dodecyl mercaptan treated mesh. (f) Efficiencies for the separation of a variety of oils and water with the PDA-NDM mesh. Adapted with permission from Y. Z. Cao, X. Y. Zhang, L. Tao, K. Li, Z. X. Xue, L. Feng and Y. Wei: Mussel-Inspired Chemistry and Michael Addition Reaction for Efficient Oil–water Separation. *ACS Appl Mater Inter* **5**, 4438 (2013). Copyright 2013 American Chemical Society.

fibres.[20] Other groups have chemically modified wire meshes with 1*H*, 1*H*, 2*H*, 2*H*-perfluoroalkyltriethoxysilane[42] or *n*-dodecyl mercaptan[43] (Figure 12.5e and f) to lower the mesh surface energy and repel water.

Several groups have also used a modified copper mesh substrate for fabricating HP/OL membranes. Hierarchical roughness was added by cathodic electrodeposition or nitric acid etching, followed by treatment with *n*-dodecanoic acid[44] or hexadecanethiol.[45] Other copper meshes have added texture and changed the surface energy in one step by depositing Sylgard 184 silicone elastomer (PDMS) with aerosol-assisted chemical vapour deposition, which formed 3–5 μm tall micropillars on the mesh surface.[46]

Another class of substrates is porous polymers where HP/OL wetting properties may be formed directly, or may also be modified through additional texture and chemical treatments. Membranes have been formed by electrospinning cellulose acetate fibres and dip-coating with a thermosetting fluorinated polybenzoxazine (F-PBZ) monomer and SiO_2 nanoparticles (Figure 12.6a and b).[47] The thermal and mechanical stability of the membranes can be further increased by exchanging poly(*m*-phenylene isophthalamide) plus multi-walled carbon nanotubes for the cellulose acetate and using a new version of F-PBZ.[48] In other work, a single-component, hierarchical poly(vinylidene fluoride) (PVDF) membrane, formed by phase inversion, was found to be capable of separating 5–20 μm sized water-in-oil emulsions, not just free oil and water.[49]

HP/OL filter papers have been developed by coating the underlying porous substrate with polystyrene and HP nanoparticles, such as PDMS-modified silica nanoparticles[50] or PTFE nanoparticles.[51] A third method uses a mixed cellulose ester (MCE) membrane to filter suspensions of single-walled carbon nanotubes (SWCNTs) to form SWCNT films (Figure 12.6c and d). The MCE membrane is dissolved afterward. These novel carbon nanotube (CNT) films met the need for achieving the thinnest membrane possible, while maintaining a useful pore size for maximum permeation rates. Very high permeation rates, up to 107 140 L m^{-2} h^{-1} bar^{-1} for the 30 nm thick film (with a surfactant-free water-in-petroleum ether emulsion), were achieved (Figure 12.6e).[52] Growing silicone nanofilaments (Figure 12.6f–h), by chemical vapour deposition,[53] or depositing metal oxide nanocrystals on textiles (and treating with octadecyl thiol)[54] has yielded useful HP/OL membranes as well.

Although many different methods have now been developed for separating oil and water mixtures with a HP/OL membrane, there are inherent difficulties with this type of wettability. First, gravity separation is prevented if water contacts the HP membrane before oil, due to its higher density. Secondly, these membranes are subject to fouling as oils adsorb to the membrane surface, which decreases the desired permeate flux.[55,56] This can lead to significant downtime, cleaning, and membrane replacement costs when these types of membranes are used in a commercial environment. To overcome these disadvantages, membranes with other selective wettabilities have also been explored, as discussed below.

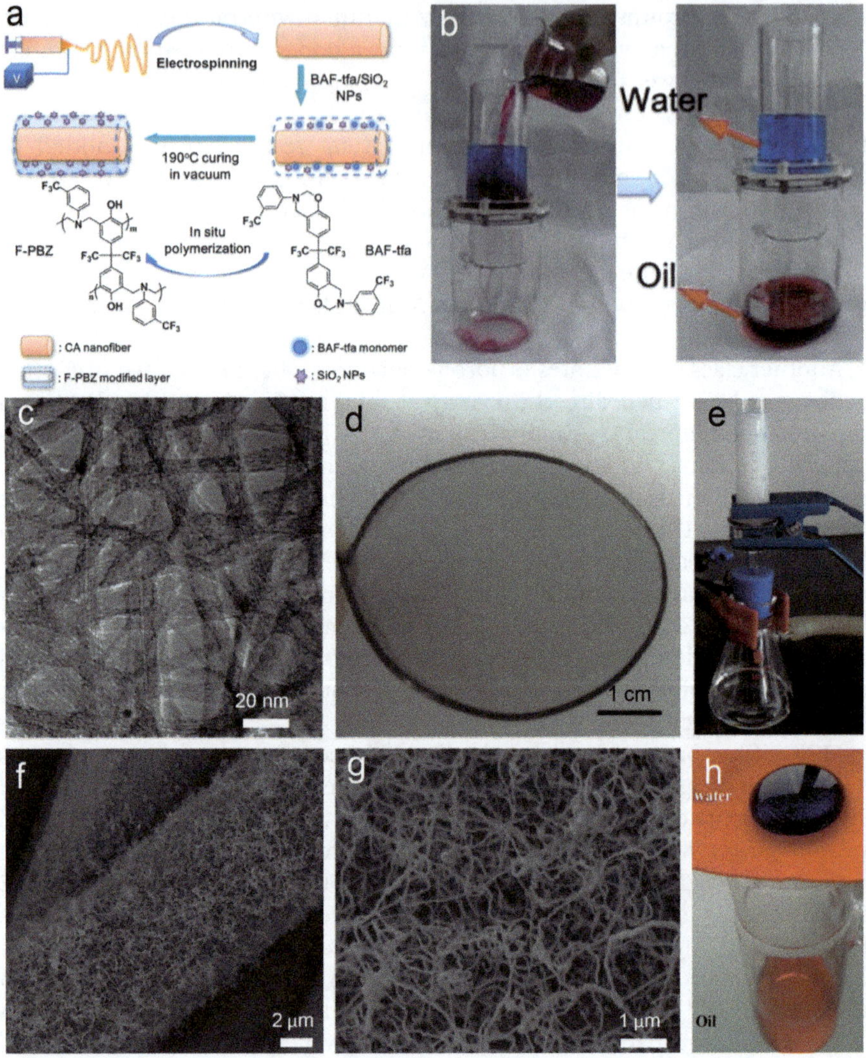

Figure 12.6 Electrospun polymer, carbon nanotube, and silicone nanofilament-based HP/OL membranes. (a) The fabrication strategy for a F-PBZ/SiO_2 nanoparticle-modified, electrospun cellulose acetate membrane and (b) the separation ability of the fabricated membrane tested with a 50% v/v mixture of dichloromethane and water. Adapted from ref. 47 with permission from The Royal Society of Chemistry. (c) A TEM image of a 70 nm thick SWCNT film showing its interlaced structure. (d) The SWCNT film supported by a steel hoop and (e) the selective permeation of oil from an emulsion using this film. Reproduced from Shi *et al.*[52] © 2013 Wiley-VCH Verlag GmbH & Co. KGaA, Weinheim with permission from John Wiley & Sons, Inc. (f–g) Silicone nanofilaments grown on a polyester textile. (h) The separation of a free octane and water mixture. Reproduced from Zhang *et al.*[53] © 2011 Wiley-VCH Verlag GmbH & Co. KGaA, Weinheim with permission from John Wiley & Sons, Inc.

12.4.2 Hydrophilic and Oleophilic Membranes

Recently, the non-wetting behaviour of oil droplets on fish scales underwater has inspired the concept of underwater superoleophobic surfaces.[57] From Young's relation (see eqn (12.1)), it is clear that HL/OL surfaces in air can become OP when underwater.[57,58] In the presence of HL rough structures, water readily wets and fills all the cavities present on the surface, leading to a composite solid–oil–water interface. Similar to the composite solid–oil–air interface formed on superoleophobic surfaces in air, this new composite interface prevents the permeation of oil droplets, yielding underwater superoleophobicity. Such superhydrophilic and underwater superoleophobic surfaces exhibit excellent oil fouling resistance when submerged in water.[57] However, these types of membranes may not be effective in stop-and-go operations where the loss of water would allow oil contamination to occur. A number of membranes that display superhydrophilicity in air and underwater superoleophobicity ($\theta^*_{oil} > 150°$) have been developed in previous work.

One of the early reports on superhydrophilic and underwater superoleophobic membranes concerned rough, polyacrylamide hydrogel-coated stainless steel meshes.[59] The hydrogel coating reduced the affinity for oil droplets, which could foul typical membranes, through a reduction in the adhesion force of an oil droplet from 46.5 ± 2.3 μN, on the uncoated stainless steel mesh, to 0.8 ± 0.3 μN for the underwater, hydrogel-coated mesh. Inspired by shrimp shells, previous work has also looked into the anti-oil-fouling behaviour of chitosan-coated rough copper meshes that could separate a range of oil–water mixtures in hypersaline and broad pH conditions after fully cross-linking the chitosan.[60] More durable, hydrogel-based membranes were fabricated by grafting polyacrylamide-*co*-poly(acrylic acid) hydrogel particles onto a poly(glycidyl methacrylate)-grafted stainless steel mesh.[61] In addition to hydrogels, various other HL materials have also been used for coating porous meshes to engender superhydrophilicity and underwater superoleophobicity. Stainless steel meshes have been coated with graphene oxide nanosheets,[62] as well as pure-silica zeolite, silicalite-1 (Figure 12.7a–c),[63] which showed high separation efficiency under repeated use (Figure 12.7d–f).[64]

In addition to metal meshes, there have been reports of polymeric membranes that are superhydrophilic in air and superoleophobic underwater. This type of selective wettability was obtained through grafting zwitterionic polyelectrolyte brushes onto PVDF microfiltration membranes,[65] grafting poly(acrylic acid) onto polypropylene microfiltration membranes, and then depositing HL, nanosized $CaCO_3$ minerals on top,[66] and co-depositing polydopamine (PDA) and polyethyleneimine (PEI) on polypropylene membranes, with (Figure 12.7g) and without nanosilica.[67,68] The polypropylene membranes[66–68] exhibited the ability to separate polydisperse emulsions (Figure 12.7h). Inorganic fibre filters have also been utilized for oil–water separation. Zwitterionic poly(sulfobetaine methacrylate) has been grafted onto glass fibre filters,[69] and 1,2-bis(triethoxysilyl)ethane and polyacrylamide have

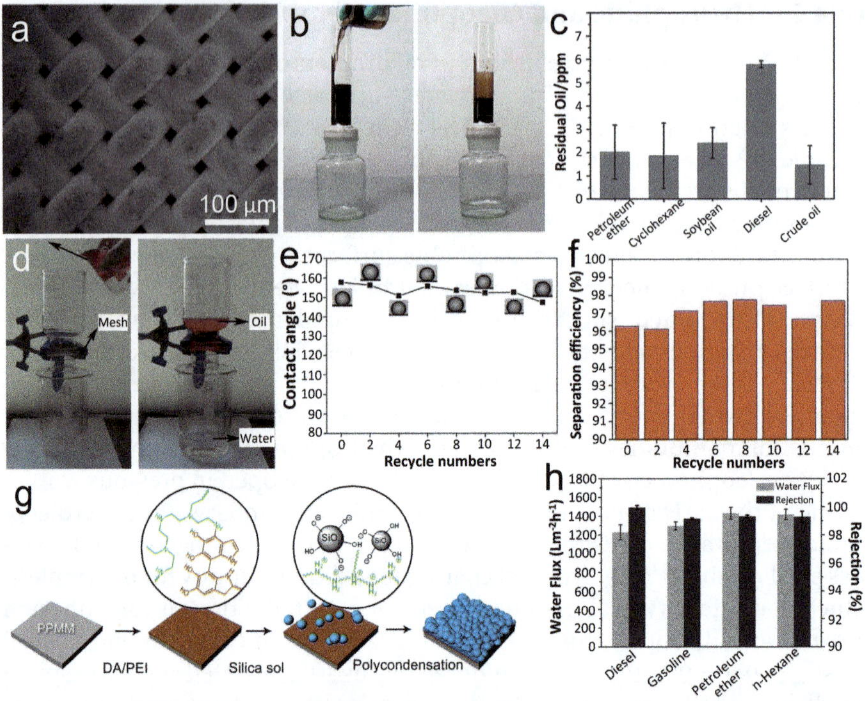

Figure 12.7 HL and underwater OP membranes. (a) An SEM image of a zeolite-coated mesh film (ZCMF-12) and (b) a demonstration of its ability to selectively remove water from crude oil. (c) The residual oil content in water for various oils after the separation. Adapted from ref. 63 with permission from The Royal Society of Chemistry. (d–f) A zeolite membrane on top of stainless steel mesh separated chloroform (dyed red) and water mixtures efficiently, while maintaining high dichloromethane contact angles over 14 separations. Reprinted from *Colloids and Surfaces A: Physicochemical and Engineering Aspects*, 444, Zeng *et al.*, Superhydrophilic and underwater superoleophobic MFI zeolite-coated film for oil/water separation, 283–288, Copyright 2014 with permission from Elsevier.[64] (g) Methodology for producing silica and PDA/PEI decorated polypropylene membranes that (h) show high water permeation, while rejecting several oils from oil-in-water emulsions. Adapted with permission from H. C. Yang, J. K. Pi, K. J. Liao, H. Huang, Q. Y. Wu, X. J. Huang and Z. K. Xu: Silica-Decorated Polypropylene Microfiltration Membranes with a Mussel-Inspired Intermediate Layer for Oil-in-Water Emulsion Separation. *ACS Appl Mater Inter* **6**, 12566 (2014). Copyright 2014 American Chemical Society.

enhanced a silica gel-modified quartz fibre mesh, which became resistant to harsh acidic and saline environments.[70]

Practical applications of HL or superhydrophilic membranes in oil–water separations are limited by contamination from low surface energy oil.[71,72] Once the HL/OL membrane is fouled by oil, it is difficult to remove the adsorbed oil. This leads to decreased separation performance, and necessitates periodic washing of the membranes, resulting in higher operating costs. To overcome this limitation, self-cleaning membranes have also been studied. The following examples are based on the ability of TiO_2 to remove contaminants under ultraviolet (UV) light. One membrane was formed through layer-by-layer (LBL) assembly of sodium silicate and TiO_2 nanoparticles on stainless steel mesh.[73] Full HL recovery of the membrane was seen after five cycles of contamination with oleic acid and then UV treatment. Another self-cleaning membrane was fabricated by the calcination of a titanium membrane to form a HL TiO_2 surface with underwater oleophobicity.[74]

Although membranes with superhydrophilic and underwater superoleophobic properties can be successfully used for gravity-driven separation of oil–water mixtures, and are more resistant to fouling, they are unsuitable for the separation of free water-in-oil or water-in-oil emulsions. This is because both oil and water easily permeate through them, unless every pore within the membrane is pre-wetted by water. Consequently, oil permeates through the membrane if even a single pore dehydrates within the superhydrophilic membrane, which can typically happen in a matter of minutes.[33]

12.4.3 Hydrophilic and Oleophobic Membranes

As discussed in previous sections, HP/OL membranes are unsuitable for most gravity-driven separations. Although HL/OL membranes are applicable for the gravity-driven separation of oil-in-water emulsions, they do not work for free oil–water or water-in-oil emulsions, unless they are repeatedly pre-wetted by water. HL/OP membranes are expected to overcome these limitations. However, it has been considered challenging to fabricate such membranes due to the surface tension of water (γ_{LV} = 72.1 mN m^{-1}) being significantly higher than that for most oils (γ_{LV} = 20–30 mN m^{-1}), which typically yields greater contact angles with water on surfaces.

In recent work,[33] hygro-responsive membranes that are superhydrophilic and superoleophobic both in air and underwater, were successfully fabricated. A polymer blend of 1*H*,1*H*,2*H*,2*H*-heptadecafluorodecyl polyhedral oligomeric silsesquioxane (f-POSS) and cross-linked polyethylene glycol diacrylate (x-PEGDA) was coated on porous substrates, including steel meshes and polyester fabric (Figure 12.8a). The selective HL/OP wettability of the membrane is attributable to the surface reconfiguration of the coating (Figure 12.8b and c). This membrane was also used in tandem with a HP/OL membrane to achieve continuous oil–water emulsion separation (Figure 12.8d).

In addition to polymer blends, synthesis of polymers possessing HL and OP constituents has also been proposed for fabricating HL/OP coating materials. A polymer with HL and OP constituents can be synthesized through the reaction of poly(diallyldimethylammonium chloride) (PDDA) with sodium perfluorooctanoate (PFO). This fabricated material mixed with silica nanoparticles was spray-casted onto various substrates, such as stainless steel meshes and

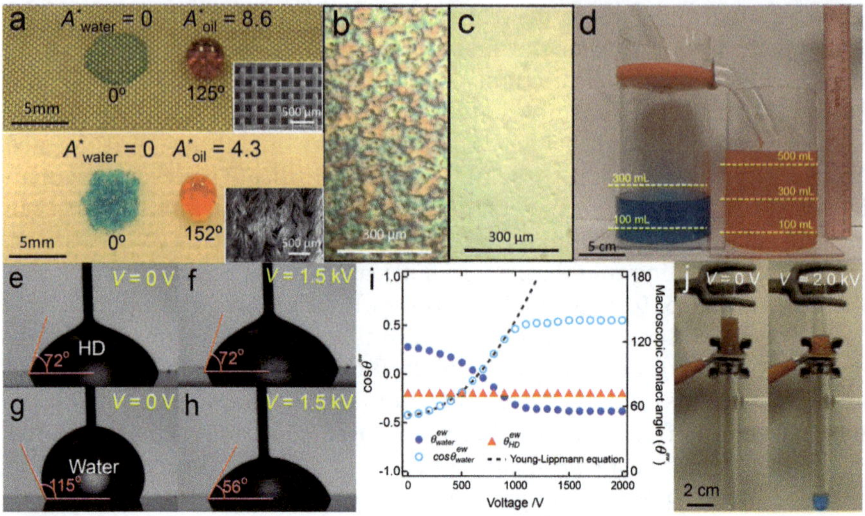

Figure 12.8 HL/OP Hygro-responsive and HP/OP electrowetting membranes. (a) Water (blue) and rapeseed oil (red) contact angles on a stainless steel mesh (top) and a polyester fabric (bottom) dip-coated in 20 wt% fluorodecyl POSS + x-PEGDA blend. (b) Optical microscopy image of a 20 wt% fluorodecyl POSS + x-PEGDA blend surface in air and (c) underwater showing the surface reconfiguration. (d) A continuous separation apparatus separated 30:70 v:v water-in-hexadecane emulsions stabilized by polysorbate80. It used a 20 wt% fluorodecyl POSS + x-PEGDA blend membrane (superhydrophilic and OP) on the bottom, and a Desmopan9370 coated sidewall membrane (HP/OL). The hexadecane and water fluxes were consistent over a period of 100 hours. Reprinted from Kota *et al.*[33] © 2012 with permission from Nature Publishing Group. (e–f) Contact angle of hexadecane on a non-textured 50 wt% fluorodecyl POSS + x-PDMS substrate was unchanged by the application of a 1.5 kV potential, while (g–h) contact angle or water decreased significantly. (i) The macroscopic contact angles for water and hexadecane on the non-textured surface as a function of applied voltage. (j) The EWOD effect was used to separate hexadecane (red) and water (blue) on demand. Adapted from Kwon *et al.*[80] © 2012 Wiley-VCH Verlag GmbH & Co. KGaA, Weinheim with permission from John Wiley & Sons, Inc.

paper, to form superhydrophilic/superoleophobic membranes.[71] Another option for fabricating HL/OP membranes was simply blending the PDDA, PFO, and silica nanoparticles and coating the mesh.[75] Membranes can also be formed from PVDF blended with additive polymers containing perfluoroalkyl polyethylene glycol surfactant chains, which exhibited anti-organic and anti-biofouling properties and were capable of emulsion separation.[76,77] A final method is to bind perfluorinated polyethylene glycol (Zonyl® FSN-100) to glass fibre membranes with a silane linker.[78] These methods all share molecules containing a low energy fluorinated segment and a polar HL segment, which permits water to permeate through the membrane while rejecting oils.

12.4.4 Hydrophobic and Oleophobic Membranes

In contrast to previously discussed membranes, HP/OP or omniphobic membranes prevent permeation of both oil and water. In order to utilize HP/OP membranes for the separation of oil–water mixtures, pressure must be selectively exerted on either the water or oil phase, leading to $P_{\text{breakthrough, water}} < P_{\text{applied}} < P_{\text{breakthrough, oil}}$ or *vice versa*. An electric field is capable of tuning the wettability of polar (or conducting) liquids, and the decrease in the macroscopic contact angle for a polar liquid droplet on a dielectric, in response to an external electric field, is known as electrowetting on a dielectric (EWOD) (Figure 12.8e–i) and is described by the Young–Lippmann equation:[79]

$$\cos\theta^{\text{ew}} = \cos\theta + \frac{\varepsilon_0\varepsilon_{\text{d}}}{2\gamma_{12}d}V^2 \tag{12.5}$$

here, θ^{ew} is the macroscopic electrowetting contact angle, θ is Young's contact angle, ε_0 is the vacuum permittivity, ε_{d} is the dielectric permittivity, γ_{12} is the interfacial tension between the liquid and surrounding medium, d is the dielectric thickness, and V is the voltage applied. Using EWOD, an on-demand oil–water separation triggered by an electric field was developed.[80] An omniphobic membrane was obtained by dip-coating nylon mesh in a blend of 50 wt% f-POSS and cross-linked polydimethylsiloxane. The membrane retained both water and oil until an external electric field was applied across the conducting liquid (*e.g.*, water) and the electrode at the membrane. The conducting liquid, initially in the Cassie–Baxter state on the porous membrane, transitioned to the Wenzel state and permeated through, while a non-conducting liquid (*e.g.* oil) did not undergo such a transition and remained above the membrane. Before transitioning to the Wenzel state, increased pressure (P_{applied}), due to the applied voltage, leads to the liquid–air interface sagging until it reaches a critical texture angle, ψ_{cr}. This critical angle, along the curvature of each cylindrical wire comprising the nylon wire mesh, is where the surface can withstand the greatest pressure P_{critical} before entering the Wenzel state. For cylindrical surface geometry, P_{critical} is given by:[80]

$$P_{\text{critical}} = \frac{\gamma_{12}\sin(\theta - \psi_{\text{cr}})}{D + R - R\sin\psi_{\text{cr}}} \tag{12.6}$$

where

$$\psi_{\text{cr}} = \theta - \cos^{-1}\left(\frac{R\sin\theta}{R + D}\right) \tag{12.7}$$

here again, R is the cylinder radius and D is half of the cylinder spacing. Utilizing this preferential transition, the on-demand separations of free oil and water, oil-in-water emulsions, and water-in-oil emulsions were demonstrated, with $\eta > 99.9\%$ separation efficiency (Figure 12.8j). Such on-demand separation could be useful for the remote operation of oil–water separation units, microfluidic valves, and lab-on-a-chip devices.

12.5 Conclusions and Future Outlook

The development of membranes with selective wettability is an ongoing process, which aims to more effectively meet today's needs for alternative and efficient oil–water separation. The numerous sources of oily wastewater, and increasingly strict environmental guidelines, necessitate a highly effective, economical, and durable membrane, with a long service life, for purifying waste streams and spills. As discussed throughout, there are four selective wettabilities to choose from, and many pathways for achieving them. The type of selective wettability membrane used, for a particular application, will depend on the waste stream composition, fouling potential, and the system employed for the separation (on-demand, gravity fed, high pressure, *etc.*). The form of oil, whether free or emulsified, will dictate the membrane pore size, and thus, is directly related to the permeation rate through the membrane. All these parameters must be taken into account and optimized to utilize membranes with selective wettability.

Overall, as discussed, a multitude of selective wettability systems have been used to successfully separate oil and water mixtures with greater than 99.9% efficiency, but the future lies in imparting these wetting properties to membranes that withstand high transmembrane pressures, have greater permeation rates of the desired liquid, are anti-fouling, and can be scalably manufactured at a reasonable cost. Developing a selective wettability membrane with all these characteristics will require creative solutions, and provides a range of intellectual and research challenges. Such membranes will help meet the growing needs for waste and byproduct treatment in a wide variety of fields.

Acknowledgements

We thank Dr Ki-Han Kim and the Office of Naval Research (ONR) for financial support under grant N00014-12-1-0874. We also thank Dr Charles Y. Lee and the Air Force Office of Scientific Research (AFOSR) for financial support under grant FA9550-10-1-0523. We also thank the National Science Foundation and the Nanomanufacturing program for supporting this work through grant #1351412. EP would like to acknowledge support through the National Science Foundation Graduate Research Fellowship under Grant No. DGE 1256260.

References

1. P. Kajitvichyanukul, Y.-T. Hung and L. K. Wang, *Handbook of Environmental Engineering: Membrane and Desalination Technologies*, The Humana Press Inc., New York, 2011, vol 13.
2. C. E. Clark and J. A. Veil, *Produced water volumes and management practices in the United States*, Argonne National Laboratory: Environmental Science Division, 2009.

3. Office of the Federal Register: Code of Federal Regulations, *Title 40 – Protection of Environment*, Washington, DC, 2014, vol. 30, part 435.13.
4. J. W. Patterson, *Industrial Wastewater Treatment Technology*, Buttersworth, Stoneham, MA, 2nd edn, 1985.
5. M. O. Adebajo, R. L. Frost, J. T. Kloprogge, O. Carmody and S. Kokot, Porous materials for oil spill cleanup: A review of synthesis and absorbing properties, *J. Porous Mater.*, 2003, **10**, 159.
6. M. Cheryan and N. Rajagopalan, Membrane processing of oily streams. Wastewater treatment and waste reduction, *J. Membr. Sci.*, 1998, **151**, 13.
7. A. A. Al-Shamrani, A. James and H. Xiao, Destabilisation of oil–water emulsions and separation by dissolved air flotation, *Water Res.*, 2002, **36**, 1503.
8. J. Rubio, M. L. Souza and R. W. Smith, Overview of flotation as a wastewater treatment technique, *Miner. Eng.*, 2002, **15**, 139.
9. T. Ichikawa, Electrical demulsification of oil-in-water emulsion, *Colloids Surf., A*, 2007, **302**, 581.
10. A. A. Al-Shamrani, A. James and H. Xiao, Separation of oil from water by dissolved air flotation, *Colloids Surf., A*, 2002, **209**, 15.
11. M. Toyoda and M. Inagaki, Heavy oil sorption using exfoliated graphite – New application of exfoliated graphite to protect heavy oil pollution, *Carbon*, 2000, **38**, 199.
12. V. K. Gupta, P. J. M. Carrott, M. M. L. R. Carrott and Suhas, Low-Cost Adsorbents: Growing Approach to Wastewater Treatmenta Review, *Crit. Rev. Environ. Sci. Technol.*, 2009, **39**, 783.
13. G. Rios, C. Pazos and J. Coca, Destabilization of cutting oil emulsions using inorganic salts as coagulants, *Colloids Surf., A*, 1998, **138**, 383.
14. L. F. Song, Flux decline in crossflow microfiltration and ultrafiltration: mechanisms and modeling of membrane fouling, *J. Membr. Sci.*, 1998, **139**, 183.
15. J. Kong and K. Li, Oil removal from oil-in-water emulsions using PVDF membranes, *Sep. Purif. Technol.*, 1999, **16**, 83.
16. G. Kwon, E. Post and A. Tuteja, Membranes with selective wettability for the separation of oil–water mixtures, *MRS Commun.*, 2015, **5**, 475.
17. M. Kai, K. Ishii, H. Tsugaya and T. Miyano, Development of polyethersulfone ultrafiltration membranes, in *Reverse Osmosis and Ultrafiltration, ACS Symposium Series*, ed. S. Sourirajan and T. Matsuura, ACS Publications, Washington, DC, 1985, pp. 21.
18. A. K. Kota, W. Choi and A. Tuteja, Superomniphobic surfaces: Design and durability, *MRS Bull.*, 2013, **38**, 383.
19. T. Young, An Essay on the Cohesion of Fluids, *Philos. Trans. R. Soc. London*, 1805, **95**, 65.
20. A. Tuteja, W. Choi, M. L. Ma, J. M. Mabry, S. A. Mazzella, G. C. Rutledge, G. H. McKinley and R. E. Cohen, Designing superoleophobic surfaces, *Science*, 2007, **318**, 1618.
21. X. J. Feng and L. Jiang, Design and creation of superwetting/antiwetting surfaces, *Adv. Mater.*, 2006, **18**, 3063.

22. A. K. Kota, G. Kwon and A. Tuteja, The design and applications of superomniphobic surfaces, *NPG Asia Mater.*, 2014, **6**, e109.
23. R. N. Wenzel, Resistance of solid surfaces to wetting by water, *Ind. Eng. Chem.*, 1936, **28**, 988.
24. A. B. D. Cassie and S. Baxter, Wettability of porous surfaces, *Trans. Faraday Soc.*, 1944, **40**, 0546.
25. W. Choi, A. Tuteja, J. M. Mabry, R. E. Cohen and G. H. McKinley, A modified Cassie–Baxter relationship to explain contact angle hysteresis and anisotropy on non-wetting textured surfaces, *J. Colloid Interface Sci.*, 2009, **339**, 208.
26. R. Johnson and R. Dettre, Wettability and contact angles, *Surf. Colloid Sci.*, 1969, **2**, 85.
27. D. Quere, Rough ideas on wetting, *Phys. A*, 2002, **313**, 32.
28. G. McHale, N. J. Shirtcliffe and M. I. Newton, Contact-angle hysteresis on super-hydrophobic surfaces, *Langmuir*, 2004, **20**, 10146.
29. A. Marmur, From hygrophilic to superhygrophobic: Theoretical conditions for making high-contact-angle surfaces from low-contact-angle materials, *Langmuir*, 2008, **24**, 7573.
30. A. Marmur, Wetting on hydrophobic rough surfaces: To be heterogeneous or not to be?, *Langmuir*, 2003, **19**, 8343.
31. A. Tuteja, W. Choi, J. M. Mabry, G. H. McKinley and R. E. Cohen, Robust omniphobic surfaces, *Proc. Natl. Acad. Sci. U. S. A.*, 2008, **105**, 18200.
32. M. Nosonovsky, Multiscale roughness and stability of superhydrophobic biomimetic interfaces, *Langmuir*, 2007, **23**, 3157.
33. A. K. Kota, G. Kwon, W. Choi, J. M. Mabry and A. Tuteja, Hygro-responsive membranes for effective oil–water separation, *Nat. Commun.*, 2012, **3**, 1025.
34. G. K. Batchelor, *An Introduction to Fluid Dynamics*, Cambridge University Press, New York, NY, 2000.
35. W. Choi, A. Tuteja, S. Chhatre, J. M. Mabry, R. E. Cohen and G. H. McKinley, Fabrics with Tunable Oleophobicity, *Adv. Mater.*, 2009, **21**, 2190.
36. A. Tuteja, W. J. Choi, G. H. McKinley, R. E. Cohen and M. F. Rubner, Design parameters for superhydrophobicity and superoleophobicity, *MRS Bull.*, 2008, **33**, 752.
37. S. S. Chhatre, W. Choi, A. Tuteja, K. C. Park, J. M. Mabry, G. H. McKinley and R. E. Cohen, Scale Dependence of Omniphobic Mesh Surfaces, *Langmuir*, 2010, **26**, 4027.
38. K. Golovin, D. H. Lee, J. M. Mabry and A. Tuteja, Transparent, Flexible, Superomniphobic Surfaces with Ultra-Low Contact Angle Hysteresis, *Angew. Chem., Int. Ed.*, 2013, **52**, 13007.
39. R. Hensel, A. Finn, R. Helbig, H. G. Braun, C. Neinhuis, W. J. Fischer and C. Werner, Biologically Inspired Omniphobic Surfaces by Reverse Imprint Lithography, *Adv. Mater.*, 2014, **26**, 2029.
40. L. Feng, Z. Y. Zhang, Z. H. Mai, Y. M. Ma, B. Q. Liu, L. Jiang and D. B. Zhu, A super-hydrophobic and super-oleophilic coating mesh film for the separation of oil and water, *Angew. Chem., Int. Ed.*, 2004, **116**, 2046.

41. J. Wu, J. Chen, K. Qasim, J. Xia, W. Lei and B. P. Wang, A hierarchical mesh film with superhydrophobic and superoleophilic properties for oil and water separation, *J. Chem. Technol. Biotechnol.*, 2012, **87**, 427.
42. Q. J. Wang, Z. Cui, Y. Mao and Q. M. Chen, Stable highly hydrophobic and oleophilic meshes for oil–water separation, *Appl. Surf. Sci.*, 2007, **253**, 9054.
43. Y. Z. Cao, X. Y. Zhang, L. Tao, K. Li, Z. X. Xue, L. Feng and Y. Wei, Mussel-Inspired Chemistry and Michael Addition Reaction for Efficient Oil/Water Separation, *ACS Appl. Mater. Interfaces*, 2013, **5**, 4438.
44. S. T. Wang, Y. L. Song and L. Jiang, Microscale and nanoscale hierarchical structured mesh films with superhydrophobic and superoleophilic properties induced by long-chain fatty acids, *Nanotechnology*, 2007, **18**, 015103.
45. C. X. Wang, T. J. Yao, J. Wu, C. Ma, Z. X. Fan, Z. Y. Wang, Y. R. Cheng, Q. Lin and B. Yang, Facile Approach in Fabricating Superhydrophobic and Superoleophilic Surface for Water and Oil Mixture Separation, *ACS Appl. Mater. Interfaces*, 2009, **1**, 2613.
46. C. R. Crick, J. A. Gibbins and I. P. Parkin, Superhydrophobic polymer-coated copper-mesh; membranes for highly efficient oil–water separation, *J. Mater. Chem. A*, 2013, **1**, 5943.
47. Y. W. Shang, Y. Si, A. Raza, L. P. Yang, X. Mao, B. Ding and J. Y. Yu, An *in situ* polymerization approach for the synthesis of superhydrophobic and superoleophilic nanofibrous membranes for oil–water separation, *Nanoscale*, 2012, **4**, 7847.
48. X. M. Tang, Y. Si, J. L. Ge, B. Ding, L. F. Liu, G. Zheng, W. J. Luo and J. Y. Yu, *In situ* polymerized superhydrophobic and superoleophilic nanofibrous membranes for gravity driven oil–water separation, *Nanoscale*, 2013, **5**, 11657.
49. W. B. Zhang, Z. Shi, F. Zhang, X. Liu, J. Jin and L. Jiang, Superhydrophobic and Superoleophilic PVDF Membranes for Effective Separation of Water-in-Oil Emulsions with High Flux, *Adv. Mater.*, 2013, **25**, 2071.
50. S. H. Wang, M. Li and Q. H. Lu, Filter Paper with Selective Absorption and Separation of Liquids that Differ in Surface Tension, *ACS Appl. Mater. Interfaces*, 2010, **2**, 677.
51. C. Du, J. D. Wang, Z. F. Chen and D. R. Chen, Durable superhydrophobic and superoleophilic filter paper for oil–water separation prepared by a colloidal deposition method, *Appl. Surf. Sci.*, 2014, **313**, 304.
52. Z. Shi, W. B. Zhang, F. Zhang, X. Liu, D. Wang, J. Jin and L. Jiang, Ultrafast Separation of Emulsified Oil/Water Mixtures by Ultrathin Free-Standing Single-Walled Carbon Nanotube Network Films, *Adv. Mater.*, 2013, **25**, 2422.
53. J. P. Zhang and S. Seeger, Polyester Materials with Superwetting Silicone Nanofilaments for Oil/Water Separation and Selective Oil Absorption, *Adv. Funct. Mater.*, 2011, **21**, 4699.
54. J. Li, L. Shi, Y. Chen, Y. B. Zhang, Z. G. Guo, B. L. Su and W. M. Liu, Stable superhydrophobic coatings from thiol-ligand nanocrystals and their application in oil/water separation, *J. Mater. Chem.*, 2012, **22**, 9774.

55. A. Maartens, E. P. Jacobs and P. Swart, UF of pulp and paper effluent: membrane fouling-prevention and cleaning, *J. Membr. Sci.*, 2002, **209**, 81.
56. B. Hu and K. Scott, Influence of membrane material and corrugation and process conditions on emulsion microfiltration, *J. Membr. Sci.*, 2007, **294**, 30.
57. M. J. Liu, S. T. Wang, Z. X. Wei, Y. L. Song and L. Jiang, Bioinspired Design of a Superoleophobic and Low Adhesive Water/Solid Interface, *Adv. Mater.*, 2009, **21**, 665.
58. Z. X. Xue, M. J. Liu and L. Jiang, Recent developments in polymeric superoleophobic surfaces, *J. Polym. Sci., Part B: Polym. Phys.*, 2012, **50**, 1209.
59. Z. X. Xue, S. T. Wang, L. Lin, L. Chen, M. J. Liu, L. Feng and L. Jiang, A Novel Superhydrophilic and Underwater Superoleophobic Hydrogel-Coated Mesh for Oil/Water Separation, *Adv. Mater.*, 2011, **23**, 4270.
60. S. Y. Zhang, F. Lu, L. Tao, N. Liu, C. R. Gao, L. Feng and Y. Wei, Bio-Inspired Anti-Oil-Fouling Chitosan-Coated Mesh for Oil/Water Separation Suitable for Broad pH Range and Hyper-Saline Environments, *ACS Appl. Mater. Interfaces*, 2013, **5**, 11971.
61. B. X. Jing, H. T. Wang, K. Y. Lin, P. J. McGinn, C. Z. Na and Y. X. Zhu, A facile method to functionalize engineering solid membrane supports for rapid and efficient oil–water separation, *Polymer*, 2013, **54**, 5771.
62. Y. Dong, J. Li, L. Shi, X. B. Wang, Z. G. Guo and W. M. Liu, Underwater superoleophobic graphene oxide coated meshes for the separation of oil and water, *Chem. Commun.*, 2014, **50**, 5586.
63. Q. Wen, J. C. Di, L. Jiang, J. H. Yu and R. R. Xu, Zeolite-coated mesh film for efficient oil–water separation, *Chem. Sci.*, 2013, **4**, 591.
64. J. W. Zeng and Z. G. Guo, Superhydrophilic and underwater superoleophobic MFI zeolite-coated film for oil/water separation, *Colloids Surf., A*, 2014, **444**, 283.
65. Y. Z. Zhu, F. Zhang, D. Wang, X. F. Pei, W. B. Zhang and J. Jin, A novel zwitterionic polyelectrolyte grafted PVDF membrane for thoroughly separating oil from water with ultrahigh efficiency, *J. Mater. Chem. A*, 2013, **1**, 5758.
66. P. C. Chen and Z. K. Xu, Mineral-Coated Polymer Membranes with Superhydrophilicity and Underwater Superoleophobicity for Effective Oil/Water Separation, *Sci. Rep.*, 2013, **3**, 2776.
67. H. C. Yang, K. J. Liao, H. Huang, Q. Y. Wu, L. S. Wan and Z. K. Xu, Mussel-inspired modification of a polymer membrane for ultra-high water permeability and oil-in-water emulsion separation, *J Mater. Chem. A*, 2014, **2**, 10225.
68. H. C. Yang, J. K. Pi, K. J. Liao, H. Huang, Q. Y. Wu, X. J. Huang and Z. K. Xu, Silica-Decorated Polypropylene Microfiltration Membranes with a Mussel-Inspired Intermediate Layer for Oil-in-Water Emulsion Separation, *ACS Appl. Mater. Interfaces*, 2014, **6**, 12566.
69. Q. S. Liu, A. A. Patel and L. Y. Liu, Superhydrophilic and Underwater Superoleophobic Poly(sulfobetaine methacrylate)-Grafted Glass Fiber Filters for Oil–Water Separation, *ACS Appl. Mater. Interfaces*, 2014, **6**, 8996.

70. Y. N. Chen, Z. X. Xue, N. Liu, F. Lu, Y. Z. Cao, Z. X. Sun and L. Feng, Fabrication of a silica gel coated quartz fiber mesh for oil–water separation under strong acidic and concentrated salt conditions, *RSC Adv.*, 2014, **4**, 11447.
71. J. Yang, Z. Z. Zhang, X. H. Xu, X. T. Zhu, X. H. Men and X. Y. Zhou, Superhydrophilic-superoleophobic coatings, *J. Mater. Chem.*, 2012, **22**, 2834.
72. J. A. Howarter and J. P. Youngblood, Self-cleaning and anti-fog surfaces *via* stimuli-responsive polymer brushes, *Adv. Mater.*, 2007, **19**, 3838.
73. L. B. Zhang, Y. J. Zhong, D. Cha and P. Wang, A self-cleaning underwater superoleophobic mesh for oil–water separation, *Sci Rep*, 2013, **3**, 2326.
74. Y. Sawai, S. Nishimoto, Y. Kameshima, E. Fujii and M. Miyake, Photoinduced Underwater Superoleophobicity of TiO2 Thin Films, *Langmuir*, 2013, **29**, 6784.
75. H. Yoon, S. H. Na, J. Y. Choi, S. S. Latthe, M. T. Swihart, S. S. Al-Deyab and S. S. Yoon, Gravity-driven hybrid membrane for oleophobic-superhydrophilic oil–water separation and water purification by graphene, *Langmuir*, 2014, **30**, 11761.
76. X. Y. Zhu, H. E. Loo and R. B. Bai, A novel membrane showing both hydrophilic and oleophobic surface properties and its non-fouling performances for potential water treatment applications, *J. Membr. Sci.*, 2013, **436**, 47.
77. X. Y. Zhu, W. T. Tu, K. H. Wee and R. B. Bai, Effective and low fouling oil/water separation by a novel hollow fiber membrane with both hydrophilic and oleophobic surface properties, *J. Membr. Sci.*, 2014, **466**, 36.
78. J. A. Howarter and J. P. Youngblood, Amphiphile grafted membranes for the separation of oil-in-water dispersions, *J. Colloid Interface Sci.*, 2009, **329**, 127.
79. B. Berge, Electrocapillarity and wetting of insulator films by water, *C. R. Acad. Sci., Ser. II*, 1993, **317**, 157.
80. G. Kwon, A. K. Kota, Y. X. Li, A. Sohani, J. M. Mabry and A. Tuteja, On-Demand Separation of Oil–Water Mixtures, *Adv. Mater.*, 2012, **24**, 3666.

CHAPTER 13

Droplet Manipulation on Liquid-Repellent Surfaces

ROBIN H. A. RAS*[a], XUELIN TIAN†[a], BO CHANG[a] AND JAAKKO V. I. TIMONEN[a,b]

[a]Aalto University School of Science, Department of Applied Physics, Puumiehenkuja 2, 02150 Espoo, Finland; [b]Harvard John A. Paulson School of Engineering and Applied Sciences, Harvard University, 02138 Cambridge MA, USA
*E-mail: robin.ras@aalto.fi

13.1 Droplet Friction

Recent years have witnessed a tremendous growth in the design and preparation of liquid-repellent surfaces due to their broad application potential in self-cleaning, anti-fouling, anti-icing, drag reduction, enhanced thermal transfer, and other applications.[1–3] This rapidly increasing research interest since the late 1990s is largely boosted by study of several natural liquid-repellent surfaces (such as lotus leaves) as well as progress in micro/nanofabrication techniques, which have led to the development of a large number of superhydrophobic surfaces. Generally, a surface is regarded as superhydrophobic if it displays a contact angle larger than 150° and the contact angle hysteresis is small so that a water droplet can readily slide away from the surface.[4] Here, the contact angle hysteresis is characterized as the difference

†Current Address: School of Materials Science and Engineering, Sun Yat-sen University, 510275 Guangzhou, China

RSC Soft Matter No. 5
Non-wettable Surfaces: Theory, Preparation, and Applications
Edited by Robin H. A. Ras and Abraham Marmur

Published by the Royal Society of Chemistry, www.rsc.org

between the advancing contact angle θ_A and the receding one θ_R, which correspond respectively to the critical contact angles to initiate advance and retreat of the solid–liquid–air triple contact line. Superhydrophobic surfaces possess two essential features, namely micro-/nanotextured surface structure and low energy surface chemistry, to allow trapping of an air layer within the texture when in contact with a liquid. The solid–liquid contact area can be minimized by introducing appropriate geometrical texture, and a droplet can then be regarded as sitting on an air cushion. This leads to largely suppressed friction between the droplet and the surface. As a consequence, droplets on such surfaces can be easily manipulated for novel technological applications by external stimuli, including gravity or magnetic and electric fields.

The ability to quantitatively describe friction force caused by contact angle hysteresis is crucial for controlled droplet manipulation since droplet motion is initiated only if the external force is large enough to overcome this hysteresis force. By studying the sliding behaviour of droplets along an inclined surface (Figure 13.1), Macdougall *et al.* found that θ_A and θ_R were important for determining the critical slope α of the surface above which the droplet moves continuously downward.[5] They observed that the contact angles at the lowest and highest positions along the droplet base were always θ_A and θ_R, respectively, and the hysteresis force, which equalled gravity $mg\sin\alpha$ at the critical inclined condition, was proportional to $\gamma(\cos\theta_R - \cos\theta_A)$, where γ is the surface tension of the liquid. This relationship was confirmed by many other investigations,[6–9] and the hysteresis force can be expressed as:

$$F_h = kr\gamma(\cos\theta_R - \cos\theta_A) \tag{13.1}$$

where k is a constant and r is the radius of the droplet base.

Numerical determination of k is not easy due to the three-dimensional geometrical character of the problem and that the contact angles along the periphery of droplet base (*i.e.* the contact line) vary continuously from θ_A to θ_R. For a circular droplet base, Extrand *et al.* analytically calculated the constant k using a linear approximation of cosines of contact angles. By assuming that $\cos\theta$ increases linearly from $\cos\theta_A$ to $\cos\theta_R$ along the contact line (from the advancing edge to the receding edge), they predicted k to be $4/\pi$.[9,10] In practice, the droplet base profile may deviate from a circular shape attributed to asymmetric deformation of the droplet at the critical configuration. For

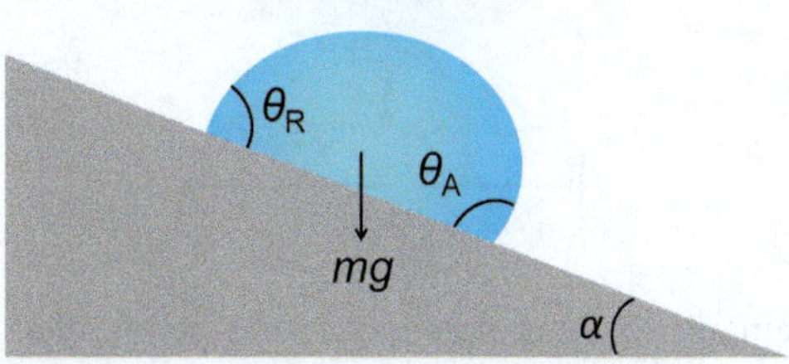

Figure 13.1 Droplet sliding down an inclined surface with slope of α.

a non-circular drop base, Furmidge found that the hysteresis force can be expressed as:

$$F_h = 2w\gamma(\cos\theta_R - \cos\theta_A) \quad (13.2)$$

where $2w$ is the width of the droplet base, *i.e.* the base dimension perpendicular to the droplet moving direction (Figure 13.2).[6] Furmidge's finding implies that along the triple contact line the contact angle variation does not follow a simple linear relationship; instead, the contact angles at the half rear portion are closer to θ_R and the contact angles at the half front portion are closer to θ_A. In fact, by assuming constant contact angles of θ_A and θ_R at the advancing and receding portions, respectively, Dussan *et al.* analytically deduced that eqn (13.2) applies to droplets of arbitrary base shapes.[7]

Eqn (13.1) or (13.2) provide an easy way to quantitatively predict the mobility of droplets on surfaces, and are thus widely used by the community. However, it is worth noting that on an inclined surface, the two contact angles at the front and back just before droplet sliding may not simultaneously be the advancing and receding angles.[11]

The magnitude of hysteresis force on superhydrophobic surfaces can be quantified using the above equations. Considering droplets with size R smaller than capillary length $l_{cap} = \sqrt{\gamma/\rho g}$ (2.7 mm for water), where ρ and g denote respectively liquid density and acceleration due to gravity, the effect of gravity on droplet morphology is then negligible. Due to the low hysteresis of superhydrophobic surfaces, it is reasonable to assume the droplet base as a circle with radius of r, which obeys the following relationship: $r = R\sin\theta$, where R is the droplet radius and θ can be approximated as $(\theta_A + \theta_R)/2$. Using

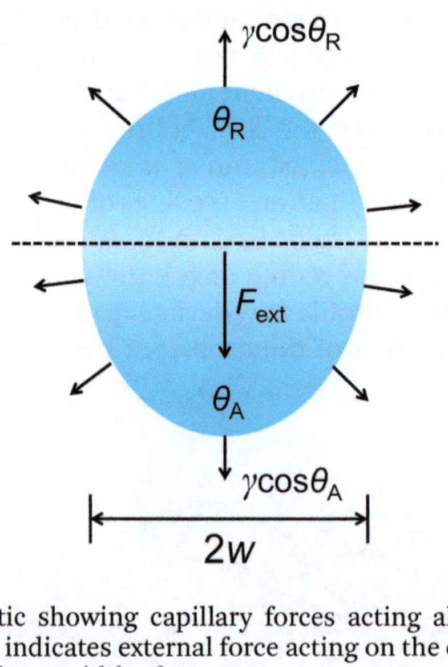

Figure 13.2 Schematic showing capillary forces acting along the triple contact line. F_{ext} indicates external force acting on the droplet. The base of the droplet has a width of $2w$.

eqn (13.2), the hysteresis force can be expressed as $F_h = 2\gamma R\sin\theta(\cos\theta_R - \cos\theta_A) \approx 2\gamma R\sin^2\theta\Delta\theta$, where $\Delta\theta = \theta_A - \theta_R$ is the contact angle hysteresis. It is clear that both contact angle and contact angle hysteresis are important in determining the hysteresis force. For a typical superhydrophobic surface with θ of 150° and $\Delta\theta$ of 10°, respectively, the hysteresis force is about 4% of $2\gamma R$. Balancing this force with gravity $4\pi R^3\rho g/3$, we can obtain the critical size $R_c = \sqrt{(3\Delta\theta/2\pi)}\sin\theta l_{cap}$ for a droplet to be able to slide away from a vertically placed surface, which is ~0.4 mm for water droplets. Below this size, droplets will get stuck on the surface. If θ is greater than 170°, even a water droplet with size down to ~0.1 mm may be able to move off the surface. Therefore, superhydrophobic surfaces allow manipulation of droplets as small as few hundred micrometres by gravity or other moderate stimuli.

Once the external force overcomes the hysteresis force and initiates droplet motion, another source of friction that acts near the solid–liquid interface, *i.e.*, viscous drag force, will play a role in determining the droplet dynamics, including its velocity and acceleration. Currently there are only a few reports studying the effect of viscous force on motion of droplets on superhydrophobic surfaces. Reyssat *et al.* studied the water droplet motion on an inclined surface with θ and $\Delta\theta$ of ~165 ± 5° and ~10 ± 5°, respectively, and found that the droplet followed a free fall law with a constant acceleration of $1/2g\sin\alpha$.[12] Within a short distance of tens of millimetres, the droplet exhibited exactly a trajectory of $x = 1/2gt^2\sin\alpha$, where x is the motion distance and t is the time. This suggests that viscous force plays a negligible role. Tracers placed inside the moving droplet only indicated translation and no rotational motion, confirming that the droplet adopted a purely slipping motion. Sakai *et al.* monitored the velocity field distribution within a water droplet that moved along superhydrophobic ZnO nanorod surfaces using particle image velocimetry technique.[13] Their investigation also indicated an entirely slipping motion of the droplets, and constant acceleration motion mode was also observed.[13,14] Though viscous dissipation may be negligible for motion of low-viscosity droplets on superhydrophobic surfaces, our group recently developed a magnetic field-induced droplet oscillation method that could measure a very low viscous force as well as the hysteresis force of ferrofluid droplets on highly superhydrophobic surfaces.[15]

Motion of viscous droplets on superhydrophobic surfaces is quite different. Richard *et al.* found that a glycerol droplet (which has a surface tension of 63 mN m^{-1}, comparable to water, but with much higher viscosity of 950 mPa s) rolled instead of slipping along an inclined superhydrophobic surface (θ and $\Delta\theta$ for glycerol are 165 ± 5° and 10°, respectively).[16] Moreover, it moved at a constant velocity almost from the beginning, contrasting the constant acceleration mode in water droplet motion. Viscous dissipation plays a significant role here, which is deduced to occur near the droplet–surface contact area with a size of $r \sim R^2/l_{cap}$. This size relationship assumes that the contact is caused by gravity-induced descent of the droplet mass centre (*i.e.* gravity-induced flattening), and is valid for droplets with radius larger than $l_{cap}\sin\theta$ but smaller than l_{cap}. The viscous force then scaled as $F_v \sim \mu(v/R)$

$(R^2/l_{cap})^2$, where μ and ν are liquid viscosity and droplet velocity, respectively. By balancing the viscous force torque of $F_v r$ with gravity torque $\rho g R^4 \sin\alpha$, the steady velocity is obtained:[16,17] $\nu \sim \dfrac{\rho g \sin\alpha}{\mu (l_{cap})^3 R}$. This relationship results in a counterintuitive fact: a smaller droplet runs faster than a bigger one since ν is proportional to $1/R$.

For a droplet with size less than $l_{cap} \sin\theta$, its contact area with a superhydrophobic surface is mainly caused by wetting-induced contact, rather than gravity-induced flattening as described above, and scales as $r \sim R\sin\theta$. The viscous force scales as $F_v \sim \mu(\nu/R)(R\sin\theta)^2$. Applying the torque balancing relationship yields a steady velocity:[12] $\nu \sim \dfrac{\rho g R^2 \sin\alpha}{\mu(\sin\theta)^3}$. This relationship applies to droplets with size less than $l_{cap} \sin\theta$, but large enough to overcome the hysteresis effect. Therefore, it is applicable for moderately superhydrophobic surfaces (to allow $l_{cap} \sin\theta$ to be reasonably large) with very low hysteresis.

Apart from superhydrophobic surfaces, a new type of liquid-repellent surfaces, namely SLIPS (slippery liquid-infused porous surfaces), has been developed very recently.[18,19] Such surfaces use a lubricating film infused within porous structures to repel other immiscible liquids (Figure 13.3). A prominent advantage of SLIPS is its intrinsic self-healing ability, a benefit of the free flow of the lubricating liquid within the porous structure. SLIPSs can show very low hysteresis to droplets, though contact angles of liquids on SLIPSs are not significant, which makes them suitable for droplet manipulation by external stimuli.

Systematic investigations on friction between droplets and SLIPSs are rare so far. Unlike solid surfaces, where hysteresis forces are governed by triple lines at the solid–liquid–gas interface, the hysteresis force for droplets on SLIPSs is likely to be governed by a liquid–liquid–gas triple line. All three interface phases are highly deformable fluids, and different hysteresis behaviour is expected for SLIPSs. It is thus interesting to investigate whether a hysteresis force relationship similar to eqn (13.1) is present for SLIPSs. It was shown that the thickness of lubricating film can significantly affect the sliding angles of droplets on SLIPSs;[20] however, how hysteresis force relates to the wetting properties of SLIPSs (*e.g.* θ_A or θ_R) still remains unclear. Another

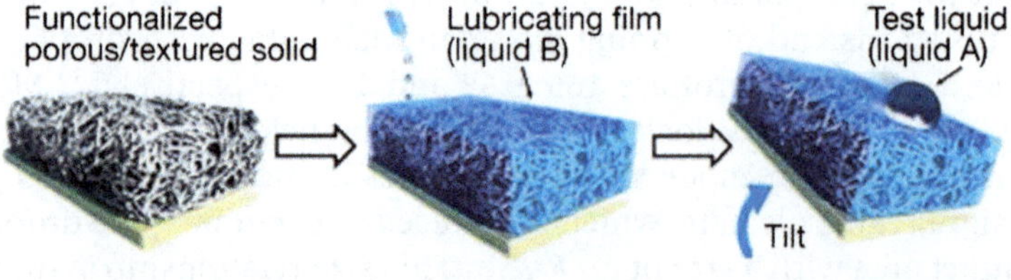

Figure 13.3 Schematic illustration of SLIPS, which uses a liquid-infused porous surface to repel another immiscible liquid. Reprinted by permission from Macmillan Publishers Ltd: *Nature*,[19] Copyright 2011.

interesting topic is to investigate viscous dissipation of droplets on SLIPSs, including how the SLIPSs wetting property, lubricating liquid viscosity, droplet viscosity, and size affect viscous dissipation behaviours. Systematic investigations of these topics are important to further our understanding on friction of droplets on SLIPSs and will improve our capability to manipulate droplets for various technological applications.

13.2 Gravity-Induced Droplet Manipulation

Precise droplet manipulation is essential in many applications, such as droplet-based microfluidics systems, drug delivery, and chemical screening. Gravity is frequently used as a driving source for droplet manipulation. Mertaniemi *et al.* demonstrated that a droplet can be transported using gravity as a driving force in open tracks on a superhydrophobic metal plate.[21] Figure 13.4a shows multiple water drops moving along the curved superhydrophobic track with a depth of 0.3 mm and width of 1.5 mm as the metal plate is tilted about 2°. The droplet moves in the direction based on the gravity gradient, and the droplet follows the track precisely. Such superhydrophobic tracks were used for rebounding droplet–droplet collisions, to enable simple Boolean logic devices operating with water droplets.[22] In addition to entirely superhydrophobic tracks, hydrophilic–superhydrophobic patterned surfaces can also be used for guiding water droplets. Seo *et al.* reported that water droplets on hydrophilic–superhydrophobic patterned surfaces move precisely along the trajectories of the tilted hydrophilic tracks (Figure 13.4b).[23] The hydrophilic water guiding tracks were obtained *via* selective patterning of the hydrophilic region on superhydrophobic silicon nanowire arrays.

Mertaniemi *et al.* also demonstrated that a sharp superhydrophobic surface positioned in the middle of the track can be used as a blade to cut droplets in half, as shown in Figure 13.4c and d.[21] Droplet splitting is also possible using droplets falling on macrotextured superhydrophobic surfaces, such as submillimetre-scale ridges[25] or wires[24] (Figure 13.4e and f). In both cases, the hydrodynamics of the droplet was significantly affected by the impact. It resulted in reduction of the contact time of bouncing drops, thereby providing a new route for design of anti-icing surfaces. Song *et al.* demonstrated that a droplet could split into multiple smaller droplets using superhydrophobic stripes on hydrophilic surfaces.[26] As a droplet falls down due to the gravity and hits a hydrophilic surface with superhydrophobic stripes, the unbalanced surface tension at hydrophilic/superhydrophobic interface causes the splitting of the droplet. By adjusting the landing position of the droplet, the droplet could be split into different sizes of the smaller droplets.

By combining gravity and hydrophilic/superhydrophobic patterned surfaces, transportation and deposition of nanolitre-sized droplets has been demonstrated utilizing sliding droplets on an inclined hydrophilic/superhydrophobic patterned surface.[27] The patterned surface consists of hydrophilic black silicon pads with superhydrophobic black silicon substrate coated with fluorocarbon polymer. The measured advancing contact angle

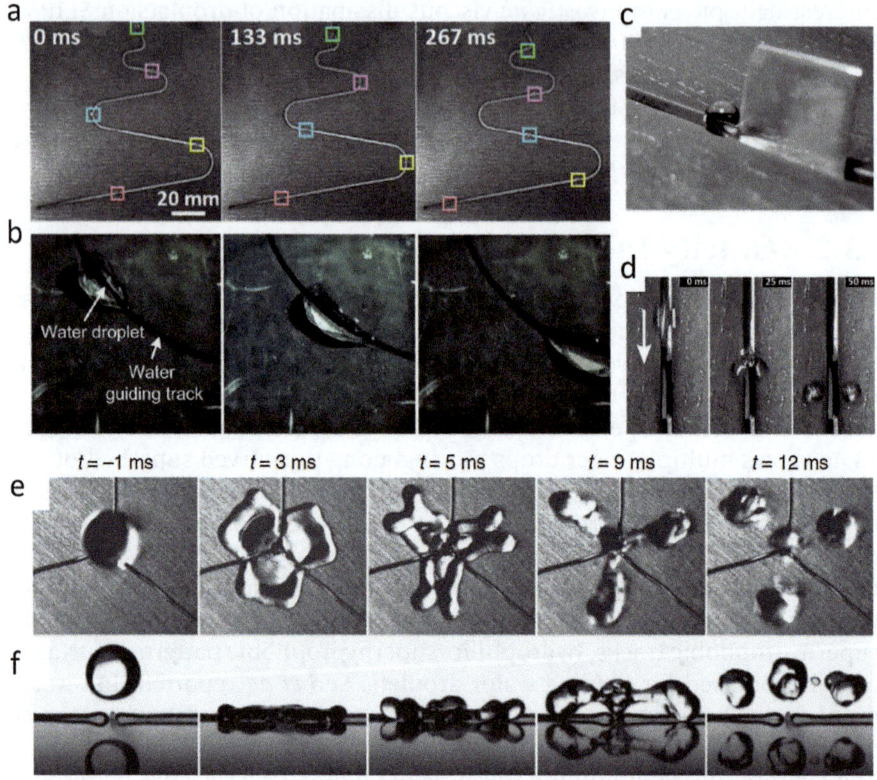

Figure 13.4 (a) Gravity-induced droplet movement in an open track on a superhydrophobic metal plate. Reproduced from ref. 21 with permission from John Wiley & Sons. Copyright © 2011 Wiley-VCH Verlag GmbH & Co. KGaA, Weinheim. (b) A water droplet moves along the trajectory of a tilted hydrophilic track on a superhydrophobic surface. Reproduced with permission from J. Seo, S. Lee, J. Lee and T. Lee, *ACS Appl. Mater. Interfaces*, 2011, **3**, 4722–4729.[23] Copyright (2011) American Chemical Society (c–d) A superhydrophobic knife can be used to split a drop into two. Reproduced from ref. 21 with permission from John Wiley & Sons. Copyright © 2011 Wiley-VCH Verlag GmbH & Co. KGaA, Weinheim. (e–f) Bouncing of a water drop on a symmetry centre of three superhydrophobic wires of radius 100 μm. Reprinted from Gauthier *et al.* 2015,[24] with permission from Nature Publishing Group, Copyright 2015.

on the substrate and the pad is 170° and 30°, respectively. The droplet deposition process is shown in Figure 13.5a, where a droplet is sliding down on an inclined hydrophilic/superhydrophobic patterned surface due to gravity. As the rear edge of the droplet transforms from the pinning state to the depinned, part of the droplet is deposited on the hydrophilic pads. The combination of a good wetting property of the pads and superhydrophobicity of the substrate, as well as reasonable gaps between the pads, can lead to deposition with uniformity better than 5% (Figure 13.5b and c). The parallel

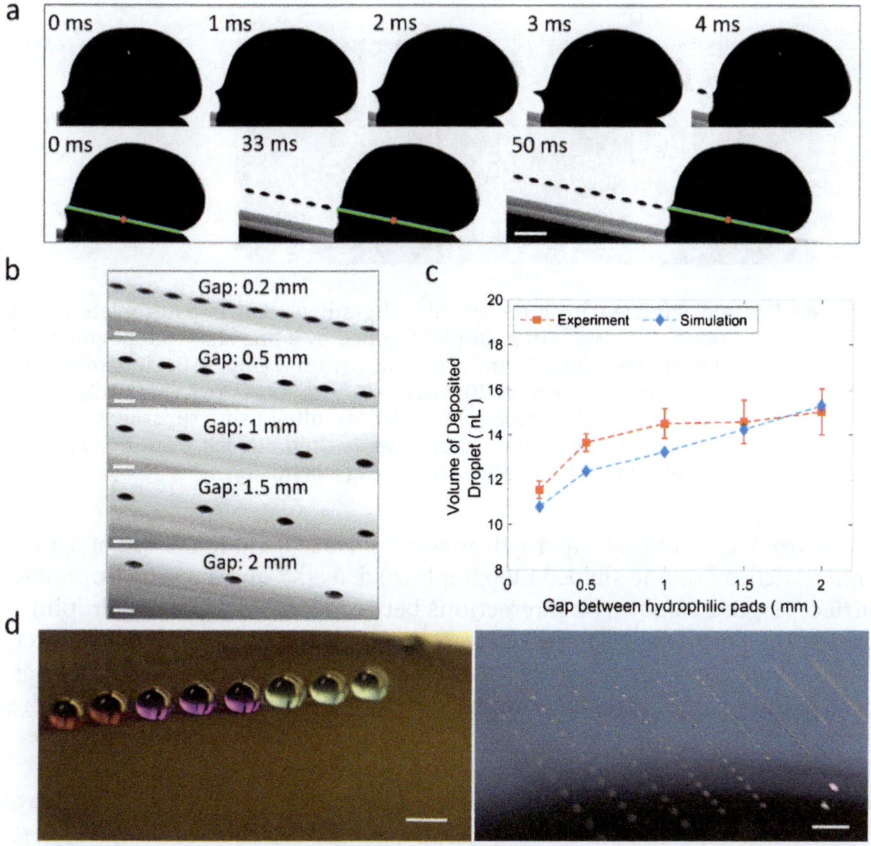

Figure 13.5 (a) Gravity-induced droplet deposition on an inclined hydrophilic/superhydrophobic patterned surface using a sliding droplet, scale bar 2 mm; (b) deposited droplets on pads of 0.5 mm × 0.5 mm and with gap of 0.2 mm, 0.5 mm, 1 mm, 1.5 mm and 2 mm, scale bars 0.5 mm; (c) volume of deposited droplet as function of gap between pads; (d) parallel droplet deposition using different coloured liquids on 0.5 × 0.5 mm pads, scale bars 1 cm. Reprinted from B. Chang, Q. Zhou, R. H. A. Ras, A. Shah, Z. Wu and K. Hjort, *Appl. Phys. Lett.*, 2016, **108**, 154102[27] with permission from AIP Publishing.

nanolitre deposition has also been demonstrated using different liquids (Figure 13.5d), which shows the potential of the proposed method for large-scale and high-density liquid deposition. Similar work was done by Kong *et al.* and they demonstrated the possibility of transporting multiple droplets in parallel and performing sequential fluidic reactions on planar plastic hydrophilic/superhydrophobic patterned sheets.[28]

Controlled trapping of sliding water droplets is feasible using designed wetting defects, as demonstrated for a hybrid surface consisting of metal domes and polymer pillar arrays (Figure 13.6a).[29] The metal domes affect the adhesion properties of water on the surface and therefore could act as a gate to trap droplets based on their mass. Other techniques used for pinning a

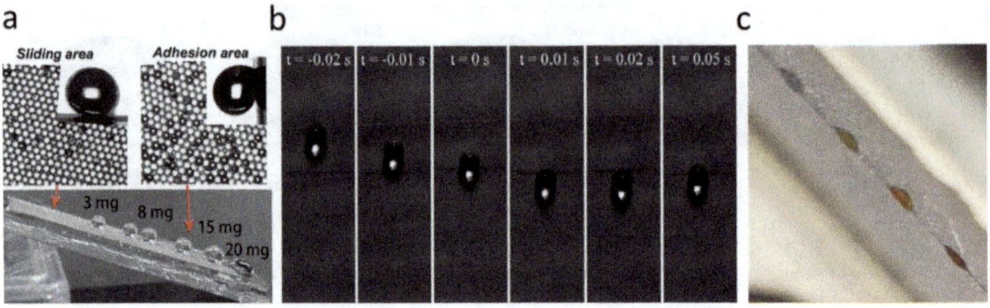

Figure 13.6 (a) Surface with a topographical gradient that traps droplets based on their mass.[30] (b) A droplet trapped by a line-shaped topographical defect. Reproduced with permission from P. Olin, S. B. Lindström and L. Wågberg, *Langmuir*, 2015, **31**, 6367–6374.[31] Copyright (2015) American Chemical Society. (c) Droplet sampling using superhydrophobic paper patterned with wettable spots. Reproduced from ref. 32 with permission from the Royal Society of Chemistry.

water droplet on a tilted superhydrophobic surface include the use of hydrophilic patches[30] or line-shaped topographical defects[31] on a superhydrophobic surface (Figure 13.6b). The interactions between a drop and the hydrophilic patch or the topographical defect creates a wetting potential and drag force to pin the droplet. Furthermore, an open-air surface microfluidic devices has been demonstrated for transport, mixing, and rapid droplet sampling.[32] The microfluidic device is made of superhydrophobic paper patterned with wettable spots using a common household inkjet printer. By placing a larger amount of droplet at the top of the tilted surface, the droplet rolls down, leaving behind a liquid sample on each of the wettable spots as shown in Figure 13.6c. Also electrically tunable wetting defects allow the controlled trapping of sliding water droplets.[33]

13.3 Magnetic Field-Induced Droplet Manipulation

Manipulation of aqueous drops on superhydrophobic surfaces with magnetic fields is an attractive alternative to manipulation with electric fields. Magnetic manipulation is straightforward with well-defined dipolar forces, whereas electric manipulation can show non-ideal behaviour due to dielectric breakdown, leaky dielectrics, and unintended charge build-up. Water itself is diamagnetic and thus experiences a force towards the direction of decreasing magnetic field strength. Unfortunately, this force is weak compared to forces that resist motion of a sessile drop, including the contact angle hysteresis and viscosity. This applies even to the very best superhydrophobic surfaces developed so far. In order to enhance the magnetic response and make magnetic manipulation feasible, a magnetic component needs to be added either to the droplets or to the superhydrophobic substrate. Different scenarios, working principles, and corresponding demonstrations given in existing literature are outlined below.

13.3.1 Magnetic Droplets Based on Non-Uniformly Dispersed Magnetic Particles

This very simple yet powerful approach utilizes ferromagnetic particles (typically carbonyl iron) that are (typically) of the order of micrometres in diameter.[34–39] Particles are mixed with the liquid, but they do not form a colloidally stable dispersion due to the strong magnetic interparticle attraction. However, particles are confined within droplets and migrate towards increasing magnetic field strength direction in external field. When they reach the droplet–air interface, they are held there by capillary forces, and transfer the magnetic force to the liquid droplet. This mechanism forms the basis for droplet manipulation (Figure 13.7a and b).

This technique is very general, and can be applied to many kinds of fluids, including biological ones.[36] Practical use has been demonstrated, *e.g.* by combining and mixing droplets with two different reagents and performing electrochemical detection of the reaction products (Figure 13.7c).[37] Also, magnetic particles can be extracted from the droplet by applying a large enough field gradient to overcome the capillary force,[40] and thus the particles can be easily recycled.

13.3.2 Magnetic Droplets Based on Uniformly Dispersed Magnetic Nanoparticles

In this approach, magnetic nanoparticles (typically iron, iron oxide, cobalt, or nickel; diameter typically from few nanometres to 10–20 nm) are uniformly dispersed throughout the droplet by introducing an interparticle repulsion

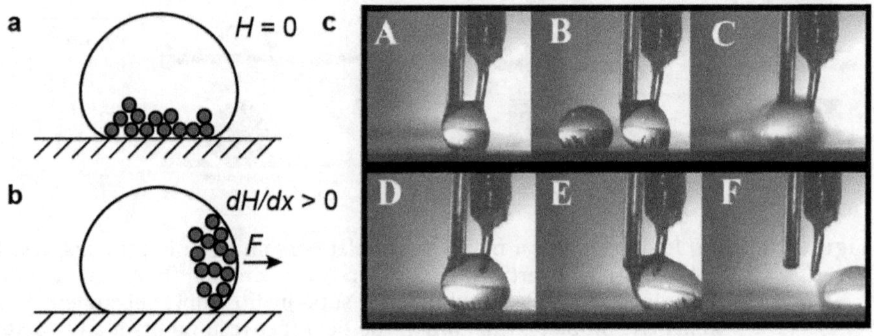

Figure 13.7 (a) A scheme of magnetic particles mixed with a droplet on a superhydrophobic surface under zero magnetic field and (b) under a magnetic field with increasing field strength towards the right. (c) Series of photographs demonstrating the use of magnetic forces to combine and mix two droplets (left: glucose oxidase in water, right: glucose in water), followed by amperometric measurement of reaction products and removal of the analysed sample in the end ((c) Reproduced from ref. 37 with permission from the Royal Society of Chemistry). Magnetic microparticles are clearly visible inside the droplets as chain-like aggregates.

by surfactants (steric stabilization) or by electrical double layer (electric stabilization) (Figure 13.8a). The particles remain (nearly) uniformly distributed even under strong magnetic field gradients due to thermal motion. These fluids are generally called ferrofluids, and the research on them has been recently reviewed.[41]

The biggest advantage compared to non-uniformly dispersed particles is the ability to convey well-defined body forces to the droplets *via* an external

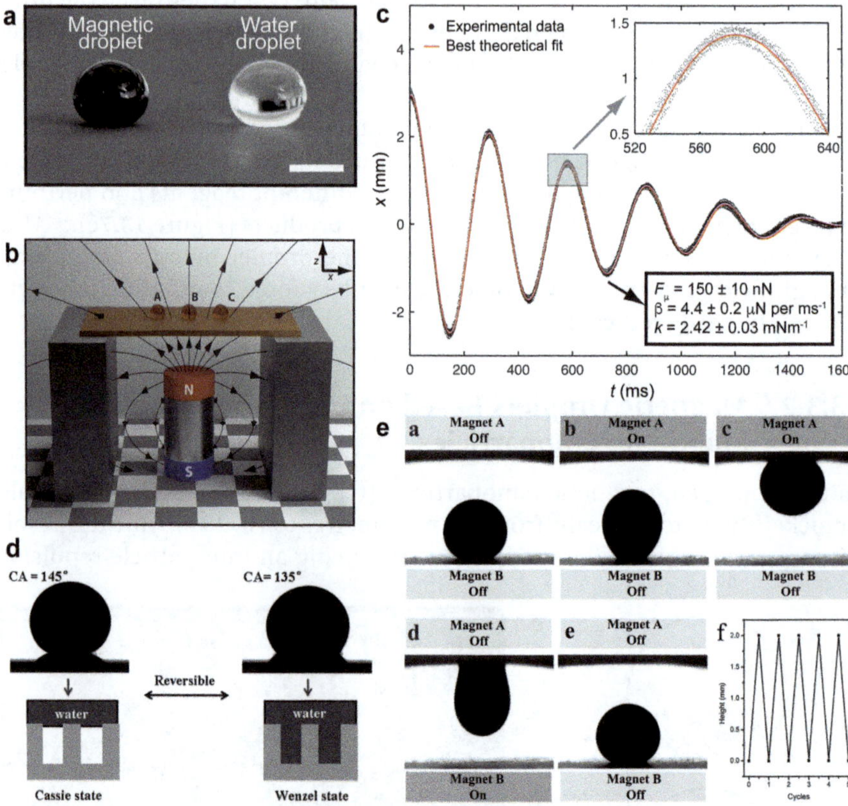

Figure 13.8 (a) Photograph of a magnetic droplet containing uniformly dispersed iron oxide nanoparticles, and a pure water droplet. (b) A scheme of a cylindrical magnet placed under a superhydrophobic substrate and a magnetic droplet oscillating in the field (between points A and C). (c) Experimental data of the oscillation of a 5 μL magnetic droplet with a theoretical fit. (d) Photographs and corresponding schemes of a magnetic droplet going through a reversible transition between Cassie and Wenzel states induced by an external magnetic field. (e) Photographs of transport of a magnetic droplet between two superhydrophobic surfaces, induced by external magnetic field. ((a–c) from ref. 15, (d) reproduced from Z. Cheng, H. Lai, N. Zhang, K. Sun and L. Jiang, *J. Phys. Chem. C*, 2012, **116**, 18796–18802.[42] Copyright (2012) American Chemical Society, and (e) reproduced from X. Hong, X. Gao and L. Jiang, *J. Am. Chem. Soc.*, 2007, **129**, 1478–1479.[43] Copyright (2007) American Chemical Society).

magnetic field. On the other hand, the biggest disadvantage is the strong optical absorbance throughout the droplet in the whole visible region by the nanoparticles (Figure 13.8a). Thus, applications requiring optical clarity (*e.g.* for optical probing of chemical reactions inside droplets) are most likely not feasible with this system. However, the well-defined body forces make these droplets ideal candidates for observing and measuring droplet-scale fluid dynamics under well-defined external forces.[15,44,45]

One interesting and readily available magnetic field geometry is produced by a cylindrical magnet placed under a superhydrophobic substrate (Figure 13.8b).[15] This induces two forces on the droplet: one perpendicular to the substrate (pulling the droplet against the surface) and another parallel to the substrate (pulling the droplet towards the symmetry axis of the cylindrical magnet). Because the perpendicular magnetic force is a constant (to a first approximation) and the parallel force has Hookean dependency of $-kr$, the droplet essentially forms a harmonic oscillator with an adjustable friction term. Indeed, magnetic droplets oscillate in this field and dissipate their energy through contact angle hysteresis and viscous dissipation (Figure 13.8c). Thanks to the adjustable perpendicular force, which functions as an effective tuneable gravitational force, the energy dissipated can be quantified as a function of normal force and resulting deformation of the droplet (three-phase boundary and contact area both increase with increasing normal force).[15] Also, controllable Cassie-to-Wenzel transition has been demonstrated with magnetic droplets in a magnetic field (Figure 13.8d).[42,46] Similarly, the magnetic body force technique can be used for measuring droplet adhesion[47] on superhydrophobic surfaces and to transport droplets between different superhydrophobic surfaces (Figure 13.8e).[43]

From the theoretical point of view, the deformation and motion of dilute magnetic droplets is reasonably well understood.[15,48] However, when the concentration of the magnetic nanoparticles is increased enough, the droplets can become unstable in external magnetic fields and divide into smaller droplets.[49] The as-formed daughter droplets can form various static and dynamic patterns under DC and AC magnetic fields.[49]

13.3.3 Magnetically Controllable Superhydrophobic Surfaces

In this interesting approach, the superhydrophobic surface itself is made responsive towards the applied magnetic field. Generally speaking, the applied magnetic field induces a deformation of the surface, which in turn induces droplet motion either through capillary force or an unbalanced component of the gravitational force along the deformed surface. In order to induce large enough deformations, the substrate needs to be reasonably soft and the magnetic elements inside the substrate reasonably strongly magnetic. This combination requires some clever engineering, but has the benefit of requiring no magnetic particles to be introduced into the liquid drops.

Taking into account the long tradition of making superhydrophobic coatings based on micropillars, it is clear that making the micropillars magnetic can offer a straightforward way to magnetically controlled superhydrophobicity. For example, magnetic microparticles have been embedded in soft poly(dimethylsiloxane) (PDMS) posts, making it possible to tilt the posts with an external magnetic field.[50] In the absence of the magnetic field, the posts remain straight up and show a sliding angle of 26° for a 10 μL water drop (static angle 146°). Application of the magnetic field, and thus tilting of the pillars, led to decrease of the sliding angle to 19° (when pillars were tilted along the sliding direction) or increase to 46° (when pillars were tilted against the sliding direction). In another work, micropillars were fabricated of pure nickel and attached to a rigid substrate through a soft PDMS layer (Figure 13.9a).[51] In this system, the pillars remain straight and rigid, and bending is made possible by the soft PDMS junction. Modest dependency of drag force on a water drop on a tilted substrate was observed as a function of magnetic field direction (Figure 13.9b and c).

Another exceptionally elegant approach is based on soft magnetic elastomers.[52] A superhydrophobic coating is applied on a magnetically responsive

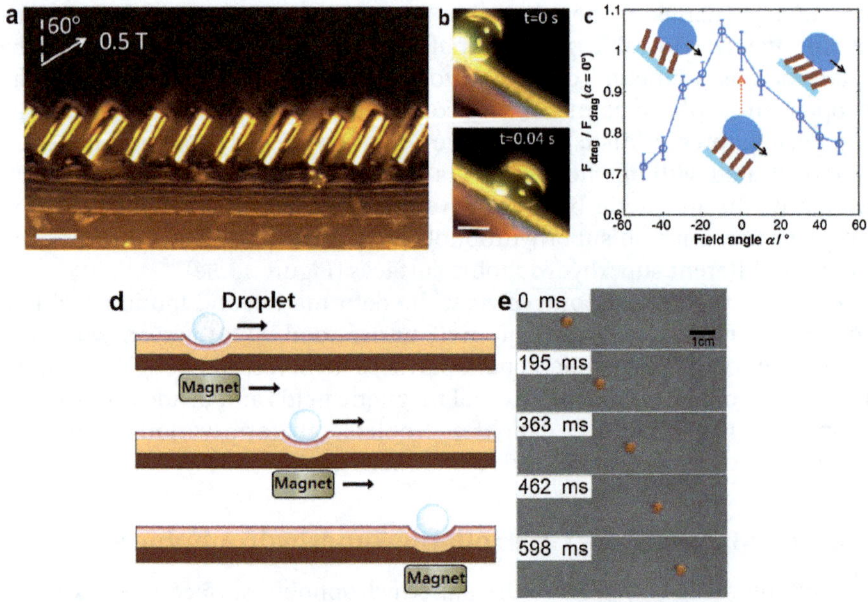

Figure 13.9 (a) An optical image of nickel micropillars tilted in a magnetic field (scale bar 50 μm) and (b) a water droplet moving on a tilted micropillars. (c) Drag force as a function of the angle of the magnetic field on top of the nickel micropillars. (d) Scheme of droplet transport on a deformable superhydrophobic surface based on soft magnetic elastomer and (e) optical images of droplet transport when the magnet is moved under the substrate towards right. ((a–c) Reproduced from ref. 51 with permission from John Wiley & Sons. Copyright © 2014 Wiley-VCH Verlag GmbH & Co. KGaA, Weinheim; (d–e) reproduced from ref. 52 with permission from John Wiley & Sons. Copyright © 2013 John Wiley & Sons, Ltd).

soft elastomer stack. Then, under local magnetic actuation with a handheld permanent magnet, the soft elastomer stack deforms and a dimple is formed on the superhydrophobic top layer (Figure 13.9d). This dimple serves as a gravitational trap for water droplets, which can be moved on the substrate simply by moving the magnet under the substrate.

Also, polystyrene fibres with embedded iron particles have been suggested to show magnetically controllable contact angle hysteresis.[53]

13.3.4 Other Systems

All systems described above have been based on "classic" lotus-mimetic superhydrophobic surfaces. In addition, there are a number of closely related systems that also show high droplet mobility under an applied magnetic field, but which do not use the lotus-like structure to demonstrate low friction. One such system consists of a porous surface infused with a lubricating liquid that is immiscible with water.[18,19] Magnetic droplets can be manipulated on these surfaces in very similar manner as on the superhydrophobic counterparts describe above.[54] However, things get more interesting if the lubricant itself is magnetic (*e.g.* contains colloidally stable magnetic nanoparticles). In that case, plain liquid droplets can be manipulated through magnetic forces that are exerted on the droplets through the lubricant that both supports and wraps the liquid droplets.[55]

High droplet mobility has also been demonstrated in so-called liquid marbles, which are aqueous droplets coated with a hydrophobic powder.[56] If the powder is made of magnetic matter (with hydrophobic coating), the liquid marbles become responsive to magnetic fields.[45,57–60] In contrast to magnetic particles dispersed in the volume of the droplet (as discussed above), the magnetic particles residing permanently on the liquid–air interface behave quite differently. For example, under zero magnetic field the marbles are uniformly covered with the magnetic particles. Under an applied field gradient the particles can be pulled along the interface to "open" the marble shell, *e.g.* to allow optical probing of the droplet contents.[57,58] The deformation and dynamics of these magnetic liquid marbles have been discussed in depth recently.[59]

13.4 Conclusions

The very low friction force between a droplet and a superhydrophobic surface provides opportunities for manipulation of the droplets. A tiny external force such as gravity or a magnetic field is sufficient to accelerate the droplet on a superhydrophobic surface and control its motion. Or, by providing an external force sufficiently large to dominate over surface tension, a droplet can split into two or more droplets. Here we have presented many examples of clever engineering of droplet systems, by using surface textures, magnetic fields or other design features. The first applications are emerging in lab-on-a-chip systems and anti-icing surfaces. This review has focused on gravity and magnetic field as driving force for the manipulation; however electric fields, vibrations, or light can also be used to drive droplets.

References

1. W. Barthlott and C. Neinhuis, *Planta*, 1997, **202**, 1–8.
2. L. Feng, S. H. Li, Y. S. Li, H. J. Li, L. J. Zhang, J. Zhai, Y. L. Song, B. Q. Liu, L. Jiang and D. B. Zhu, *Adv. Mater.*, 2002, **14**, 1857–1860.
3. A. Tuteja, W. Choi, G. H. McKinley, R. E. Cohen and M. F. Rubner, *MRS Bull.*, 2008, **33**, 752–758.
4. A. Nakajima, *NPG Asia Mater.*, 2011, **3**, 49–56.
5. G. Macdougall and C. Ockrent, *Proc. R. Soc. London, Ser. A*, 1942, **180**, 151.
6. C. G. Furmidge, *J. Colloid Sci.*, 1962, **17**, 309–324.
7. E. B. Dussan and R. T.-P. Chow, *J. Fluid Mech.*, 1983, **137**, 1–29.
8. E. Pierce, F. J. Carmona and A. Amirfazli, *Colloids Surf., A*, 2008, **323**, 73–82.
9. C. W. Extrand and A. N. Gent, *J. Colloid Interface Sci.*, 1990, **138**, 431–442.
10. C. W. Extrand and Y. Kumagai, *J. Colloid Interface Sci.*, 1995, **170**, 515–521.
11. B. Krasovitski and A. Marmur, *Langmuir*, 2005, **21**, 3881–3885.
12. M. Reyssat, D. Richard, C. Clanet and D. Quéré, *Faraday Discuss.*, 2010, **146**, 19–33.
13. M. Sakai, H. Kono, A. Nakajima, X. Zhang, H. Sakai, M. Abe and A. Fujishima, *Langmuir*, 2009, **25**, 14182–14186.
14. M. Sakai, J.-H. Song, N. Yoshida, S. Suzuki, Y. Kameshima and A. Nakajima, *Langmuir*, 2006, **22**, 4906–4909.
15. J. V. I. Timonen, M. Latikka, O. Ikkala and R. H. A. Ras, *Nat. Commun.*, 2013, **4**, 2398.
16. D. Richard and D. Quéré, *Europhys. Lett.*, 1999, **48**, 286–291.
17. L. Mahadevan and Y. Pomeau, *Phys. Fluids*, 1999, **11**, 2449–2453.
18. A. Lafuma and D. Quéré, *Europhys. Lett.*, 2011, **96**, 56001.
19. T.-S. Wong, S. H. Kang, S. K. Y. Tang, E. J. Smythe, B. D. Hatton, A. Grinthal and J. Aizenberg, *Nature*, 2011, **477**, 443–447.
20. X. Yao, Y. Hu, A. Grinthal, T.-S. Wong, L. Mahadevan and J. Aizenberg, *Nat. Mater.*, 2013, **12**, 529–534.
21. H. Mertaniemi, V. Jokinen, L. Sainiemi, S. Franssila, A. Marmur, O. Ikkala and R. H. A. Ras, *Adv. Mater.*, 2011, **23**, 2911–2914.
22. H. Mertaniemi, R. Forchheimer, O. Ikkala and R. H. A. Ras, *Adv. Mater.*, 2012, **24**, 5738–5743.
23. J. Seo, S. Lee, J. Lee and T. Lee, *ACS Appl. Mater. Interfaces*, 2011, **3**, 4722–4729.
24. A. Gauthier, S. Symon, C. Clanet and D. Quéré, *Nat. Commun.*, 2015, **6**, 8001.
25. J. C. Bird, R. Dhiman, H.-M. Kwon and K. K. Varanasi, *Nature*, 2013, **503**, 385–388.
26. D. Song, B. Song, H. Hu, X. Du and F. Zhou, *Phys. Chem. Chem. Phys.*, 2015, **17**, 13800–13803.

27. B. Chang, Q. Zhou, R. H. A. Ras, A. Shah, Z. Wu and K. Hjort, *Appl. Phys. Lett.*, 2016, **108**, 154102.
28. T. Kong, R. Brien, Z. Njus, U. Kalwa and S. Pandey, *Lab Chip*, 2016, **16**, 1861–1872.
29. D. Ishii and M. Shimomura, *Chem. Mater.*, 2013, **25**, 509–513.
30. S.-J. Hong, C.-C. Chang, T.-H. Chou, Y.-J. Sheng and H.-K. Tsao, *J. Phys. Chem. C*, 2012, **116**, 26487–26495.
31. P. Olin, S. B. Lindström and L. Wågberg, *Langmuir*, 2015, **31**, 6367–6374.
32. M. Elsharkawy, T. M. Schutzius and C. M. Megaridis, *Lab Chip*, 2014, **14**, 1168.
33. D. A. T. Mannetje, S. Ghosh, R. Lagraauw, S. Otten, A. Pit, C. Berendsen, J. Zeegers, D. van den Ende and F. Mugele, *Nat. Commun.*, 2014, **5**, 1–7.
34. A. Egatz-Gómez, S. Melle, A. A. Garcia, S. A. Lindsay, M. Marquez, P. Domínguez-García, M. A. Rubio, S. T. Picraux, J. L. Taraci, T. Clement, D. Yang, M. A. Hayes and D. Gust, *Appl. Phys. Lett.*, 2006, **89**, 034106.
35. A. Egatz-Gómez, J. Schneider, P. Aella, D. Yang, P. Domínguez-García, S. Lindsay, S. T. Picraux, M. A. Rubio, S. Melle, M. Marquez and A. A. Garcia, *Appl. Surf. Sci.*, 2007, **254**, 330–334.
36. A. A. Garcia, A. Egatz-Gómez, S. A. Lindsay, P. Domínguez-García, S. Melle, M. Marquez, M. A. Rubio, S. T. Picraux, D. Yang, P. Aella, M. A. Hayes, D. Gust, S. Loyprasert, T. Vazquez-Alvarez and J. Wang, *J. Magn. Magn. Mater.*, 2007, **311**, 238–243.
37. S. Lindsay, T. Vázquez, A. Egatz-Gómez, S. Loyprasert, A. A. Garcia and J. Wang, *Analyst*, 2007, **132**, 412.
38. K. A. Cavicchi, A. S. Zalusky, M. A. Hillmyer and T. P. Lodge, *Macromol. Rapid Commun.*, 2004, **25**, 704–709.
39. J. Schneider, A. Egatz-Gómez, S. Melle, S. Lindsay, P. Domínguez-García, M. A. Rubio, M. Marquez and A. A. Garcia, *Colloids Surf., A*, 2008, **323**, 19–27.
40. Z. Long, A. M. Shetty, M. J. Solomon and R. G. Larson, *Lab Chip*, 2009, **9**, 1567.
41. I. Torres-Díaz and C. Rinaldi, *Soft Matter*, 2014, **10**, 8584–8602.
42. Z. Cheng, H. Lai, N. Zhang, K. Sun and L. Jiang, *J. Phys. Chem. C*, 2012, **116**, 18796–18802.
43. X. Hong, X. Gao and L. Jiang, *J. Am. Chem. Soc.*, 2007, **129**, 1478–1479.
44. Z.-G. Guo, F. Zhou, J.-C. Hao, Y.-M. Liang, W.-M. Liu and W. T. S. Huck, *Appl. Phys. Lett.*, 2006, **89**, 081911.
45. E. Bormashenko, R. Pogreb, Y. Bormashenko, A. Musin and T. Stein, *Langmuir*, 2008, **24**, 12119–12122.
46. P. Poesio and E. N. Wang, *Exp. Therm. Fluid Sci.*, 2014, **57**, 353–357.
47. Z. Cheng, L. Feng and L. Jiang, *Adv. Funct. Mater.*, 2008, **18**, 3219–3225.
48. G.-P. Zhu, N.-T. Nguyen, R. V. Ramanujan and X.-Y. Huang, *Langmuir*, 2011, **27**, 14834–14841.
49. J. V. I. Timonen, M. Latikka, L. Leibler, R. H. A. Ras and O. Ikkala, *Science*, 2013, **341**, 253–257.

50. D.-M. Drotlef, P. Blümler, P. Papadopoulos and A. del Campo, *ACS Appl. Mater. Interfaces*, 2014, **6**, 8702–8707.
51. Y. Zhu, D. S. Antao, R. Xiao and E. N. Wang, *Adv. Mater.*, 2014, **26**, 6442–6446.
52. K. S. Seo, R. Wi, S. G. Im and D. H. Kim, *Polym. Adv. Technol.*, 2013, **24**, 1075–1080.
53. T. Ho, N. Ghochaghi and G. Tepper, *J. Appl. Polym. Sci.*, 2013, **130**, 2352–2358.
54. L. Chen, A. Geissler, E. Bonaccurso and K. Zhang, *ACS Appl. Mater. Interfaces*, 2014, **6**, 6969–6976.
55. K. S. Khalil, S. R. Mahmoudi, N. Abu-dheir and K. K. Varanasi, *Appl. Phys. Lett.*, 2014, **105**, 041604.
56. P. Aussillous and D. Quéré, *Nature*, 2001, **411**, 924–927.
57. Y. Zhao, J. Fang, H. Wang, X. Wang and T. Lin, *Adv. Mater.*, 2010, **22**, 707–710.
58. Y. Zhao, Z. Xu, M. Parhizkar, J. Fang, X. Wang and T. Lin, *Microfluid. Nanofluid.*, 2012, **13**, 555–564.
59. N.-T. Nguyen, *Langmuir*, 2013, **29**, 13982–13989.
60. F. Li, W. Wu, A. Ning and J. Wang, *Int. J. Chem. React. Eng.*, 2015, **13**, 113.

Subject Index